This book is dedicated to the men and women of the Australian Force Somalia who served in Somalia and adjacent waters in 1992-93, especially Shannon McAliney, who made the ultimate sacrifice, as well as those who sustained physical and mental injuries as a result of their service there.

First published in 2018 by Barrallier Books Pty Ltd,
trading as Echo Books

Registered Office: 35-37 Gordon Avenue, West Geelong, Victoria 3220, Australia.

www.echobooks.com.au

National Library of Australia Cataloguing-in-Publication entry.

Creator: Breen, Bob, author.

Title:A Little Bit of Hope: Australian Force Somalia 1993

Edition: Second edition

ISBN: 9780995367746 (hardback)

Notes: Includes bibliographical references and index.

Subjects: International relief--Somalia. Australia--Armed Forces--Foreign service--Somalia. Australia--Relations--Somalia. Somalia--Social conditions--1960-

Dewey Number: 363.349880967730994

Book layout and design by Peter Gamble, Canberra.
Set in Garamond Premier Pro Light Display, 12/17 and Minerva Small Caps.

www.echobooks.com.au

Front cover image: Gary Ramage photograph of Lance Corporal Shannon McAliney at a food distribution point escorting an elderly woman to protect her from robbery.

A LITTLE BIT OF HOPE

AUSTRALIAN
FORCE
SOMALIA
1993

BOB BREEN

ECHO BOOKS

CONTENTS

Abbreviations

A

ABC   Australian Broadcasting Commission

ABCA   America Britain Canada Australia

ACOPS   Assistant Chief of Operations

ADF   Australian Defence Force

ADFA   Australian Defence Force Academy

ADMINORD   Administration Order (G)*

AFS   Australian Force–Somalia

AHQ   Air Headquarters

AICF   Action Internationale Contre La Faim (G)

AIRCDRE   Air Commodore [RAAF]

AM   Member of the Order of Australia

AME   Aero-Medical Evacuation (G)

ANZAC   Australian and New Zealand Army Corps

AO   Area of Operations (G)

AO   Officer of the Order of Australia

APC   Armoured Personnel Carrier (G)

AS   Australian

ASF   Auxiliary Security Force [Somali] (G)

AUSMIPS   Australian Services Materiel Issue and Movement Priority System
0A   zero alpha (G)

B

3 BASB   3rd Brigade Administrative Support Battalion
BDE   Brigade (G)
BG   Brigadier General [US]
BHQ   Battalion Headquarters
BMSS   Box Metal Stores Shipping
BN   Battalion (G)
BRIG   Brigadier [Army]

C

CAFS   Commander Australian Force—Somalia
CARE   Relief agency(G)
CAPT   Captain [Army and Navy]
CDF   Chief of the Defence Force
CDRE   Commodore [Navy]
CENCOM   Central Command [US]
CGS   Chief of the General Staff [AS/UK]
CI   Counter Intelligence (G)
CINCPAC   Commander in Chief –Pacific [US]
CMDR   Commander [Navy]
CMOT   Civilian and Military Operations Team (G)
CNN   Cable News Network.
CO   Commanding Officer
COL   Colonel [Army].
COY   Company [Army] (G)
CPL   Corporal [Army]
CPOYS   Chief Petty Officer Yeoman Signals [Navy]
CRS   Catholic Relief Service (G)
CSC   Conspicuous Service Cross [AS]

| CSM | Conspicuous Service Medal [AS] |
| CSM | Company Sergeant Major (G) |

D

DFSU	Deployed Forces Support Unit (G)
DIO	Defence Intelligence Organisation (G)
DJOPS	Director—Joint Operations (G)
DMEO	Disposal of Malfunctioned Explosive Ordnance (G)
DP	Disruptive Pattern

F

| FAD 1 | Force Activity Designator 1 (G) |
| FFR | Fitted For Radio (G) |

G

GEN	General [Army]
GPCAPT	Group Captain [Air Force]
GPS	Global Positioning System (G)

H

HF	High Frequency (G)
HMAS	Her Majesty's Australian Ship
HQ	Headquarters
HRS	Humanitarian Relief Sector (G)

I

2IC	Second in Command
ICRC	International Committee of the Red Cross (G)
ID	Identification
IMC	International Medical Corps (G)
INT	Intelligence
IO	Intelligence Officer (G)
ISO	International Shipping Organisation

J

JAPG	Joint Administrative Planning Group (G)
JFHQ	Joint Force Headquarters (G)
JOPG	Joint Operations Planning Group (G)
JOPS 1	Joint Operations and Policy Staff (Grade 1) (G)
JOPS 2	Joint Operations and Policy Staff (Grade 2) (G)

K

KIA	Killed in Action

L

LARCV	Lighter, Amphibian Resupply Cargo (G)
LC AUST	Land Commander—Australia (G)
LCDR	Lieutenant Commander [Navy]
LCM 8	Landing Craft Mechanised—Series Eight (G)
LCPL	Lance Corporal [Army]
LEUT	Lieutenant [Navy]
LHQ	Land Headquarters
LOG	Logistic
LSF	Logistic Support Force (G)
LT	Lieutenant [Army]
LTCOL	Lieutenant Colonel [Army]
LTGEN	Lieutenant General [Army]

M

MAJ	Major [Army]
MAJGEN	Major General [Army]
MASH	Mobile Army Surgical Hospital (G)
MCAUST	Maritime Commander Australia
MHQ	Maritime Headquarters
MLO	Mortar Line Officer (G)
MSF	Medecins Sans Frontieres (G)
MSU	Media Support Unit (G)

N

NCO	Non-Commissioned Officer [Army]
NGO	Non-Government Organisations (G)
NVG	Night Vision Goggles (G)

O

OBE	Order of the British Empire
OC	Officer Commanding
ODF	Operational Deployment Force (G)
OFOF	Orders for Opening Fire (G)
OP	Operation
OPSO	Operations Officer (G)

P

PAL	Patrol Ambush Light (G)
PI	Public Information
PL	Platoon [Army] (G)
PR	Public Relations
PRO	Public Relations Officer
PTE	Private [Army]
PX	Post Exchange [US]

Q

QRF	Quick Reaction Force (G)
QM	Quartermaster [Army]

R

1 RAR	1st Battalion, the Royal Australian Regiment
RADM	Rear Admiral [Navy]
RAN	Royal Australian Navy
RCL	Recoilless Rifle (G)
RCT	Regimental Combat Team (USMC)
Recon	Reconnaissance

RHB	Reinforcement Holding Branch (G)
RHC	Reinforcement Holding Company
ROE	Rules of Engagement (G)
RSL	Returned and Services League of Australia

S

SAD	Ship's Army Detachment (G)
SDM	Somali Democratic Movement (G)
SEAL	Sea-Air-Land [US]
SF	Special Forces
SGT	Sergeant [Army]
SK 50	Sea King helicopter (G)
SLA	Somali Liberation Army (G)
SNA	Somali National Alliance (G)
SNF	Somali National Front (G)
SNM	Somali National Movement (G)
SO1	Staff Officer—Grade 1
SPM	Somali Patriotic Movement (G)
Spt	Support
SQNLDR	Squadron Leader [Air Force]
SSDF	Somali Salvation Democratic Front (G)
SSGT	Staff Sergeant [Army]

T

TI	Thermal Imager (G)

U

UD	Unauthorised Discharge (G)
UN	United Nations
US	United States
UNHCR	United Nations High Commission for Refugees (G)
UNICEF	United Nations International Childrens Emergency Fund (G)

UNITAF	Unified Task Force
UNOSOM	United Nations Operations in Somalia (G)
USC	United Somali Congress (G)
USMC	United States Marine Corps

W

WFP	World Food Program (G)
WGCDR	Wing Commander [Air Force]
WIA	Wounded in Action
WO1	Warrant Officer Class 1 [Army]
WO2	Warrant Officer Class 2 [Army]

NOTE:

*The symbol (G) after an entry means that there is a definition of the entry in the Glossary.

PREFACE

This book is a story, an analysis, a reminder and an acknowledgment.

The storyline runs along the cutting edges of two operational worlds. The first world encompassed the Baidoa Humanitarian Relief Sector in southern Somalia where young Australian commanders, their subordinates, as well as those who supported them in the field, worked eye-ball to eye-ball with the Somali populace. This cutting edge determined the success of Operation *Solace*. The second world comprised the various headquarters that commanded and supported those who served in the first world. It was individuals at this cutting edge who planned, deployed, commanded, sustained and redeployed the Australian Force-Somalia overseas and back again.

The analysis of lessons from Operation *Solace* at the end of each chapter identifies decisions, factors and events that contributed to success, as well as to frustration and failure. The aim is to illuminate the planning and conduct of future ADF off-shore peace operations in the developing world. This analysis should be a stimulus for change, shed light on the realities of peace operations and preserve corporate memory.

The book is also a reminder to the ADF that failure to learn from the lessons of the past may result in unnecessary frustration and failure being repeated. Veterans from the 1 RAR Group that deployed to Vietnam in 1965 and a 1 RAR Company Group that deployed to the waters off Fiji in 1987 will be disappointed to see some of the frustration and unnecessary pressures they endured repeated in this story of another 1 RAR Group in 1993

Finally, this book is the 'flesh and blood' of Operation *Solace*. It acknowledges the men and women who planned, raised and dispatched the Australian Force-Somalia, those who served in Somalia and those who supported them while they were there. It reports back on what was done, affirms those who did the work, and hopes to educate and inspire those who will face the challenges of peace operations in the future.

Bob Breen,

Sydney, 1997

In 2017, twenty years after this book was first published and almost twenty five years since the US-led Unified Task Force intervened to protect the distribution of humanitarian aid, Somalia is still a troubled land and parts of Africa are still impoverished and politically unstable. New groups of marauding terrorists, some claiming Islamic destinies, threaten life and property.

Africa's strategic importance has waned and humanitarian interventions problematic since this book was written. The war against terror in the Eurasia, the rise of China and India and new Russian assertiveness dominate strategic matters.

While United Nations military intervention into Africa is less likely the book has enduring lessons for Australia and Western armed forces intervening anywhere in the world, especially where what has become known as hybrid war is being waged.

In 1993 Australia and its allies were unprepared for what they encountered in Somalia. Fortunately, the Australians left before the United

States was humiliated later in 1993 This book has been republished in order to remind the Australian Defence Force and its allies that some things do not change while the evolution of hybrid war in places like the Ukraine and the Middle East demand change.

Bob Breen
Canberra, 2017

ACKNOWLEDGMENTS

The seed for this book was planted at a New Year's Eve celebration to welcome in 1993 at Colonel Kerry Gallagher's married quarter at Victoria Barracks in Sydney. After watching fireworks exploding around the Centre Point Tower from the upstairs veranda, I was speaking with two other guests, Major General Murray Blake, the Land Commander, and Lieutenant Colonel Paul Retter, Blake's chief planning officer, about the departure of 1st Battalion, the Royal Australian Regiment Group (1 RAR Gp) to Somalia in two weeks time. I mentioned that I had written to Lieutenant Colonel David Hurley, the Commanding Officer, encouraging him to remind subordinates to keep personal diaries and for his staff to collate key documents so I could write a book about the Group's experiences in Somalia. My first book about the operations of another 1 RAR Gp in Vietnam in 1965/66 had been written in 1988 after several years of part-time research. I wanted to avoid the difficulties I had endured finding first-hand primary source material for the earlier book by getting in on the ground floor of this 1 RAR Gp story.

Four months later on Thursday morning before the ANZAC Day long weekend, Kerry Gallagher telephoned my office. He asked if I was prepared to re-enlist in the Regular Army for four weeks and deploy to Somalia in six days time to conduct field research for a book on Operation *Solace*.

I do not think my family really believed that I had agreed to go until I arrived home the next evening with a Kevlar helmet, field webbing and a flak jacket. I flew to Somalia on 29 April on a RAAF C130 transport aircraft, forever grateful to Major General Murray Blake, Brigadier Iain Macinnis, his Chief of Staff, and Colonels Kerry Gallagher, Colonel (Training), and Phil McNamara, Colonel (Operations), for getting me on that flight.

In Somalia, I found that members of the Australian Forces-Somalia (AFS) were ready and willing to tell me about their experiences and to share their perspectives on Operation *Solace*. I was humbled by their cooperation, commitment and trust. I am grateful to Colonel Bill Mellor and Lieutenant Colonel David Hurley for allowing me the access to troops and activities that made my research trip a success. For them, the first two weeks of May 1993 were very busy, finalising plans for returning the AFS to Australia as well as having Lieutenant Colonel Paul Retter's Operational Study Team, Lieutenant Colonel John Kelley's psychologists and me, Land Command's newly-appointed Combat Historian, interviewing and holding discussions with their subordinates and tagging onto activities.

In a perfect world, I would have returned from Somalia in May 1993 and begun to write this book full-time. However the *ad hoc* circumstances of my deployment were followed by *ad hoc* and intermittent employment and remuneration arrangements. Despite the best intentions of senior military staff at Land Headquarters, who wished to create a permanent position for a Combat Historian to research and write about operations, resources for full-time employment were not available after a suitable position was created, but subsequently transferred to Canberra. Consequently, I had to juggle my time between writing the book and running a busy training consultancy. The key supporters during this time were Major General Blake's successors, Major Generals Peter Arnison and Frank Hickling, Brigadier Macinnis's successors Brigadiers Tim Ford, Peter Kilpatrick and Paul O'Sullivan and Colonel Phil McNamara's successors Colonels Gerry Warner and Mike Hyde. Concurrently, I also received the support and encouragement of several peers, Paul Retter, Richard Greville, Peter Schmidt, John Culleton and Gary Martin.

Of the prime movers behind this book, I would like to make special mention of Colonel Kerry Gallagher, now out of the Army, and Colonel Mike Hyde. Kerry was the originator and Mike was the inheritor; however, both were convinced that the ADF should research, incorporate and share its operational experiences with the Australian public in general and with members of the ADF in particular. Kerry, supported by Phil McNamara, initiated proposals for my deployment to Somalia and set the scene for my subsequent employment as a Combat Historian on ADF operations in Rwanda, Egypt, Israel and Mozambique. Mike Hyde not only inherited a Combat Historian with an unfinished book in 1996 but also the problem of finding resources to employ him and maintain operational research and analysis services at Land Headquarters within a climate of financial constraint and unforeseen cost cutting. He won the battle to have the book completed and published without significant alteration, and also successfully negotiated the provision of operational research and analysis services until the end of 1997.

The final production of the book was a team effort. Garth Pratten from the Military History Section of the Doctrine Centre at Headquarters Training Command edited and advised on the book. Captain Karen Barnett and Mrs Jean Stiller from the Publications Section of the Doctrine Centre continued the editorial process, formatted the book and proof read its contents. The photographs of George Gittoes, the still images taken from video footage shot by Lisa Keen and her staff at the Electronic Media Unit in Canberra and processed by Gerry Crown the Audio Visual Section at Georges Heights as well as the photographs of staff photographers at the *Townsville Bulletin*, Corporal Gary Ramage and several others have brought the book to life with the images of Australians at work in Somalia. The maps were prepared by Sergeant Andrew Westover and Corporal Maria O'Donohue of Engineer Branch at Land Headquarters. The nominal roll was produced over many months by Major Stuart Dodds, who wanted to show where everyone served in Somalia, not just that they were there. Debbie Robinson at the Directorate of Infantry collated and set up the data base for the roll.

My final acknowledgment goes to the men and women who served in Somalia and those who supported them during Operation *Solace* from off-shore, and from Kenya and Australia. Their input, their trust, and their expectations, as well as their encouragement, have inspired me to keep going and to see the book through to its completion.

1   BACK TO THE DARK AGES

Me and my clan against the world,
Me and my brother against the clan,
Me against my brother.

Somali Proverb

Early on the morning of 20 October 1992, a twin-engined King Air light aircraft from Nairobi in Kenya landed at Mogadishu, the capital of Somalia. It was one of the scores of aircraft chartered by international aid agencies, called Non-Government Organisations (NGO), to shuttle staff and stores around the Horn of Africa. Hitching a ride on the 16-seater were two Australian Army officers, Lieutenant Colonel Bill Nagy and Captain Paul Angelatos.[1] They were conducting a reconnaissance for an Australian service contingent scheduled for deployment to Somalia in a few weeks time. The Australian Government had agreed to deploy a 30-strong movement control unit to Somalia to support a United Nations mission, known as United Nations Operations in Somalia (UNOSOM), whose mission would be to protect the distribution of humanitarian aid. The Australian unit's task would be to coordinate the movement of UNOSOM personnel and materiel for six months.[2]

As the King Air touched down, Nagy and Angelatos looked out expectantly. The runway was modern and well-maintained but ran amidst piles of rotting rubbish and junk, run-down buildings and the wreckage of burnt-out vehicles and old jet fuselages. At the south-eastern end of the runway were abandoned Somali Air Force hangers that still housed a number of broken-down Russian-made MIG jet aircraft. To the west, a line of dirty-white buildings on a ridge marked the edge of the city. To the east, a line of sand dunes ran parallel to the runway, overlooking the airport on one side and the Indian Ocean on the other.

After the plane stopped in front of a dilapidated terminal building, Nagy and Angelatos were met by a UNOSOM escort officer who was travelling in a Ute accompanied by a group of Somali gunmen riding in the tray of another Ute. This utility and its occupants looked like they had come from the set of one of Mel Gibson's *Mad Max* movies. The Somalis, dressed in torn T-shirts, baggy trousers and sandals, and adorned with necklaces and colourful bandannas, were armed with an assortment of knives and weapons, most favouring the Russian-made AK 47 assault rifle. The utility, known as a 'technical', was battered, dirty and daubed with Somali graffiti. Welded into the middle of its tray was a metal rod, standing as high as a man's chest, on which was mounted a .30 calibre machine gun. A standing man could swivel the machine gun around on a pintle and fire it in all directions.

Somali Gunmen riding in the back of a 'Technical' 1993
(ABC News)

Nagy and Angelatos noticed that there were many more 'technicals' with groups of heavily-armed Somali gunmen perched on their trays roaring around the airport kicking up long plumes of dust.

The Australians were driven to HQ UNOSOM, a large, leased mansion surrounded by a high wall festooned with powerful security lights.[3] They reported to the Canadian Chief of Staff, Colonel Cox. He greeted them cordially and invited them to be seated in a briefing room. He then gave them a bleak assessment of the situation in Somalia. Two years of civil war had destroyed the infrastructure and the economy, and driven hundreds of thousands of Somalis from their homes. Control of southern Somalia was still being contested by four warlords who commanded well-equipped but poorly-maintained militia armies. The areas where the warlords did not exert their authority were controlled by urban criminals and nomadic bandits.

In southern Somalia, the warlords' armies, based on major Somali clan groups, were no longer campaigning against each other as formed units. However, there were still numerous minor military actions and acts of political terrorism, robbery and murder. The roads were unsafe. Bandits were ambushing vehicles and travellers regularly, killing and maiming unarmed civilians after robbing them of their meagre possessions and cash.

Mogadishu had been partitioned by two warlords whose forces stared at each other over a mini Berlin Wall of barbed wire, barricades and gutted buildings, called the Green Line, that had been brokered by the UN a year ago. Each warlord directed a political and military campaign of terror. Political factions, as well as criminal gangs, shot at each other every night in a violent struggle for supremacy.

In this lawless, violent environment, over four million Somalis were starving and dying from disease in southern Somalia because insufficient humanitarian aid was getting through to villages and refugee camps. Somali gunmen preyed on humanitarian aid as it arrived, in transit and at distribution points. UN and NGO staff were being intimidated into paying hundreds of US dollars daily to armed extortionists to protect relief operations and to rapacious Somali gunmen for personal protection.

Cox told Nagy and Angelatos that the UN intervention into Somalia faced several significant challenges. Politically, UNOSOM did not have the unanimous support of the Somali warlords. Each warlord jostled for favour with the UN at the expense of their opponents, but UNOSOM was unable to get them all to agree to the deployment of foreign troops. No one really knew what authority UNOSOM could exert in Somalia to protect the distribution of aid. Several days before Nagy and Angelatos arrived, the Pakistani UNOSOM Force Commander, Brigadier General Shaheen, had:

> ... returned to UN New York to sort out a number of points. Apparently he was having some difficulty getting his message across to UN New York by phone/fax and felt a personal approach was the only way to resolve issues, which included: nature of mandate (scope of military operations).[4]

The military position of UNOSOM was perilous. The only armed UNOSOM unit on the ground in Somalia was a lightly-armed, 500-man Pakistani battalion, holed up in a barbed wire compound at Mogadishu Airport. Outnumbered and out-gunned, they were in a Mexican stand off with Somali gunmen who menaced airport operations. The beleaguered Pakistanis were a long way from home with only a Morse Code radio link back to Pakistan. There was no UN or Pakistani supply system supporting them. They bought their food from the city markets and boiled water, drawn from local wells, to drink. Even after 20 minutes boiling and the addition of cordial or fruit, the water still tasted foul.[5]

Other UNOSOM operations had also come to a standstill under pressure from Somali gunmen. Unarmed military observers had abandoned a monitoring program in the city after two observers had been robbed and shot. No UNOSOM personnel moved at night for fear of being caught in the crossfire between rival Somali political factions and criminal gangs, beaten up or taken hostage.

After Cox's comprehensive briefing and several hours of further discussions with UNOSOM staff, Nagy and Angelatos consolidated their notes and shared their impressions of the day. UNOSOM staff told Nagy several times that the Somali warlords controlling the city could 'take out' the Pakistanis and HQ UNOSOM anytime. He wondered whether Australia should be involved with this faltering UN intervention. Late that night, they sought a few hours sleep after a day of intense concentration. They were exhausted by the oppressive heat, fears for their personal safety and the jet lag from a long journey that had begun in Sydney three days before.

It was not easy to sleep soundly in Mogadishu in October 1992 Nagy noticed that security of the accommodation compound was provided by hired Somali guards and that UNOSOM military and civilian staff kept their weapons and ammunition out of sight in order not 'to send the wrong message to the locals'.[6] The only difference between the attitude of the Somali UNOSOM security guards and the gunmen threatening the Pakistanis at the airport was a daily cash payment of US dollars.

Outside, he could hear the sounds of a restless population and a war-torn city. The lights of UNOSOM and NGO compounds were the only ones on after dark. Large crowds milled around these compounds throughout the night chewing a mild narcotic plant called *khaat* (pronounced 'cat' and similar in appearance to the greenery growing from the top of a mature carrot),[7] and socialising. Aside from the oppressive heat and the chatter and shouts of the crowds, the night was punctuated by a cacophony of grenade explosions, rifle shots and bursts of automatic fire. In the distance, lines of tracer bullets arced up into the sky. The throaty, growling sounds of numerous 'technicals' marauding through the city streets completed Mogadishu's anarchic lullaby.

The next day Nagy and Angelatos went on a tour of the airport, port and inner city area in a vehicle driven by a Somali, accompanied by a technical; both vehicles flew a small UN flag from their mud guards. Though it was mandatory for hired Somali gunmen to accompany UNOSOM personnel and visitors moving on foot or in vehicles, these precautions had not helped the two UNOSOM observers who had been shot a few days before. Nagy and Angelatos took little comfort from the wolfish grins of their cocky body guards.

As they toured the inner city, the impact of the civil war was obvious. Almost every building was roofless and gutted. The remaining walls were pocked with bullet holes and larger holes made by heavier weapons and blasts from artillery and tank rounds. Wiring, windows, pipes and doors had been ripped out or blown out long ago. There were no public utilities; people defecated randomly and rubbish rotted in the streets. There was no running water or electricity. Hundreds of children, with no school to go to, played in the filthy streets. Young children called and waved as UN vehicles passed by, their hands outstretched for food. Teenagers threw stones.

Despite the devastation and media reports of widespread starvation, Nagy noticed a large, thriving market for food and all sorts of other commodities. Plenty of meat was being butchered for sale on blood-encrusted hessian matts. There was an abundance of grain, vegetables, fruit, cattle, camels, goats and chickens for sale. He wrote in his reconnaissance report, 'If you have the money you can buy anything ... the majority of the food has been bought from the looters by the black market.'[8]

Late that afternoon, the two Australians caught the UN air shuttle back to Nairobi. Nagy sighed with relief and felt the adrenalin subside as the aircraft took off. He contemplated the anarchy he had witnessed and the uncertain future of UNOSOM. He had been told that most Somalis yearned for peace and stability. The average Somali was sick and tired of violence, and did not support the warlords. What pressed on Nagy's mind was how this supposedly peace-loving society could destroy itself. Why were several million Somalis about to die because fellow Somalis would not allow humanitarian aid to get through to them?

The story of Somalia's journey to anarchy began in 1960. About two thirds of Africa was being decolonised under UN supervision. These were the heady, optimistic days of African independence. Somalia, a new

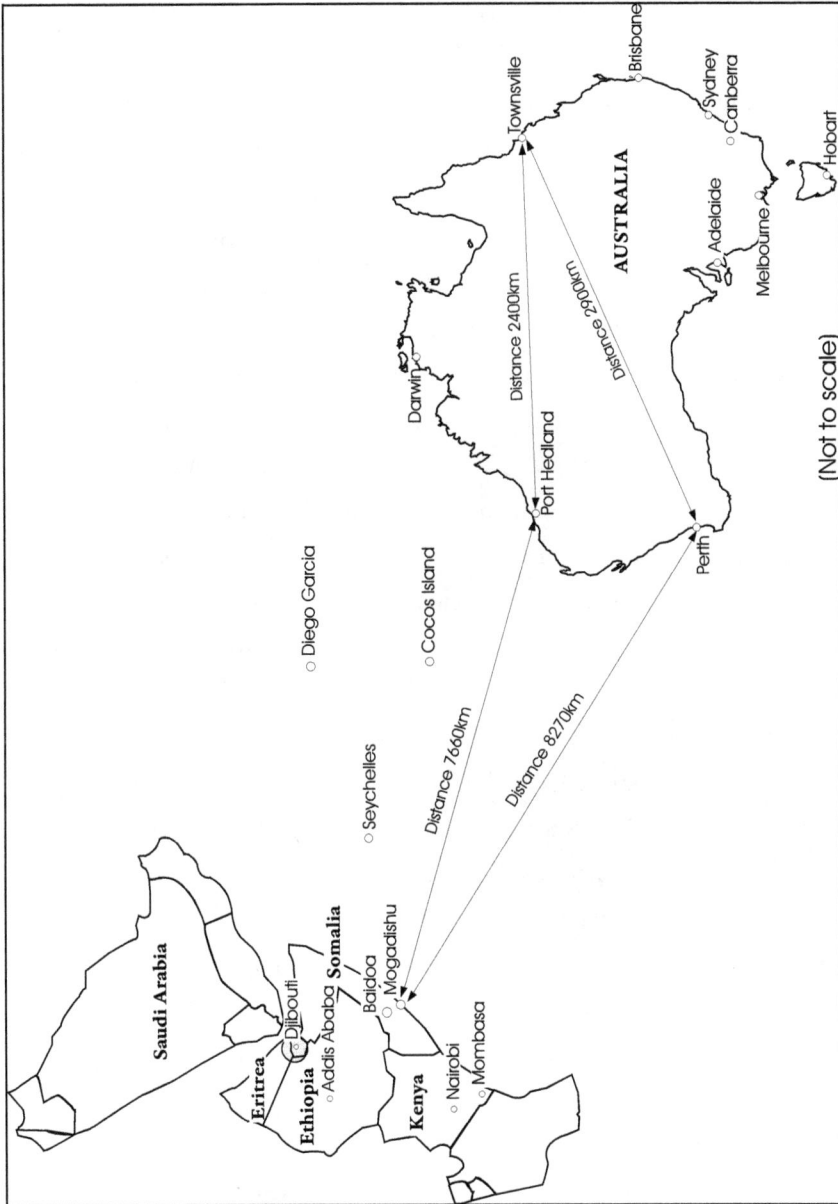

Map 1: Theatre of Operations–Operation Solace

boomerang-shaped state located on the jutting edge of the Horn of Africa, was formed from the former British Somaliland Protectorate in the north and Italian Somaliland in the east and south.

To the west, Islamic Somalia shared a border with its ancient adversary, Coptic Christian Ethiopia, and to the south another colonial border was shared with another traditional rival, Kenya, a former British colony. In the north-west, nestled between the old Italian colony of Eritrea and Somalia was Djibouti, a tiny colony established by the French as a conduit for Ethiopia's export trade through French-built rail links and a port facility on the Red Sea.

Before Somali independence in 1960, the British presided over, but did not nurture, a primitive pastoral economy exporting camels, cattle, sheep and goats locally and to Arab countries, mainly Saudi Arabia. Britain's Somaliland Protectorate was a butcher's shop for British military forces in Aden, across the Red Sea. Life in the north involved seasonal migrations of Somali clan groups with herds of cattle, goats, camels and other livestock in search of grazing areas. Aside from two brief wet seasons, the northern plains were swept by hot, dry and dusty winds and temperatures averaged in the high 30°sC and low 40s°C for most of the year.

In the south, the Italians had nurtured and invested in a vast tropical market garden embraced by the Shabeelle and Juba Rivers. The area not only provided fresh foodstuffs for regional consumption but also yielded export surpluses of bananas, maize, millet, rice and citrus fruits.

The southern agricultural area of Somalia was not only the 'breadbasket' but also the centre of commerce, culture and learning. Mogadishu was the capital and a major port. There were other ports and commercial centres further south along the coast. The largest city at the head of the Juba and Shabeelle Rivers was Kismaayo. Inland from Mogadishu was the city of Baidoa, a rural centre for agricultural commerce and the site of a university.

The major mistake made by the UN during negotiations with Britain, France, Italy, Ethiopia and Kenya was to create the new state of Somalia inside old colonial borders. By not eliminating all the arbitrary borders made by previous British, Italian and French colonial rulers, three Somali minorities continued to live under the jurisdiction of Ethiopia, Kenya and Djibouti. About two and a half million Somalis lived in Ethiopia's Ogaadeen region. This region had been given to Ethiopia by the British in 1897 during a period of colonial 'horse trading'. A further 250 000 Somalis lived in the northern part of Kenya and an influential mercantile clan of about 100 000 Somalis lived in Djibouti. Thus, the Somalis, one of the most homogeneous people in Africa, remained 'partitioned into mini-lands'.[9]

The UN set out to establish democracy in Somalia as the culmination of the process of de-colonisation. However, Somali culture presented many challenges for UN policy makers as they facilitated the development of a Western-style democratic government. Somalis shared common customs, origins, history, religion and language but were divided into fiercely independent and competitive clans. In pre-colonial times, 'Somalis lived in a world of egalitarian anarchy, a world of camel husbandry and clan-families as liable to be at war with one another as to assemble under an acacia tree in order to exchange oral poetic contests that sometimes lasted for days.'[10]

For obvious reasons, the British and Italian Governments had not encouraged democratic processes during the period of colonial rule. After independence, the major Somali clans were represented by Western-style political parties, formed with UN sponsorship and encouragement, but attitudes had not changed. During the nine years following independence in 1960, the political parties representing these clans remained balanced to share power; not in harmony, but in deep, mutual distrust.

The common ground shared by the rival clans within Somalia was a desire to unite all Somalis into 'Greater Somalia'. The clans wanted to free Somali minorities from the jurisdiction of neighbouring countries by reclaiming the land they occupied. As a consequence, the Somali Government diverted substantial national resources away from nation-building and productive economic activity into support for Somali guerrilla groups in neighbouring countries and the establishment of a formidable national army.

By 1969, deep divisions persisted in the national government, despite the broad representation of northern and southern clans. Acts of political violence, sabotage and assassination increased. In March, a total of 64 political parties, representing numerous clans and sub-clans, contested a national election, marred by many acts of terrorism. A north-south coalition led by southerner, Dr Abdi Rashid Ali Shermaarke, and northerner, Mohammed H.I. Igaal, was returned to power. An uneasy calm settled over the country.

This calm was shattered by an assassin's bullet in July 1969 when Shermaarke was gunned down by a soldier from a rival clan while he was reviewing a military parade in the north. Igaal hurried back from a state visit to the United States where he had been arranging an aid program. In the ensuing months, the political haggling and violence associated with trying to find a new president undermined Somalia's democratic institutions. This squabbling created an opportunity for an ambitious army officer, Mohammed Siad Barre, to secretly form an alliance between the army and the police to seize power in a bloodless coup.

For the next eight years, Barre built up his personal power through the patronage of his Mareehan Clan and by playing the other clans off against each other. He also recruited, trained, and equipped a modern national army and air force with the support of the Soviet Union. During the 1970s, Somalia became a Russian client state building military power and experimenting with scientific socialism.

In August 1977, the Somalis in the Ogaadeen region rose in revolt after the overthrow and assassination of the Ethiopian Emperor, Haile Selassie. Despite being outnumbered by the Ethiopian Army three to one, Barre launched an invasion of the Ogaadeen with 35 000 regular and 15 000 militia troops. The Somalis achieved surprise and recaptured 90 percent of the disputed territory before the Ethiopians were able to stop their advance and force them into defensive positions.

During the ensuing stalemate, the Ethiopian Government offered the Soviet Union, a strategic alliance in exchange for military support and the abandonment of Somalia as a client-state. After signing a secret agreement, the Soviet Union sent 1.5 million US dollars worth of weapons, vehicles and equipment to Ethiopia in February and March 1978. This military aid was accompanied by two Soviet generals, 1 500 Soviet advisers and 11 000 Cuban troops. The Soviet advisers with the Somali Army deserted and carried their maps of the Somali dispositions in the Ogaadeen to Ethiopian territory.

Within weeks of the rearmament of the Ethiopian Army and the arrival of Cuban troops, the Somali Army was driven from the Ogaadeen. This humiliating defeat caused dissent among Somali officers. Predictably, the dissent followed clan lines and officers from the Majeerteen Clan, previously a prominent clan in the civilian Governments of the 1960s, conspired against Barre.

In April 1978, Barre ordered the execution of 17 Majeerteen officers after a failed coup attempt. He then launched a savage attack on the Majeerteen Clan throughout the country. Elite units, such as the specially-recruited and trained 'Red Hats', laid waste to the pastoral areas in the north occupied by the Majeerteen Clan. Barre's forces killed over 2 000 people, wounded many more and raped hundreds of Majeerteen women, the ultimate degradation in Islamic culture. They destroyed centuries-old watering wells and reservoirs, and killed tens of thousands of camels, cattle, sheep and goats. This attack on the northern Majeerteen Clan by the southern-based Mareehaan-led military units, loyal to Barre, set the agenda for future civil war and secession.

North from where the Majeerteen Clan had been ravished, the Isaaq Clan owned pastoral lands and had built several cities. The Isaaq occupied Hargeisa, the second largest city in Somalia, Bur'o, an interior pastoral centre, and the port of Berbera. Since independence in 1960, the Isaaq had felt deprived both as a clan and as a region.[11] In the south, traditional rivals, the Daarood and Hawiye clans, had dominated the civilian Governments of the 1960s and ensured that national resources were directed to the developed agricultural heartland rather than the less-developed northern pastoral zone.

Somali Gunmen chewing Khaat, a mildly narcotic plant. Khaat heightend the senses and reduced the appetite.
Image: ADF Electronic Media Unit

During the 1970s under the southerner, Barre, this imbalance had continued.

The persecution of the northern Majeerteen by the Somali Armed Forces led by southern clansmen brought the long-held grievances of the Isaaq to a head. In 1981, the Somali National Movement (SNM), led by 400 to 500 Isaaq exiles was formed in London. They proclaimed that their purpose was to free Somali from Barre, 'the military tyrant'. Two years before, Colonel Abdullaahi Yusuf, a Majeerteen officer who had escaped Barre's post-coup purge a few years earlier formed the Somali Salvation Democratic Front (SSDF) in the Ogaadeen. The SSDF and the SNM became powerful political forces, based on traditional northern clan-families, dedicated to overthrowing Barre.

The year of 1988 was both a political watershed and a time of bloodshed for Somalia. The SNM launched a general offensive aimed at securing the Isaaq northern heartland, based on the cities of Hargeisa, Bur'o and Berbera. The offensive was accompanied by an uprising of the Isaaq civilian population. In a matter of weeks, however, SNM forces were defeated and driven from the area under a hail of artillery fire and aerial bombardment from Barre's well-equipped Somali Armed Forces.[12]

Once again, Barre took revenge. During his counterattack, savage reprisals were taken against Isaaq civilians. The wholesale destruction of property and livestock was repeated and the systematic rape of Isaaq women was conducted on an even larger scale than the mass rape of Marjeerteen women a decade before. There were numerous incidents of Isaaq men, women and children being rounded up and having their throats cut.

The scale and ferocity of the destruction was exemplified in the razing of the cities of Hargeisa and Bur'o. Both cities were reduced to rubble and not reoccupied. Barre's troops and the Somali Air Force killed over 5 000 Isaaqs, and wounded and injured tens of thousands more during fighting and reprisals. The fighting and the razing of cities and villages displaced 300 000 Isaaqs. They fled inland and began to die from starvation, thirst and disease in squalid refugee camps, constructed from bracken and scavenged waste materials.

In the twenty years since he took power, Barre's dictatorship had robbed the Somali people of their national wealth and standard of living. In the mid-80s, principal economic indicators revealed that 40 percent of urban Somalis and 70 percent of rural Somalis lived in 'absolute poverty'. By the late 1980s, these percentages had risen significantly. In Mogadishu alone, over 70 per cent of the population lived in slums and squatter camps.[13]

It was now the turn of the Hawiye Clan from the south-central region of Somalia around Mogadishu and inland up the Shabeelle River Valley to oppose Barre. A group of Hawiye exiles in Rome formed the United Somali Congress (USC) and set an agenda to overthrow Barre. Having established his method of quelling rebellion with the Majeerteen and Isaaq clans, Barre ordered his Red Hats to ravage the lands and property of the Hawiye Clan even before they had been able to organise any significant political or military action against him. In so doing, he accelerated the formation of Hawiye guerrilla units, dedicated to avenge the killing, rape and destruction. Barre had also precipitated the formation of a coalition of clans who combined to attack Mogadishu and throw him out of power.

In January 1991, Barre fled from Mogadishu after a two-week assault by the coalition of clans fighting under the USC banner. He looted the Government treasury and, accompanied by his remaining supporters, drove south to his Maaraheen clan stronghold at Garbhaarey, north-west of Kismaayo, in a long column of armoured vehicles and trucks. Subsequently, his son, General Maslah and his son-in-law Colonel Mohamed Hersi Morgan went on an international spending spree to buy new weapons, vehicles, equipment and ammunition. For his part, Barre fled to Nigeria where he

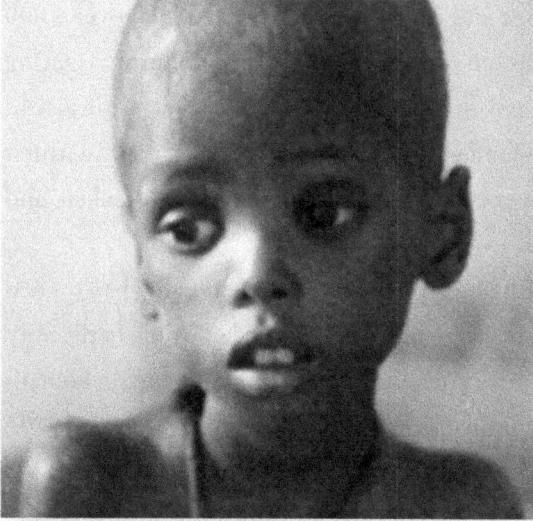

The face of starving Somalia
Image: ADF Electronic Media Unit

retired to a life of luxury.

On 21 March 1991, Simon Manthrope of the *Montreal Gazette* reported:

> This is Mogadishu six weeks after President Mohammed Siad Barre obliterated it with his artillery in face of the rebel assault on his palace and then loaded a tank with all the gold in the national bank and sped off to his region in the south-west.
>
> This is a town with no law, no government, no water, no electricity, no fuel, no food. There is *Khaat* (or cathaedulis, a narcotic herb that Somalis are fond of eating) which depresses the appetite and the trade flourishes ... And there are guns.
>
> In Somalia, there is an unspoken acceptance that the eight rebel groups that toppled Siad [Barre] will not meet at the negotiating table to draw up a new constitution and pave the way for elections. They will meet on the battlefield, probably in protracted and irresolute war.[14]

Soon after Barre's departure, a USC faction declared Ali Mahdi Mohamed, a member of the Hawiye Clan, President of Somalia. This hurried action, without consultation with the other factions of the USC coalition of clans, meant that Mahdi did not have a mandate outside his clan. Consequently, the other clans opposed him. Despite calling for a national conference to negotiate a political agreement for power sharing, Ali Mahdi remained isolated. Opposing him from another faction of the USC was General Mohamed Farah Aideed.

In the north, the SNM, representing the Isaaq Clan, convened a northern clans conference at Berbera to develop a regional understanding. In effect, the Isaaq clan was organising a northern secession from southern Somalia. They managed to develop a consensus among the northern clans based on their hatred of southerners. Later in 1991, a new northern Somaliland Republic was declared under the Presidency of Ahmad Ali Tur.

By 1992, the struggle for power in southern Somalia focused around the strategic gateways of Mogadishu and Kismaayo. General Aideed gathered around him dissenting factions of the USC to oppose Ali Mahdi in Mogadishu. The fight for Kismaayo was waged between forces loyal to Barre, under the command of Barre's son-in-law, newly-promoted General Morgan, and a coalition of southern Daarood Clan forces under the command of General Omar Jess, a former minister in Barre's government.

During the first six months of 1992, the combination of civil war, anarchy in the countryside and a prolonged drought resulted in an estimated two million Somalis fleeing from their homes into remote areas of Somalia, and into Kenya, Ethiopia and Djibouti. The Australian equivalent would be for the people living in the coastal cities and towns of Queensland to flee inland to the remote regions of the Northern Territory, north-eastern South Australia and north-western New South Wales.

Like Australia, the further inland one travels in Somalia the less food, facilities and water are available. Thousands of refugees congregated around wells, near rivers or where aid organisations had set up camps to accommodate refugees and distribute food and other necessaries. The refugee problem was not new to Somalia. The Ogaadeen War in 1978 had displaced over one million refugees into camps along the Ethiopia-Somalia border. The Ethiopians did not want them back and the Somali Government could not feed them or provide land for their sustenance. A group of aid agencies had been supporting the displaced Somalis in these camps for over ten years.

The cycle of severe droughts combined with Barre's purges of rival clans in the late 1980s had increased the number of refugees significantly. The outbreak of civil war in 1991 and another drought doubled the refugee population and turned most of southern Somalia, the colonial agricultural breadbasket, into a desolate, dusty wasteland dotted with refugee humpies. Men survived by taking up arms, fighting against rival clan armies or bandit gangs, and stealing what they could find. Women, children and the elderly not supported by armed husbands, fathers, sons or male relatives, simply starved to death or lived in refugee camps accepting handouts.

By July 1992, the survival of over four million Somalis depended on the distribution of food and the provision of clean water and medical services by UN agencies and NGOs. Having defeated one dictator, southern Somalis were now prey to four would-be dictators. To achieve their ends, Ali Mahdi, Aideed, Morgan and Jess, now dubbed by the Western media as 'warlords', were prepared to allow hundreds of thousands of men, women and children to die from starvation and disease. Traditional clan rivalries now found expression in the systematic looting and hoarding of food and other life-sustaining commodities.

The unanswered question was whether the UN, representing the ideals of international cooperation and humanitarianism, could intervene in Somalia successfully and restore the flow of humanitarian aid to those most in need, and facilitate political reconciliation and a restoration of law and order.[15]

The architect for armed international intervention into Somalia was Peter Hanson, a former Assistant Secretary General with the UN, who visited Somalia in August 1992 and estimated that 'as many as 4.5 million

Somalis are in desperate need of food'. He wrote of a 'vicious cycle of insecurity and hunger' that could only be broken by the urgent deployment of UN security forces. He recommended a 'comprehensive program of action covering humanitarian relief, the cessation of hostilities, reduction of organised and unorganised violence and national reconciliation'. He envisaged local police forces being established under UN arrangements to restore law and order.[16]

Hanson's report set an agenda that not only addressed the immediate requirement to save lives by protecting the distribution of humanitarian aid but also committed the UN to facilitating national reconciliation and restoring law and order. This agenda echoed the idealism that had accompanied the UN pressure to give Somalia independence in 1960. Unlike 1960, Western-style democratic institutions were going to be forced on the Somalis under the guns of foreign military units.

On 28 August, four days after Hanson published his report, the UN Security Council passed Resolution 775 proposing the deployment of five infantry battalions to Somalia to protect the distribution of humanitarian aid. The UN began circulating a proposed organisation and charter to member nations and UNOSOM was born. A few weeks later HQ UNOSOM and a battalion of Pakistanis deployed to Mogadishu.

Prompted by UN interest in involving Australian troops, military planners in Australia put several proposals to the Defence Minister, Senator Robert Ray. The UN was guided towards requesting the Australian Government to deploy a 30-strong movement control unit (MCU).[17] After Cabinet considered the UN request, Ray announced the commitment of the MCU on 14 October.

Meanwhile, as the UN finalised arrangements for UNOSOM with donor nations, the situation in Somalia worsened. On 12 October, Kismaayo port closed because every ship containing relief supplies was being looted immediately after docking. At the airport armed clansmen were looting over 50 per cent of relief supplies brought in by air and denied landing rights to UN and NGO aircraft unless they received cash payments.

In the inland city of Baidoa, amidst anarchy and widespread lawlessness, up to 100 people a day were dying on the streets. Their bodies were being loaded onto carts and trucks and dumped into mass graves. Over 12 000 bodies lay in the softer soil of the banks of the river which ran through the city, awaiting the next wet season to be exhumed by running water to spread disease.

The Australian involvement in Somalia began six days after Senator Ray's announcement with the reconnaissance by Bill Nagy and Paul Angelatos. Despite Nagy's concerns about the security of HQ UNOSOM, the military observers and the Pakistani battalion at the airport, he had recommended that the Advance Party of the Australian MCU, led by Major Greg Jackson, continue its deployment to Mogadishu on schedule.[18] Jackson arrived with two others on 28 October. By 14 November, he was operating under very difficult conditions with 10 Australian personnel planning for the deployment of UNOSOM units to Somalia.

Of particular concern to Jackson was the working conditions of his Advance Party. Despite assurances given to Nagy and Angelatos by UNOSOM staff during their reconnaissance, the Australians were not allocated vehicles for their own transport and had no reliable communications system to support their coordination work.

In absence of any amenities from UNOSOM, Jackson asked Land Headquarters (LHQ) in Australia to send portable refrigerators and generators so his troops could have cold water and soft drinks, as well as reliable lighting at night. His requests were denied several times but he stated:

> I will continue to ask for refrigeration. If for no other reason than the maintenance of morale. Immediate access to cold water, even cold soft drinks, may be one of the few luxuries that can be provided to the contingent.[19]

In contrast to the situation endured by Jackson and his staff at the airport, Australians working for UN agencies and NGOs in Mogadishu enjoyed far better working conditions and participated in a lively Western social circuit. Jackson and his subordinates eagerly accepted invitations to dinner and air-conditioned social functions where fresh Western food was served and alcohol was available.[20] The Australian Army had an austere approach to supporting service contingents in developing countries whereas UN agencies and NGOs had a policy of ensuring that their expatriate personnel lived as comfortably as circumstances permitted. These agencies felt that the extra cost of flying in mail, fresh food and alcohol, and setting up refrigeration and air-conditioning, was a worthwhile investment in the mental and physical health of their employees who faced the horrors and dangers of starving, war-torn Somali society every day.[21]

Meanwhile, clan battles and the looting of humanitarian supplies continued in Mogadishu and the southern parts of Somalia throughout November unchecked by UNOSOM. By mid-November 1992, Dr Jeylani Dini of the International Committee of the Red Cross reported that 840 people were dying each month in Kismaayo, half of them under the age of five. The 40 Red Cross feeding centres could only offer one meal a day to the estimated 300 000 refugees huddled in camps around the city.[22]

Malcolm Fraser, the President of CARE International, reported:

> During a recent fact finding mission to Somalia, I witnessed extreme depths of human suffering, against a back drop of naked anarchy, wanton destruction and total collapse of social, economic and political structures. ... The UN must accept the challenge of Somalia with determination and perseverance. ... The consequences of failure do not warrant contemplation.[23]

About the same time Mr Ismat Kittani, an Iraqi diplomat, was appointed by the UN as Special Envoy to Somalia. Soon after his arrival in Mogadishu on 8 November, he told the Western media:

> The resolutions of the Security Council and the mandate talk about 'authorities', Somali authorities. Authority in Somalia, in the sense

that we know it anywhere else, does not exist. ... In some cases the [local] authority is no more than three or four people with guns.[24]

Every day Western news services were flooded with stories and images of Somali women, children and old people starving to death while gunmen rode around on their technicals and the warlords bickered with the UN over what needed to be done. World public opinion, stimulated by the media exposure of the plight of the Somali people, grew, favouring a decisive military intervention.

On 15 November, Greg Jackson's situation report described an escalation of hostilities in Mogadishu. Aideed's forces were stepping up their harassment of UNOSOM vehicles and personnel. In one 90-minute engagement between Aideed's technicals and the technicals hired by UNOSOM for protection, rifles, machine guns and mortars were used to send UNOSOM technicals scurrying back to UNOSOM compounds. Aideed was on the record stating that his forces would oppose any attempt to deploy more foreign troops to Somalia.

Humanitarian operations in Mogadishu were also under increased threat. Vehicles belonging to aid agencies were being hijacked and employees robbed. Ships and aircraft carrying humanitarian aid were being fired on and looted. Somali guards, hired by the expatriate NGO managers, looted relief supplies from warehouses, broke into offices in search of cash and valuables, and robbed expatriate employees of their cash and belongings.

As UNOSOM's predicament deteriorated in November, UN leaders in New York began to develop options for breaking the deadlock in Somalia. Colonel Bruce Osborn,[25] the Australian Defence Attache with the UN in New York, kept Headquarters Australian Defence Force (HQ ADF) abreast of developments and predicted that Australia would be one of the nations requested to contribute to an increased international effort in Somalia.

On 25 November a breakthrough occurred. Lawrence Eagleburger, the acting US Secretary of State, visited the UN Secretary General, Dr Boutros

Boutros-Ghali, to advise him that the US were prepared to lead and sustain a multi-national peace enforcement operation into Somalia to protect the distribution of humanitarian relief operations. Eagleburger promised a force headquarters, 30 000 US troops and logistic support.

Prompted by Eagleburger's visit, Boutros-Ghali wrote to US President-elect, Bill Clinton, three days later. The options he discussed in his letter were to persevere with the UNOSOM *status quo*, withdraw from Somalia completely, conduct a show-of-force operation with the Pakistanis in Mogadishu or conduct a large scale US-led or UN-led peace enforcement operation to protect the distribution of humanitarian aid and to 'promote national reconciliation and reconstruction'. He also called for the disarming of the clan armies and the confiscation of heavy weapons during peace enforcement operations to set the scene for renewed UNOSOM peace keeping efforts to resolve Somalia's political problems.

Independent of UN deliberations and discussion of options, American military planners told the world of their intentions. On 29 November, the US Armed Forces Central Command, based at MacDill Air Force Base, Tampa Bay, Florida, declared that a US Marine Air-Ground Task Force of 1 800 men, currently sailing off the southern Indian coast, had been ordered to change course for Somalia. Their orders were to spear-head an assault on Mogadishu.

The Marine Task Force, called the Tripoli Unit, after its flagship USS *Tripoli*, was a specialised unit trained for humanitarian missions. The helicopter assets of the Tripoli Unit could transport a battalion of Marines in a single lift accompanied by four AH-1 Huey Cobra attack helicopters for fire support. The primary focus of the Tripoli Unit's training had been the evacuation of American civilians from areas where they were endangered, either as a consequence of a breakdown in law and order or a natural disaster. The unit had been deployed to Mogadishu in January 1991 when the lawlessness accompanying the overthrow of Siad Barre had threatened the lives of 300 American citizens, mostly diplomatic and NGO staff, and their families. The Marines had protected American life and property until

Lieutenant Colonel Ray Martin, Acting Director, Joint Plans HQ ADF, *during the planning of Operation Solace*
Image: Ray Martin

Lieutenant Colonel Paul Ritter, Acting Colonel Operations and key planner at Land Headquarters for Operation Solace, the ADF's *participation in the US-led Operation Restore Hope*
Image: P B Retter

Aideed and Ali Mahdi had agreed to their UN-brokered ceasefire on either side of the Green Line.

Earlier in November Bruce Osborn's advice had initiated contingency planning at HQ ADF in Canberra and at LHQ in Sydney.[26] This type of planning had become routine because of an increase in the tempo and complexity of Australian military support to UN-sanctioned operations and missions overseas during 1991 and early in 1992 The planning staff at both headquarters had provided options for Cabinet consideration, and had subsequently planned and mounted Australian service contingents to serve with the UN Transitional Authority in Cambodia, the UN Mine Clearance Training Team in Pakistan, the Multi-National Force and Observers in the Sinai and UNOSOM.

Lieutenant Colonel Ray Martin in Canberra and Lieutenant Colonel Paul Retter in Sydney were given responsibility for the detailed preparation of options for ADF support to a large scale US-led military intervention into Somalia.[27] Retter recalled:

Ray and I had worked together for some time and had built up the trust required for cooperative planning as well as a healthy scepticism about the possibilities of seeing our plans get up. In late November our discussions were not optimistic, but we felt from the beginning that the chances of a combat force getting away [overseas] were higher than they had ever been in our time in the contingency planning game.[28]

The key to Martin and Retter's success in contingency planning was the experience of their staff. Major Dan Ryan was 'the whiz kid' of contingency planning and brief-writing at HQ ADF.[29] Ryan had mastered the use of computer technology to manipulate data, consolidate information and produce concise, high quality staff work. He and Martin had eight UN operations abroad, as well as a number of operational activities and exercises in Australia and overseas, to report on each day. They were also 'on call' anytime during the day or night to prepare land operations briefs at short notice. On top of these duties they had to consolidate contingency planning information developed by Retter's staff in Sydney for presentation to General Peter Gration, the Chief of the Defence Force.[30] Dan Ryan was on a keyboard treadmill every day, sometimes in 12 hour stretches, to meet reporting and briefing deadlines. For his part, Martin was a formidable communicator, able to absorb complex detail and present the various options effectively and intelligently to senior military officers, Department representatives and politicians.[31]

Retter was supported by Lieutenant Commander David White, a Navy officer seconded to LHQ, and Major Melva Crouch, a movements specialist.[32] White and Crouch operated around the clock to produce contingency plans. By the time they were required to prepare contingencies for deploying various combinations of units and sub-units to Somalia, they had a vast store of information and experience to draw on.

Initially, Martin and Retter's contingency planning for Somalia was severely hampered. They had little information to work with and were not permitted to involve Maritime or Air Headquarters in Sydney or subordinate headquarters, such as HQ 1st Division in Brisbane, in their planning considerations.[33] Consequently, they could not discuss the options for deploying Army assets by sea and air to Somalia with their Navy and Air Force counterparts, or other Army headquarters staff.[34] Strict secrecy also reduced their ability to gain access to crucial planning information from units within the Army.[35]

Both men knew that the timing of this operation was going to provide a major challenge. Army formations and units were about to stand down for the annual Christmas-New Year leave period. During the same period, there would be a high turnover of personnel moving to and from units as a consequence of the Army's annual posting cycle. If one of their options for an increased commitment to a US-led operation in Somalia was accepted by the Cabinet, they knew that it would be a challenge to recall personnel from leave and deploy an expeditionary force 11 000 kilometres across the Indian Ocean to Somalia during the Christmas-New Year period.[36]

The only combat formation available for overseas deployment at short notice was 3rd Brigade, based in Townsville, which was constituted as an Operational Deployment Force (ODF). The brigade was due to stand down for Christmas leave on 4 December. However, 1st Battalion, the Royal Australian Regiment (1 RAR) would be on call with a company group ready to move quickly and the remainder of the battalion to follow soon after.

Fortunately, 1 RAR had practised rapid deployment by air and sea in 1992 In March most of the battalion had deployed by sea in HMAS *Tobruk* to Melville Island, located in the north west of the Northern Territory. During this exercise, 1 RAR soldiers had been practised in service protected evacuation training. This type of training involved setting up vehicle check points, controlling crowds and operating under strict rules of engagement (ROE) and orders for opening fire (OFOF) for the use of weapons.

Observing strict secrecy, Retter, White and Crouch developed 11 options and assessed the suitability of each for application in Somalia. The options ranged from deploying medical, supply, transport, engineer and maintenance sub-units of no more than 200 personnel to sending a full strength infantry battalion group of just over 1 000 troops.[37]

On 3 December, the UN Security Council passed Resolution 794 which endorsed, under Chapter 7 of the UN Charter, the conduct of a US-led multi-national peace enforcement operation into Somalia to be called Operation *Restore Hope*. The mission was to establish a secure environment for the distribution of humanitarian aid as soon as possible. While Resolution 794 was being drafted, China and several African nations had vigorously opposed American command of Operation *Restore Hope*.[38] A day had been spent haggling behind closed doors until the wording was amended to play down American command and increase UN oversight of the operation.

The Security Council concurred with the appointment of a US Marine Corps officer, Lieutenant General Robert Johnston, General Norman Schwarzkopf's Chief of Staff and operational spokesman during the Gulf War, to command the operation. However, unlike Schwarzkopf, Johnston would have an *ad hoc* Security Council Commission to oversee the operation, a UN liaison office attached to his headquarters and a requirement to provide Boutros-Ghali with periodic progress reports.[39]

Expatriate NGO managers and employees in Somalia were enthusiastic about the forthcoming military intervention. The looting of relief supplies made them angry and frustrated. 'We want them to come now, not tomorrow, now. We are fed up with the war and these stupid boys with guns who have made our lives a misery for two years. This horror story has gone on for too long and the time has come to break the cycle of crime and take away the guns at the point of a gun.'[40]

On same day Operation *Restore Hope* was announced, a *Sydney Morning Herald* editorial summarised American intentions for Somalia:

> Once inside Somalia, therefore, US troops will probably set about securing delivery points for relief supplies, punching safe supply routes to the country's drought-stricken interior and providing protection for food distribution centres. They won't be in the business of solving Somalia's underlying political and economic problems. Their efforts could make two valuable contributions to ending the suffering there: ensuring that relief supplies get through to the needy and depreciating the value of food to Somalia's extortionists by increasing its supply

throughout the country. That may not be the stuff of which humane world orders are made. But it is an overdue act of humanity towards the victims of Somalia's hell on earth.[41]

Prompted by a message from Osborn in New York that the Australian Government was about to be asked to support both Operation *Restore Hope* as well as reinforce UNOSOM, Retter's team considered providing units and personnel for both operations. Major General Murray Blake, the Land Commander, recalled:

> At that time there was some uncertainty about the Government's intent. The policy options were to provide short-term larger scale humanitarian relief or longer- term, smaller-scale nation-building support to a new UNOSOM II. I did not think we were going to participate in both to a significant level because of cost. In respect to the nation building option, I was pressing for a kind of military assistance group based on a rifle company providing security for a range of specialist services—mainly engineer and medical. I saw this unit of about 200 personnel going into a region, presumably after stability had been restored, and getting on with support tasks for the local nation building effort.[42]

To support Operation *Restore Hope* in the short term, Blake decided he would only offer two infantry battalion group options.[43] Advised by Retter's planning team, he based his assessment on the enforcement nature of Operation *Restore Hope*, the speed of deployment required, the military threat in Somalia and the knowledge that after deployment the Americans planned to reduce their forces in Somalia. Retter recalled:

> From the beginning we understood that the Americans planned to kick the door down in Somalia and, to use US Marine parlance, 'take down' Mogadishu and the major regional towns. The Marines were then going to hand over to coalition forces who would establish themselves throughout the country to protect the distribution of humanitarian aid. The Americans did not plan to stick around. We had to deploy a force that could take care of itself and fight hard to protect itself if that became necessary.[44]

In the weekend newspapers on 5-6 December the Australian public learned more about American plans for the military intervention into Somalia and that several countries around the world, including Belgium, Morocco, France, Pakistan and Italy, had offered infantry battalions.[45] It was this sort of information that confirmed in the minds of Martin and Retter that infantry battalion group options for providing Australian support to Operation *Restore Hope* were appropriate.

On Monday morning 7 December, General Colin Powell, Chairman of the US Joint Chiefs of Staff, conducted a press conference in Washington to brief media representatives on the forthcoming operation. He stated that it would take about three weeks to get troops into position in Somalia and then a further six weeks to subdue the country and create a secure environment for the distribution of humanitarian aid.

Initially, the US Marines would seize airports and the docks in Mogadishu and Kismaayo. Twenty six thousand American and coalition troops would follow and consolidate in those two metropolitan areas before clearing inland routes and securing food distribution centres. In the final phase of Operation *Restore Hope*, the Unified Task Force (hereafter UNITAF), as the *Restore Hope* formation was to be called, would hand over to UNOSOM units for a continuation of peace keeping operations. Based on Powell's timings and the likelihood of the Marines landing within the week, he estimated that UNITAF would complete its mission by mid-February 1993

While Powell delivered his briefing to the international media, senior military officers in Canberra waited for a formal request from the Americans for Australian military support to UNITAF. On the same day, gunmen in Somalia went on the rampage to capture as much relief supplies as they could before *Restore Hope* began. In Baidoa, an important distribution point for several aid agencies, including CARE Australia, looting at gunpoint forced the evacuation of all but two expatriate staff. Rick Grant, a spokesman for CARE Australia, reported, 'Security has fallen apart. We tried to evacuate everybody, but two had to stay because they figured that they would be killed if they tried to leave.

Thugs in Baidoa have threatened that they will beat up the staff, loot their belongings and rape the female relief workers.'[46] On the previous Friday Somali gunmen had attacked and robbed a CARE worker of $US 20 000 in cash.

Unbeknown to HQ ADF at the time, the formal request for Australian support had arrived at the American Embassy in Canberra two days before on 5 December. Unfortunately, it had been redirected internally and not sent to the Department of Foreign Affairs and Trade immediately. General Peter Gration discovered that the New Zealand Defence Force had agreed to provide air transport support to UNITAF. Presumably, this support had been agreed to after an American request. Gration wondered why the Australian Government had not been approached. After a discreet, informal inquiry through military diplomatic channels, the American request was found and the American Ambassador, Mel Sembler, arranged for a meeting with Senator Gareth Evans, the Minister for Foreign Affairs and Trade, on Tuesday 8 December to discuss the request.[47] Soon after meeting Sembler, Evans called a meeting with the Minister of Defence, Senator Robert Ray to examine the American proposals.

On the day Sembler met with Evans, General Gration and his senior staff were examining the battalion group options developed by Retter's planning team and refined by Ray Martin and Dan Ryan. General Blake was very keen for Gration to recommend to Robert Ray that an infantry-based force from the ODF should be deployed:

> The difference between this request for support and others we had received before was the real possibility of getting part of the ODF away. That was significant because the commanders in the ODF had been saying for sometime that they had been training hard, were fully equipped and on the shortest notice to go. Why not us? It was becoming a morale problem and everyone from the CGS [Chief of the General Staff, Lieutenant General John Grey] down was getting the message. I had discussed this with the CDF [Gration] and the CGS [Grey].[48]

Meanwhile, as the Australian generals considered their options, the Americans were conducting a psychological operations campaign to let the

Somali warlords know that the Marines would be arriving soon. Mr Dick Cheney, the US Defence Secretary, had decided that it was better to create fear in the hearts of Somalis by showing what was in store for them rather than cloaking the operation in secrecy. He stated emphatically, 'We are not going to wait for people to shoot at us'. The American media showed pictures of heavily-armed troops from the 10th Mountain Division, a rapid deployment formation, based at Fort Drum in New York, and the 1st US Marine Division, based in Camp Pendleton, California, making their final preparations for deployment.

The signals from Somalia were mixed. At the lower levels, within the armed clans, and among urban criminals and the many roving bandits, there was talk of killing American troops and resisting the invasion. In contrast, the warlords stated that they welcomed the arrival of UNITAF. Robert Oakley, a former US Ambassador to Somalia and now Special Envoy appointed by out-going President George Bush, met with the two warlords controlling Mogadishu. Both Aideed and Ali Mahdi assured him that they would not resist a beach assault by American Marines. Commentators suggested that this show of support was calculated to gain favour with the Americans at the expense of rival political opponents.[49]

Several realities remained constant throughout the exchange of words in the press and on television. Every Somali male over the age of 12 years had a personal weapon, and every armed faction had access to heavy machine guns and anti-armour weapons mounted on vehicles. The larger clan armies had artillery, mortars and tanks at their disposal. Hundreds of thousands of armed Somalis had experienced two years of war—they were ready and able to kill. Major General Shaheen, the UNOSOM commander, estimated that there were 15 000 heavily-armed Somalis in Mogadishu supported by 150 vehicles mounted with heavy weapons, such as .50 calibre machine guns, and an endless supply of ammunition.[50]

The first signs that American tactics were working was when Western journalists reported that hundreds of Somali gunmen were leaving

Mogadishu in long convoys of technicals and trucks. There were reports that weapons were being stockpiled in secret locations. The city had an unprecedented quiet night on 8 December.

It was a lull before a storm of military activity. Oakley announced that Marines would land the next day. He assured journalists that the arrival would not involve 'all guns blazing' and would be spread out over several days. Journalists obliged him by reporting the arrival of American air traffic controllers at Mogadishu airport and an American direction to aid agencies that there would be a suspension of civilian aircraft flying into Mogadishu until all the Marines had arrived safely. This was turning out to be an extremely well-publicised military operation. The imminence of the operation was also telegraphed to the residents of Mogadishu by several F-14 Tomcat reconnaissance jets conducting low level photographic runs over Mogadishu throughout the day.

In Australia, since the receipt of the American request and the acceptance by Gration of the infantry battalion group option, Paul Retter, Melva Crouch and David White were examining the feasibility of concentrating, equipping, training and deploying an infantry battalion group based on 1 RAR with APC support to Somalia. The guidance from Canberra was that a force of no more than about 900 personnel had to be in Somalia by mid-January 1993 to take over from departing US Marine units. The operation would have a duration of 17 weeks and no longer.

The selection of 17 weeks as the duration of Operation *Solace* was a judgment by General Gration, in consultation with Lieutenant General Grey and Major General Blake. A finite commitment made good political and financial sense. General Colin Powell, had estimated that UNITAF would create the secure environment for the distribution of humanitarian aid in about two months. Gration, Grey and Blake doubled the American estimate based on their appreciation of the time it would take to deploy UNITAF, establish a presence and hand over to UNOSOM units; hence two months became four months.[51]

The two other factors influencing the choice of the duration of the Australian commitment were cost and the perception by media commentators that Australia would be vulnerable with the majority of 3rd Brigade deployed in Africa. Gration assessed that the Government would only accept the deployment of a larger force on the condition that the duration of its operations was limited to keep costs down and was politically compatible with achieving humanitarian objectives, not the beginning of a longer term military commitment to the UN's agenda in Somalia.[52]

In the early hours of 9 December (local time), US Marines landed in Mogadishu. The time differences enabled the American public to get a ringside seat through broadcasts on CNN About two hours before the landing hovercraft arrived, groups of heavily-armed American Navy Sea-Air-Land (SEAL) special forces teams landed silently in rubber dinghies. They were surrounded immediately by journalists and camera crews with an assortment of bright, white lights. This made the SEALS targets for any enterprising Somali gunman hiding in the dark watching the show. The discomfort of the SEALS with their blackened faces and bulky personal equipment was obvious.

After the first group of SEALS shook off eager reporters with gruff, 'No comment, No comment'! and disappeared into the night towards pre-arranged rendezvous points around the airport, another rubber dinghy arrived with a Marine officer who warned the gathering of journalists to pull back from the water line so they would not be injured by the following landing craft. The lone officer was then drowned out by the sound of several hovercraft containing Marines and vehicles speeding towards the beach.[53]

LESSONS

Somalia's story since 1960 is a useful case study of post-independence self-destruction in the Third World as well as how international military intervention can occur when Western public opinion is aroused. After being granted UN-brokered independence, Somalia was not the only nation in Africa and the developing world to be left with a legacy of divisive colonial borders and a well-intended experiment in Western-style democracy. Some have become broken-back states with ailing economies, high levels of international debt, low standards of living and persistent inter-tribal and criminal violence. Like Somalia, the road they have travelled since independence has been signposted with political instability, dictatorship, tribal conflict, Cold War politics and civil war.

The story of Somalia provides an insight into the origins and nature of the violence that can stunt as well as traumatise a whole society—possibly a modern form of the Dark Ages, possibly more accurately dubbed the Dark Decades. Like several other Third World nations, Somalia's dark decades were characterised by ancient barbarism and modern weapons technology. The armies who looted, raped and pillaged in Somalia were motorised and armed with artillery, tanks and aircraft, not travelling on horseback, armed with swords, spears and arrows. Persistent barbarism, combined with modern lethality, increased the numbers of Somalis who would suffer and made the devastation of Somalia's social structure, as well as its infrastructure, inevitable.

The UN, international financial institutions and NGOs work hard to pull these broken-back states out of their Dark Decades. Unfortunately, many of them have the potential to deteriorate further to the point where international military intervention will be called for to protect on-going international humanitarian efforts and save thousands of lives. The near-intervention of a Canadian-led multi-national force into eastern Zaire in November 1996 to facilitate the distribution of humanitarian aid to thousands of Rwandan refugees is a case in point.

International media coverage seems to constitute one of the most influential factors prompting Governments to send the young men and women of their armed forces into hell's kitchens like Somalia. The harrowing images and emotional stories of those suffering during the famine in Somalia provoked a sense of outrage that nothing was being done to stop the warlords and their cohorts looting humanitarian supplies. Consequently, the NGOs, assisted by the media, not only launched appeals for funds in many Western countries but also called for military intervention.

The Ethiopian famine in the 1980s is another example of the influence of media coverage. The international popular music industry, mobilised by Irish vocalist, Bob Geldorf, stimulated publicity about the plight of thousands of starving Ethiopians. Millions of dollars were raised through concerts, donations and record sales. Fortunately, military intervention was not called for to protect the distribution of the resultant humanitarian aid.

The Ethiopian and Somali examples, as well as the images of starving or physically-threatened refugees broadcast before the military interventions on behalf of fleeing Kurds in northern Iraq in 1991 and Hutu refugees in Rwanda in 1994, prompts an hypothesis, 'If the international media were kept out and as a consequence unable to broadcast images of humanitarian disasters, would there be sufficient public opinion mobilised to pressure Governments into committing troops?'

The absence of harrowing images from eastern Zaire in November 1996 supports this hypothesis. The UN's efforts to get a sufficient commitment from member states for military intervention seemed to be stymied by a lack of images of thousands of suffering Rwandan refugees. There was also a problem of verifying how many refugees were wandering around in the forests because no journalists and camera crews were permitted to move into the area and report back.

The lack of international support for intervention into eastern Zaire was also influenced by the absence of a substantial American commitment to deploy US armed forces. The stories of the formations of UNOSOM and UNITAF illustrates the difference that US commitment can make to international military interventions. Under the pressure of international public opinion and informed by the Hanson Report, the UN Security Council passed Resolution 775 in late August 1992 proposing the protection of the distribution of humanitarian aid by deploying five international infantry battalions to Somalia as soon as possible. Considering the well-publicised dire circumstances of over 4 million Somalis, the response of member nations was slow. It took over six weeks for a group of nations to agree to donate sufficient forces to fill out the proposed organisation of UNOSOM. Like other nations, the Australian Government took its time to consider a number of options before agreeing to provide a small movement control unit. On 20 October the ADF sent Bill Nagy and Paul Angelatos to prepare the way for the deployment of the movement control unit to Somalia, only to discover that UNOSOM was politically and militarily snookered in Mogadishu.

UNOSOM faltered because negotiations were not backed up with sufficient force. The combination of diplomacy, the deployment of military observers to supervise a truce in 1991 and the prospect of lightly-armed foreign forces protecting the distribution of humanitarian aid in 1992 did not persuade Somali warlords to give UNOSOM a chance. They survived through political cunning, as well as the threat and application of force.

UNOSOM negotiators found that there was little prospect of Somali warlords responding co-operatively or compassionately to appeals to allow foreign troops into southern Somalia to interfere with their looting of humanitarian relief supplies and save the lives of thousands of their compatriots unless they were faced with overwhelming force.

Eventually, the US Government decided not to stand by and let the UN fail in Somalia in order to prevent a monumental human catastrophe and to respond to strong international support for military intervention. Despite being in the last weeks of his administration, President George Bush committed 30 000 US troops, a fleet of USAID-managed transport aircraft and tens of millions of dollars to enforce peace in Somalia and fly in thousands of tonnes of emergency relief supplies. The subsequent landing of US Marines on Green Beach in the early hours of 9 December was a defining moment for American foreign policy. For the first time a large number of American ground troops were being committed to enforcing peace and protecting the distribution of humanitarian aid in Africa. With the Americans in the lead, Australia and over 20 other nations quickly pledged their support and encouraged US requests for their assistance.

Aside from underwriting the international intervention into Somalia with the promise of troops, aircraft, relief supplies and funds, President Bush and his Chairman of the Joint Chiefs of Staff, General Colin Powell, reiterated the US Government's 'no casualties' policy. President Bush was not going to follow the UN into Somalia with too little, too late and put the lives of American and allied service personnel at risk. The US-led UNITAF was going into Somalia with overwhelming force that had characterised American military interventions into the micro-states of Panama and Grenada in the 1980s, and the prosecution of the Gulf War in 1991. The Somali warlords who had toyed arrogantly with UNOSOM were now compelled to take notice of UNITAF. Images of the Hollywood-inspired arrival of the cavalry fitted American intent. Senior US Government officials, like Mr Dick Cheney, the Secretary of Defence, made no secret of their intention to take over southern Somalia, save the lives of thousands of beleaguered Somalis and blow anyone out of the saddle who opposed them.

Until Operation *Restore Hope* was announced the Australian Government had pursued a policy of supporting UN peace keeping operations with small specialist contingents. This policy had the benefit of putting a reasonable price on the Government's claim that Australia was a responsible member nation of the UN. The prospect of participation in Operation *Restore Hope* triggered a more traditional Australian strategic policy. Like their conservative predecessors in 1950 and 1965 who had committed infantry battalions to the Korean and the Vietnam wars, the Labor Government was attracted to the strategic benefits of making a more substantial commitment to a military intervention into Somalia now that Australia's major international ally was in the lead.

In great secrecy Generals Peter Gration, John Grey and Murray Blake authorised ADF planners to develop a proposal for the deployment of about a 1000-strong battalion group, including a small national headquarters, for 17 weeks. They assessed that the political will was probably there for a commitment of this size because the Americans were going in, a Federal election was in the offing and Australian public opinion was supportive.

They also felt that this proposal made good military sense. The situation in Somalia was volatile, the mission was peace enforcement, not peacekeeping, and the Americans were planning to withdraw their units as international forces arrived. If the warlords decided to confront the international battalions after American combat units left as planned in January, then an Australian infantry battalion, supported by APCs, would have the ground mobility and fire power to take care of itself.

In addition to external factors influencing the selection of a battalion group option, there was also internal pressure from within the Army to deploy a combat force from the ODF. Since its inception in early 1980s, this formation had been kept at high states of readiness without 'getting a guernsey' when the ADF deployed contingents overseas for peacekeeping operations. The generals agreed that the opportunity to support the US in Somalia gave a battalion group from the ODF a chance to test itself on operations.

Though they may not have known it then, the deployment of a combat force of this size overseas for four months would also test the ADF's ability to plan, prepare for, conduct and support off-shore land operations.

2   GETTING THERE

*'What is certain is that conflict will continue in the world,
and the best way to deal with it will continue to be through
negotiation. Sadly that negotiation will frequently only
be possible when the protagonists have been held apart by
armed and objective nations acting as peacekeepers. Even
more sadly, there will occasionally be times when objective
nations will have to impose peace through the application of*
UN *sanctioned force ...'*

General de Chasterlain,
the Canadian Chief of the Defence Force Staff,
January 1991

On 8 December, the day before the US Marines landed in Mogadishu, Major Glenn Crosland, 2IC 1 RAR, received an unexpected and unusual telephone call from Colonel Peter Sibree, Colonel Operations, Land HQ.[1] Having failed to make contact with Lieutenant Colonel David Hurley, CO 1 RAR,[2] who was on local leave, Sibree directed Crosland to have Hurley go to HQ 3rd Brigade as soon as possible and call him on the secure telephone used for confidential discussions. Crosland did this but remained suspicious even after Hurley told him subsequently that the call from Sibree was routine, just a pre-Christmas update on 'world-wide issues'.[3] The next day, among front page news of the successful, unopposed Marine landing in Mogadishu, Cameron Stewart, a journalist from *The Australian*, broke the news that the Federal Government was giving serious consideration to a US request for 'Australia to provide a large number of troops to Somalia as a part of OP *Restore Hope*.' [4]

Soon after reading about the Marine landing in Mogadishu and the US request to the Australian Government, Crosland discovered that Hurley had come in again to talk to Sibree on the secure telephone. He had a sinking feeling in his stomach that he would be left behind if 1 RAR deployed to Somalia because he had recently accepted a redundancy package from the Army and was due for discharge in February.[5]

Later that day, he received a call from Hurley to come to HQ 3rd Brigade for a briefing. On arrival he found Hurley in the company of Colonel Neil Weekes, acting Brigade Commander, Lieutenant Colonel Don Murray, CO 4th Field Regiment, Royal Australian Artillery, Lieutenant Colonel Mike Saw, CO 3rd Brigade Administrative Support Battalion and Major Gary Banister AM, Deputy Quartermaster Assistant Adjutant General (DQ), 3rd Brigade.[6] Hurley announced that 1 RAR with attached sub-units might deploy to Somalia in January in support of OP *Restore Hope*.[7] Crosland was told to prepare for a trip to Sydney with Mike Saw and Gary Banister to attend a planning conference at Land HQ on Friday 11 December.[8]

Meanwhile, Colonel Bill Mellor, the Chief of Staff, Headquarters 1st Division in Brisbane, learned from his commander, Major General Peter Arnison, that he was to be the national commander of the Australian Force-Somalia (CAFS), if the Cabinet approved the proposals being put before them the following Tuesday. Mellor was also told that Lieutenant Colonel Graeme Woolnough, a newly-arrived staff officer at the Headquarters would 'stand up' his headquarters and be his logistics liaison officer.[9]

On Friday 11 December Paul Retter convened a Joint Operations Planning Group (JOPG) meeting in Sydney.[10] This planning group knew that the 1 RAR Group would be allocated a Humanitarian Relief Sector in Somalia and its mission would be to protect the distribution of humanitarian aid by NGOs to the populace. They knew from intelligence and media sources, as well as first-hand reports from Greg Jackson in Mogadishu, that the Somali warlords had large numbers of lightly-armed militia and 'technicals' at their disposal.[11] Although the Marines had landed in Mogadishu unopposed, Retter's group planned for worst case scenarios. If the warlords decided to oppose the Unified Task Force (hereafter UNITAF), they had sufficient artillery, heavy weapons mounted on vehicles, armed troops and ammunition to cause casualties.

The JOPG made other assumptions so they could produce a realistic deployment plan for presentation to General Blake the next day. They assumed that only ADF strategic assets, RAN Ships *Jervis Bay* and *Tobruk*, and RAAF 707 and C-130 Hercules aircraft would be available to move the AFS.[12] The main body of the AFS had to be in Somalia in four weeks time with 14 days resupply stowed on *Jervis Bay* and 30 days 'follow-on' stocks in certain commodities stowed on both HMAS *Jervis Bay* and HMAS *Tobruk*. Leutenant Tony Powell from *Jervis Bay* and Warrant Officer Peter Macdonald from *Tobruk* assisted David White and Melva Crouch to formulate loading plans for both ships.[13]

Water and fuel resupply were key issues. The JOPG had to strike a balance between the quantity of ammunition, vehicles and other equipment to be stowed on *Jervis Bay* and *Tobruk* and quantities of water and fuel. The question was, 'How far would US forces, who were planning to reduce their presence after international forces arrived in Somalia, be able to support the 1 RAR Group ?' For the time being, the JOPG assumed that the Americans would provide essential commodities such as water, fuel, food and certain classes of ammunition after the first 30 days.[14]

The provision of aviation support was another key issue. Retter recalled:

> We made a conscious decision not to consider aviation assets [helicopters] because we knew that the inclusion of helicopters to support the force would necessitate reductions in numbers of Infantry and APCs. We were limited to sending just over 900 personnel. Given the large number of US Army and Marine Corps aircraft expected to be deployed into the theatre [Somalia], we anticipated receiving sufficient support from them. Had we gone with helicopters and their supporting people, the 900 figure would have been blown out the window. There seemed to be a desire in Canberra to keep the force under a thousand [personnel]- some sort of magical number.[15]

General Gration, after receiving advice from a member of his planning staff, assessed that the 'magical number' of 1000 personnel operating for 17 weeks at a total cost of $20 million constituted a proposal he could sell to Cabinet.[16] Grey and Blake accepted this assessment. Blake was also conscious that, in the time frame of just over four weeks, whatever force was deployed had to stow its weapons, vehicles, equipment and reserve stocks on *Jervis Bay* and *Tobruk*.[17]

While Retter's planning group in Sydney were putting the final touches on documents for Blake's approval on Saturday 12 December, the Australian public was being prepared for a sizeable Australian commitment to Somalia. Tony Wright, of the *Sydney Morning Herald*, reported the content of the discussions between Ambassador Sembler and Senator Evans. Quoting 'sources', he summarised Sembler's preference for a battalion-sized force but acceptance of a lesser-sized force if that was all Australia could spare. Wright's claim that Sembler had presented Evans with a 'shopping list' of options and had expressed a preference for a battalion group was not supported by military planners in Canberra.[18]

On Saturday 12 December Retter briefed General Blake on numbers of personnel, weapons, equipment and vehicles that would comprise the AFS. He then elaborated on the concept for deployment of the AFS to Somalia and the resupply of the AFS after getting there. After some lengthy discussions, Blake approved the proposals and they were forwarded to Canberra to become the basis for Defence Department proposals being prepared for submission to the Cabinet on the following Tuesday. Ray Martin, now the acting-Director of Joint Operations, and Dan Ryan spent a busy Saturday night and Sunday developing a draft Cabinet submission.[19]

The important job of estimating the cost of Blake's proposals fell to Bob McDonald, Deputy Director-Operational Costing, at Army Office in Canberra. McDonald was a veteran of 34 years in the Defence Department and had worked in the Army's Resources Planning Section for 17 years. He attended a Joint Administrative Planning Group (JAPG) meeting on 9 December chaired by Colonel Chris Stephens.[20] The meeting began at 10am. At 1pm McDonald was told to go away and have financial estimates for OP *Solace* ready by 3pm. He thought this was a ridiculous time frame. In any case, these estimates could not be made until General Blake approved Retter's final plans and administrative concepts. McDonald waited until the final proposals were available on 12 December to develop his estimates.

For the next two days McDonald approached several agencies for information upon which to base his financial estimates. All he knew for sure was the composition of the AFS and the duration of the commitment. Staff at Army Office, Land HQ and HQ Logistic Command could not tell him what extra payments that would be made to those serving in Somalia under amended conditions of service, the estimated usage rates for stores and spare parts, the estimated costs of repairs for vehicles and equipment, and what extra costs might be incurred to meet the operational needs of the AFS. No senior staff appeared to have been recalled from leave to provide this sort of information. McDonald was left to speak with junior officers 'minding the shop' over the Christmas-New Year period.

After McDonald received General Blake's approved plan on 12 December, he and a colleague, Ms Fiona Hanks, sat back to back in his small office for almost 48 hours straight preparing the financial estimates by hand with calculators. They had to have them ready for inclusion with Ray Martin's draft Cabinet submission.[21] McDonald and Hanks met the deadline and he delivered his final estimate of $19, 557, 000 personally to the Defence Minister's office. The member of Robert Ray's staff he handed his estimates to said, 'That's good, it's under 20 million.'[22]

On the afternoon of 12 December, after approving Retter's proposals, General Blake had assessed that the Government would probably baulk at the cost as well as the responsibility for sending Australian combat troops overseas for four months. He told Retter to take a week off because the odds were three to one against the proposal getting up. Retter took his family to Terrigal, two hours drive north of Sydney. He felt professionally satisfied that in the time available he and his staff had prepared a feasible, well-argued proposal. Personally, he thought it would be just one of the many exercises in contingency planning he had completed over the past year and, like most of them, it would come to nothing.

While Ray Martin and Dan Ryan prepared a Cabinet submission based on General Blake's proposals. Key personnel who would be responsible for loading and despatching the 1 RAR Battalion Group to Somalia met in Townsville. This planning group was chaired by Gary Banister, who had returned from Retter's meeting in Sydney on Friday and briefing Mellor and Woolnough in Brisbane on Saturday afternoon. Others in attendance were David Hurley, Major Greg Birch, OC 2 Field Supply Battalion, the unit holding the bulk of ODF stocks, and three staff Captains from HQ 3rd Brigade, Mick Fulham, Ian Young and Nick Roundtree.[23]

Fulham and Young were told that they had been selected to serve in Somalia with Colonel Bill Mellor's national headquarters. Both were delighted to be going overseas but were puzzled that a headquarters group was not coming from HQ 3rd Brigade in Townsville which commanded the units and sub-units making up the AFS. The provision of such a headquarters was already in the Standing Operating Procedures of the ODF for situations when one of the two ODF infantry battalions was deployed.[24]

At this meeting Hurley stated that the 1 RAR Group would be recalled in three phases once the official announcement of the deployment was made in Canberra.[25] He did not want to recall everyone immediately because the deployment was not until mid-January, deciding that as many families as possible should have loved ones at home for Christmas rather than waiting around the barracks in Townsville while plans were being made.[26]

The weekend newspapers of 12/13 December featured the operations in Somalia prominently. All the news was good. The Mogadishu-based warlords agreed in a joint communique brokered by Robert Oakley to declare an immediate ceasefire and to withdraw their troops and 'technicals' from the city. The most significant breakthrough was the call for a 'unity committee' to meet and begin solving Somalia's political impasse. For the international media, Aideed and Ali Mahdi embraced and shook hands with a beaming Oakley in the background. In one 'photo opportunity' Oakley, representing the United States, had recognised two of Somalia's most ruthless warlords as the two legitimate contenders for power in southern Somalia.[27]

On Monday 14 December Chris Stephen's JAPG met again in Canberra. However, it was too late to change anything. From Bob McDonald's point of view, the Army was locked into financial estimates that were based on preliminary planning information.[28] Later that day, General Gration and his senior operations adviser, Air Vice Marshall Les Fisher,[29] endorsed

Lieutenant Colonel David Hurely, CO 1 RAR Group, at a media briefing in Townsville, 15 December 1992
Image: *Townsville Bulletin*, Michael Chambers

Ray Martin's Cabinet submission and copies were forwarded to Senator Robert Ray for consideration. Gration and Ray then spoke at 4 30 pm and Ray approved the submission. At 6 30 pm Gration briefed the Prime Minister, Paul Keating, the Minister for Finance, Ralph Willis, the Minister for Foreign Affairs and Trade, Gareth Evans and Robert Ray. [30] The weekend's good news from Somalia and a forthcoming election in March may have influenced the Ministers who listened to Gration's briefing that evening.

On the morning of 15 December Paul Retter picked up a newspaper and read that the Government had approved the deployment of the 1 RAR Group to Somalia. Some 36 hours later, with a mixture of regret at the interruption of a family holiday and excitement at the prospect of contributing to the largest deployment of Australian troops overseas since the Vietnam War, he drove back to Sydney. [31] In the meantime, David White and Melva Crouch were already putting the final touches to the Land HQ Operations Order and were initiating the deployment of the AFS to Somalia.

The announcement on 15 December triggered strong media reactions throughout Australia. The news was on the front pages of all newspapers and headed radio and television news reports. The deployment was perceived as a humanitarian mission with life-saving benefits for the Somali people. Donations of over $ 500 000 to an ABC Radio National-Community Aid Abroad fund-raiser for Africa over the previous weekend were proof of the pre-Christmas sentiment of the Australian public. With an election due in March 1993, sources close the incumbent Labor Government felt that sending troops to Somalia was both 'electorally attractive' and consistent with Australia's policy to assist UN-approved operations. [32] The fact that the US was commanding and under-writing the operation may have also made the deployment of Australian troops to Somalia strategically attractive.

The Prime Minister, Paul Keating, and the Leader of the Opposition, John Hewson, emphasised the humanitarian purpose of the commitment and its short duration. Delivering promised bipartisanship, Hewson said:

There can be no doubt about Somalia's urgent requirement for a massive international effort to provide and to ensure the delivery of humanitarian relief ... I take this opportunity of fully supporting the Government's decision. It is an important and appropriate one, not just in terms of responding to the needs of the Somali people but also in terms of supporting an expanded United Nations role in the post-Cold War world.[33]

Alpha Company, 1 RAR, the on-line ODF company, commanded by Major Doug Fraser,[34] had deployed into the Bluewater State Forest near Townsville to brush up on their patrolling and tactical procedures on Monday 14 December. Fraser recalled that it was difficult for his men to concentrate on their work knowing that the Government was considering sending a battalion group to Somalia. He had cautioned each of his platoons to get on with the job and stick to the exercise program he had devised. In reality, members of each of Fraser's platoons were huddled around their radios waiting for the announcement. In the early hours of 15 December ABC Radio broke the news of the 1 RAR Group deployment. The dark, quiet of the bush was punctuated with wild shouts, 'You beauty! You bloody beauty !'[35]

For Fraser's diggers and those in Townsville who now knew that they were likely to be deployed with the 1 RAR Group, the reaction to the news was consistent. The vast majority felt that they were about to put into practice their years of training; the culmination of their aspirations for operational service. Though not well articulated, there was also a strong feeling that they were about to embark upon a dangerous military quest and represent their country in an historic overseas operation.[36] Many of the infantrymen from 1 RAR and cavalrymen from B Squadron, 3rd/4th Cavalry Regiment, hoped OP *Solace* would give them the opportunity to be tested in combat. The infantrymen looked at the Somali gunmen shown on the television news reports with new interest and the cavalrymen examined their technical vehicles.[37]

For the spouses and partners of members of the 1 RAR Group the announcement came as an unexpected shock. Many families were on holidays, scattered all over Australia visiting friends and relatives and enjoying a break after a demanding year where units and sub units had been away on exercises and training courses, on and off, for about six months. Those with school-age children were waiting for the end of the school year at the end of the week before leaving on holidays. These plans were cancelled. There were immediate concerns about when their husbands, wives and partners would have to return to work and whether there would be a 'closed camp'.[38] Casual interest in television, radio and newspaper stories about the situation in Somalia turned into intense hour-by-hour scrutiny.[39]

One of the major challenges facing the commanders of the 1 RAR Group was who to take to Somalia and who to leave behind because of the annual posting cycle. On 15 December many officers, Warrant Officers (WO), senior NCOs and soldiers had been posted out to other units and were in the process of leaving, or had just left Townsville. Just as many personnel were inbound to Townsville to take up new appointments. A balance had to be struck between maintaining continuity and experience in units and sub-units by retaining some of those who had been posted out, and being fair to those who had arrived, or were about to arrive in Townsville keenly anticipating being sent overseas on active service. The directions of many military careers were decided in the days after the official announcement. No formal policy was decided on or promulgated. In the end it came down to a mixture of personal preference by commanders and whether those posted in or out had packed up their furniture and effects, and had either departed, arrived or were in transit.[40]

A great deal of bitterness was created when members of 1 RAR who thought their unit of 713 personnel was going to Somalia were told that over 60 personnel had been stripped from 1 RAR's establishment to fit an arbitrary figure of 650 set by higher headquarters staff officers to keep the overall strength of the AFS under 930 personnel. Hurley had the painful task of telling each company commander what numbers they had to lose to 'down-size' for deployment. Hardest hit was Administration Company which was reduced from 120 to 78 personnel.[41] In the following days Hurley had to cancel the deployment of more members of his unit as extra specialist personnel were added to the AFS and the size of its National Headquarters was increased.

Meanwhile in Somalia strong support for military intervention among aid agencies had turned to impatience. There were calls for more assertive action by Johnston and his Marines. Mr Russ Kerr, World Vision's Vice-President for Relief, was angry about the time it was taking for Marines to move inland to relieve the situation in Baidoa, the epicentre of starvation in the 1993 agricultural hinterland. He described the town as being under siege,

> We're living in fear of our lives. There was a gun battle only three
> kilometres from our centre at Baidoa, and earlier a local man was
> shot. The delay in sending troops to Baidoa, where up to a 100 bodies
> are being picked up from the streets every day, is criminal. [42]

On the day of the announcement to commit Australian troops to Somalia, the Australian connection to Baidoa received further publicity. Reports were received about Australian aid workers being evacuated and others being trapped and waiting anxiously for the US Marines to restore law and order. Inter-clan gun battles and looting were on the increase in and around the town. CARE Australia's three women relief workers in Baidoa, including former Prime Minister Malcolm Fraser's daughter, Phoebe, had been evacuated. Two other Australians, Lockton Morrissey, an ex-member of the Australian Army's elite Special Air Service Regiment, and Richard Chambers, were receiving death threats daily but still operating the CARE International relief program from the CARE compound in town.

Humanitarian operations in Baidoa were being crippled by greed and violence. At Baidoa Airfield only a trickle of relief supplies were getting through because the area was under the control of armed Somali extortionists. No supplies were getting through by road because of the risk of ambush.[43]

A spokesperson from CARE International, Cynthia Osterman, said,

> There is a feeling of disappointment that the military moved into Mogadishu so fast and now don't seem to be doing much more. This is no army they would be up against—it is a bunch of rag-tag thugs—and, by delaying, the Marines are leaving our people [in Baidoa] in great jeopardy.[44]

General Johnston countered these criticisms by stating that,

> We will be in Baidoa soon, but I am not yet prepared to send my troops until I have enough [of them in position] to make certain that they can decisively and immediately impose security in that town.[45]

On 16 December the day after the Government's announcement of the deployment of the 1 RAR Group to Somalia, General Gration and Major General Blake had met in Canberra.[46] They discussed what needed to be done to ensure that the 1 RAR Group was well employed on OP *Restore Hope* and how Australian military and political interests would be protected while Australian units were under the operational control of UNITAF. Not since 1965 had the ADF had to make arrangements for protecting Australian political and military interests while Australian troops served under foreign command. In that year another 1 RAR Group had been cobbled together at short notice and deployed overseas in four weeks to operate under the command of US 173rd Airborne Brigade (Separate) in southern Vietnam.

Gration and Blake had already agreed that an Australian National Headquarters would be set up under the command of Colonel Bill Mellor, who would be called, Commander, Australian Force—Somalia (CAFS). Gration and Blake were confident that Mellor, a gifted communicator, would provide a strong, authoritative voice among the donor nations to UNITAF.[47] However, they disagreed about the size and composition of Mellor's headquarters. Blake sought more personnel for the headquarters, but Gration wanted the numbers capped at ten, as had been agreed by the Defence Minister.[48]

Both generals agreed that Mellor and his staff would be a liaison team. Gration felt that it was possible that Australian officers might be able to serve on the American headquarters following a UN model of constituting multi-national headquarters staffs. Neither envisaged Mellor commanding the 1 RAR Group tactically. The Americans would have operational control.

Mellor's mission was to monitor UNITAF operations in general and Australian operations in particular, co-ordinate visits and media activities, be attentive to political sensitivities and report back to Australia daily. Gration insisted that Mellor only needed a staff of nine to fulfil these functions.

Both Gration and Blake wanted Bill Mellor and a small liaison team to fly to Mogadishu as soon as possible to negotiate with the Americans for a suitable area of operations for the 1 RAR Group and to determine what support Australian forces could expect from UNITAF logistic system. Gration argued that it would be 'first in best dressed' when it came to the Americans allocating tasks, areas of operation and resources to arriving units from donor nations. Blake agreed, but he did not want to rush Mellor, insisting that he be briefed by agencies in Canberra and by him personally before leaving. He wanted Mellor to have a first-hand understanding of the political sensitivities in Canberra of OP *Solace* and Blake's personal expectations of the way the Australians would operate.

Two days later, on 18 December, Mellor received a round of briefings in Canberra. He appreciated being brought up to date on the situation in Somalia, a country he knew nothing about aside from what had been reported in the media.[49] Blake met him at Sydney Airport on his way back to Brisbane. Blake emphasised the political implications for the 1 RAR Group of the forthcoming Australian Federal election, the sensitivity of the Australian public to casualties, the need to keep him informed and the challenges being faced to sustain the 1 RAR Group in Somalia. Blake concluded by assuring Mellor of his personal support and his commitment to 'backing the man on the ground' when it came to deciding on the merits of different courses of action.[50]

Blake made it clear to Mellor that he would hold him personally responsible for the success of OP *Solace* as the national commander in Somalia.[51] Like Colonel David Jackson, Military Attache in Saigon, South Vietnam, in 1965, Mellor was a senior Australian officer in an area of operations with complex and challenging responsibilities.[52] Like Jackson, Mellor was given the dual role of being a military liaison officer to a senior American headquarters and also being a 'watch dog' on the fortunes of a 1 RAR Group under the American command.[53]

On the day that Gration and Blake held their discussions, the 3rd/9th US Marine Battalion of about 700 men and an advance party of 142 French Foreign Legionnaires drove into Baidoa in a convoy of 70 vehicles, supported from the air by Cobra helicopter gunships. For the Somalis helicopters were a symbol of power. The local population turned out to cheer and wave to the Americans and Frenchmen in jubilant scenes reminiscent of the liberation of European cities in World War II. A sign saying, 'We are happy for the intervention' [sic] was suspended across one of the main streets. [54]

A large media contingent accompanying the Marines filed numerous stories about the sights, the smell and the sounds of what they dubbed 'The City of Death'. Australian journalist, Sue Neales, observed a 'body' truck making its rounds one morning to pick up those who had succumbed during the night. The sight of so many children's bodies was deeply moving. That day the skeletal remains of a family of six had been found in a roofless house. They had crawled in from some outlying district weeks before, were forgotten and died slowly and painfully in each others' arms.[55]

On 17 December Land HQ issued the OP Order for OP *Solace* after two days of hard work by David White and Melva Crouch.[56] General Blake directed that HQ 1st Division, 3rd Brigade's superior headquarters in Brisbane, stay out of OP *Solace*. He told General Arnison to concentrate on reconstituting the assets of 3rd Brigade to maintain the operational readiness of the ODF.[57] Blake wanted Land HQ to mount the operation and to deal directly with HQ 3rd Brigade who would dispatch the AFS from Townsville.

In reality Blake's staff in Sydney dealt directly with David Hurley and his staff because the previous Brigade Commander, Brigadier Mick Keating, his Brigade Major and other key 3rd Brigade staff had been posted out.[58] The most senior officer providing continuity at Brigade Headquarters was the Brigade's DQ, Gary Banister who was packing up for a move to Queenscliff in Victoria to attend the Australian Command and Staff College for a year.

The new Commander, Brigadier Peter Abigail, flew up from Canberra to Townsville with the previous Brigade Major, Major Mark Kelly, to see what needed to be done.[59] He spoke briefly with Hurley and directed Kelly and Banister to produce Operations and Administration Orders for despatching the AFS. Everything was in hand so Abigail assessed that neither he nor his staff needed to closely supervise Hurley's management of pre-embarkation training and administration. Hurley was an experienced ODF battalion commander who had supervised several sea and air deployments in 1992. Abigail directed Hurley to get on with it, but emphasised that, if Hurley needed any advice or further assistance, he and his staff would be on hand.[60]

Hurley's preparations for deploying the 1 RAR Group to Somalia were greatly simplified by being able to by-pass HQ 3rd Brigade and HQ 1st Division. In addition, the absence on leave or on posting of senior officers in the chain of command meant that three Lieutenant Colonels ran the show: Hurley in Townsville acting on behalf of his Brigade Commander; Retter in Sydney acting as Colonel Operations; [61] and Martin in Canberra acting as Director of Joint Operations.[62] In their acting appointments, Retter worked directly to Major General Blake, and Martin worked directly to General Gration.

The absence of senior officers on leave or on posting also had its downside. Major John Caligari, Hurley's Operations Officer, and 1 RAR's Quartermaster, Major Rod McLeod, faced the challenge of getting an Army on leave, a week before Christmas Day, to respond quickly to the needs of the 1 RAR Group. Caligari wrote in his diary, 'Despite the urgency and priority of this deployment, no one of any suitable rank is manning any of

Commander Kevin Taylor,
CO of HMAS Tobruk on the
bridge, off the coast of Somalia
in March 1993
Image: G. Gittoes

the agencies likely to be able to help us. Ho hum.'[63] There was no formal recall of personnel by higher headquarters. There were key officers and staff, especially in logistic units, who needed to be recalled so that swift action could be taken to 'top up' the 1 RAR Group with stocks to take to Somalia. The absence of many Majors and Captains on leave also meant that detailed staff work was not able to be done.

Another consequence was the absence of information about Somalia. Caligari and his operations and Intelligence staff had to rely on news reports and the local Townsville libraries for their information. No one at the Defence Intelligence Organisation in Canberra was recalled to support the 1 RAR Group.[64]

By this time the commanding officers of both *Jervis Bay* and *Tobruk* had recalled their crews from leave and were preparing to support OP *Solace*.[65] Like Army units in Townsville, there had been a significant turnover of personnel early in December. The hurried recall of crew added to confusion about who would be sailing for Somalia and who would continue with their postings to and from their ships.[66]

Jervis Bay, commanded by Commander Errol Morgan, completed preparations for the deployment to Somalia in four days and sailed from Sydney for Townsville on 19 December. Fortunately *Jervis Bay* was already working up at sea off Sydney, after a refit, with most of her crew aboard when Morgan received his orders.[67]

Tobruk was not so lucky. When the order came for deployment to Somalia on 15 December, *Tobruk* had been 'in pieces and alongside' at Garden Island for six weeks.[68]

Newly-arrived Commander Kevin Taylor was told to be ready to sail before the end of December. Just over two months earlier *Tobruk's* main port engine camshaft had disintegrated on a voyage from Melbourne to Perth. Major repairs were made in Melbourne before the ship limped back to its home base in Sydney for an engine rebuild, and an extended repair and maintenance period, intended to last until 20 January 1993

Taylor and his crew had to put *Tobruk* back together and expedite the refit program. Maintenance tasks that had been scheduled over the next six weeks had to be completed in ten days and some tasks had to be cancelled. Many of the remaining tasks involved reassembling complex equipment that had been pulled apart for detailed servicing. Concurrently, the ship had to be loaded with provisions for a six month deployment and provide for an increase in the ship's company. With the addition of a SK50 Sea King Helicopter with 16 personnel from 817 Helicopter Squadron and additional medical personnel, his ship's company swelled to 220 personnel A banner saying '*Bound for Somalia*' was secured on the guard rails of the vehicle deck, not as an advertisement for *Tobruk's* humanitarian mission, but to explain to the citizens of Woolloomooloo, the suburb adjacent to the wharf, why the ship was bathed in light all night and sounds of repairs and loading were going on 24 hours every day.[69]

In Taylor's opinion, when the order for deployment to Somalia came, *Tobruk* and its crew needed a rest after spending over 8 months at sea in 1992 away from their home port of Sydney.[70] He was proud of the way the crew responded to the requirement to return from leave at short notice, to reassemble the ship's equipment and prepare for deployment. Despite the prospects of another six month period of separation from their families, they worked hard in shifts around the clock. The morale of the crew rose steadily, spurred by the media coverage of OP *Restore Hope*. Personally, Taylor was excited about being given the opportunity to 'go operational' so soon after assuming command.[71]

Taylor's own preparations for *Tobruk's* deployment were hampered by having to serve on the Board of a Court Martial.[72] He had to rely on his Executive Officer, Lieutenant Commander Brian Cowden, to prepare the ship by Christmas.[73] In addition there was a great deal of uncertainty about the nature and duration of *Tobruk's* support for OP *Solace*. Taylor was 'totally frustrated with the planning processes' prior to the Government announcement on 15 December. Many staff officers had gone on leave and he had difficulty in establishing what was required of *Tobruk* during OP *Solace*.

During the days after the Government announcement, Gary Banister at HQ 3rd Brigade and Rod McLeod at 1 RAR had been finding it difficult to get the Army's supply system to stock up the 1 RAR Group, especially with spare parts. Within 24 hours of the announcement, Banister and McLeod had hit the logistic system with supply demands to ensure the 1 RAR Group was equipped and stocked up for 30 days of independent operations and the battalion Support Group (BSG) deploying with the 1 RAR Group would have 30 days further stocks. In theory this should not have been difficult because the ODF had stocks set aside for short notice deployments.

Banister had sent a document specifying the logistic requirements for OP *Solace* to Land HQ on 16 December.[74] He based them on the composition of the 1 RAR Group, his intimate knowledge of force logistic needs developed in Townsville over two years, the available scant intelligence reports and the likely environmental conditions within Somalia.[75] On the same day McLeod sent a batch of 1 000 Priority One demands for stores directly to Greg Birch's 2 Field Supply Battalion for satisfaction in seven days.[76] By the end of the week McLeod and Harry O'Brien, his veteran Regimental Quarter Master Sergeant, were sending scores of demands for stores every day.

Majors Craig Mills, Mark Baker, a British exchange officer, and David Bucholtz at Land HQ were in the logistic hot seats.[77] Baker assessed Banister's logistic requirements as being a 'wish list'. Administration Branch staff then cut back the quantities of several items significantly, including hand-held Global Positioning Systems (navigation equipment), Night Vision Goggles, Thermal Imagers, PACE 10 data processing equipment, squad radios, lap top computers and other technical items.[78]

The culling of Banister's 'wish list' led to a vigorous debate between the staff officers in Townsville and staff officers in higher headquarters. Banister argued that the 1 RAR Group should be deployed overseas with as much competitive edge as possible in navigation, night vision equipment, communications and computer support.[79] Some officers in LandHQ, Materiel Division in Canberra and HQ Logistic Command in Melbourne were cautious about sending these expensive technical assets to Africa and stripping Australian-based units of their capabilities.[80]

Logistic staff officers on higher headquarters demanded further justification from Banister, questioning why technical items, such as lap-top computers and squad radios should be purchased. Subsequently, Banister's requests and justifications were pushed higher and higher for resolution. Eventually, the Deputy Chief of the General Staff, Major General Geoff Carter, reviewed Banister's justifications and authorised the purchase of additional items of computer and communications equipment for OP *Solace*.[81]

McLeod's 1 000 demands submitted within 24 hours of the Government announcement had also caused problems for the supply system. Many were for spare parts that had not been available previously because of a lack of funds. Others were for stocks that had been earmarked for other ODF contingencies and there was some tardiness in releasing these stocks for OP *Solace*. Junior staff officers, who were the only personnel available at supply agencies, had not been informed officially that OP *Solace* had a higher priority for stocks than these contingencies despite Banister's request for the highest priority, called Force Activity Designator One (FAD 1), within 12 hours of the Government announcement.[82]

During this busy time, logistics staff at Land HQ were planning an independent Point of Entry for the resupply of the AFS across the beach near Mogadishu. The planning for this Point of Entry had originated from a Maritime Headquarters resupply concept based on assumptions that there would be insufficient port facilities at Mogadishu, and that water, fuel and other classes of supply would have to be provided by *Tobruk* sailing to and from Mombassa in Kenya.[83]

On 22 December Maritime HQ had asked Land HQ to deploy amphibious craft and personnel from the Army's 10 Terminal Regiment, located at Woolwich in Sydney, onto *Tobruk*. Their mission would be to operate a ship-to-shore ferry service from *Tobruk* at sea to a supply dump on the beach near Mogadishu. Presumably, the 1 RAR Group would have had responsibility for setting up, operating and protecting this dump. Vale envisaged trucks from the 1 RAR Group picking up stocks from this beachside distribution point and transporting them to the BSG at Baidoa.[84]

Brian Vale, faced with contradictory information about the reliability and security of the port at Mogadishu as well as the extent of support available from US supply agencies as American UNITAF units withdrew in January, had decided to plan for the Maritime HQ logistic option. He had not been dissuaded from taking this decision despite receiving a draft cross-servicing agreement from US Central Command in Tampa Bay on 17 December agreeing to provide water, fuel, ammunition and repair and maintenance services in accordance with cross servicing agreements signed in 1990. Vale took a cautious approach and accepted Australian Navy advice that the 1 RAR Group would have to be resupplied across the beach near Mogadishu.

At around this time Taylor was told that *Tobruk* would be a logistic platform for OP *Solace*, providing supplies of water, fuel and other commodities across the beach near Mogadishu. To achieve this he was

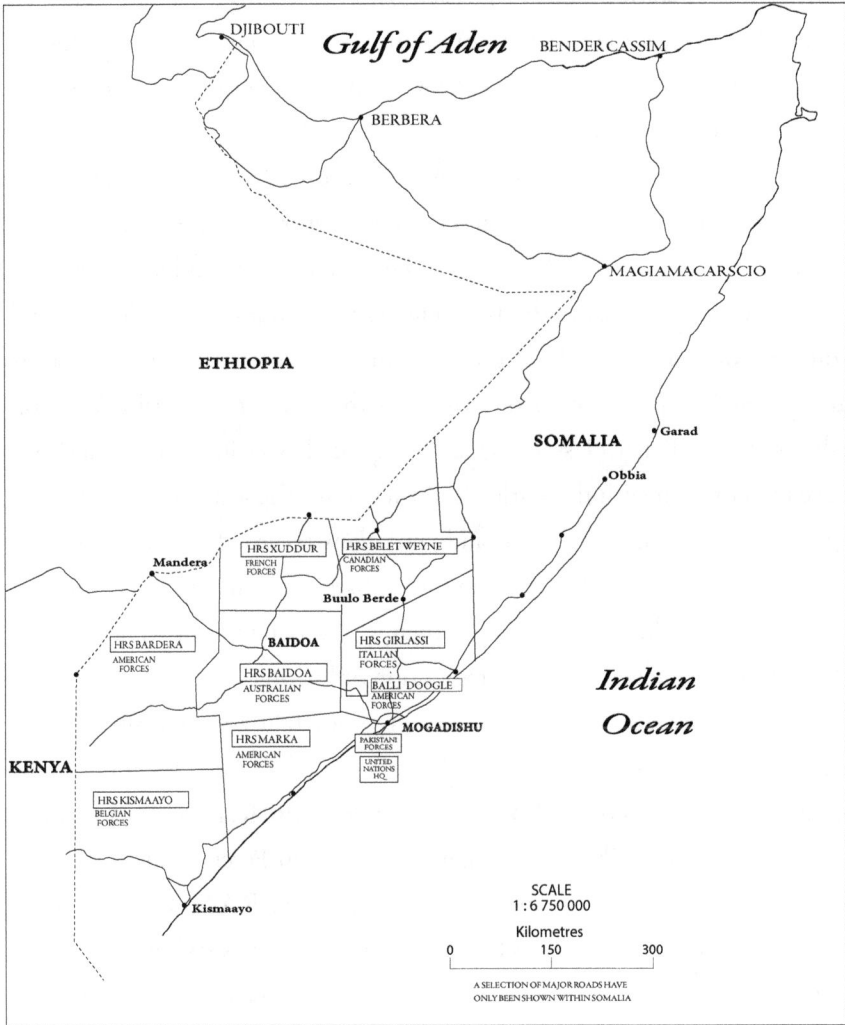

DJIBOUTI

Gulf of Aden   BENDER CASSIM

BERBERA

MAGIAMACARSCIO

ETHIOPIA

SOMALIA   •Garad

•Obbia

Mandera• | HRS XUDDUR |
FRENCH FORCES

| HRS BELET WEYNE |
CANADIAN FORCES

Buulo Berde•

| HRS BARDERA |
AMERICAN FORCES

BAIDOA | HRS GIRLASSI |
ITALIAN FORCES

| HRS BAIDOA |
AUSTRALIAN FORCES

| BALLI DOOGLE |
AMERICAN FORCES

Indian Ocean

| HRS MARKA |
AMERICAN FORCES

| PAKISTANI FORCES |
•MOGADISHU

| UNITED NATIONS HQ |

KENYA

| HRS KISMAAYO |
BELGIAN FORCES

SCALE
1 : 6 750 000

Kilometres

0   150   300

A SELECTION OF MAJOR ROADS HAVE
ONLY BEEN SHOWN WITHIN SOMALIA

•Kismaayo

Map 2: Locations of Humanitarian Relief Sectors

warned out to load amphibious craft and take personnel from 10 Terminal Regiment aboard. He was happy to provide the planning support for the amphibious operation but not pleased with some Army staff officers lack of understanding about the limitations his ship would have to load vehicles, equipment and stores for the 1 RAR Group after loading the wherewithal for amphibious operations.[85]

On the day *Jervis Bay* sailed from Sydney on 19 December, newspaper reports about OP *Restore Hope* described the return of armed gunmen and technicals to the streets of Mogadishu. The Somali warlords had interpreted the American reticence to disarm them, despite requests from the UN and some donor countries to UNITAF, as permission to continue to carry arms and get back to the business of terrorising their opponents. An American-educated Somali, Mr Hassan Togani, was quoted as saying, 'The Americans have to get tough with the warlords. If they don't, the warlords will just play games.'[86] As violence increased and rival clans returned to attacking each other, it became clear that the Americans had to persuade the warlords to impound their vehicles and stockpile their weapons. It was uncertain whether this could be achieved through negotiation or by force.

Set against the background of this uncertain security situation in Mogadishu, an Australian National Liaison Team left Australia for Somalia in an Air Force Falcon 900 on 21 December. The Team was made up of Mellor accompanied by Woolnough, Majors Geoff Peterson and Cameron Powrie, Leutenant Commander Andy Naughton, RAN, Captain Nigel Catchlove and Leutenant Tony Powell, RAN.[87] Petersen's task was to link the Australians into the US intelligence network. His first priority was to assess the threat to the 1 RAR Group and establish a steady flow of intelligence from HQ UNITAF to Mellor's headquarters. Powrie's task was to arrange for medical support for the 1 RAR Group and assess the preventative medical measures that would need to be taken to maintain the health and well-being of Australian personnel before and after deployment to Somalia. Catchlove was a Defence Public Relations representative.[88] Naughton was on the trip to assess the port facilities at Mogadishu and make arrangements for the arrival of *Jervis Bay* and *Tobruk*. He was to complete these tasks in a day and then fly back to Australia in the Falcon. Powell remained in Mogadishu to co-ordinate the reception of *Jervis Bay* and *Tobruk*.

Mellor and his team were met at Mogadishu Airport by Greg Jackson and a US Marine Corps Liaison Officer, Captain Mike Owen, who put a driver and a Hum Vee vehicle at Mellor's disposal. By this time UNOSOM operations had ceased while OP *Restore Hope* dominated the political and military scene in Somalia. Most UNOSOM staff had left Somalia on Christmas leave. Jackson had decided to make himself and his staff available to assist Mellor and his team settle in and introduce them to key staff at HQ UNITAF.[89]

Gration had been right about a 'first in best dressed approach' by the Americans to the allocation of tasks and resources to inbound coalition units. Brigadier General Anthony Zinni, General Johnston's senior operations officer, offered the Baidoa Humanitarian Relief Sector to Mellor on the day Mellor arrived. Baidoa just happened to be the next Sector to be allocated. The day before, the Belgians had been allocated the Kismaayo Sector.[90] The day after, Zinni allocated Sectors to other contingents that would not have suited the Australians as well as Baidoa.[91] On Christmas Eve, after discussions with Mellor, Blake agreed to accept the Baidoa Sector as the Australian area of operations.

Meanwhile, in Canberra, Senator Robert Ray, possibly with an eye on the forthcoming March Federal election, ordered General Gration to coordinate maximum media coverage for OP *Solace*. The ADF had a policy of 'professional coexistence with the Media and the Military' which included providing transport, accommodation, security and access to journalists, camera crews and photographers. Brigadier Adrian D'Hage, the Director General of Public Information,[92] had written that,

> ... despite the fact that the Media and the Military are following
> basically different and often contradictory agenda, there is a good

deal of common ground. ... Public support is essential to the nation's prosecution of war, and this will be strongly influenced by the Media's reporting of the conflict. Contradictory agenda or not, the Military and the Media need each other. [93]

D'Hage directed Ms Lisa Keen, the Manager of his Directorate's Electronic Media Unit, to raise, and command a Media Support Unit (MSU). Fortunately, the Public Relations Directorate had produced guidance on the establishment of these units to support operations and one had supported the Kangaroo 92 Exercise in north-western Australia in March 1992. However, Keen faced a challenge to have a MSU ready for deployment to Somalia in three weeks time with Hurley's advance party. All of the vehicles, stores and equipment looked fine on paper but, where would they come from and how would they get to Townsville in time to be loaded on *Jervis Bay* by Christmas Eve?

The highest priority for supply was given to the MSU from the beginning.[94] Craig Mills received strict instructions from General Grey's office in Canberra to direct 3rd Brigade staff in Townsville to load the vehicles and stores for the MSU onto *Jervis Bay* on 24 December at all costs. Hurley and his staff were caught unawares. Hurley recalled,

> The thing about the media [MSU] was that demands came out of the blue. We had not dealt with this sort of thing before at battalion-level. We were just told to provide space and dedicate vehicles while we were trying to stuff all our own stores aboard *Jervis Bay* and *Tobruk*.[95]

On 18 December Captain Mick Fulham from HQ 3rd Brigade informed Mills that, 'The loading of MSU [vehicles and] stores will effect the 1 RAR Group's ability to sustain itself for 30 days.'[96] On 22 December Bucholtz advised Fulham that Land HQ had been able to negotiate a 'much reduced' list of vehicles and stores for the MSU, but 'urgent action needed to be taken to include stores and vehicles on *Jervis Bay*.' He went on to say that, 'It is accepted [by Land HQ] that this will cause disruption. However ALL [original emphasis] items MUST [original emphasis] sail on same [*Jervis Bay*].'[97]

First contingent from the 1
RAR Group to leave for
Somalia from Townsville on
Christmas Eve 1992, aboard
HMAS Jervis Bay.
Image: Royal Australian
Navy

Concentrating vehicles and stores on the docks ready for loading on *Jervis Bay* and *Tobruk* was a major challenge for the staff at HQ 3rd Brigade and the 1 RAR Group. *Jervis Bay* was due to dock on 22 December and *Tobruk* a week later. Leutenant Tony Powell had arrived to assist the loading of *Jervis Bay* and Sergeant Ungus McMillan also joined Hurley's planning team to advise on loading *Tobruk*.[98]

The days before *Jervis Bay* docked were chaotic. Despite previous experience loading 1 RAR's vehicles and stores onto *Tobruk* for field exercises, the co-ordination of the concentration of stores onto the docks and finalising a loading schedule for both ships was a shambles. Loads of stores were arriving in Townsville in semi trailers and by air around the clock. Until semi-trailers and aircraft were unloaded, no one in Townsville knew what stores had arrived. Similarly, it was difficult to track stores in transit and ascertain quickly what stores were overdue.[99] It was also very expensive to move tonnes of stores to Townsville from the major supply depots in Albury, Sydney, Brisbane and western New South Wales to top up the 1 RAR Group and reconstitute ODF stocks at short notice just before Christmas.

Captain John Atkinson, recently appointed to command the 8th Movement Unit in Townsville, arrived in Townsville on 19 December to find

Major Doug Fraser, OC A Company,
1 RAR Group, aboard HMAS Jervis Bay,
on the way to Somalia in early 1993
Image: NT Services Ltd by DJ Fraser.

Banister's staff at 3rd Brigade trying to confirm what needed to be loaded for the six supporting sub units comprising the 1 RAR Group and Glenn Crosland's staff concentrating on ensuring sufficient vehicles and stores were being loaded for 1 RAR.[100] On top of this lack of co-ordination, the allocation of vehicles and loading of stores for the MSU had not been finalised between Mick Fulham, and David Bucholtz and Lisa Keen.

On 22 December, the day *Jervis Bay* docked in Townsville, Banister reminded Bucholtz that OP *Solace* had still not received a high priority status for the issue of stores and he had not been advised what Army supply depots were to be the sources of classes of supply for the operation after the 1 RAR Group had arrived in Somalia.[101] The next day, after Bucholtz had made a formal request to Canberra, the 1 RAR Group were given the highest priority for the issue of stores from Army stocks. Thus, eight days after the Government announcement, the ADF logistic system was officially focused on meeting the needs of the 1 RAR Group as Banister raced against time to get stocks to the docks.

Fortunately the Navy had not been as tardy. Morgan's and Taylor's preparations were made easier because Maritime HQ gave *Jervis Bay* and

Colonel Bill Mellor, Commander of the
Australian Force–Somalia, pictured in
Mogadishu in early 1993
Colonel Mellor led a National Liaison Team
to Mogadishu in December 1992 to set up
arrangements for the 1 RAR Group.
Image: ADF Electronic Media Unit,
David Brill

Tobruk top priority in the Fleet for the issue of stores immediately it was known that they would deploy to Somalia. Both Morgan and Taylor would have found it very difficult to load all the stores required and sail on time without this timely action.[102]

The loading of *Jervis Bay* over the next two days became a 'hand to mouth' exercise. Previous planning had assumed that stocks would be available for loading from the wharf, guided by agreed priorities and loading schedules. In reality, stocks were loaded as they arrived in unpredictable order from the Army's supply system.[103] Hurley accurately described loading *Jervis Bay* as 'stuffing' stores aboard. Soldiers formed human chains and manhandled boxes into every spare space. All of the APCs and other vehicles were stuffed with stores and sand bags, and rolls of barbed wire were stacked on top.

Hurley decided to embark Major Doug Fraser's A Company, reinforced by Infantry Corps members of the 1 RAR Band, as well as a group of drivers from the Transport Platoon, a detachment from B Squadron and a mix of 65 vehicles, 17 tonnes of ammunition and 183 tonnes of general stores.[104] The Navy were keen to load *Jervis Bay* and sail for Mogadishu as soon as possible. General Blake had concurred with the Navy's departure date of Christmas Eve. On reflection, he regretted not ordering that the departure to be postponed until 26 December so that members of A Company and others about to leave for Somalia could spend

Lieutenant Colonel Graeme Woolnough, Chief of Staff, HQ Australian Force, Somalia, pictured in Mogadishu early in 1993 Lieutenant Colonel Woolnough accompanied the Australian Liaison Team to set up logistic arrangements in Somalia for the 1 RAR Group.
Image: ADF Electronic Media Unit, David Brill

Christmas Day with families and loved ones. A postponement of 24 hours for this sentimental reason would have made no difference to the arrival time of *Jervis Bay* in Mogadishu.[105]

The scenes on the Townsville docks on Christmas Eve were reminiscent of the many emotional farewells of expeditionary forces during the World Wars. Families, relatives and friends waved to their loved ones who looked down from the decks of *Jervis Bay*. The 1 RAR Band played regimental and patriotic music. Hundreds of locals, a bevy of VIPs and a large media contingent turned up to farewell the soldiers and sailors of Australia's latest expeditionary force.

The day before *Jervis Bay* departed Dr Boutros-Ghali released a report criticising President Bush's reluctance to order General Johnston to disarm the clan armies. If the Somali warlords' forces were left intact when UNITAF withdrew in a few weeks time, Boutros-Ghali feared that UNOSOM II troops would find it difficult to protect the distribution of humanitarian relief. 'It would be tragedy if the premature departure of UNITAF were to plunge Somalia back into anarchy and starvation, and destroy the fragile political process of recent weeks.'[106]

The issue of whether or not to disarm the clans went to the heart of UNITAF's mission in Somalia. President Bush, his Special Envoy, Robert Oakley, and General Colin Powell had stressed that OP *Restore Hope* was a humanitarian mission of limited duration to create a secure environment for the distribution of humanitarian aid. They did not want UNITAF forces to take on the armed clans who would no doubt resist any systematic attempt to disarm them.

The differences between UN, and US political leaders and US military commanders revolved around the interpretation of the term, 'create a secure environment'. For Boutros-Ghali the term meant disarming the clan armies, and removing their ability to fight among themselves and intimidate UNOSOM troops as they had done in previous months. Bush, Oakley and Powell feared that the US would be drawn into a long-term, expensive security commitment if they gave Johnston the task of disarming the warlords' militia as part of the mission to create a secure environment. Senator Robert Ray supported the UN position on disarmament, 'Our general attitude is, if UNOSOM, the group that is to follow OP *Restore Hope*, is to be successful, some attempt must be made to disarm the warring factions.'[107]

Senator Ray had sent a signal to the Army that the 1 RAR Group might be involved in disarming the clan armies. There was concern among military planners in Canberra and Sydney that, if the UN interpretation of 'create a secure environment' prevailed, probably as a result of pressure from American and Western public opinion, and Johnston was directed to disarm the Somali militia armies, the fragile agreement with the warlords to stand aside and permit the flow of humanitarian aid would disappear. If the warlords thought that they were going to lose the military assets which were the source of their power, the military situation in Somalia might quickly escalate into semi-conventional war with the potential for significant Somali and UNITAF casualties.[108]

Meanwhile in Mogadishu, Mellor and his staff were finding it difficult to obtain information despite the assistance of Greg Jackson, his staff, and their Marine Liaison Officer. OP *Restore Hope* was in full swing but there was a planning stalemate between UNITAF and the UN on transitional arrangements between OP *Restore Hope* and a new, larger-scale UNOSOM II.

OP *Solace* had a duration of 17 weeks which meant that the 1 RAR Group's stay in Somalia would overlap OP *Restore Hope* and UNOSOM II: an overlap from peace enforcement to peace keeping. Zinni told Mellor that the Bush administration preferred that all American combat units be out of Somalia by 20 January in time for the inauguration of President Bill Clinton. However, this political agenda did not account for the time it would take for international units to take over from US Marine and Army units deployed for OP *Restore Hope*. There was also pressure on the US not to leave Somalia until UNOSOM II was ready to take over. With UNOSOM staff on leave until mid-January and the UN staff in New York finding it difficult to attract international support for UNOSOM II until the results of OP *Restore Hope* were evident, Zinni was pessimistic about the prospects of US Army units being clear of Somalia by 20 January.[109]

This delay in clarifying transitional arrangements left Mellor and Woolnough without any reliable information on the extent of the resupply and other support that would be available to Australian forces in Somalia from UNITAF or UN sources. Accordingly, Brian Vale and his staff planned for the Australians, to be 'self sustaining after 20 January because US involvement cannot be guaranteed' and convened a Joint Administrative Planning Group (JAPG) on 5 January 1993 to develop an Administrative Order for OP *Solace*.'[110]

For the time being Australian staff in Townsville and in Mogadishu were guided by Banister's Administration Order produced on 19 December, four days after the Government announcement.[111] By the time Brian Vale and the logisticians from higher headquarters planned to provide administrative guidance, *Jervis Bay* and *Tobruk* would be at sea, Hurley's advance party would have left and the main body of the AFS would be on their way.

On 27-28 December Mellor, Petersen and Catchlove travelled by road to Baidoa. They were able to visit the town area and conduct an aerial reconnaissance by helicopter. The Marines patrolled the urban area of Baidoa by day and night, on foot and in vehicles. They escorted food convoys and protected food and water distribution points as well as searching buildings and vehicles for weapons and ammunition. Mellor reported that it would be the 1 RAR Group's responsibility to secure the countryside of the Baidoa Sector as a further step in creating a secure environment for the distribution of humanitarian relief.[112]

While Mellor and his group were in Baidoa, Hurley issued orders in Townsville. The Intelligence Officer, Captain Jim Burns,[113] briefed those attending on the situation in Somalia using a map [circa 1942],[114] concluding that the threat from the clan armies was high[115] Major Steve McDonald, OC Administration Company, wrote in his diary, 'Good int[elligence] brief from Jim Burns. Seems the threat from clans is higher than we thought with [a] possibility of artillery, 120 mm mortars, tanks etc. [This] shows that the situation in Somalia has potential to 'hot up'.[116]

Events in Mogadishu seemed to contradict Burns' threat assessment. On 29 December competing warlords, Aideed and Mahdi, organised a large peace rally in Mogadishu. A crowd of several thousand Somalis observed Aideed and Mahdi embrace, and then lead them hand-in-hand along the Green Line. This was assessed to be an astute political gesture by Aideed and Mahdi to reassure the US of their peaceful intentions before a New Year visit from President George Bush to Somalia. An uneasy calm settled over the war-torn metropolis.[117]

While the warlords embraced in a gesture of peace, they still maintained an arsenal for war. This made it difficult to assess the possible military threat against the 1 RAR Group after deployment to Baidoa. General Blake recalled,

> It was frustrating not to have a clear intelligence picture of the likely threat to our forces once they deployed to Somalia. For example, we did not know whether the warlords had merely cached [hidden] their weapons and driven their 'technicals' across the border. If so, they could just wait until the US Marines and the bulk of US Army combat units had left and then mount sizeable operations against coalition forces. Because of this uncertainty, and despite little

evidence to support our concerns, we had to allow for the possibility that our force could be taken on conventionally.[118]

At battalion level, Burns' briefing showed that Hurley and his staff were expecting a conventional armed threat from forces loyal to various warlords. Accordingly, *Jervis Bay* had been 'loaded for bear' with sufficient mortar and heavy machine gun ammunition to meet such a threat. Rather than anticipating deployment to a lightly-defended airfield to protect the distribution of humanitarian aid, 1 RAR prepared to occupy and develop 'Fire Support Base Baidoa', a defensive position able to withstand conventional armed assault.

Meanwhile *Tobruk* had sailed on time from Sydney for Townsville on 26 December amidst the colourful, festive backdrop of scores of pleasure craft and yachts preparing for the start of the annual 'Sydney to Hobart' yacht race. Over the next four days personnel from the Sea Training Group (STG) from Fleet HQ put *Tobruk's* crew through their paces with 'a demanding around-the-clock exercise routine of fire, flood, toxic gas, man-overboard, action and defence stations, and aircraft crash-on-deck incidents'. Though this was a stressful period without much sleep for Taylor and his crew, the STG exercises became a common challenge to bond the newly-posted and seconded personnel with the experienced members of the crew. Over one third of those aboard had never been to sea before. *Tobruk* arrived in Townsville at 7:30pm on 29 December as a 'worked-up' crew, tired from being awake since 3:30am that morning when the last STG exercise was conducted, but happy that they had been declared 'Mission Capable.'[119]

The task of co-ordinating the loading of *Tobruk* fell to the newly-arrived commander of the Ships Army Detachment (SAD), Major Paul Le Large, his experienced Ship Sergeant Major, Peter Macdonald, Sergeant Ungus McMillan, and Corporals 'Dinger' Bell and Darryl Burke's five Army freight handlers as well as the Ship's Quarter Master Sergeant and the SAD Clerk.[120] Shortly after 8pm this team, assisted by volunteers from the Ship's company, began a 50 hour marathon to load 93 vehicles and trailers, 52 tonnes of ammunition and 70 tonnes of general cargo, and embark 75 Army personnel under the command of Lieutenant Janelle Lawson.[121]

Le Large quickly discarded the load lists and loading schedules he had received in Sydney four days before. On arrival they received a completely different load list and loading plan. There were tonnes of stores on the wharf that had not been loaded onto *Jervis Bay* and there were almost hourly amendments to *Tobruk's* load lists after loading began to accommodate arriving consignments of stores. Once again a 'hand to mouth' situation prevailed as cargo was loaded and more cargo arrived continuously during the loading period. In a perfect world *Tobruk* should have been 'combat loaded', i.e. stores needed for immediate operations in Somalia should have been loaded last. All Le Large could do was load the stores unable to be fitted onto *Jervis Bay* and then load the remainder of stores in the order they arrived at the wharf.

Loading *Tobruk* was as manpower-intensive as the loading of *Jervis Bay* had been before Christmas. Often pallets of stores were broken up and boxes stuffed into every bit of spare space available.[122] The Army did not have sufficient standard shipping containers to efficiently load either ship.[123] Like their forefathers in World War II, a large contingent of soldiers from 1 RAR, B Squadron, 3/4 Cavalry Regiment and 3rd Brigade Administrative Support Battalion (3 BASB) formed a human chain to manhandle boxes up gangplanks into the holds.

After being awake since 3 30am the previous morning, Le Large ordered a two hour break at 5am and resumed loading at 7am. At 2.15pm there was an unexpected three hour respite from loading after a toxic gas leak from the ship's sewerage system was detected. The ship had to be evacuated quickly under the direction of Brian Cowden and a Damage Control Team who then ventilated the ship and repaired the sewerage system. This was dangerous, dirty work. Sailors from the Damage Control Team performed exceptionally well during this emergency and there were no injuries.[124]

After Cowden gave the 'All clear' at 5 30pm , Le Large and his men, helped by sailors from *Tobruk,* and soldiers from 1 RAR, 3/4 Cavalry Regiment and 3 BASB, continued loading the ship throughout the night without a break until 7am that morning. Two hours later Rear Admiral Robert Walls, the Maritime Commander, came aboard *Tobruk* to address the ship's company before their departure.[125] Paul Le Large and his 'freighties' were 'flaked out in their racks below; quite oblivious to all the official farewells and tears.'[126] *Tobruk* was loaded to the 'gunwales' but had bulked out. Over 50 pallets of stores and six specialist vehicles remained on the wharf as mute testimony to the gap between the intentions of military planners and the final outcome. Consignments of stores continued to trickle in over the next few days.[127]

As *Tobruk* sailed for Darwin on 31 December, Mellor and his staff in Mogadishu were using their time to meet key UNITAF staff officers, work out how things were operating and establish useful relationships. Woolnough was active trying to set up arrangements for resupply. He had been briefed by Brian Vale and Craig Mills at Land HQ that the 1 RAR Group would deploy with sufficient stocks to operate for 30 days and the BSG would have another 30 days stocks. He also knew that Brigadier Bill Traynor from HQ ADF was negotiating cross-servicing arrangements between the US Commander In Chief, Central Command, and the Australian Chief of Logistics.[128]

Towards the end of December Blake and Retter realised that no one from 1 RAR had been deployed to Somalia to conduct a reconnaissance and prepare the way for Hurley's advance party that was due to leave in just over a weeks time. [129] On 28 December Steve McDonald was given 24 hours to pack and fly to Canberra for briefings before catching a flight to Mogadishu.[130] Three days later, in the early hours of New Years Day, he was sewing an Australian flag logo on the shoulder of his shirt in a hotel room in Nairobi. The next day he landed in Mogadishu. He recalled,

> It was like being plonked into the middle of a movie set. CH 53 [helicopters] went past us as we were on late finals [before landing].

Tents, Marines, bulldozers, trucks, aircraft of every type. It was big !
A dozen ships offshore. The airport was a hive of activity.[131]

McDonald had been on staff at HQ 3rd Brigade before posting to 1 RAR. He brought with him a sound knowledge of the reception requirements for the units being deployed from Townsville. Based on his experience in the ODF, McDonald suggested refinements to the plans made by Jackson's UNOSOM staff. By this time Mellor and his staff had been allocated office and accommodation facilities in the US Embassy compound.

McDonald took a few days to acclimatise to the heat and the 'crack-thump' of rounds hitting the walls of the Embassy Compound each night.[132] There was a significant gap between Aideed's and Mahdi's peace-making gestures for the international media and the reality of their continuing violent struggle to gain supremacy in Mogadishu.

Soon after McDonald was deployed, Bill Mellor requested that liaison officers from the 1 RAR Group be sent to Somalia to facilitate the hand over arrangements between the US Marines and the 1 RAR Group in Baidoa. Major Greg Hurcum, OC Support Company, 1 RAR, and Major Dick Stanhope, Battery Commander, 107 Field Battery, were given 24 hours to pack, attend briefings in Canberra and then fly to Mogadishu to link up with the Marines in Baidoa.[133] Both families were up early on 31 December preparing for departure from Townsville airport at 6am. Hurcum wrote, 'Up at 4am very sad and upset. It is very hard to leave Jenny and Lauren. Met Dick, Haylie and family at the airport. Flight left at 6am. Said our farewells with tears and a heavy heart. Thank God there were no media to intrude.'[134] After an interesting journey via Tokyo, London and Nairobi, Hurcum and Stanhope arrived in Mogadishu on 4 January.

Meanwhile, by 31 December the 1 RAR Group was complete in Townsville.[135] Despite 1 RAR's official high state of operational readiness, the unit had to be reinforced with 56 soldiers from the Townsville-based 2nd/4th Battalion, the Royal Australian Regiment (2/4 RAR) and Major Doug Fraser's A Company had been reinforced by Infantry Corps members of the 1 RAR Band before embarking on *Jervis Bay*. This late reinforcement added to the impact of personnel turbulence in 1 RAR caused by the end-of-year posting cycle and 'down-sizing' to 650 personnel. Fortunately, Lieutenant Colonel Pat McIntosh, the Commanding Officer of 2/4 RAR, insisted that his Battalion's best soldiers get their opportunity for overseas operational service. 1 RAR benefited greatly from this injection of well-trained and disciplined soldiers. McIntosh's professionalism also gave David Hurley and his staff much more flexibility when 1 RAR personnel were unable to be deployed because of medical, disciplinary or compassionate circumstances.

David Hurley addressed a muster parade of the 1 RAR Group on New Year's Day. He highlighted the humanitarian nature of OP *Solace* as well as the likelihood of armed opposition. From 2 January all members of 1 RAR were to carry weapons around the barracks area with a live 'round up the spout.' From 8 January everyone was to wear flak jackets around the battalion area to acclimatise them to the heat and tape two First Aid Dressings to their webbing.[136] After Hurley's address the 1 RAR Group dispersed to begin 14 intensive days of training and personal administration.

Major Colin Chidgey, OC Reinforcement Holding Branch, and his staff co-ordinated the 1 RAR Group's pre-deployment training.[137] Chidgey was responsible for bringing personnel and resources to Townsville from around the ADF and elsewhere for this training. A major challenge was that the 1 RAR Group had loaded most of its training stores on *Jervis Bay* and *Tobruk*. Fortunately, Pat McIntosh 'came to the party' once again and loaned significant quantities of 2/4 RAR's stores to support the 1 RAR Group's pre-deployment training. The external training support provided by Chidgey's unit and 2/4 RAR freed Hurley, his company commanders and his headquarter staff to concentrate on overall preparations for future operations in Somalia.

Hurley and his planning group had plenty to think about. After assessing the scarce information available on the situation in Somalia and their likely mission in the Baidoa Sector, they decided to rotate the four rifle companies and platoons from Support Company through four operational tasks; protecting the Baidoa Airfield, protecting the Baidoa metropolitan area, protecting humanitarian aid convoys and patrolling the countryside. In addition, they decided to constitute a Quick Reaction Force of one platoon mounted in trucks to support troops deployed into Baidoa.

As operational planning progressed pre-deployment training and administration was in full swing. Aside from local field firing exercises, zeroing weapons and crowd control training, the most effective and important training sessions were delivered by the 3rd Brigade's Legal Officer, Captain Bruce 'Ossie' Oswald, who covered Rules Of Engagement (ROE), Orders For Opening Fire (OFOF) and Geneva Conventions on Human Rights.[138] Many of the infantrymen and cavalrymen of the 1 RAR Group were thinking in terms of conventional operations and envisaged fighting groups of Somali clansmen mounted on technicals; images that had dominated media coverage from Somalia over the past weeks.

Oswald, with Hurley's encouragement, set out to change the temperament of members of the 1 RAR Group. Fortunately he had been training soldiers in the ODF in the use of ROE and OFOF on exercises throughout 1992 He emphasised the more complex nature of forthcoming operations and legal responsibilities to use deadly force only when directly threatened. Oswald spoke in terms that were easily understood by the soldiers. He illustrated his teaching points by conducting a number of demonstrations showing hypothetical incidents that might occur in Somalia, and following up the demonstrations by facilitating discussion about what had gone on.

Long lectures by representatives from the Defence Intelligence Organisation in Canberra were less effective in preparing members of the 1 RAR Group for service in Somalia. Cataloguing the military capabilities of the warlords raised expectations among members of the 1 RAR Group they would be involved in conventional, war-like operations. Complex and inaccurate explanations of Somali culture and the Muslim religion confused many soldiers and created inaccurate expectations of Somali behaviour. The soldiers were much more appreciative of the time they spent hearing from Phoebe Fraser from CARE Australia who had just spent time in Baidoa. Like Oswald, she gave commonsense advice and described the reality of the threat from Somali gunmen and technicals.[139]

Each evening members of the 1 RAR Group read newspaper reports about the dangerous and uncertain situation in Somalia. At this time, Boutros-Ghali was attempting to promote the UN agenda for Somalia. A personal three hour visit to Mogadishu on 3 January had aroused Somali protests. Aideed's supporters gathered outside UNOSOM Headquarters accusing the UN of bias and meddling in local political affairs.[140] This was an uncertain prelude to a meeting of 14 clan chiefs in Addis Ababa, Ethiopia, on 4-5 January called by Boutros-Ghali.

Boutros-Ghali was in a difficult political position. On one hand he was calling on American troops to disarm the militia armies, and on the other was trying to get the clans together to discuss an agenda and schedule for national reconciliation. One journalist commented, 'Dr Boutros-Ghali has offended just about everybody, in itself a mark of some distinction in the normally polite world of diplomacy.'[141]

On 5 January Dr Sean Devereux, an Irish doctor working for UNICEF, was gunned down in the port of Kismaayo by several Somali guards he was trying to sack. Devereux was an experienced UN manager who had returned to Kismaayo after working for the UN in Liberia. Colonel Fred Peck, the US Marine spokesman, said '[Devereux] told his security guards that he was no longer going to pay them inflated wages. As he was leaving after making this announcement he was shot in the back several times and killed.'[142] Within a few days UNICEF withdrew its expatriate staff from Kismaayo. This was an ominous precedent as Hurley and his staff planned for the security of expatriate NGO staff in Baidoa.

Later on the day that Devereux was shot and as clan leaders gathered in Addis Ababa to begin the UN-brokered national reconciliation conference, Johnston announced the timetable for the withdrawal of UNITAF. He planned to hand over command of military operations to UNOSOM

Major Geoff Peterson, SO2 Operations and
Intelligence, the Australina Force–Somalia.
Pictured in Mogadishu in early 1993

Image: ADF Electronic Media Unit,
David Brill

II on 20 January to coincide with the inauguration of President-elect Bill Clinton. After the hand over, his troops would begin a progressive withdrawal over a period of 60-90 days. [143] The US Armed Forces were very keen to leave Somalia for military as well as political reasons. They were not confident that political reconciliation was possible. Colonel Jock Murray,[144] the Australian Liaison Officer at Central Command in Tampa reported that,

> Prospect for political reconciliation, a prerequisite for any level of stability, remains poor. The view is widely held [at CENTCOM] that clans will appear to cooperate until the bulk of UNITAF departs, then return to the status quo.[145]

Reacting to the gathering of clan chiefs in Addis Ababa and the prospect of a return to UN command in a few weeks time, a spokesperson for General Aideed, Mr Mohamed Awale, told reporters that Aideed, 'expected nothing to come out of this meeting, which is too short and is taking place too soon.' He added that UN operations in Somalia were 'defective.' [146]

Johnston's announcement meant that the 1 RAR Group would arrive in Baidoa as American UNITAF units were beginning to withdraw. The Australians would cover the withdrawal by occupying the Baidoa Sector and continuing security operations until UNOSOM II units were ready to assume control. The major uncertainty was how much logistic support would still be available to the Australians during the American withdrawal.

A JAPG, convened by Brian Vale and chaired by Brigadier Bill Traynor, met in Sydney on 5 January to finalise the logistic and administrative arrangements for OP *Solace*. Key staff officers from all three Services were present as well as staff from major Army logistic headquarters. The Americans had advised the ADF that UNITAF logistic units would, 'Provide supplies and services according to capability, but not to the detriment of US forces.'[147] Despite this assurance, the meeting was characterised by pessimism and uncertainty. Brian Vale pointed out that despite Mellor's assessment that the US forces would remain in Somalia long enough to support the 1 RAR Group, he felt that there were no guarantees. Those attending then discussed how a Message Demand System would be used to resupply the 1 RAR Group.

In simple terms, the Message Demand System involved Rod McLeod in Baidoa raising demands for items and sending them via Major Mark Harnwell's BSG to Mellor's headquarters with a priority for satisfaction varying from seven to 28 days.[148] Woolnough and his logistic staff, Major David Creagh and Captains Mick Fulham and Mark McKeon would decide whether the demands were to be satisfied from the UNITAF resupply system in Somalia, through local purchase by McKeon, who would act as a Local Purchase Officer in Nairobi, or from Logistic Command units in Australia.

In Australia, Colonel Tim Winter, the senior Logistics Operations officer at HQ Logistic Command in Melbourne and his staff, were to supervise the satisfaction of McLeod's demands from Logistic Command units. Winter directed his senior operations officer, Lieutenant Colonel Tony Ayerbe, to raise HQ Movement Control to co-ordinate the movement of stores to Mogadishu.[149] All stores for the AFS were to be sent to Moorebank Freight Terminal in Sydney. From there they would be delivered to where ever the staff of HQ Movement Control decided they would be loaded for transportation to Mogadishu.[150]

Traynor closed the meeting by pointing out that Brian Vale and his staff at Land HQ, who act as the link between the Australian logistics system and Mellor's headquarters in Mogadishu, would have to maintain close co-operative arrangements with his staff at HQ ADF, as well as movements staff at Air and Maritime HQ and logistic staff at HQ Logistic Command.[151]

After this meeting Vale's staff continued to work on an Administrative Order for OP *Solace*. On 7 January Traynor signed the cross-servicing agreement on behalf of the Army Chief of Logistics with the US Central Command.[152] Plans to resupply the 1 RAR Group across the beach near Mogadishu were cancelled. Australian forces deployed overseas would once again rely on the American logistics system for the resupply of basic commodities and any items of supply that were common to the American and Australian military inventory.

On 5 January, the same day Traynor and ADF logisticians were meeting in Sydney, Mellor, accompanied by Petersen, Hurcum, McDonald and Stanhope, left Mogadishu with a UNITAF resupply convoy bound for Baidoa. The move was an impressive display of firepower. Gunships flew in vigilant circles above the convoy and Hum Vees and other escort vehicles bristled with gun barrels pointed in every direction. The trip took seven hours and the convoy passed through country that 'looks like Queensland or north-western New South Wales scrub country. ... monotonous and flat as a pancake.'[153]

The Marines of the 3rd/9th Battalion, 7th Regimental Combat Team, were very pleased to see the 1 RAR officers because their arrival was tangible evidence that they would be leaving soon. After several briefings on the situation in the Baidoa Sector, the Australians went into town and received a briefing from a very tense Lockton Morrissey from CARE Australia. Despite the arrival of the Marines, Morrissey was still receiving death threats because he was cutting the wages of the Somali drivers and guards who had been charging exorbitant prices for their services when there was no law and order. Now that UNITAF had taken over, Morrissey felt that some of the Somali gunmen should be laid off and wages should be cut. He was fearful of being gunned down like Sean Devereux in Kismaayo. Morrissey emphasised that the Marines had not co-operated well with the aid agencies and had been heavy-handed with local Somalis, including trusted Somali employees. Marines searched NGO vehicles, residential compounds and warehouses without informing expatriate NGO managers.[154]

On the morning of 6 January Mellor and Petersen returned to Mogadishu with a convoy. Hurcum, McDonald and Stanhope went to a 7am briefing at the Marine Headquarters at the airfield. After that they spent the rest of the day inspecting the airfield area from the ground and from the air. The co-operation from the Marines was first class. Stanhope also met with Colonel Werner Hillmer, who was the Marine Corps officer in charge of civil and military operations within the Baidoa Sector. Hilmer commanded six teams each made up of at least one officer, a junior NCO, an engineer and a Special Forces representative. Hilmer informed him that several committees had already been established to co-ordinate UNITAF activities with local Somali authorities and aid agencies. Stanhope or a member of his team was invited to attend a meeting of the Security Committee daily, the Regional Relief Committee weekly, the NGO Committee three times per week and

Captain Mick Fulham, SO3 Logistics,
HQ Australian Force–Somalia, saying goodbye
to his daughter before departing with the Advance
Party on 8 January 1993
Image: Defence Public Information,
Queensland. Gary Ramage.

Major David 'Spike' McKaskill, OC B Squadron, 3rd/4th Cavalry Regiment, pictured dockside in Townsville with the Defence Minister, Senator Robert Ray, on Christmas Eve 1992
Image: Royal Australian Navy

the Elders of Baidoa Committee every second day.[155]

Later that day McDonald hitched a flight back to Mogadishu in a Marine Sea Stallion helicopter carrying a visiting Botswanian general. He wanted to make some notes and draw sketches before contacting Hurley early the next day on the secure telephone link and sending him a sketch showing the layout of the Baidoa Airfield and the proposed locations he, Hurcum and Stanhope were recommending for the sub-units of the 1 RAR Group.

By this time McDonald and Hurcum had formed harsh first impressions of Somalia,

> ... it is clear that the Somalis caused their famine through fighting and destroying their infrastructure. It has little to do with drought and lack of food. They are an aggressive, selfish race. ... 'Thank you' is not in their vocabulary. [156]
>
> The problems here [in Somalia] are not from famine but from thugs and bandits. They [the Somalis] have no family infrastructure. Survival of the fittest is the only law. I am in awe of the welfare [NGO] people. I could not do their job. They get no thanks from the Somalis. This is a country without hope or pity.[157]

While Mellor, Petersen and McDonald were returning to Mogadishu General Aideed's troops fired on a Marine patrol and American convoys in Mogadishu. This escalation of violence formed a pattern of provocation. In previous days Aideed had refused entry by Marine patrols into an area of the city where he had warehouses full of weapons, equipment and ammunition. General Johnston had told all clan chiefs in Mogadishu to remove their arsenals from the city several days before. He now decided to get tough with General Aideed and ordered a large scale ground operation with Cobra helicopter support against Aideed's cantonment for early next morning.

That evening Mogadishu erupted in several gun battles as UNITAF convoys and patrols exchanged fire with Somali gunmen. The Embassy compound was subjected to Rocket Propelled Grenade, heavy machine gun and small arms fire. Next morning McDonald saw 12.7 mm heavy machine gun and small arms bullet holes in the newly-built ply-wood Mess Hall inside the compound.[158]

At first light on 7 January Johnston's operation began with deadly accuracy when Marine snipers 'head shot' each member of the Somali crew of an anti-aircraft gun.[159] From then on television audiences, tuned into CNN, around the world watched as the Cobras hovered in an aerial firing line and sent rockets crashing into Aideed's warehouses. On the ground US Marines fired hard-hitting, anti-armour rockets at outposts of Aideed's forces. It was an impressive display. The Marines sustained light casualties and about 30 Somalis were reported to have been killed and scores of others wounded.[160]

With the images of the fighting in Mogadishu in their heads, the 1 RAR Group Advance Party comprising Hurley, members of his headquarter staff, Major David 'Spike' McKaskill, the OC B Squadron, the remainder of Mellor's staff, and key administrative personnel left Townsville on 8 January in two RAAF C 130 Hercules transport aircraft. A third C 130 carrying Lisa Keen's MSU with about 23 journalists and assistants met the Advance Party in Port Hedland, on the Western Australian coast.[161]

Hurley's group had been farewelled by the Minister for Defence Science and Personnel, Gordon Bilney, who told reporters that the Australians would be able to fire in self defence and could be involved in 'a shoot up' after they arrived in Mogadishu.[162]

Hurley and his advance party arrived in Baidoa on 11 January (local time). Hurley recalled,

> I certainly did not expect to see what I saw at Baidoa when I got there. I do not know whether my mind had been elsewhere or I had totally misread the information I had received. It could be entirely my fault that I had not picked up on what people were saying.[163]

Hurley had expected that his troops would occupy reasonably well-organised, but lightly-fortified defensive positions, designed to deter Somalis from infiltration into the airfield area and to prevent unauthorised vehicle access. What he saw was a congregation of American units housed in tents amidst a military junk yard overrun by hundreds of Somali adults and children, either milling about, scavenging, or begging.[164]

For the next four days the Advance Party worked hard to prepare for the arrival of Fraser's Alpha Company group sailing on *Jervis Bay* and the QANTAS 747 flights that would begin to shuttle the remainder of the 1 RAR Group into Mogadishu from 15 January onwards. The Australians averaged four hours sleep each night as they planned to replace the Marines, attended briefings and conferences, inspected the area and established a small headquarters.

On 11 January there appeared to be a breakthrough in the quest for peace in Somalia. Clan leaders met in Addis Ababa and agreed to a ceasefire to begin at midnight that night. There was also an agreement to place all heavy weapons under the control of a neutral UN monitoring authority and to disarm.

Taken on face value, the hostilities in Somalia were over for the time being and the country was beginning the process of disarmament and reconciliation. Hurley and his staff were sceptical. In the following days neither the UNOSOM office in Mogadishu or Robert Oakley announced how armed clansmen would be informed about the ceasefire, or when, by whom or how the heavy weapons would be collected, or who would supervise the further disarmament of the clan armies.[165]

On 14 January (local time) *Jervis Bay* docked at Mogadishu and began unloading. Greg Jackson's Australian UNOSOM personnel came down to help Steve McDonald co-ordinate the unloading and securing of stores. 'Unloading of JB [*Jervis Bay*] went very smoothly thanks to the hard work of the Termites (Ships Army Detachment), our UNOSOM movers plus Rod McLeod. Alpha Company provided work parties along with the ship's company and all worked well.'[166]

Jervis Bay's voyage to Somalia had not been uneventful. After several days at sea Doug Fraser's contingent had been looking forward to a 24 hour stopover at Diego Garcia, a British-owned island, located in the middle of the Indian Ocean and leased by the US as a strategic air and submarine base. A combination of the sense of relief at being off the ship and having access to deceptively potent local drinks, called Mae Taes, contributed to several embarrassing incidents, culminating in a fight among a group of Australian soldiers at a bus stop. As a consequence of the fight, one soldier's jaw was broken. The injured soldier was evacuated to the US base at Guam and eventually, with his jaw wired up, he was returned to Australia.

On 15 January the first QANTAS 747 aircraft landed at Mogadishu with Major Jim Simpson's Bravo Company, half of Major Mick Moon's, Charlie Company, most of Steve McDonald's Administration Company, half of Greg Hurcum's Support Company and Battalion Headquarters staff.[167] There was some delay in unloading baggage and moving the Australians into a transit area at the airport because US trucks were not available.[168] Several Australian journalists who had deployed with the MSU were there seeking interviews and taking photographs.

The Australians had been standing on the airport tarmac for two hours waiting for transport when there was a flurry of shots 'far off in the city'. Several Marines began running towards the firing, cocking their weapons.

HMAS *Tobruk arriving in Mogadishu, on 20 January 1993, loaded to the gunwales.*
Image: PW Macdonald

The Australians soldiers were not amused because their weapons and ammunition were still in the hold of the QANTAS Jumbo sealed tight in 'bubble wrap'. When the sound of shooting continued, a Marine shouted to the Australians, 'Take cover! 'The vast majority of soldiers 'stood their ground' and looked towards the sound of the shooting.

About 20 Australians followed the Marine's advice and found cover behind a low wall in the vicinity. This was an exciting reception for the 'keyed up' diggers. Journalists had an unexpected bonus for their story of the arrival of the first flight of Australians and reported that the contingent had been fired on at the airport.[169]

The sound of firing finished a few minutes later and the Australians were driven to transit accommodation near the airport tarmac. McDonald recalled,

> I was travelling with the first vehicle-load of our soldiers to arrive at the transit camp. We were initially refused access by armed Moroccan soldiers who were still occupying the camp and spoke no English. ... After a difficult exchange with a Moroccan officer in my school boy French, we were eventually allowed to enter the camp. We found that the Moroccans had been shitting indiscriminately, including in the tents earmarked for 1 RAR.[170]

This was an unfortunate beginning for 350 tired Australians who then had to wait a further two hours before clean toilet units and a fresh water trailer arrived so they could settle in and wash themselves. In the meantime they had the unpleasant task of cleaning up their tent lines. Colonel Philip McNamara, the newly-arrived Colonel Operations at LandHQ, who had accompanied the first contingent recalled that, 'The reception arrangements left a lot to be desired. Transport did not work well. The standard of the transit camp was disgraceful.' [171]

The next day the final QANTAS flights arrived with Major Anthony 'Ant' Blumer's, Delta Company, Major David 'Spike' McKaskill's, B Squadron, Major Mark Harnwell's Battalion Support Group (BSG), Major Dick Stanhope's Civil and Military Operations Team (CMOT), Lieutenant Bill Bowyer's engineers and the remainder of Moon's and Hurcum's companies.[172] These contingents were accommodated in transit lines for 24 hours and then ferried by road to Baidoa. McKaskill's cavalrymen preferred to sleep with their APCs on the wharf than occupy the shabby transit lines. They drove their APCs to Baidoa the next day without a breakdown in a 13 hour test of the endurance of men and machines. McKaskill attributed the success of this move to the meticulous work done to refurbish the APCs in Townsville before departure and running the engines every two days during the sea voyage.[173]

The safe and efficient arrival of the main body of the 1 RAR Group in Mogadishu was a credit to the Flight Operations staff of QANTAS airlines. They had responded very professionally, like their predecessors had for short-notice deployments of Australian troops to Japan in 1950 and to Vietnam in 1965. The QANTAS Captains and flight crews had made the flights comfortable for the troops and had remained on the tarmac despite delays in off-loading their aircraft's cargo and the dangers of the Mogadishu airport.

After unloading at Mogadishu, Errol Morgan sailed *Jervis Bay* out to sea for a rendezvous with *Tobruk*. On 17 January Kevin Taylor and key members of his crew flew by Sea King helicopter to *Jervis Bay* for briefings about the conditions in Mogadishu from Morgan and his staff. Two days later *Tobruk* was anchored off Mogadishu among UNITAF's armada of 30 ships, near USS *Tripoli*, UNITAF's command ship and helicopter carrier.

On 20 January it was *Tobruk's* turn to use the Roll-On-Roll-Off wharf facilities. After a tricky manoeuvre, known as a Mediterranean Moor, which involved *Tobruk* swinging around on her anchor chains to present her stern at right angles to the wharf, unloading began. The Australian ship was dwarfed in between the 55 000 tonne USS *Belatrix*, using the south pier to off-load US Army vehicles and stores, and the equally large Marine Corps Pre-positioning Ship '*The Private Franklin J. Phillips*' taking aboard Marine vehicles and stores.

The unloading area was barricaded in by a two-tiered rectangular wall of ISO shipping containers. US Marines and Australian diggers patrolled along the top of these containers to deter opportunistic gunmen and to ensure that local scavengers did not steal supplies. There were hundreds of children and youths milling about looking for opportunities. 'Kids are kids the world over ... these ones were obviously under-privileged and likeable ... Their bigger brothers up the hill a bit, facing the 1 RAR and Marine patrols, were definitely not to be liked or trusted.'[174]

Unloading was completed at 3am on 21 January after a '22 hour day' for Le Large's 'freighties', members of the Ship's company and diggers from Blumer's Delta company who began operations in Somalia as wharf labourers. Unfortunately while Blumer's soldiers worked hard in the hot Somali sun in their unaccustomed role as wharfies, the thieves of Mogadishu were at work. Despite tight security measures, someone managed to get in and steal a Steyr rifle. The loss of a rifle on operations is a serious military offence in the Australian Army. The soldier who had been issued the rifle was charged automatically.

Members of D Company were very angry about the theft of the rifle because it lessened the reputation of the company early in the operation. Like householders who have been burgled, they were angry with the unseen Somali perpetrators. First impressions were lasting. Whilst under constant surveillance by groups of scavenging Somali youths and older males, members of D Company assessed that Somali males could not be trusted and should be dealt with firmly when they failed to comply with directions.

Errol Morgan and Kevin Taylor were proud of the performance of their ships and crews during the deployment to Somalia. At short notice and with demanding deadlines, they had recalled their crews, prepared their ships and sailed to Townsville to load personnel, vehicles and stores. Their Ships Army Detachments had adapted to continual changes in loading schedules and worked around the clock in Townsville and Mogadishu to load and unload quickly. *Jervis Bay* was well on her way to Diego Garcia when *Tobruk* finished unloading in Mogadishu on 21 January. Morgan and his crew were headed from there to Singapore and then onto Australia. Taylor told his crew on 21 January that he had received orders that *Tobruk* would remain in support of their countrymen and women of the AFS offshore. *Tobruk* made ready to take her place among the UNITAF armada.

After a small celebration of two soft drinks on the stern door of *Tobruk* with his diggers, Warrant Officer Peter Macdonald, retired to his cabin, two decks above,

> I reflected that we in the Ships Army Detachment were a lot more comfortable tonight than our recently disembarked 1 RAR shipmates, with our hot showers, clean clothes and fairly soft bunk mattresses. ... soon we would cast off from the uncertainties of Mogadishu and steam away over the horizon to a safer and more exciting port of call, and come and go with impudent indifference, whilst they would now be committed to a dangerous and uncertain future. ... almost 24 years earlier I had been a young National Serviceman in 5 RAR and had watched my troopship (HMAS *Sydney*) and its escorting destroyer (HMAS *Derwent*) steaming off to the south away from the port of Vung Tau. I had felt then exactly the same as no doubt the young 1 RAR soldiers were feeling this day.[175]

LESSONS

Within the ADF, getting the 1 RAR Group to Somalia over the Christmas-New Year period of 1992-93 proved to be a very busy time for a few while many were on leave or in transit between postings. The challenge was to get the 1 RAR Group to Somalia in four weeks with stocks to operate independently in Somalia for 30 days. The recall, administration and training of personnel in four weeks was a relatively straightforward process compared to concentrating stocks in Townsville in two weeks and loading *Tobruk* and *Jervis Bay*. This phase of OP *Solace* contained important lessons for those planning future short notice overseas deployments. This phase also set the scene for subsequent difficulties with command and control arrangements and resupply.

Before discussing pre-deployment preparations, there is much to be learned from the process to obtain Government approval. Several factors worked in favour of Generals Gration's, Grey's and Blake's assessment that the time was right to get an ODF battalion group away for operations in Africa for four months. The first factor was the experience of staffs at Land HQ, HQ ADF, the Defence Department's International Policy Division and the Resources Planning section at Army Office who had prepared several proposals for Cabinet consideration over the past 12 months.

Several proposals had resulted in deployments of Australian service contingents overseas to serve with the UN. By the time the Marines landed in Mogadishu on 9 December, Paul Retter, David White and Melva Crouch at Land HQ had developed a detailed battalion group proposal with supporting documentation. In the next four days the proposal was refined after consultation with representatives from 3rd Brigade and approved by General Blake. Two days later Ray Martin and Dan Ryan submitted a draft Cabinet Submission to the Defence Department's International Policy Division. After an intense 48 hours, Bob McDonald's and Fiona Hanks' cost estimates caught up with the submission at the Minister's office. In the end, the ADF, with Minister Robert Ray's endorsement, presented a well-argued 20 million dollar proposal to the kitchen Cabinet.

Aside from having experienced staff to produce high quality staff work quickly, the timing and the cost of the proposal were both fortuitous. The news from Somalia over the weekend prior to the Cabinet meeting on 14 December was good. Optimism was reported to be high for successful political reconciliation in Somalia and the level of reported violence was low.

The warlords were jostling for favour with UN Special Envoy, Robert Oakley, hoping to gain a political edge over their opponents during UNITAF's occupation of southern Somalia. For the time being, the likelihood of serious armed opposition to UNITAF seemed remote.

With a Federal election only a few months away, the Labor Government was also jostling with their political opponents for favour. The Prime Minister and his senior Ministers and advisers probably assessed that the political and strategic advantages of supporting the US on a high profile international humanitarian operation outweighed the cost of a battalion group operating in Africa for just over four months. Prime Minister Keating and his colleagues also considered the submission when the price was right. Costs had been calculated on preliminary planning information. The estimates increased substantially after allowances for service in Somalia were calculated and further planning identified additional costs. Within the context of Departmental rivalry for funds, Treasury and Department of Finance got a bargain. Any costs over and above the supplementation of $20 million would have to be met by the ADF from current budget allocations or painstakingly justified.

This unintended under-costing highlighted the dilemma the ADF faces when putting proposals to the Government at short notice. From the Government's perspective, the Minister and Cabinet will not make a decision until the cost, the numbers of personnel, the dangers, and the duration and scope of an operation are presented to them. From the ADF's perspective, the duration and scope of overseas operations, the structure of the force to be deployed and the dangers to ADF personnel are difficult to identify when information is scarce and the situation is volatile. This was the quandary when it came to providing early cost estimates for a deployment of a battalion group to Somalia. On one hand cost estimates developed to meet a tight decision deadline could only be based on preliminary planning information and the uncertainties of a developing situation in Somalia. On the other hand, increasing the cost estimates in anticipation of these uncertainties could put the price up too high for Government approval.

After the Government announcement on 15 December, preparations began in earnest with the shrouds of secrecy pulled away. Maritime Headquarters ordered *Jervis Bay* and *Tobruk* to sail to Townsville from Sydney to pick up personnel, vehicles, equipment and stores. *Jervis Bay* stocked up and sailed in four days. *Tobruk* followed five days later on Boxing Day after 10 days of around-the-clock preparations that included reassembling an engine and other major components, accelerating a repair and maintenance program and enduring a gruelling testing program at sea to achieve 'Mission Capable' status a few hours before arrival in Townsville.

The lessons from the Navy's deployment were that the earliest possible warning should be given to the COs of ships that are earmarked to support the deployment of forces overseas at short notice. Fortunately, Commander Kevin Taylor, the newly posted CO of *Tobruk,* was tipped off early when officially he should not have been told of that contingency planning had begun in November for a major deployment to Somalia. Without this timely warning, *Tobruk* may not have been on-line in time to load troops and materiel in Townsville within 10 days of the Government's announcement.

It is instructive to compare the stories of preparations of the 1 RAR Group in 1992 and another 1 RAR Group in 1965 that was also given four weeks after a Government announcement to depart by sea and air overseas. In 1965 the 1 RAR Group's mission was to operate as the third manoeuvre battalion of the US 173rd Airborne Brigade (Separate) in South Vietnam. Despite almost thirty years separating the two deployments, there were many similarities. Taken together, both stories suggest that the Army of the 1960s and the ADF of the early 1990s still had room to improve the process of getting an expeditionary force ready for overseas deployment. The similarities also suggest that there are consistent features of short notice deployments worth noting when planning for them in the future.

One of the first issues influencing preparations for the deployment of both 1 RAR Groups was the imposition of strict secrecy. In 1965 CO 1 RAR,

Lieutenant Colonel Ivan 'Lou' Brumfield, and a few of his senior staff were told in the strictest secrecy at the end of February that 1 RAR was likely to be deployed to South Vietnam in a few months time. This coincided with the Australian Government's expectation that the US would accept the offer of a battalion group to serve alongside US ground forces about to be committed to secure major airbases in South Vietnam. In 1992, Lieutenant Colonel David Hurley was told on 8 December that the ADF was putting a proposal to the Government for his battalion and a number of sub-units to deploy to Somalia in six weeks time. With hindsight, he could have been told three weeks earlier to coincide with the realisation at Land HQ that the only unit capable of short notice deployment was 1 RAR, supported by elements from the ODF, or several sub units of the ODF protected by a company group from 1 RAR.

With hindsight, the advantages of imposing strict secrecy on the preparations of the two 1 RAR Groups before the Government announcements of their deployments were outweighed by disadvantages. In 1965, the advantages of telling Brumfield and his senior staff were that they had two months to collect their own thoughts, conduct research, plan the manning and stocking of the battalion and prepare Standard Operating Procedures and administrative instructions. The disadvantages were that, because supporting logistic units were not told of 1 RAR's pending deployment, 1 RAR received no priority for personnel, vehicles, weapons, equipment and stocks. Nor was the unit accorded any priority for training resources. In effect, the Army paid lip service to 1 RAR's authorised level of operational readiness to support Australian commitments to South East Asian Treaty Organisation (SEATO) military contingencies. Brumfield was unable to stock up to authorised levels or top up on personnel. He was also not given the resources or flexibility to train for operations in South Vietnam. Brumfield was left to train his under-stocked, under-strength and newly-reorganised battalion for conventional warfare against a notional enemy in a local State Forest using a limited supply of blank ammunition, road transport and helicopter support.

After the Government announcement on 29 April 1965, Brumfield and his senior staff were mentally prepared for what had to be done to train and equip 1 RAR before departure in four weeks time, but, because no one else in the Army was prepared, the next four weeks was spent catching up on administration, taking on reinforcements and trying to get the Army's logistic system to concentrate sufficient stocks on the wharfs at Garden Island on Sydney Harbour to load HMAS *Sydney*. Fortunately, troops were given the opportunity to train with live ammunition for operations in South Vietnam subsequently. Those sailing on *Sydney* trained in transit and the whole Battalion Group were able to train together for two weeks after arrival at Bien Hoa Airbase.

David Hurley and his staff were given a week to collect their thoughts, conduct research, plan the manning and stocking of the 1 RAR Group and prepare operational and administrative instructions before the Government announcement on 15 December. Like Brumfield and his staff 27 years before, no action could be taken to stock up, conduct relevant training or consolidate manning until after the Government announcement. Once again the fact that 1 RAR was on the highest level of readiness in the Army for overseas deployment did not mean that stocks or numbers of personnel were at their authorised levels.

Fortunately, members of 1 RAR in 1992 was administratively better prepared than their predecessors in 1965 and there was time for two weeks of 'special-to-mission' training in Australia before beginning operations in Somalia. The major disadvantage of not telling Hurley, his senior staff and key logistics officers when the 1 RAR Group was earmarked for deployment four weeks earlier was the reduction in time for the logistic system to be alerted to 1 RAR's deficiencies in vehicles, weapons, equipment and general stores—a thousand demands for resupply hit the logistic system within 72 hours. Earlier warning and setting a higher priority for resupply would have allowed stocks to be concentrated progressively and inexpensively, and more time for Hurley's administrative staff to sort out what needed to be loaded on *Tobruk* and *Jervis Bay*.

There is a risk that warning senior staff from a number of units and headquarters early in the contingency planning process could result in the premature disclosure of preparations before political processes have been worked through by the Australian Government and other nations considering participation in a multi-national force. Past experiences with short notice deployments of Australian expeditionary forces overseas, such as 3 RAR to Korea in 1950, a 1 RAR Group to South Vietnam in 1965, and a 1 RAR Company Group to the waters off Fiji in 1987, shows that information is invariably leaked intentionally or unintentionally by Governments, not by military planning staffs or commanders. Based on previous experience, early advice to commanders and key staff from units earmarked for deployment and their supporting supply depots and movement agencies is worth the risk.

The recall of personnel in December 1992 for service in Somalia was successful. In 16 days over 900 personnel concentrated in Townsville and began pre-deployment training and administration. This was a notable achievement because many members of the 1 RAR Group and their families were either on their way out of Townsville, or were inbound with posting orders from around Australia as a consequence of the Army's block leave arrangements and traditional end-of-year posting cycle. Consequently, there was some confusion during the recall about who would be deployed. During this time the postings of some inbound personnel were hastened or cancelled, the postings of some outbound personnel were cancelled or confirmed.

Many commanders and subordinates met for the first time amidst the jubilation of those who had been included and the disappointment of others who were excluded. In addition to the personnel turbulence caused by the annual posting cycle, end-of-year changes to appointments within the 1 RAR Group and the arrival of 56 reinforcements from 2/4 RAR contributed further turbulence.

One of the lessons from the period of concentration and pre-deployment preparations for both 1 RAR Groups, and the deployments of 3 RAR in 1950 and a 1 RAR Company Group in 1987 is that personnel turbulence is the norm for short notice deployments. In a small Army this situation is offset to some degree because personnel in each Corps and rank have trained and been posted together, and know each other or have mutual friends that accelerate the establishment of close working relationships. However, for units on short notices to move there is might be some advantages in staggering block leave periods and posting cycles for married and single personnel. Incentives could be provided for single personnel to take leave and move on posting outside the periods of school holidays and over the Christmas/New Year period.

Neighbouring units are significantly effected by short notice deployments. The experiences of the 1 RAR Groups in 1965 and 1992 show that the equivalent of another battalion group had to be cannibalised and temporarily immobilised to facilitate their deployment overseas. For example, in addition to providing 56 riflemen and a number of critical items on loan to 1 RAR, 2/4 RAR loaned vehicles, heavy weapons, equipment and stores to 1 RAR for pre-deployment training because 1 RAR's vehicles, heavy weapons, equipment and stores had been loaded on *Tobruk* and *Jervis Bay*. The other sub-units of the 1 RAR Group had to borrow from parent units to stock up as well as to conduct pre-deployment training. In 1965 1 RAR had depended on neighbouring 5 RAR for substantial reinforcement and the transfer of serviceable vehicles, weapons, equipment and stores prior to deployment to South Vietnam.

In 1992 the Army's logistic and movement systems were put under extreme pressure just before Christmas. A combination of the late notice, the absence on leave of many key logistic personnel, long term deficiencies in stock holdings for the ODF in Townsville and a delay in authorising the highest priority for issuing stocks to the 1 RAR Group contributed to an uncoordinated and slow response. The first consequence was an expensive concentration of stocks in Townsville at short notice by air and road. The second was the 'hand-to-mouth' loading of *Jervis Bay* and *Tobruk* without sufficient shipping containers. The third was that the ships sailed, leaving over 50 pallets of stores and six vehicles on the wharf, while more stores continued to dribble in. These vehicles and stores did not find their way to Somalia, leaving the 1 RAR Group short of stock and mobility, and setting the scene for expensive, urgent and *ad hoc* resupply.

The size of the media presence was a major difference between the deployments in 1965 and 1993 The late inclusion of a Media Support Unit with specialist vehicles and stores disrupted the 1 RAR Group's preparations and added another complication to loading *Jervis Bay*. In 1965 a handful of journalists had accompanied the 1 RAR Group by sea on HMAS *Sydney* to South Vietnam. In 1993 25 journalists, members camera crews and photographers deployed to cover the arrival and settling in of the 1 RAR Group into Somalia. The lesson for the ADF is to anticipate that a deployed force will have to move, administer and protect contingents of media representatives during operations.

Two lessons emerge from the logistic preparations of the two 1 RAR Groups. The first is that there appears to be a gap between the declared high states of operational readiness for units earmarked for short notice deployment and the reality of the actual readiness and responsiveness of the logistic systems supporting them. In 1965 and 1992 1 RAR was the unit on the highest state of readiness for deployment in the Army. When it came time for deployment, long term deficiencies in stock holdings and maintenance schedules were exposed, neighbouring units had to be cannibalised for serviceable vehicles, weapons, equipment and stores, and supply depots and movement agencies found it difficult to concentrate stocks for onward movement to the area of operations in time to load ships efficiently.

The second lesson is that sufficient time and prompt clarification of priorities are required to make appropriate logistic preparations for short notice deployments and to permit the efficient concentration and loading of materiel. This time was not afforded to the Army's logistic systems in 1965 or in 1992, nor were priorities for logistic effort clarified quickly.

Key logisticians must be informed as soon as possible, priorities of effort specified by higher headquarters and operations and logistic planning staff should be co-located while the force is being concentrated, stocked up, trained, loaded and deployed.

In the end, the AFS arrived in Somalia on time, and unloaded personnel and stores on schedule. QANTAS Airlines proved once again to be the quickest means of moving large numbers of Australian military personnel and their personal equipment overseas at short notice. After some reception difficulties, the main body of 1 RAR Group moved through a transit camp in Mogadishu, arriving at Baidoa in time to take over from the Marines who departed quickly to ensure they were out of Somalia by the time President Clinton was inaugurated on 20 January. Based on outcomes, but not on the efficiency of some processes, this was a significant achievement, worthy of the praise afforded to the ADF at the time.

Of interest during the six weeks from 8 December 1992 until the arrival of the main body of the 1 RAR Group in Baidoa on 17 January 1993 were the pressures put on General Robert Johnston to increase the scope of UNITAF operations while also having to satisfy domestic American political agendas. These pressures illustrate the potential for mission creep and compromising military judgement in the future as well as the temptation to use the military forces deployed to create a secure environment for the distribution of humanitarian aid to take more assertive action against the forces that created the insecure environment in the first place.

From the beginning General Johnston was put under pressure from UN agencies, NGOs and the media, to deploy forces inland as quickly as possible. Johnston insisted that he would build up his forces in Mogadishu before advancing inland. He was being asked to suspend his military judgement in order to save lives. This sort of pressure to save lives by acting quickly, especially if there has been some perceived delay in deploying military forces in the first place, will apply in future operations. Like Johnston, future commanders will have to balance their responsibility for the lives of their troops against the call for risks to be taken to save the lives of starving or physically threatened women and children.

The UN exerted significant pressure on Johnston to order UNITAF units to disarm the warlords while UN diplomats tried to facilitate political reconciliation. Dr Boutros-Ghali was concerned that after UNITAF left Somalia in a couple of months time, the warlords would resume their intimidation of UNOSOM units. The difficulty for Johnston and the US Government was that the warlords were unlikely to give up the source of their military power and political survival without a fight. In effect, Johnston was being invited by Dr Boutros-Ghali to risk restarting the civil war and having to wage a full military campaign to defeat the warlords. This may have incurred high civilian casualties and having to operate across international borders to destroy the warlords' arsenals of tanks, artillery, mortars, heavy weapons and technicals.

This paradoxical approach of calling for UNITAF to strip the warlords of their military power on one hand while negotiating with them for peace on the other seemed to echo one of the adages of the Vietnam War—'If you have them by the balls, their hearts and minds will follow.' In the final analysis, the UN could not have it both ways and achieve both objectives simultaneously. At best the objectives were sequential, not concurrent. At worst, the simultaneous pursuit of both objectives might have consolidated Somali opposition to UN intervention into Somali affairs and given the warlords the last laugh.

Finally, General Johnston also operated under the pressure of having to get US units in to secure southern Somalia as well as getting them out of the country by the time President-elect Bill Clinton was inaugurated on 20 January. In the end the appearance of getting out on time was achieved by publicising the relief-in-place of US Marine units by international combat units. In reality, the Marines left, but US Army logistics, communications and other support units stayed in Somalia. In addition, US Army combat and Special Forces units arrived to fill the gap between what the international community could provide and the US Armed Service's assessment of what was required for OP *Restore Hope* to succeed. Despite the change of Presidents, the US Government remained committed to creating a secure environment before handing command of operations back to a larger-scale UNOSOM II. The need for the Americans to leave their logistic, communications and other support units in Somalia emphasised once again the importance of the US underwriting large scale international military interventions logistically for them to be viable.

3 SECURING BAIDOA

*The local people are generally friendly and don't appear
concerned about our patrol. The streets are full of strong
pungent smells. There is rubbish and rubble in many places.
It is uncomfortable and hot patrolling in webbing and a flak
jacket, wondering if a bandit is going to pop up and shoot you.*

Captain Andrew Somerville,
3 February 1993

After their arrival in Somalia, the 1 RAR Group had the mission to create a secure environment for the distribution of humanitarian aid in the Baidoa Sector. To fulfil this mission Hurley and his staff determined that there were four simultaneous tasks. The first was to protect the battalion's base. This was a straightforward task of constructing a barbed wire fence around the perimeter of the airfield and ensuring that the entire perimeter was able to be observed from a cordon of manned observation towers. Any movement was monitored by pairs of sentries in the towers who communicated with troops manning checkpoints. Those on the checkpoints controlled the movement of vehicles and personnel. A small Quick Reaction Force of six to eight men with a Land Rover was on standby 24 hours a day to respond to the approach of armed individuals or groups, or any other crises. In addition to their own personnel and property, the Australians protected collocated and transiting UNITAF personnel and their property, and humanitarian supplies stored at the airfield. In effect, UNITAF commandeered the airfield from the Somalis who had controlled its operations, giving the Australians jurisdiction over the entire area and authorising David Hurley to set the rules.

The second task was to create a secure environment in the Baidoa metropolitan area to protect humanitarian operations. This entailed protecting expatriate aid agency staff, their accommodation areas, their vehicles and other aid agency property. This was not as straight-forward as protecting the airfield. Militarily, Hurley had to assess the threat to the UN aid agencies and the NGOs, allocate sufficient resources to meet the threat, impose a regime of patrolling and building and house searches as well as select key areas to occupy to maintain an effective level of deterrence. Politically, Hurley had to decide on what regime of liaison and coordination was required with UN and NGO representatives, as well as local Somali elders, community groups and political factions, to allow his commanders to operate safely and effectively in town.

The third task was to maintain sufficient presence in the countryside of the Baidoa Sector to deter anyone contemplating interfering with the distribution of humanitarian aid. Once again this was a politico-military task. Hurley and his staff had to assess the threat, allocate resources and conduct operations that would deter interference. He also had to decide on what needed to be done to gain the cooperation of local Somalis whom the Australians would depend on to provide information on individuals and groups that posed a danger to humanitarian operations.

The fourth task was the close protection of food convoys and other humanitarian activities in the countryside. Militarily, this was relatively easy to achieve by the provision of armed escorts vehicles. However, administratively, the coordination with UN and NGO representatives to ensure that sufficient troops and vehicles were on-line for every convoy

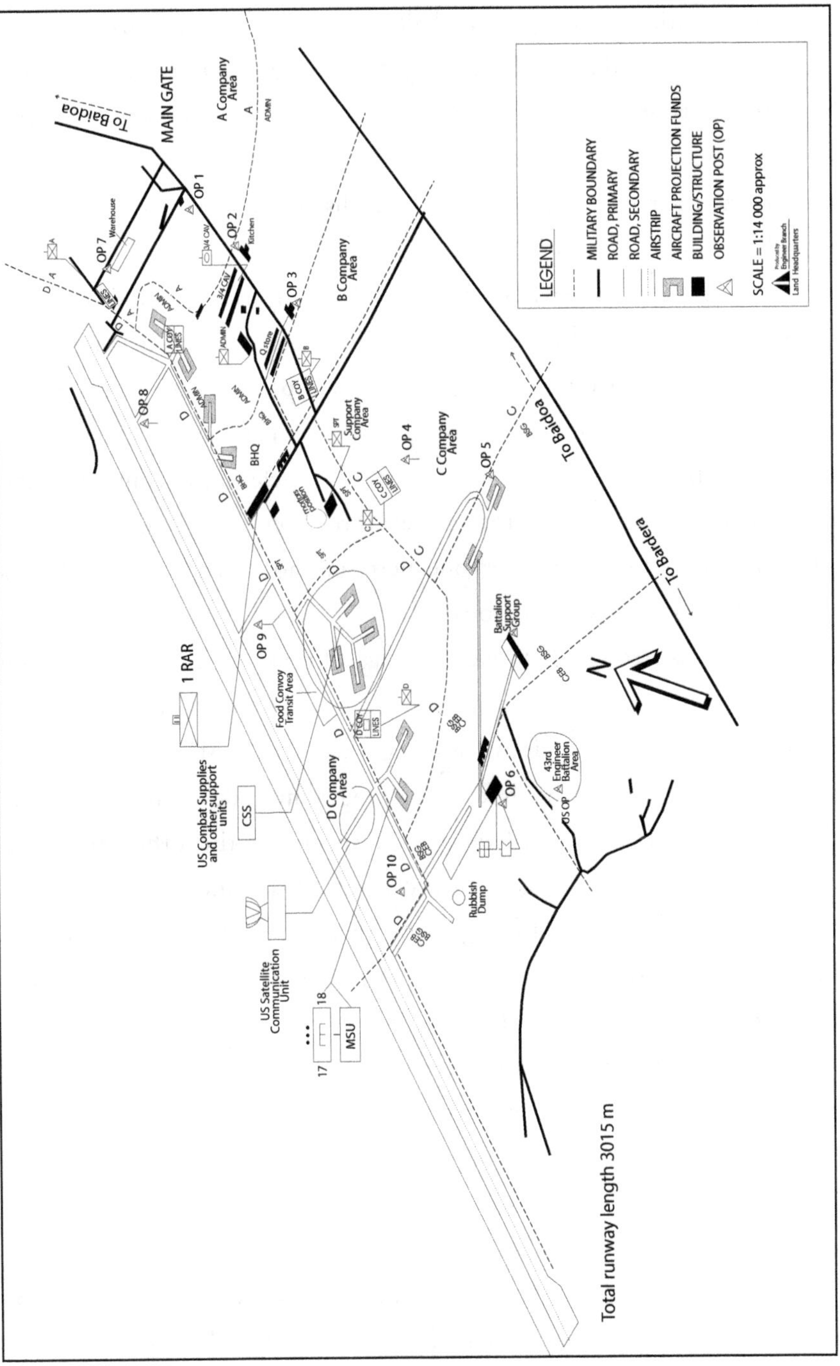

Map 3: 1 RAR Group positions, Baidoa Airfield, Somalia, 1993.

and humanitarian activity was a challenge. The additional challenge the Australians accepted beyond escorting the convoys was to ensure that food and other supplies were distributed in an orderly, equitable way once they arrived at distribution points. From the beginning the Australians assessed that it was their responsibility to set up distribution points and control crowds in conjunction with NGO staff.

Back in Australia, David Hurley and John Caligari had decided to assign the rifle companies bolstered by Reconnaissance, Mortar and Direct Fire Support Weapons platoons from Support Company to each of these four tasks on a nine-day rotation. Consequently company commanders could rotate their three platoons through a three-day cycle within companies and platoon commanders could rotate their three sections on one-day cycles. The intention was to keep everyone alert and motivated by giving them a variety of tasks in a number of locations.

There was strongly-held view among all ranks of the 1 RAR Group that they were in Somalia to do some good. There was a spectrum of interpretations of what 'doing some good' meant. At one end of the spectrum were those who felt that 'doing some good' meant creating deterrence through a heavily-armed presence. At the other end were those who felt that this meant operating as trained infantry and cavalry to eliminate hostile armed groups who threatened the lives of innocent Somalis, their property and their entitlements to humanitarian aid. On arrival most members of the 1 RAR Group were at the 'find 'em and fight 'em' end of the spectrum. There was a strong desire to confront any hostile groups or individuals who were preying on Somali citizens or stealing humanitarian supplies. The troops were very keen to test themselves in combat to validate their training and answer the centuries-old warrior's question of how they would react in life and death situations.

Baidoa was a modern Dark Ages town, gutted by a civil war, ruled by the gun, corrupt by Western-standards and socially dysfunctional. Power was distributed among gangsters, corrupt tribal elders and violent Western-style

Street scene Baidoa February 1993
Image: ADF Electronic Media Unit.

political factions, fronting for centuries-old clan affiliations. Under the blaze of the guns of those wrestling for power and booty, people led a dog-eat-dog lifestyle, fighting for resources to survive against starvation and disease. When the Marines arrived in December 1992, about 20 000 people cowered in their homes, scuttling out now and again for food and water. About a hundred bodies were being picked up and buried each day and an estimated 40 000 thousand refugees struggled for survival in squalor on the outskirts of town or living and sleeping in the streets, defenceless prey for criminals and bandits. David Hurley wrote later about what he saw when he arrived in January 1993:

> Most of the government buildings in our area were completely destroyed. Anything that wasn't fixed had been stripped from other buildings—roofing, windows, doors, wiring. The main shopping area was littered—a lot of wreckage, old car bodies, just like a big garbage dump. No power. Very few people on the streets at all. No ordinary citizen dared venture out on the streets at night and only a few would come out in the daytime. Surrounding the town were these humpy villages with refugees living basically under piles of rubbish, twigs with grass interlaced covered with plastic bags or plastic sheeting, or cardboard—whatever they could get hold of.[1]

In the absence of agrarian commerce, the economy of Baidoa was based on trade in stolen humanitarian aid, *khaat*, guns, sex and audio cassettes of Western music. Despite being nominally a Muslim town, every night the

Khaat Market in Baidoa throbbed with sex, drugs, rock 'n roll and reggae as well as reverberating with gunfire and punctuated by cries of those being mugged. Somali men, high on *khaat*, would fire bursts from their rifles and pistols into the air, indiscriminately at buildings or, after provocation, at each other. Groups of bandits were always on the lookout for prey and regularly robbed anyone foolish or unlucky enough to be in the wrong place at the wrong time. Every night the local hospitals run by the International Medical Corps and the Medicens Sans Frontieres treated scores of Somalis suffering from gunshot wounds and the results of severe beatings.

The fight against famine in Baidoa had been won by January and many refugees were returning to their homes in the countryside. However, several UN agencies and NGOs continued to work in the area. Dick Stanhope's summary written soon after arrival appeared in an Army report later:

> The relief organisations represented [in Baidoa] included CARE Australia, World Vision, Medicens Sans Frontieres-Holland, CONCERN, GOAL, International Medical Corps, International Committee of the Red Cross and the Catholic Relief Service. Each had a different mandate and level of funding. The UN was represented by the World Food Program, UNICEF and UNOSOM. There were about 102 expatriate NGO managers and workers in Baidoa, in 43 different compounds, 20 of which were living accommodation. Each NGO varied in size, but employed a large number of Somalis as drivers, security guards and workers.[2]

Most NGOs were based in a line of compounds along what became locally known as, 'NGO Road'. Expatriate NGO staff were mostly Australian, American, Irish, Dutch, British or French, many with experience in other parts of Africa, the Middle East, India and in Asia. Most were single and their ages varied from early 20s to late 50s. There was a fairly even mix of genders because of the employment of female nurses in local medical facilities. Expatriate staff had a variety of motives for being in Baidoa, ranging from religious or ideological beliefs in helping people in the Third World to strong personal desires to make money quickly.

From the beginning, the US Marines had came down heavily on the people and property in Baidoa and surrounding villages. In the days after occupying the town in December 1992, they 'busted'[3] and searched buildings and houses, confiscated weapons and patrolled on foot and in vehicles, questioning, searching and roughly handling any Somali male who looked suspicious. The Marines gained the physical and psychological upper hand quickly but created resentment among many Somalis, especially local criminals and Somali youths who were used to being a law unto themselves on the streets of Baidoa. From the Marine point of view, they had 'kicked Somali butt.'

On 27 December 1992, the 3rd/9th Marine Battalion, a brother battalion from the 7th Regimental Combat Team, relieved the 2nd/9th Battalion. The fresh Marines kept up the pressure in Baidoa, but quickly became frustrated and bored. Maintaining security in Baidoa for three weeks until the Australians arrived sapped their energy and patience. They were trained to act as shock troops, not as policemen in a hot, dusty, stinking urban environment among a hostile populace. Young men and teenage boys did not help the temperament of the Marines by throwing stones and spitting at them as they patrolled.

Soon after David Hurley and his advance party arrived on 11 January, the Marines began to take casualties in Mogadishu. Somali gunmen, many of them quite young, had begun shooting at Marine patrols employing hit and run tactics. Most of the time the ambushers fired bursts quickly and inaccurately, and then ran away, using narrow streets and crowds as cover for their escape.

On 12 January, hurried sprays of bullets found their mark. A marine was killed and another shot in the head by gunmen in separate incidents. Somali gunmen threw grenades during one of these contacts.[4] This marked an ominous escalation of hostilities. In Baidoa, later that same day a Marine patrol detained a 15 year old youth armed with an AK 47 assault rifle after he had stolen an NGO vehicle at gunpoint. Two days later a group of Somali youths stoned a patrol, injuring another marine.

Possibly spurred by the death of the marine in Mogadishu, and the other casualties taken in the preceding days, Marine commanders decided to give the Somalis in the Baidoa area one last taste of Marine Corps muscle before the Australians took over. In their own parlance, the Marines 'took down' Buurhakaba and several other towns and villages in the Baidoa Sector in airmobile assaults, and conducted rigorous searches of homes, buildings and compounds.

This violent farewell from the 'Hammer Heads'[5] of the 3rd/9th Battalion stirred up Baidoa. Property owners were angry about the damage caused to houses and businesses during searches. The expatriate NGO staff were angry with the Marines for being so violent and not cooperating with them. Local criminals were angry that their lucrative extortion operations had been interrupted for several weeks and that their arms and vehicles were being confiscated.

After the stage had been set so violently by the Marines for the Australian takeover, the Somalis did not know what to expect next. Those who had observed members of the Australian advance party at the airfield and during subsequent reconnaissance visits into town would have noticed that Australians wore different uniforms, carried short-barrelled Steyr rifles and spoke English. Initially they reported back to friends and relatives in Baidoa that the Australians were white, European-looking, English speakers like the Americans, but impersonal and aloof—just like the French.[6]

Aside from the ruckus caused by the Marines, the Australians had to contend with another security crisis as they prepared to take over. On 15 January, a Swiss NGO employee of the Red Cross at Berdaale was shot in the back of the head and killed. The motive was robbery and a large amount of US currency was taken. Three of the six Somali gunmen involved in this murder and robbery were Red Cross employees, another had been a patient at the Red Cross medical facility. This murder increased anxiety among expatriate NGO staff in Baidoa. They were disillusioned with the Marines who had concentrated on escorting convoys and confiscating weapons, and had not provided nocturnal security for them. There was talk of some NGOs ceasing relief operations in Baidoa and pulling out.[7]

Responding to the threat of an NGO walk out, David Hurley and his advisers planned to create a secure environment in Baidoa 24 hours a day.[8] Unlike the Marines, they intended to provide close physical security for expatriate NGO managers and all UN and NGO compounds and warehouses by sending patrols into the streets of Baidoa around the clock. They also intended to implement the UNITAF policy of confiscating all arms and ammunition found and registering the weapons carried by the guards employed by the UN agencies and NGOs.

The Australians faced three armed groups in Baidoa who depended on criminal activity for their livelihoods and presented threats to the security of humanitarian aid operations. The first group were local thugs, extortionists, arms dealers and drug runners. The second group were nomadic bandits who visited Baidoa to rob, rape and loot. The third group were some of the guards employed by less-experienced NGOs. Many of these guards were not local Baidoans and, because they were armed and lived in NGO compounds, posed a particularly difficult security problem for the Australians. Few could be trusted and some were suspected of colluding among themselves to steal goods from NGO compounds and warehouses at night when they were supposedly on guard. Based on recent killings of expatriate NGO managers in Kismaayo and Berdaale, expatriate NGO managers in Baidoa were more likely to be shot by these men than by local criminals or bandits.

On the morning of 16 January, Alpha Company flew in from Mogadishu after disembarking from *Jervis Bay*. Their reception in Baidoa was chaotic and disorganised—a frustrating beginning to their service in Somalia. The situation was compounded by the presence of a scrum of journalists seeking interviews and taking photographs and video footage.

Doug Fraser had expected to take over airport perimeter security positions from the Marines at midnight. He was now told to set up vehicle checkpoints on the approaches to the airfield at the same time and send out his first patrol into Baidoa at 5:30 am next morning.[9] A long, busy, chaotic day was now becoming a sleepless 24 hours. Fraser quickly arranged for those who would lead the first Australian patrols next morning to accompany the last Marine patrols into town that night so they would get a feel for the situation. He then prepared and issued new orders to his platoon commanders and gave several interviews to journalists who continually asked him what dangers he and his men anticipated in Baidoa. Later, he was told that a group of journalists and several camera crews would accompany his first patrol into town and footage of that patrol would be on all the Australian national television network news programs next evening.

Many diggers from Alpha Company experienced feelings of great excitement that night as they listened to Baidoa crackle with gunfire. Lieutenant Bob Worswick, the Commander of 2 Platoon, who accompanied one of the Marine patrols into Baidoa that night, was struck by the hatred in Somali eyes as the Marines drove past them.[10] He deduced that the Somalis knew that the Marines were there to do a job, but not to help.[11] No one at the airport, or any of the Marines patrolling in Baidoa noticed one flurry of shots among all the others. Three Somali gunmen attacked and looted the Medicines Sans Frontieres (MSF) NGO compound and escaped with a large quantity of US currency.

It fell to Corporal Terry 'Burger' Conner,[12] from Worswick's Platoon, to lead the first Australian patrol into Baidoa just before first light on 17 January. He had been out most of the night with a Marine patrol but was still alert and keen. Worswick, who had also had a busy sleepless night patrolling with the Marines, waited with Conner until 5 30 am for the journalists to appear. Word came to him that they had slept in. With a feeling of relief, he ordered Conner to proceed as planned.[13]

> The first impressions of town in the morning were completely different to what was expected. As you left the main gate for the first time and allowed the action of your rifle to chamber a live round,[14] the reality of the situation suddenly hit you. Everything you had trained

for was 'Now', and outside was the unknown. The first operational patrol since Vietnam. The effect on the soldiers was overwhelming: eyes widened trying to see into the morning darkness, ears were pricked up, heads cocked listening for the slightest sound but hearing nothing but your heart pounding, hands tightening on pistol grips, and sweat pouring from every pore as the soldiers' instinct took over and they became 'switched on'. We had only gone 20 metres outside the gate, there were three hours and another 12 kilometres ahead.[15]

As Conner's men headed into town, the news of the robbery at the MSF compound spread among the NGO community. The Marines were no longer responsible for security in town so the Australians picked up the consequences of this incident. Later that morning a delegation from the NGO Committee turned up for a meeting with Dick Stanhope. The combination of the murder of the Red Cross manager in Berdaale and the MSF robbery had brought months of anxiety and tension to a head. All the expatriate NGO staff in Baidoa had signed a letter which had been sent to the HQ UNITAF in Mogadishu stating that they would leave Baidoa immediately if one of them was killed or wounded.[16] When this was reported to Hurley, he knew that if he was unable to protect the NGO compounds and expatriate personnel his mission would fail.[17]

Meanwhile, Conner's patrol was well on its way. Worswick wrote later:

As the sun came up the reality of town was realised. The streets were alive with people slowly awakening from a night's sleep, like creatures coming out of their holes. The foul stench that violated your sense of smell came from the filth littering the streets; rubbish, human excrement and animal waste ...[18]

The next patrol into Baidoa put the patrol commander, Corporal Dion Jobson under considerable pressure.[19] Jobson and his men were accompanied by Doug Fraser, a team of journalists, photographers, a Media Liaison Officer and a video camera crew. Every time Jobson gestured or called to his men, camera shutters would click rapidly, a video cameraman with a sound recordist scrambling behind him would

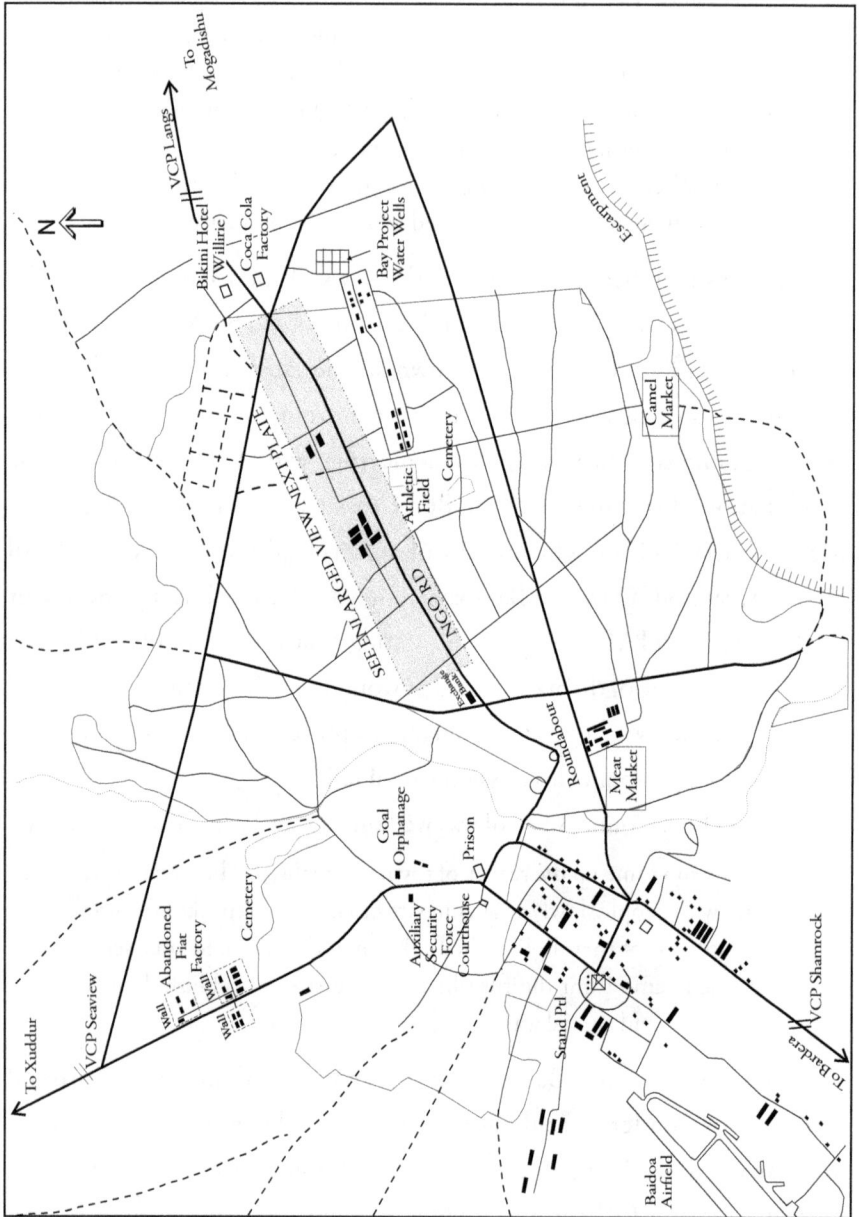

Map 4: Baidoa township and some locations—not to scale

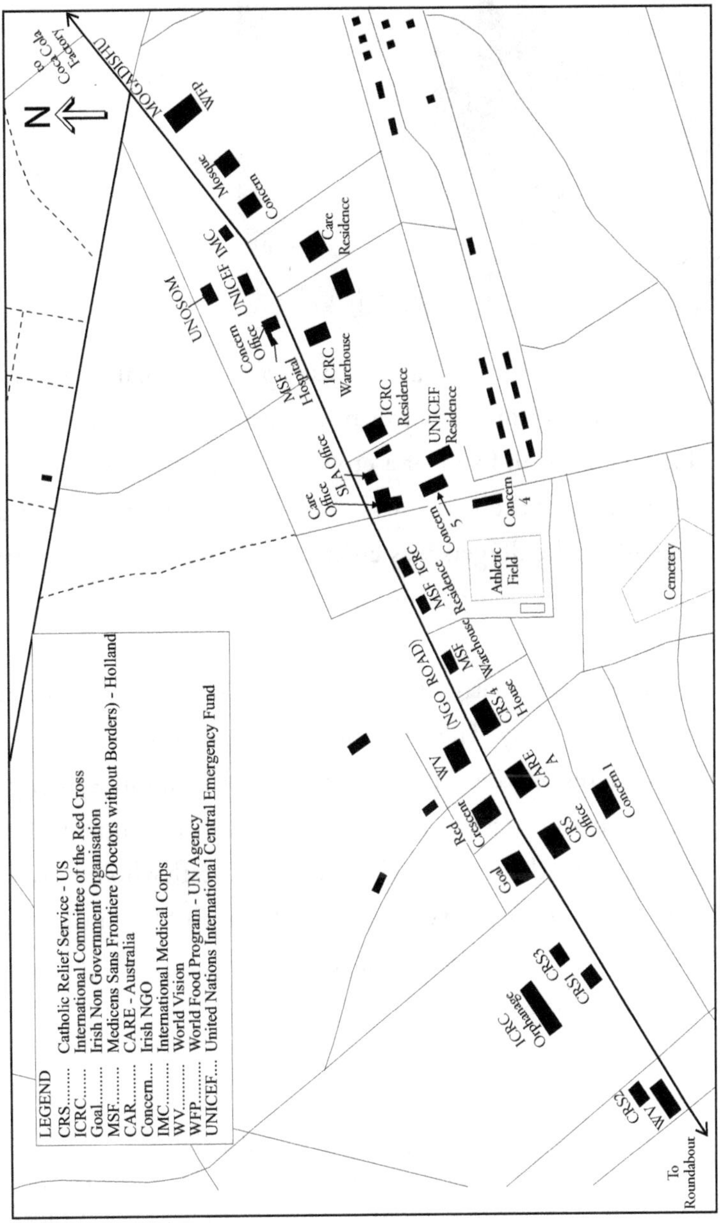

Map 5: A schematic diagram of NGO Road Baidoa, showing NGO compounds

*An Australian patrol
moves through the
narrow alleyways of
Baidoa in
February 1993*
Image: G Gittoes

move into a better position to get a 'close up' and journalists would dart here and there to get into a better position to see what was going on. All of this happened in front of a crowd of amused and contemptuous Somalis. Fortunately the patrol was completed without incident and a positive image of the diggers was created back in Australia on national television network news programs.

That night a single shot was fired at one of Fraser's patrols at about 10:45 pm.[20] The patrol 'froze' and listened for 15 minutes before continuing on their way. Their response was limited by a ban on using illumination flares in town because of concerns about starting fires among the stick and hessian huts. The incident was handled routinely by the patrol and by the battalion Headquarters staff, who reported it to John Caligari, and to the Duty Officer at HQ 10th Mountain Division, 1 RAR's superior American headquarters in Mogadishu. The contest for supremacy in Baidoa had begun.

By 20 January, after three days of almost continuous patrolling, Worswick's men were exhausted:

> The first patrols were marathons. Dressed in flak jacket and helmet, the diggers had to walk one kilometre before they left the perimeter, then cover 12 to 15 kilometres [21] in temperatures in excess of 40C. To sleep after a patrol was near impossible. A company of men [over 100] was crammed into a roofless building in sweltering conditions. After returning from patrol there was the compulsory debrief, then a

briefing for the next patrol which had to be passed on to the section as formal orders. In reality, the men were lucky to get three or four hours sleep between patrols. The effects were noticeable after a day and blatant after two.

For a commander it was awe inspiring. Men were physically wrecked, eyes reddened from lack of sleep, yet the moment they left the gate, they switched back on, alert to the slightest noise or movement. Reporters shat themselves and were physically wrecked after one patrol. The diggers were back out a few hours later, weighed down by a flak jacket, webbing and a helmet.

The diggers wanted to dominate the town. My sections competed against each other regarding the number of patrols they conducted, number of weapons confiscated and number of shots fired at them. They were true testimony to the ANZAC legend.[22]

The initial impact of the intense patrolling program conducted by the Australians was good. A favourable response was noticed:

Baidoans are impressed that the Australian forces are conducting foot patrols by day and night. They are in favour of us further prosecuting operations against bandits and are excited by prospects of Australian forces being involved in operations to outlying villages.[23]

The policy regarding the confiscation of weapons was a source of confusion and frustration in these early days. Fraser was constantly seeking clarification of the policy when his patrols reported that Battalion Headquarters were giving inconsistent advice. At this time there was no effective weapon registration system in town. Guards employed by the NGOs were particularly vocal if an Australian patrol tried to take their weapons from them. The Australians were unable to determine who were guards and who were bandits. They had already been told that Somalis carrying weapons could be both.

After some contradictory guidance from Battalion Headquarters, patrol commanders were told to confiscate any weapon they found.[24] The diggers liked this approach because any weapon confiscated was one less weapon that could be used against them or their mates. Every weapon confiscated also added to their score back at the company. The diggers were also curious and fascinated with the different types of weapons. They had only handled these weapons briefly during training. Several diggers coveted the weapons as attractive trophies to commemorate their operations in Somalia.[25]

One of the major difficulties with this confiscation policy was that every household, shop and business in town contained at least one weapon for self-protection. Local families and businesses had them to protect themselves from bandits and criminals. NGO staff were also concerned that their guards kept their weapons to protect lives and property. Hurley developed a registration policy for NGO guards and allowed them to carry weapons. However, he continued with an 'all-weapons' confiscation policy for the local Somalis for the time being. For their part, the Somalis became more adept at hiding and moving weapons to keep them from being discovered by Australian patrols. For every one confiscated there was another available from the arms market in town.

On the morning of 17 January, Major Jim Simpson's Bravo Company had arrived at the airfield ready to relieve Fraser's men around the perimeter the next day so they could concentrate on patrolling in Baidoa. On 18 January Simpson deployed a platoon to the main gate and a platoon to what became known as the Warehouse, and he split the remaining platoon into its three sections to occupy observation posts. There were no problems keeping the diggers alert on sentry. A Somali gunman fired three shots at Corporal Wayne Prosser's section at Observation Post 1 on the first night.[26] Two shots were fired at the diggers guarding the Warehouse two nights later.[27] There were armed infiltrators roaming around every night and into the early hours of the morning.

There were no fortifications built or barbed wire barriers erected during these early days to protect the Australian camp. Though each sentry could not be issued a set, night vision goggles provided a significant edge for the diggers at night over Somali infiltrators around the airfield perimeter. Some of the 15 observation posts set up around the perimeter were also supported by thermal imaging devices which depicted images of humans, animals and vehicles based on the heat they emitted.[28] The Australians could observe infiltrators, 'get the drop on them' and frighten 'the living daylights out of them' by loud verbal challenges and firing warning shots if necessary.[29]

Strict ROE and OFOF governed the actions of the diggers when they discovered armed Somalis near their camp. They had to challenge them to stop in Somali and were only to fire warning shots if their verbal challenges were ignored. If weapons were raised and aimed at them, the Australians could shoot to kill. The time between a challenge, a warning shot and a requirement to shoot to kill could be a matter of split seconds. The diggers had been trained over and over again in these procedures before leaving Australia.

The ROE and OFOF made the diggers more cautious about opening fire than they would have been on conventional operations. This probably helped avoid accidents as the battalion settled in, especially when one section from B Company swept in the wrong direction on a routine clearing patrol on 22 January, ending up in front of another section's observation post in the dark without communications. Initially, they were identified through a thermal imager as an unknown group of armed men approaching the perimeter. The ROE and the night challenging procedure meant that they were able to move back inside the perimeter safely. [30]

On 21 January, Fraser's men moved from their cramped, roofless building into tent lines near the main road into the airfield. Morale was high as the diggers were finally able to sleep under canvas in their platoon groups. The tents had been abandoned by the Marines because they were old stock that would have been too much trouble to clean to quarantine standards and pack. Sufficient tentage to house all members of 1 RAR had not been brought because Battalion planning officers had not anticipated setting up a tented camp. They had expected that the tactical situation in Somalia would require the development of a conventional defensive position. In such a position, diggers would be expected to live in two-man tents near their fighting trenches or in sleeping bays below ground extending from their trenches.

Lieutenant Peter Connolly 2IC Mortar Platoon, questioning a Somali in February 1993 Connolly was one of the early commanders of 1 RAR Group's Quick Reaction Force.

Living conditions improved but were still cramped, hot and austere. Sections were accommodated in platoon-sized tents (30 man) in long rows of stretchers on dirt floors opposite each other. Each officer, NCO and soldier had about 200 cm of living space around their stretchers. There was no privacy and no electric lights. The heat demanded that sides of the tents were rolled up which meant that dust bellowed in over the sleeping areas. Rats, snakes, insects, scorpions and different varieties of African spider, were co-tenants.

The 1 RAR Group's resources were stretched as Hurley assigned his sub units to the tasks of escorting convoys, patrolling in Baidoa, securing the airfield and patrolling the countryside. He soon realised that all sub-units, including the BSG, would have to be responsible for the security of their areas of the airfield against Somali infiltrators. He also decided to create a small command post in his headquarters to coordinate airfield security and to supplement airfield defence with a section-sized quick reaction force (QRF) mounted in a Land Rover, called 'Mobile One'.

He directed Steve McDonald to provide members of the Supply, Catering, Technical Support, Medical and Transport platoons of Administration Company to constitute Mobile One. This force was on call 24 hours a day for emergencies and performed routine escort and administrative tas ks around the clock as well. The workload of members of Administration Company increased considerably. After a 16 hour working day, cooks, clerks, storemen, mechanics, drivers and medical assistants would spend two hours on duty at night manning observation posts or serve as members of Mobile One. At other times they assisted with building fortifications and providing armed escorts.

Right from the start Hurley had established another QRF of one platoon mounted in Unimog four tonne trucks to back up his patrols in town and to react to incidents there. On 21 January, the QRF task was given to Mortar Platoon, under the command of their Mortar Line Officer, Lieutenant Peter Connolly. [31] After a three hour briefing which identified 10 'hot' spots in town, Connolly briefed his men and settled back to see what fate had in store:

> We had been sent to the north of town to locate a sniper pair [32] who had not reported in on their schedule when we were reacted by 0A [Battalion Headquarters] to 'the sound of gun shots somewhere in the centre of town'. I chose a rough track going in the direction of the shots and hoped for the best. We screamed down it trying not to hit the street stalls. We dismounted just short of the main roundabout. Luckily I had been given a very good interpreter. I asked him to find out what had happened and he said the shots had been fired over 500 metres to the west at the ICRC [Red Cross] compound. We stayed on foot moving at high pace with the trucks following. The streets were very crowded and I felt very insecure. We reached the compound to find a bullet-riddled metal gate in a high wall. Kids were telling us that guns were in there and we could see men with AK 47s behind the gate. As far as I knew the shots had come from the compound—were they good guys or bad guys ? (there is no such distinction in Baidoa it would seem).
>
> I dropped into a fire position near Mitch. [33] Two thirds of his section were there, and the rest, with Johno's section,[34] were around the corner. Adrenalin pounded through me. Safety catch off. Is this really happening?
>
> One of the fellas moved closer to the gate with no cover, holding up his rifle ready to shoot.[35] There was movement at the gate—an unarmed man appeared just as the digger screamed, 'I've got a sight picture'[36]... I have never been so scared as at that instant when I thought one of my men was going to kill an innocent human being. I nearly lost it as I shouted, 'I don't give a fuck what you can see—don't shoot!

> I told all of the armed men in the compound to put all of their weapons on the ground outside the gate. When they appeared to have complied with my demand, I moved forward, asking them why did they have so many weapons and what were they shooting at? A man came forward and explained that they were NGO 'security guards' and they had just shot an intruder.
>
> About this time an Irish NGO worker, Tom O'Neill, turned up. He was very amiable and understanding. When I asked him if he trusted the guards he confided that he thought they rigged the robberies by night. He was quite embittered with them—'a pack of bastards who bite the hand that feeds them. Take their weapons and don't give them back, mate.' He also said that they could have plenty more hidden in the compound'. [37]

Connolly took O'Neill's advice and confiscated the weapons carried by the guards and found a further six weapons hidden in the compound. He reported this back to Battalion Headquarters and waited for further orders.

Connolly was in a difficult situation. All the weapons he had confiscated were unlicensed and the hidden weapons probably belonged to bandits. However, if he took all the weapons away, the Red Cross Compound would be insecure, the expatriate NGO staff would be unprotected and another clan group could move in and threaten the guards he had disarmed. He requested permission to mount an all-night picket in the compound until the situation could be sorted out.

After some delay Battalion Headquarters approved his seizure of the weapons but refused his request to mount an all-night picket in the compound. On return to the airfield he got into trouble for not issuing a receipt:

> This had been the biggest weapon take in the battalion so far. However, I forgot to give them a receipt, our communications had been poor, so I got into trouble at BHQ.[38] I later gave Tom a block receipt and he said, 'Yeah, ta, mate, I'll file it.

Connolly's men were called into town again early next morning:

> At 0200 hrs we were reacted to the CARE compound in NGO Road.[39] A security guard had fired 10 shots at an intruder (another likely story).

Once again it took me ages to find the silly bloody place and then about 15 minutes for the guards to open the door and get an Australian CARE worker to talk to us. ... When we got in they said, 'Boy are we glad to see you'. I said, 'That's fine, please tell me what happened and have you licensed all your weapons?' Their story lacked credibility (no cartridge cases). ... I told Peter [Kieseker] that he should get the weapons of the guards registered. Peter said, 'Yes, but you can't trust any Somalis'.[40] I said, 'I'm well aware of that, but unless you take responsibility for the controlled use of their weapons, I'll take them away'.

Connolly had one task that morning:

At 0500 hrs I was woken again and told to get the platoon ready. Two 'squads' of 'enemy' has just passed a sniper pair [41] on a track north of town; spaced, no noise, track discipline, sense of purpose and armed. This was very scary—we'd never seen anything like this before and I was preparing to deal with this group at night. Eventually the OPSO (MAJ Caligari) called it off.[42]

When I handed over at 0800 hrs to Todd Everett, I was a wreck.[43] I had had very little sleep, been on reaction most of the day and night (normally you spend most of the time waiting to be reacted) or writing lengthy incident reports and doing debriefs and seeing the IO [Intelligence Officer].[44] Above all, the adrenalin had totally exhausted me. The highs were followed by lows requiring deep sleep (which wasn't available!).[45]

The sighting of armed groups moving in a military manner increased the threat to the Australians significantly. It also reinforced perceptions among the Australians that their patrols would eventually have to fight trained Somali militia groups. Lieutenant Andrew Pritchard wrote in his diary, 'There had been a number of sightings of armed men and we will be patrolling into Baidoa next. This has raised the spirits of all—any chance to get amongst it is met with eagerness.'[46]

As Hurley and his staff coordinated an intense 24 hour patrol program into Baidoa, the diggers got to know a little more about what was going on in town, tightening their grip, but not without some humour:

The patrols themselves had their ups and downs ... The meat market was the worst area to be during a day patrol as the smell of rotting meat permeated from the corrugated iron market stalls that were seething with flies. Corporal Conner made it a point to always stop for a navigation check [there] as it was guaranteed to make Private Kirk Went[47] start dry retching—anything for a laugh. The locals usually provided laughs for the men on patrol. Private Dale Gollop made the members of his section laugh when he started chatting to the local ladies and was proposed to.[48] He politely refused and she 'went to pieces', bursting into tears that were uncontrollable, much to his embarrassment.[49]

Building up a picture of the threat to life and property in Baidoa was more akin to undercover work by police investigators than conventional tactical intelligence gathering through patrol reports, satellite imagery and radio intercepts. Geoff Petersen at HQ AFS provided the following information to the 1 RAR Group after talking with two expatriate NGO staff in Mogadishu:

> [Most armed groups and individuals] are not militarily trained, but are familiar with fighting.[50]

> They are not accurate shots, but most personal weapons are well maintained and used at close quarters. Moreover, volume fire (hose piping) is a common way of overcoming inaccuracy. They conduct ambushes but normally pick soft targets. They tend to like to stand up to fight. . . . [They] act like gangs. They raid fixed facilities, normally at night. Random shooting is an ever-present threat but sniping [is] not a common thing. Gangs prefer to use bravado or volume fire. Occasionally an individual [will] do the illogical and engage personnel for no reason.[51]

This anecdotal information was useful when information was scarce but subsequently Petersen sent information mostly gleaned from US sources because HQ AFS did not have the capability to gather intelligence independently. This information tended to be pitched at too high a military and political level to assist Hurley's staff to build up a picture of what was happening in Baidoa. However, it was useful information for higher headquarters and intelligence agencies in Australia and constituted an important part of HQ AFS 's role in keeping them informed about the situation in Somalia.[52]

By far the most useful way of getting to know what was going on in Baidoa during the first critical weeks was through the investigations of Staff Sergeant Brendon Thomson and Sergeant Wayne Douglas, Australian Intelligence Corps NCOs.[53] Thomson and Douglas had been included in the AFS almost as an afterthought. Hurley had to give up two 1 RAR positions to include them and their presence was initially resented by some. One of Hurley's headquarters officers told them, 'You're like squirrels in the forest. You gather your secrets like nuts, hoarding them and not telling anyone anything.' He warned them that they would be closely watched and not to try any 'spook' tricks.[54]

After some initial confusion among Battalion Headquarters staff about their role, Thomson and Douglas began gathering information about Somalis from Somalis. Using the most trusted US Marine interpreter to assist them, they employed three local Somalis to act as interpreters. They then moved among the locals, talking to people and establishing a rapport with political and community groups. Their mission was to build up a picture of who posed a threat to the success of the Australian mission in Baidoa and how they operated.

In January Thomson and Douglas operated with American counterintelligence personnel. One of their tasks was to ascertain the attitude of the Somalis to Australian troops. Some of the local Somalis were intimidated by the stern, unsmiling diggers. Thomson and Douglas advised that, if the diggers relaxed, local Somalis would be more likely to provide the Australians with information about local military, political and criminal activities. A steady flow of reliable information would make the Australian security operations more effective and safer. If they were treated fairly, most local Somalis would develop a neutral or positive attitude to Australians.[55] Fortunately, during training in Australia the diggers were told to adopt the

Army's own leadership motto of 'Firm, fair and friendly, but not familiar'.[56] There were many other references to this motto during interviews conducted by the author in Somalia. David Hurley's personal motto was, 'Gentle in manner, resolute in deed'.[57]

Thomson and Douglas discovered that many local Somalis were keen for the Australians to confront local criminals and bandit groups. They promised information on criminal activities as long as Australian patrols responded quickly. They were reluctant to provide information if there was any chance of reprisal. Thomson and Douglas concluded that patrols had to keep in close contact with the Somali population to encourage cooperation and to develop trust. To achieve this relationship, they recommended hiring more interpreters.

Gathering information from the locals was dangerous work. Typically, Thomson and Douglas travelled with a security team of at least two other soldiers. Occasionally they were allocated diggers from the rifle companies, but generally it was left to them to arrange their own security. In the main they used clerks, drivers, cooks, mechanics and storemen. Consequently, administrative personnel were given opportunities to get away from their duties at the airfield and do infantry work. These security duties became so popular a waiting list was established to give a chance to everyone who wanted to get into town.[58]

Hurley and his staff quickly recognised the value of gathering information by deploying CI Teams into Baidoa with interpreters. By the end of January three more Intelligence Corps NCOs, as well as a Warrant Officer Class One, were deployed from Australia and more interpreters were hired.[59] This was timely reinforcement. The three CI Teams were able to quickly identify the nature of the banditry in Baidoa using local American and Somali sources.

They discovered that bandit groups operated by day and night in the side streets off main roads, armed with weapons that were strapped under their arm pits beneath traditional Somali robes, that resembled long kaftans.

Typically, the wooden butts were cut down to fit more comfortably and for concealment. Fortunately for the Australians, the modifications to rifle butts made it difficult for the bandits to aim and fire their weapons accurately.

The bandits, in groups of three to six men, moved about town during the day gathering information on possible targets of opportunity, often waiting for their prey to walk into a side street where they were mugged. The proceeds were then given to female accomplices who would carry them from the area. Locals claimed that many bandits were NGO guards, a logical charge given that they had the weapons, knowledge and opportunity to get up to no good.

Banditry was not restricted to the streets in town. A focal point for bandit activity was the Sheikh Oasharow Cemetery, three kilometres south of Baidoa. The Cemetery was astride a popular route to the Baidoa markets used by camel herders and other people with something to sell or a need to buy food and other necessities. Bandits would rob passers-by in the morning, rest up in a nearby creek line during the middle of the day, and then strike again as people returned from the markets late in the afternoon. While they rested or were robbing people, sentries were posted to alert them if an Australian patrol was in the area. Children were often paid to perform these duties. Late at night bandits would go into town for *khaat* and sex, before returning to the Cemetery in the early hours of the morning for another day's villainy.[60] Sometimes bandits would move further afield and roam among the villages surrounding Baidoa, robbing villagers, extorting money at water points and forcing locals to prepare food for them.[61]

The CI Teams became the most important source of intelligence for Hurley and his staff after a few weeks. Their acceptance by 1 RAR headquarters staff could be traced by observing changes in their accommodation and working conditions. Initially Thomson and Douglas were given a tent 80 metres from Battalion Headquarters next to the latrines. Directions came after a week or so for the tent to be moved to the vehicle car park, beside the Headquarters building. Several days later, after more Intelligence Corps

NCOs arrived from Australia, the CI Teams were told to erect the tent as an annex of the battalion Intelligence Section, abutting the Headquarters building. After another week had passed, they were moved in with the Intelligence Section and shared working space in the Headquarters building. Several days after that, they moved into their own office four doors away from Hurley's office. Ultimately, they found themselves in an office next to Caligari and two doors from Hurley.[62]

In a land ruled by the gun, David Hurley effectively became the military governor of the city of Baidoa and the surrounding areas. From the Somali point of view, the Australians were probably viewed as a well-armed foreign clan, coming in and taking over at gunpoint. The locals in Baidoa knew how to survive these changes. Children, who used to call out, 'Americans Number One' and whistle the tune and sing 'Jingle Bells', learned from the Marines at Christmas several weeks before, now called out 'Australia! Australia! Australians Number One', before asking for 'Gifty! Gifty!', and holding out their hands.

Among these survivors were two of Baidoa's most notorious criminals. The first, Hassan Gutaale Abdul, had controlled Baidoa airfield with 132 gunmen and a number of technicals before the arrival of the Marines. He had

The childred of Baidoa. 'G'day mate'; 'Australians number one'; 'Gifty'.

set up a lucrative extortion operation as the self-appointed 'Head of Airfield Security', demanding cash payments in US dollars from the UN and NGOs in exchange for their use of the airfield and protection from other bandits. It was alleged that Gutaale[63] used to enjoy driving a Fiat armoured car around Baidoa and intentionally running over and killing or maiming people sleeping on the footpaths. Gutaale offered his services to the Australians to help secure the airfield and ward off any opposing military forces and bandits by 'killing them immediately.'[64] The other major criminal leader in Baidoa was Hussein Barre Warsame,[65] a 49 year old veteran of the dictator Barre's Army and an organiser for the United Somali Congress (USC) faction headed by General Aideed. Warsame organised banditry, political murders and looting while Gutaale's criminal associates extorted money and probably did their share of looting as well. These two villains were both Aideed supporters and agreed to keep out of each other's way while they conducted their respective criminal activities.

Thomson and Douglas interviewed Warsame the next day. He assured the Australians that he had significant political support in Baidoa and would assist the Australians to capture 'bad men' who were intimidating and robbing the locals.[66] The Australians had no illusions about Gutaale or Warsame, both were identified as influential criminals.

Gutaale and Warsame were not the only local criminals offering their services to the Australians. On 20 January, a delegation of local elders affiliated with Aideed's USC faction came to the battalion Headquarters to speak with Dick Stanhope. Their purpose was to increase their local status and influence by associating themselves with the Australians. In reality they had minimal local support and were well known criminals and stand-over men. They were evasive when asked questions about their activities but were forthcoming when asked to provide information about other criminals in town.[67]

One of Gutaale and Warsame's major criminal rivals in the Baidoa area was Salat Mohamid Ibrahim, known as Gaaney. Gaaney was a very tall Somali bandit whose right arm had been amputated below the elbow. He led a gang of about 20 men and was feared by the locals. Gaaney was a freelance criminal who answered to no one and did not appear to have any political affiliations. He was affected by severe mood swings, 'going berserk' and assaulting individuals in his vicinity without provocation. Aside from Gutaale, Warsame and Gaaney and their cohorts, the Australians faced a number of unidentified bandit groups, all of whom were potential threats to humanitarian activities and aid agency staff and posed some danger to Australian personnel.

Aside from armed groups who were likely to take action to deter the Australians from interfering with their criminal activities, most local Somali householders and owners of businesses had access to arms and would use them to protect their lives and property. In the dark it was difficult to determine who was attempting to enter buildings. The Australians had to be careful that they were not mistaken for criminals when bursting in for house and building searches. There were a few close calls.

One close call occurred during a night patrol on 22 January, Peter Connolly and a group of mortarmen had searched the Mosque area but had not seen anything suspicious. However, when they tried to climb the walls of the Mosque and look in, a group of Somalis began throwing stones at them— not a pleasant experience in the dark. Some time later Connolly and Lance Corporal Tim Bell were investigating a strange light flashing on and off in a nearby building. They stood against the wall of the building either side of the doorway, ready to swing inwards together and enter. Connolly wrote in his diary later:

> Suddenly there was a very loud noise above me—like a Besser brick dropped onto a corrugated iron roof and then quickly falling on its side. ... It wasn't until later that I replayed the sound in my mind—it was somehow familiar. The stop butts at the range at ADFA![68] It was a crack/thump.[69] It had been right over our heads. Why didn't they kill one of us? They could have if they had wanted to. Maybe we were into something sensitive and they wanted to draw us away. If so, they had been very successful. If they were trying to kill us, then they were bloody awful shots.[70]

During the first weeks Australian patrols were being 'needled' and tested. Somali males liked to provoke the diggers with aggressive looks, spitting in their direction and throwing stones. There were several scuffles between Somali males and diggers patrolling in town. These physical confrontations were generally over as quickly as they started. No Somali was foolish enough to continue fighting after being grabbed and held by the heavily-armed soldiers. Somali gunmen also fired at guard posts along the airfield perimeter to further unsettle the Australians. All the diggers could do was maintain their vigilance, trust their sandbag walls and hope that none of the shots found their mark.

After a few weeks the vast majority of ordinary Somali citizens became used to Australian patrols and warmed to the 'Firm, Fair and Friendly' approach of the diggers. Captain Andy Somerville wrote in his diary:

> The local people are generally friendly and don't appear concerned about our patrol. The streets are full of strong pungent smells and it is not uncommon to see a local squat in the gutter and defecate. The streets are generally tidy although there is rubbish and rubble in many places. It is uncomfortable and hot patrolling in webbing, flak jacket and carrying a radio wondering if a bandit is going to pop up and shoot you. I find it difficult shouldering my rifle and taking a sight picture. I believe we should keep the jackets for crowd control use only. The locals think we can't be killed when we wear them, so they don't try to do anything. This appears to be the prime reason we continue to wear them.[71]

By the end of January, Somali gunmen started to become bolder and more dangerous. On 29 January, Corporal Tom Aitken[72] was leading a night patrol across the bridge on the road that led into town from the airfield. He saw two rapid muzzle flashes and returned a burst of fire instinctively. One of the rounds fired at the patrol impacted a few metres away.[73] Aitken waited for several minutes and conducted a search of the area where the shots had come from but found nothing.

In the early hours of 30 January a patrol led by Corporal Mark Braeckmans[74] approached the sound of people fighting and yelling. A man on a roof fired a shot at them. After they moved under cover, their assailant continued to fire shots in their direction. Just as suddenly as he opened fire, the gunman stopped firing and fled. This incident was later reported in the Australian press as an ambush which annoyed the diggers involved because they felt it implied that they had not been vigilant.[75]

Not all shots being exchanged in Baidoa were between Somalis and Australians. On 29 January, during a house search, a digger fired his rifle accidentally.[76] An Australian machine gunner near him fired a burst of rounds instinctively at the muzzle flash of his mate's unauthorised discharge. One round from this burst of fire impacted less than 10 centimetres from the soldier's body.[77]

The settling in period for the Australians patrolling in Baidoa and guarding the airfield was tight, tense and dangerous. The presence of journalists, camera crews and photographers on many patrols added to the pressure. The diggers were understandably wary as they adjusted to the operational conditions. Local Somali youths tested their temperament and gunmen tested their discipline under fire. To their credit the Australians maintained their ROE in the face of these provocations, did not become 'trigger happy', and exercised discretion in any physical confrontations brought on by foolish Somali youths. However, their patience was being tested and there was talk of a need to be more assertive to make Somalis think twice about provoking Australians on patrol and taking pot shots at them.

On 29 January, General Johnston visited Baidoa accompanied by Colonel Mellor. He was impressed by the tempo of Australian patrolling and received positive reports from NGO representatives in town. Relief supplies were being distributed without interference and there had been an overall feeling that the Australians were keeping local criminals and nomadic bandits at bay for the time being. However, the tempo of operations was taking its toll. Major Mick Moon wrote in his diary, 'Many of the lads very tired now. I can't see us keeping up this pace for too long.'[78]

The living conditions the diggers endured at the airfield contributed significantly to their fatigue and stress. All the companies were under canvas but still had no electric light or floor boards. They lived and worked in the dirt. Sleep was difficult during the day because of the heat and noise of the camp. At night, while they prepared for patrols and rested in semi-darkness punctuated by a few kerosene lamps, they observed the well-lit areas of American and Australian logistic units and listened to the sounds of recreation. Resentment grew between the rifle companies and Support and Administration Company and the Australian and American support units in Baidoa over the differences in living conditions. The BSG was housed in a large aircraft hanger and modern tents on concrete 'hard standing' areas. They and the US Army units had created shaded recreation areas and seemed to be enjoying working hours that permitted them free nights and plenty of sleep. Weary infantrymen coming in from arduous patrols observed groups of Americans and Australian logistic personnel playing volleyball in T-shirts and shorts. Unlike the Australians, American personnel had access to cold soft drinks and snacks in the evenings, played music, watched videos and read magazines.

Aside from a Salvation Army 'Hop In' area at Administration Company that had tea, coffee, cordial and biscuits, and stationary for letter writing and a few magazines, there were no canteen facilities for the diggers to buy cold soft drinks, cigarettes, sweets or snacks, or an area to take time out to relax. Typically, soldiers flopped on their stretchers after shedding their webbing and flak jackets and snatched a few hours sleep before attending their next set of patrol orders or a rough hand woke them to put their flak jacket and webbing back on to mount sentry on the perimeter or to man a checkpoint.

While the patrols in town continued the dangerous business of asserting their presence and deterring criminals and bandits and enduring the needling of local youths, the Australians had a major challenge keeping themselves, American personnel and their property safe at the airfield. Armed intruders were observed near the perimeter every night. Most would flee quickly when challenged or after warning shots had been fired over their heads. Several times they would fire back quickly before running away. They were targeting the supply dumps, garbage tips, US engineer stocks of building supplies, shipping containers alongside the runway where relief supplies were stored, and Australian and US stockpiles of food, fuel and other stores. The vacated living areas of those who were out on operations were also targets for pilfering.

During conventional operations armed intruders approaching a defended camp or tactical position would have been shot on sight. Under the ROE in Baidoa, taking action like this against armed Somalis may have constituted unlawful homicide, i.e. murder or manslaughter.

David Hurley directed Greg Hurcum to deter the movement of suspicious vehicles at night in the vicinity of the airfield and nearby villages by establishing road blocks.[79] Subsequently Stuart Dodds was ordered to conduct a 'white light' ambush on a track on the night 22-23 January between 11 pm and 3 am. The target was a vehicle that had been observed moving in the area suspiciously over several nights. Dodds positioned two APCs, commanded by Corporal Luke Entink and Lance Corporal Paul Quentin,[80] along a section of the track. He recalled:

> They advised me I was to conduct an ambush that evening. Many thoughts went through my mind including that I could not comprehend how the situation had escalated so quickly and I would be tasked to kill anyone moving along a set track. They then changed their minds and called it a vehicle road block.[81]

Dodds' ambush plan was to have the APCs simultaneously block the vehicle and flood it with light from spot lights and headlights when it arrived in the ambush area. Once the vehicle was stopped and illuminated, Dodds planned to order the occupants out so that they and the vehicle could be searched. He knew that conducting a 'white light' ambush was a dangerous business. If the occupants of the vehicle were armed and decided to fire at the lights, Dodds would have to ensure his men were ready to return fire rapidly. As it turned out no vehicles drove into Dodds' ambush that night. He wrote later, 'If the vehicle had of appeared, we would have stopped it with the APC. I had it cleared [with Hurcum] that if they [the occupants of the vehicle] fired, we could use all our weapons to destroy it.'[82]

Dodds' concern that 'white light' ambushes were testing the limits of the ROE deepened when one of the Support Company Platoon Commanders, Lieutenant Graham Steer, reported to him that he had been encouraged to tell members of his Reconnaissance Platoon to regard armed Somalis infiltrating into the airfield area as hostile and to shoot them without warning. On 1 February his men had challenged seven armed Somalis trying to enter the BSG area, capturing one infiltrator. The others withdrew. There had been frequent sightings of armed Somalis in the area since then, including what appeared to be groups of marauding stragglers from the warlord's armies moving like trained infantry. Dodds wrote:[83]

> My concerns also revolved around the fact that [the story of] the first Somali we killed would be in all the papers at home and both the media and the hierarchy would want to know all the details. If it came out that their actions were not within the ROE then there could be a lot of trouble and charges laid for an illegal kill. I knew if this were the case ... Steery and his boys [would be left] to swing. I advised him to confirm his ROE with the OPSO [Operations Officer].
>
> I went to the 1700 hrs CO Conference still perplexed that our ROE could change so dramatically and concerned for both the ramifications on Recon Platoon and for the rest of the battalion by us escalating the situation by firing the first shots and making the first kill.[84] We still did not know who these people were and what potential they had to retaliate against us. I expressed my concerns to Jim [IO] after the Conference who had the same reservations and disbelief.[85] He told me to see the CO after the Conference. . . . The CO was adamant that Steer had to fire only warning shots at the infiltrators, if they then returned fire or aimed their weapons, then they could be shot.[86]

Meanwhile, Steer, following directions given to him, issued orders to his Reconnaissance Platoon to conduct a white light ambush south-west of the airfield where groups of armed Somalis had been infiltrating in 'a very military-like manner.'[87] Like Dodds, Sergeant Wayne Thomson, Steer's Platoon Sergeant, thought that the sudden change to the ROE was unusual.[88] He was directed to issue Claymore anti-personnel mines to members of the platoon and prepare for a conventional-style ambush just in case the armed Somalis returned fire when challenged.

Before leaving the battalion area, Steer received a message over the radio from Battalion Headquarters that the Claymores were not to be taken because of their lethality and the ROE. However, his platoon was to set up for a conventional ambush as a precaution against attack. After setting up the ambush, Thomson was told that the platoon

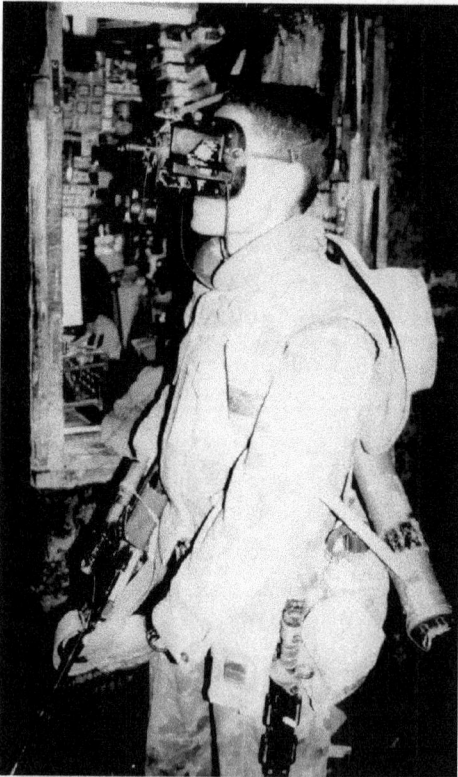

Night vision goggles gave the Australians a competitive edge in Baidoa and in protecting the Baidoa Airfield from infiltrators. Pictured is a digger wearing them near the Khaat Market on 28 March 1993
Image: G Gittoes

had received another radio message—the trip flares were to be retrieved. He was told report back to Battalion Headquarters to pick up long-handled 'Big Jim' torches that were to be used to illuminate infiltrating Somalis instead of the trip flares. No one was to fire on infiltrating Somalis until they had been illuminated by the torches and given the chance to lay down their arms after being challenged. Thomson was unimpressed with this order. Though temporarily surprised and blinded by the torch light, he expected that the Somalis would fire instinctively at those shining the torches. He asked himself, 'How would the Somalis know that after being challenged and illuminated by torchlight they were not going to be shot immediately?'

Fortunately the chances of armed Somalis surprising the Australians at night or triggering off a so called white light ambush were low. The Reconnaissance Platoon's edge at night were the thermal imagers mounted on observation posts near the ambush site. Several members of the Platoon also wore night vision goggles[89] and there were sufficient Motorola hand-held radios for the teams in the ambush to communicate among themselves and with sentries in the observation posts. The thermal imagers were capable of showing if approaching Somalis were carrying weapons or had them pointed in an aggressive manner.

At about 11 pm the first group of armed and unarmed infiltrators for the night approached where the Reconnaissance Platoon lay in wait. Thomson wrote later:

> They had been doing this regularly over the last few nights—stealing tyres, rations and fuel. We activated two patrols from our reaction force and caught one detainee—the others were too quick and got away. They were confident that they would not be caught. Previously when anyone had moved out to apprehend them, they simply lay down under the camel bush about 30-40 metres from the perimeter and hid. They were quite amazed when two patrols moved straight onto them, guided by the observers with Thermal Imagers. The boys enjoyed working with the high tech equipment and felt they were achieving something by catching these thieves.'[90]

At about 4:30 am the next morning members of the Reconnaissance Platoon caught another armed Somali on the side of the road leading out to the dump. Thomson, who led the reaction patrol that apprehended him, wrote later:

> A number of them took off straight along the road as soon as we grabbed the first one. I think they were now aware that we could locate their positions at night. I was chasing a couple along the road but they were unusually quick. I got the shits yelling at them to 'stop' in Somali. They simply ignored me. I propped and fired a 3 to 5 round burst high over the right shoulder of the closest one. Even in the night I could see his teeth as he turned around and smiled at me on the run. I guess they knew our ROE better than we did![91]

Hurley reacted decisively to the growing rumours that the ROE were to be changed by calling meetings of officers, WOs and NCOs to repeat the ROE and to emphasise the importance of using minimum force. He moved around and visited his troops whenever he could to gauge their temperament, to encourage them to maintain their concentration and to follow the ROE. He issued formal warnings that any Australian who was found guilty of physically abusing Somali citizens would be punished and sent back to Australia in disgrace.[92] He wrote at the time:

> We are starting to have some problems with the soldiers' attitude to the Somalis. I think that it stems from the detention policy which does not allow us to imprison bandits at the moment. The soldiers get very frustrated seeing our hard work go to waste when bandits are released, so they are probably taking out their frustrations on other people. I am trying to 'nip it in the bud' before it becomes serious.[93]

The detention arrangements for armed infiltrators, presumed to be bandits or local criminals, and verified criminals and bandits apprehended in town or in the countryside was a particularly sore point with many members of the 1 RAR Group. They felt that their job was to catch them and it was someone else's job to keep them locked up. After 24 hours of detention at the airfield under the supervision of the 1 RAR Group RSM, Greg Chamberlain, and his Regimental Police, all detainees, under UNITAF policy, had to be turned over to Somali police in town. Though many were handed over with quite serious charges laid against them by the Australians and, despite statements from witnesses to prove the charges, all were released within 24 hours.

Adding to the problems of reinforcing the requirement to use their weapons in accordance with the ROE among his soldiers, Hurley was becoming concerned about the number of unauthorised discharges of weapons. Whether through fatigue, over-familiarity or weapon handling procedures, or a combination of all three factors, there was a spate of UD of weapons early in February. Two incidents occurred in A Company while weapons were being cleaned and one in the living area of the QRF soldiers behind the battalion Headquarters. On 3 February 'Ant' Blumer wrote:

> Boy what a pearler of a morning ! PTE _had a three round UD in the QRF tent nearly killing one or two soldiers. All this happened 30 metres from BHQ [Battalion Headquarters]. Quite unbelievable ... Such a silly and senseless thing to do and mercifully nothing was worse for wear other than three holes in the tent. I think luck is running out. I hope that no one is hurt in any of these incidents.[94]

Hurley came down heavily on all offenders. All those firing their weapons without authorisation were charged, fined between $800 and $1000, severely reprimanded and spent 14 days in detention.

Meanwhile, violence was escalating in town again. On 2 February, a woman was shot dead and her son was shot in the leg in the market area after an argument with local standover men. She had refused to give up selling her wares from an 'unauthorised' position. This incident and several other reports of intimidation, murder and robbery showed that the people of Baidoa were still living under the rule of the gun.

Counterintelligence reports confirmed that NGO guards were involved in some of this criminal activity. They and visiting bandits competed to extort money from women who were collecting water at the Bay Project, an NGO-funded pump site. They were also implicated in night time robberies and looting of aid agency compounds and warehouses.[95] Bandits regarded the Bay Project as a 'soft mark' and had been raping young women as well as extorting their money.[96]

The guards at the *Medicens Sans Frontieres* (MSF) hospital and compound were singled out by many informants as being involved in banditry. They were reported to be led by a 'big thug', Abdullahi Mohamed Idins, known as Lesto, who was the MSF Administration Officer. Lesto was allegedly responsible for hiding weapons in the Out-Patients' building to avoid confiscation.[97]

On 3 February 17 bandits took over the Bay Project water point and harassed refugees into giving them money for water. The harassment stopped when Australian patrols were in the area but resumed after they left. The bandits were able to hide their weapons under their armpits and disperse among the hundreds of people milling around the area. There were several rapes reported later. The Somali citizens were too intimidated by the presence of the bandits and fearful of their retaliation to identify them when the diggers moved among them. The movement of Australian patrols was also being monitored closely by a network of Somali 'cockatoos', many of whom were children, who used lights at night and hand signals by day to warn bandits and local criminals of the approach of Australian patrols.[98]

Without information and under surveillance, the patrolling Australians were unable to effectively deter criminal activity or violence against citizens. Expatriate NGO staff may have reported to General Johnston during his recent visit that they felt safer with the Australians around. However, there were still problems with the guards some of the NGOs had hired. The Australians needed timely information from the locals to apprehend criminals and bandits. The locals were still cautious about giving them timely information for fear of reprisals.

At the airfield, scavenging bandits and locals were still a major security problem. A group of six Somalis had been able to creep into Charlie Company lines and steal personal items.[99] Many infiltrators had been detained but they reappeared again a few days later after being released by the ineffectual Somali police in town. This 'revolving door' at the police station was tempting some members of the 1 RAR Group to mete out their own physical punishments to detainees to deter them from returning to the airfield. For their part, the RSM and his Regimental Police made sure that Somalis caught stealing or trespassing were detained according to UNITAF policy but were not made too comfortable while in detention.

Compounding the frustrations and hard work being put in by the Australians trying to secure Baidoa and seal the airfield off from infiltrators, was another couple UDs to start the month of February off. In addition, mail had not arrived from Australia, vehicle tyres were being constantly punctured by camel thorns and the 1 RAR kitchen burnt down. This rather spectacular interruption to the routine of life at the airfield was described by 1 RAR Caterer, WO2 Allan Stow later:

> In the afternoon of Feb 4th 93 I was in my room, just off the main kitchen when I heard what sounded like a pressure release valve from a M2A Burner Unit expelling pressure. The noise lasted longer that normally expected and this came with an increasing commotion. . . . within seconds I heard someone yell, 'Get the hell out of here!' I ran to the doorway to see four people running flat out towards me and behind them a giant fireball. Someone yelled, 'The gas bottle's on fire'. I looked to see that the surrounding tent was already aflame, and it was pointless trying to extinguish it.
>
> The first priority was to get everyone out and count heads. ... Everybody was moved back to approximately 100 metres from the building, and could do nothing but stare in disbelief at what was our kitchen, well and truly alight. We watched helplessly as the fire spread to the canvas covering the sleeping area for the cooks and then dropped to engulf their Army kit and personal belongings. ... Ammunition began to explode, slowly at first, and then increasing to what sounded like continuing automatic fire. Then the 16x32 food preparation tent caught fire, bursting into a raging blaze. Helplessly we watched for a least a half an hour until the flames died down.
>
> The first impressions of the remains of a fire are horrifying. ... The sight of the cooks sifting through the ashes to see if there was anything left of their personal belongings helped to magnify the extent of the damage,

knowing that personal belongings cannot be replaced, things like photos and letters from home, that in a place like this hold more value than money or any other belongings. ... The smell of the smouldering remains is something that will long be remembered. Along with coming to grips with losing the kitchen, comes the questions, enquiries and investigation of what happened, keeping the horrid memory fresh in your mind.[100]

The day after the kitchen burnt down, the run of unfortunate incidents continued when a group of Australians were stoned by an angry crowd and had a grenade thrown at them in town. The incident began as a routine task of picking up an abandoned ISO container in NGO Road.[101] A flatbed truck, an APC fitters vehicle, commanded by Corporal Paul 'Bart' Bartlett and driven by Craftsman Craig Lindsay,[102] and a forklift were sent into town to salvage the container and bring it back to the airfield where it was to be used by the Australians for storage. Watched by a small crowd, the task initially went according to plan. Then a group of young men began throwing rocks at the Australians from among the crowd.

When the Australians resumed loading the container the young men began to stir the crowd up with loud accusations of theft and more rocks were thrown at the forklift and the truck. The Australians stopped work and radioed for help to control the crowd. The QRF, mounted in trucks, was given the task of moving quickly to the site and taking control of the situation. On the way into town a Somali truck collided with one of the Australian trucks. Lieutenant Bob Worswick, the commander of the QRF wrote:

> We were very fortunate as both vehicles were travelling at speed. If not for evasive action taken by our driver, there would have been a serious injury or worse. After checking that all my men were OK (one minor injury) we left the irate Somali and drove to the scene.[103]

The crowd near the ISO container had swelled to over two hundred shouting, rock-throwing people. The young men were hard at work among them creating the impression that the Australians were stealing the ISO container from its rightful owner. By this time Corporal Wayne Prosser[104] who was commanding a foot patrol in town had joined the group around the ISO container after observing the forklift being stoned. The crowd had retreated back from the area but continued to throw rocks, many the size of half house bricks.

The QRF arrived and Worswick directed his Platoon Sergeant, Greg Burns,[105] to deploy his men in a half-moon human cordon between the ISO container and the crowd. Earlier he had directed the forklift driver to stay where he was until the area was secure. 'The crowd began surging and a well-dressed man (trousers, long sleeved shirt and leather shoes) stepped forward and said he was the owner of the ISO container.' Worswick invited him to return with the Australians back to the airfield to discuss his claims more fully through an interpreter. He refused and Worswick continued to negotiate with him. He wrote later:

> By this stage rocks were starting to be thrown at members of the platoon. Unbeknown to me, the forklift driver had begun to advance on the container without being signalled by myself. This is what had excited the crowd. With that I stopped my conversation with the man and faced the music with the platoon. The closer the forklift came the more rocks became aerial until the air was full of debris.
>
> Concurrently the crowd had closed right up on the platoon, pushing us and spitting. I could see some ringleaders amongst them working them into a frenzy much the same as you would see rowdy fans at a football match being egged on by individuals.

Worswick, responding to a request by Corporal Shaun Moore,[106] one of his section commanders, ordered his troops to fix bayonets. The diggers flourished their bayonets close to the angry faces of the crowd, 'staring them down.' Worswick wrote:

> This had the desired effect as the crowd were unsettled by the cold steel now pointed in their direction and retreated a pace, giving us some breathing space. Looking amongst the crowd there was pure hatred in their eyes, but for the life of me I don't know why they were so worked up. Even the women in their bright native garments were amongst them, hurling rocks and abuse and spitting.

Worswick decided to tough it out and not stop the forklift driver picking up the container. The forklift was now being showered in rocks as the driver manoeuvred it close to the container. Greg Burns requested permission from Worswick for he and a few men to run out from the cordon and deal with the troublemakers. Worswick gave the go ahead and Burns and his group crashed into the throng and forced the troublemakers to retreat. This action distracted the crowd long enough for the container to be picked up by the forklift.

As the forklift driver balanced the container on the forks just above the ground, a Somali youth burst through the cordon and jumped up at the container, causing it to sway on the forks. Worswick directed two diggers to prevent him from continuing with this dangerous attempt to dislodge the container. He was dragged off the container and the forklift pulled back and began to drive away towards the airfield.

Worswick ordered his men to mount up on their vehicles and the vehicles began to back away from the crowd, heading for the airfield. The young men waited amidst the crowd with rocks in their hands. As the last QRF truck turned to leave, one last volley of rocks arced into the air. Among the rocks was a Czech RG42 hand grenade. It hit the ground, bounced and rolled unnoticed into a road ditch five to ten metres to the left rear of the last QRF vehicle. It was a lucky bounce. The grenade exploded in the ditch without injuring anyone. After detaining a Somali for a short time who was suspected of throwing the grenade, Worswick ordered his men to release him, his men remounted their trucks and drove back to the airfield.[107]

This incident was reported to HQ 10th Mountain Division and HQ AFS in Mogadishu. There was some vigorous debate between the two Australian headquarters staff about how well the incident had been handled. On advice from a member of the Elders Council in Baidoa, Hurley ordered that the ISO container be returned the next day. Worswick and his men, who were guarding the front gate area at the time, were disappointed that the hard-won container was returning to Baidoa.[108] Hurley and his staff assessed that picking up the ISO container had given local criminals an opportunity to stir up resentment against the Australians. The CI Teams reported later that they had received information Gaaney had thrown the grenade and that he and his cohorts would make further attempts to stir up public opinion against the Australians.[109]

The first week of February continued to be punctuated with violence, a bus and a truck were ambushed by bandits in separate incidents causing over 20 civilian casualties, six CARE Australia nurses were detained by bandits at Goof Guduud, intimidated and later released, and the numbers of armed Somalis trying to infiltrate into the camp area increased. The Australian vehicle fleet was also in trouble. In addition to scores of Land Rover tyres being punctured by camel thorn, immobilising vehicles until punctures in the inner tubes were repaired, there was a shortage of spare parts. Major Spike McKaskill reported that unless his APCs were supplied with additional spare parts, track pads and track link soon, most of his vehicles would have to be taken off the road by the end of the month.

As the Australians became more familiar with what they were up against in Baidoa, differences of opinion emerged among them on how to conduct operations. Using the airfield as a base and patrolling into town like conventional infantry was not producing results. Many Australians felt that they were establishing a presence but were not achieving the surprise or the stealth necessary to catch criminals and bandits off guard or deter their activities. There was a strong feeling that patrolling around town in front of the public was not enough. Commanders at all levels discussed how operate more effectively. Two broad schools of thought began to emerge. One school advocated an assertive and aggressive approach to operations that emphasised confiscating as many weapons as possible, hunting down criminals and bandits, and confronting them. The other school advocated deterring criminals and bandits by maintaining an armed presence in Baidoa 24 hours a day by occupying some compounds, accompanying NGO activities and reacting quickly to incidents. The origins of these schools of thought reflected the personalities of individuals and their expectations of what needed to be done to ensure that Australian operations were deemed successful, both in Somalia and back in Australia.

On conventional operations against an armed and well-defined enemy, the success of an infantry battalion group would be measured by the numbers of casualties inflicted on enemy forces, the capture of enemy personnel and assets, and the size and tactical importance of areas and objectives captured and held. With these criteria in mind, many members of the 1 RAR Group assessed that the Baidoa Humanitarian Relief Sector was an assigned area of operations entitling them to eliminate any armed individuals and groups who posed a threat to them, NGO operations or the civilian population. They wanted the focus of operations to be on finding and fighting these individuals and groups. Those supporting a proactive, assertive approach argued, firstly, that this was an approach that retained the tactical initiative and kept Somali criminals on the back foot, and, secondly, that it concentrated on the functions that infantry were trained to do.[110] Furthermore, the Somalis in the Baidoa area would appreciate the eradication of elements from their society that were causing them so much personal grief and sorrow. They felt that operating more like infantry and less like police in the four months available to the 1 RAR Group would have a more beneficial impact on what most Australians agreed was a 'four month blink' in African history. Major Ant Blumer wrote later:

> We should concentrate on areas that we are trained for, areas in which we can do something constructive. This was in contrast to creating police forces and judicial systems that had no basis in regional, social or cultural values.[111]

Others in the battalion took a more philosophical and, in some cases, more cynical view. They confined their measurement of success to protecting humanitarian operations and deterring criminal and bandit activity. They assessed that their four-month presence in the Baidoa region would not solve Somali society's problems with clan rivalries and criminal activity. They advocated a defensive and protective approach during operations that would not endanger Australian personnel by provoking violent retaliation from criminals and bandits who were familiar with the area, protected by members of their clan and had probably assessed that, if they were patient, the Australians would leave and they could resume their activities.

David Hurley faced a complex challenge as the tactical commander, conscious of these two competing schools of thought. He knew that his personal and professional example would determine the approach to operations and the temperament of his battalion. His dilemma was: if he or his troops were perceived to be weak, defensive and indecisive, Somali criminals, bandits and militia units would resume their activities and try to reassert their control over the population and return to intimidating the aid agencies. However, if he authorised conventional operations, the level of violence in the Baidoa area would escalate and there would be a consequent increase in the likelihood of Somali and Australian casualties. He wrote at the time:

> My concerns as a commander are my failure to really know my soldiers and thereby gain their respect. I sense a degree of respect in my conversations [with them], but how far it goes I do not know. There is often a sense of inadequacy of not doing enough to meet the soldiers expectations of this deployment. Should I be pressing harder against bandit activity—but on what intelligence [do I base my operations]? Should I be hitting more houses in town to find weapons—but why not wait until the ASF [Somali Auxiliary Security Force] is in place to assist me.[112]

In response to the persistent violence in Baidoa and continuing fears for the safety of expatriate staff in their compounds on NGO Road, Hurley decided that his soldiers would live and work where the problem was and not move back and forward from the relative safety of the airfield where his patrols could be easily observed and their movements monitored. He directed that several NGO compounds were to be occupied 24 hours a day and that patrols would be launched from them and not the airfield. Hurley's objectives were to protect the compounds and the expatriate personnel from local criminals and bandits, as well as from their own guards. By monitoring the guards' movements, he also hoped to reduce the level of criminal activity in Baidoa. By starting patrols from a number of locations in town he also intended to put criminals and bandits on the back foot and have troops on hand to respond quickly to incidents.[113]

Occupying NGO compounds was a dangerous business. The first danger was from the NGO guards who could not be trusted. The second was from grenades thrown over the walls of the compounds from adjacent streets. The diggers had to live in courtyard areas and on verandas if there were no rooms available inside buildings to accommodate them. At night, or even by day, it would be very easy to lob a grenade over the wall of the compound into a group of Australians. Like the expatriate NGO managers, the Australians were now 'putting their bodies on the line' 24 hours a day.

There were several benefits for the diggers occupying compounds. They were able to set themselves up in reasonable comfort and occasionally enjoy some amenities not available at the airfield. Lieutenant Andrew Pritchard wrote at the time:

> We relieved 4 Platoon in AO FALCON, a separate urban AO [Area of Operations] in the centre of town. My HQ was located in the GOAL compound—an Irish agency that provides medical aid rather than food. Most of the workers are pretty young nurses from Ireland. Definitely out of place in downtown Baidoa but a very nice change. ... [I met] a Scottish electrician who was in town to make repairs. He operates out of Kenya-for the highest bidder. After a friendly chat he disappeared inside the compound and reappeared with three beers. A beer !! Unbelievable,—in the middle of Africa with a chilled Tusker beer in my hand. Later this guy showed me an empty carton of Four X–his favourite beer from Australia. Apparently the plane that resupplies the aid workers is able to provide them with anything–a bit different from the planes resupplying us!

We heard of a satellite telephone link from the CARE Australia aid agency. A few of the diggers investigated this and were able to call back to Australia at $ US 20 a minute. Even at this price at least a half a dozen got through and thought it was well worth it. A pity our Army does not think of these things. Spend millions on the operation and save money on phone calls that would be so good for the morale of the diggers. ... Our orders group at 1900 hrs this evening was spiced up with the discovery of a slippery two foot long brown snake under the table. After a short but violent search, the snake met its fate.

... We received a number of newspapers from Australia today. A pity they were ten days old—better than nothing. *Townsville Bulletin* had been sending 20 to 50 papers per day since the deployment but they have not been getting through.[114]

Many infantry officers and soldiers found the role of guarding NGO compounds and patrolling through the streets of Baidoa not true infantry work. They assessed that they were being used as security guards for the NGOs and the de facto police force for the citizens of Baidoa. They longed for the opportunity to fight as infantry against an armed enemy. Andrew Pritchard wrote:

The guys are tired. Their frustration is mounting again. I sometimes think that it would be easier in some ways if an open war was declared. At least we could feel free to act as an Army should rather than as a police force which is what it feels like at the moment.[115]

In addition to occupying NGO compounds and starting patrols in town, Hurley decided to increase pressure on local criminals and bandits by authorising more house and building searches as long as these activities were prompted by reliable information on where to look. Lieutenant Bill Bowyer constituted several Engineer High Risk Search Teams, including his three explosive detection dogs, to support house and building searches. The Somalis were afraid of the Australian dogs. Their presence created a useful deterrent to provocation by Somali males and added to the initial shock when the Australians conducted surprise house and building searches.[116]

Another important contribution to the security was the slow patrolling of APCs with search lights around the streets of Baidoa. Typically, these activities were called 'Presence Patrols' and were conducted one and a half hours before midnight and about an hour after midnight. Expatriate managers commented to Spike McKaskill and his men that these patrols may have been noisy but were very comforting for those in the compounds. These patrols were also comforting for the infantry who knew that there were two APCs on the move, ready to reinforce them and provide armoured protection, white light and extra firepower if they were in a fire fight with bandits in the city.

Set against background of increasing pressure on criminals and bandits in Baidoa, the needling of Australian forces continued. Possibly prompted by criminals, a child rolled a grenade under an APC commanded by Lance Corporal Adam Golding and driven by Trooper Lindsay Beveridge on 14 February[117]. Fortunately, it failed to explode. There were also incidents of hit and run rock throwing by Somali youths.[118] Thomson and Douglas reported signs that Hurley's change in tactics was causing consternation among local criminals and visiting bandits.[119]

At night Baidoa continued to be a crazy, violent place. On 14 February Private Michael Sloman,[120] a forward scout, was attacked by a young Somali man with a knife who stabbed at Sloman, slashing his flak jacket. After the frenzied attack, lasting only a few seconds, the man ran off and escaped into the crowd.[121] Incidents like this and persistent provocation taught the diggers to be watchful and to expect anything. The environment of intimidation in Baidoa sharpened their instincts. They learned to react quickly and decisively in a variety of provocative situations.

Patrol commanders maintained tight discipline over their diggers when they were provoked. However, an unofficial rule developed among some patrol commanders that any Somali male identified as having spat on or at a digger could be restrained and given 'a talking to'. Typically, these retaliations were quick and lasted a few seconds. Offenders were let go immediately, stunned and suffering no more than a shock at being grabbed, roughly handled and subjected to a tirade. This form of retaliation became

Private Michael Sloman, a forward scout, talking to camera about being attacked by a Somali male who stabbed him, tearing his flak jacket. in Baidoa on 14 February 1993 He was not wounded in the attack.
ADF Electronic Media Unit.

known among the diggers as 'adjusting Somali attitudes'.

Many junior NCOs and soldiers felt that it was important to retaliate physically when they were subjected to provocation, such as being spat on or having a rock thrown at them. They argued that they had to react assertively and strongly in face of these attempted humiliations or they would lose respect. If they lost respect by not reacting decisively to the actions of local larrikins, then they felt this perceived weakness would increase the chances of criminals and bandits becoming bold enough to take more and more pot shots at them. The diggers adopted the attitude that their lives were on the line and they would exercise their rights to 'square up' if they were being treated disrespectfully.[122]

The practice of 'adjusting Somali attitudes' was actively discouraged by all company commanders and most platoon commanders, and led to many forthright warnings to NCOs and soldiers. Doug Fraser preferred his men 'to defuse a potentially hostile situation by initiating an exchange of greetings in the Somali language.' In his opinion 'This appeared to be just as effective as any other method used and generally led to a lessening of tensions in the town.'[123] Blumer commented later that:

> The reality is that young men, [Somali or Australian,] become emotionally involved in such situations ... We were lucky that we were trained sufficiently well to recognise the signs [of coming conflict] and realised that not losing control was 90 percent of problem resolution.

> I believe that junior commanders did extremely well in this regard and
> watched those who were moving close to the line of intolerance.[124]

Around this time, the CI Teams were providing information about a weapons market in Baidoa. They recommended that the area be cordoned off, weapons confiscated and dealers imprisoned. Hurley was faced with a dilemma. Militarily, mounting a large scale operation and closing down the weapons market would be useful. He knew that his diggers would be pleased with such an operation. Politically, he knew that this sort of operation would antagonise the locals. He also knew that there were thousands of weapons in the Baidoa Sector. Interfering with transactions at the weapons market and confiscating the weapons found there would redirect resources from other operations and have a minimal impact on the numbers of weapons available to criminals and bandits. He did not have the resources to close down the trade in guns day after day.

By this time he had begun to learn that even as the de facto military governor, there were limitations to his power and resources. In theory, his heavily armed troops could be used to interfere with any aspect of Somali life in Baidoa. For example, he could mount operations in Baidoa to close down the trade in weapons, khaat, grain, stolen humanitarian supplies and sex. By threatening gun runners, drug dealers, corrupt elders and criminals running brothels with continuous harassment unless they left Baidoa he might rid the town of many undesirables over time. Hypothetically, he could argue that all of these measures would enhance security for the distribution of humanitarian relief, increase the safety of his troops and benefit Somali society.

Hurley assessed, though, that he did not have the resources or the time to 'clean up Somali society'. The keys were deterrence, fairness and decisive action in response to direct threats against life and property. Too many enemies would be made by mounting operations to interfere with local commerce. He had to be selective about what pressure he allowed his troops to exert. For the time being, the occupation of several NGO compounds, the patrolling program, maintaining a QRF, and conducting well-targeted house and building searches were all that his resources would allow.[125]

By the middle of February Hurley's tactics were working. The Australians had taken much of the initiative away from criminals and bandits. Timely information from the locals was also coming in more frequently. They seemed to be trust the Australians not to compromise them and would take decisive action against wrong doing. On the afternoon of 16 February, a patrol from the QRF, comprising Lieutenant Anthony Swinsburg and his signaller, and two sections led by Corporals Gavin Baker and Richard Chapman, sped into

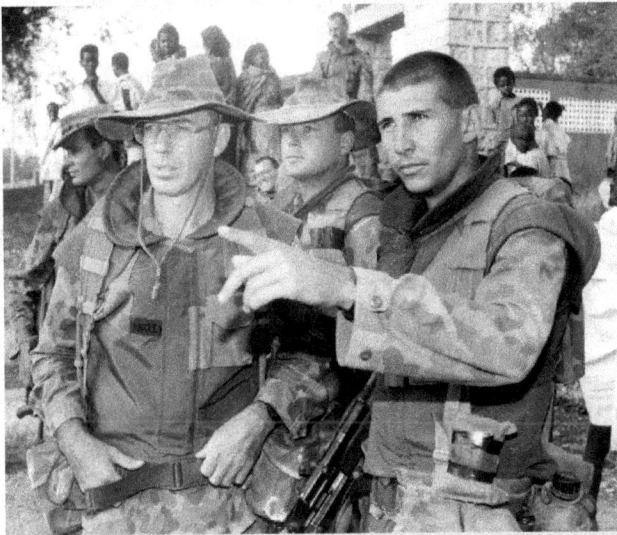

Lieutenant Colonel David Hurley (L) being briefed by Lieutenant Anthony Swinsburg, C Company, on an incident that occurred on 16 February. Major Mick Moon, Swinsburg's company commander, is standing behind them.
Image: ADF Electronic Media Unit, Gary Ramage.

town after a Somali reported that his wife and one of his teenage daughters were being abducted by bandits.[126] After deploying Baker's and Chapman's sections into a creek line east of the Governor's House to search for the woman and her daughter, Swinsburg waited with his signaller, Private Tony Williams[127], near a tree overlooking the area.

A few minutes later Baker spotted the woman and her child being led away by an armed bandit. He called out to the bandit in Somali, telling him to stop and throw down his rifle. The bandit turned immediately and opened fire. Baker returned fire as the bandit ran off into nearby camel thorn bushes and the woman and her daughter ducked and ran in the opposite direction. Swinsburg wrote later:

> Whilst I maintained communications with my two sections, the gunman Bakes' section were chasing had swung around and occupied the Governor's House. We only became aware of this as they fired at me and Willy, the rounds impacting into a tree which was only centimetres between us. I returned several rounds to the location of the muzzle shots [sic] but I don't think I hit anything.[128]

Baker reacted to the fire from the Governor's House by ordering a group of his diggers to cover him while he and the remainder of his section stormed into the building. This quick action caught four bandits by surprise. They surrendered, handing over four rifles, one with a warm barrel and fresh carbon residue. The woman and her daughter were reunited with the rest of their family later.[129]

Launching patrols from the compounds in town also started to provide the surprise and stealth necessary to deter criminal activity. In the early hours of next morning a patrol led by Corporal Adrian Hodges,[130] from 9 Platoon, C Company, deployed in the vicinity of the bridge near the pump house beside the river bed. He and his men were moving from west to east when Hodges observed some suspicious movement around the bridge area and decided to spread his patrol out to investigate. He wanted to cross the bridge with members of his patrol able to support each other in case they were engaged from the eastern side of the bridge.[131]

He sent his scouts, Sergeant Steve Boye and Private Robert McCormack, south of the bridge into the river bed and his machine gunners, led by Lance Corporal Andrew Combs, north of the bridge.[132]

Corporal Adrian Hodges talking to
camera about a contact his patrol had on
17 February 1993 in Baidao
Image: ADF Electronic Media Unit.

He and his signaller, Private James Payne, decided to move across the bridge under the cover of the other members of the patrol, Privates Adrian Wilson, David Searle and Paul Ingram, who were covering the rest of the section from west of the bridge at the intersection of a road and a track.[133]

In the pump house, a group of bandits were forcing a local technician to help them dismantle the water pump that drew water from three wells sunk on the eastern and western river banks. A bandit noticed Hodges and Payne crossing the bridge and fired three shots at them. Hodges and Payne returned fire and the scouts moved back to the stairs leading to the bridge on the western side of the river bed. A second bandit fired a burst from the pump house. Suddenly, a bandit carrying an M1 assault rifle decided to make a run for it across the bridge between the pump house and Hodges, who had gone to ground. Hodges fired at him but missed. The startled bandit dropped his weapon and sprinted away unharmed.

There were now sporadic exchanges of fire between the bandits in the pump house and Hodges' diggers. During a lull in the firing, two bandits ran from the pump house. One thought better of this attempt to escape and ran back into the pump house again. The other sprinted towards the intersection. The scouts called out in Somali for him to stop. When he continued to run, Hodges, Boye and McCormack fired simultaneously and the Somali was brought down with severe wounds to his stomach and legs.

Meanwhile, Andrew Combs decided to move his machine gunners towards the intersection to get into a better position to support Hodges and the scouts. A group of Somalis began to move towards them from the pump house. It was very difficult to ascertain what was happening in the darkness. Combs ordered the group to stop and throw down their weapons. When they still kept coming he ordered a gunner, Private Matt Dyer, to fire a warning burst above the heads of the group.[134] After the burst, they all stopped except for one man who kept creeping towards them. Combs called out in Somali for him to stop. When he didn't do so, Combs shot him, hitting him in the right shoulder and chest.

Eight minutes elapsed between the first shots being fired at Hodges and the end of the contact.[135] Hodges told his company commander later that he thought he was going to be killed during the contact but found that his training took over once he knew what was going on.[136] In those eight minutes two Somalis had been shot, a third had been wounded by departing bandits, a fourth Somali had been detained and several had escaped under the bridge during the contact, with at least one escaping bandit leaving a blood trail from a wound.

Back at the airfield, Private Kevin Low, a medical assistant, and Private Chris Hagedoorn, a driver, were sitting in an ambulance waiting for the order to move into town.[137] Later they moved down to 'marry up' with the QRF vehicles that were about to leave to secure the area where Hodge's section had been in contact.

The ambulance and QRF vehicles arrived at 2:17 am and Low and Hagsdoorn prepared the wounded Somalis for evacuation back to the airfield. About 15 minutes later the three Somalis were evacuated to the airfield and Major Darrell Duncan, the doctor with the BSG Treatment Section, began treating them.[138] The Somali who had been shot trying to escape from the pump house was dead on arrival. He had been shot through the lower abdomen and twice through the lower legs. The Somali who had been shot in the shoulder and chest was stabilised and prepared for Aero-Medical Evacuation (AME). The technician had been shot three times in the legs by the bandits, one bullet had passed through his kneecap, a favourite bandit act of violence. He was also stabilised and prepared for AME. Both were flown to Mogadishu two hours later and survived. The Somali with the chest wound had one of his lungs and an arm removed. The technician was crippled for life.

Duncan, Low and Hagedoorn were satisfied with their busy morning's work. However, there were problems with the AME and questions were raised about whether a doctor and another medical assistant should have gone with the ambulance into town.[139] The AME helicopter landed at 4:20 am to pick up the two critically wounded Somalis. Battalion Headquarters had sent a Warning Order to put a helicopter on standby just before 2 am. The request for urgent medical evacuation had been sent through at 3:10 am. The Australians had been promised that AME support would be less than two hours away, on call 24 hours a day.[140] There had been no other medical evacuations that morning, so the delay in the arrival of the helicopter was a source of concern.[141]

News of the fire fight at the pump house spread quickly around Baidoa and was broadcast by Radio Australia at 11 am local time. Hodges and Combs were interviewed by several journalists. Later CI reports revealed that Gaaney had been wounded in the head in the fire fight and had narrowly avoided capture.[142] He had probably escaped with his companions under the bridge.

That night John Caligari warned all patrol commanders to be vigilant because of fears of reprisals now that the Australians had killed and seriously wounded members of a bandit gang. This warning was timely because there was trouble at the CONCERN compound in the early hours of the next morning. At 2:30 am armed Somalis broke into a house in one of the CONCERN compounds and stole a large quantity of cash. An expatriate manager told the patrol commander who arrived a few minutes after the robbery had occurred that he suspected that

Somali guards employed by CONCERN were accomplices. He asked if the Australians could disarm the guards and take over the protection of the compound.

Caligari directed that the guards were to be disarmed and sacked. He obtained approval from Hurley to deploy a section to occupy the compound. [143] An hour later, Corporal Bill Perkins was leading his section along an alley near the Khaat Market. Private Jason Flatley, the forward scout, discovered a man in the doorway of a warehouse. [144] Flately told him to move on. A moment later he heard the sound of weapons being cocked. Flately had a split second to decide whether to fire in the direction of the sounds, or hold his fire and investigate. Something about the man he spoke to and the coincidental sound of weapons cocking told Flatley that he was about to be shot at. He opened fire.

Flatley's burst of fire was followed immediately by bursts of fire from at least two Somali gunmen. Private Chris Day, the patrol's signaller, was shot in the shoulder:

> ... as I dived for cover, I was contacted by a second gunman who was in a small alcove between two buildings ... one of the rounds hit me and the rest of the burst went over my head. I couldn't see him but he was quite close because his muzzle flash went in my face. So I fired several rounds at him and he did not return fire so I thought I had got him. I yelled out to my 2IC, Gary Lively, that I'd been hit but I was OK. [145]

In the same burst of fire that hit Day, a bullet hit Flatley's pistol grip—just above his hand, another passed through his trouser leg—just behind his calf muscle, and a third bullet hit his night vision goggles that were hanging from his neck. He and Day were stunned temporarily but were still alive. Before their unseen assailants could fire again and finish Flatley and Day off in another hail of bullets, Perkins and the other members of the patrol returned fire in the direction of the muzzle flashes. A long burst of fire from the machine gunner, Private Jason Blakeman, returned the initiative to the Australians. [146]

Chris Day maintained radio contact with Battalion Headquarters despite his wound. He sent the all-important message, 'Contact—Wait Out' and reported what was going on after that. The cavalry were on their way. Two APCs, commanded by Lance Corporal Dave Williams and Trooper Dave Bell, located in town, drove immediately to the sound of the gunfire and monitored Day's description of events over the radio.[147] Williams and Bell positioned their vehicles to extract Perkins and his men if needed.

Perkins was not sure how many gunmen he faced and how well they were positioned to shoot if he ordered his men to advance. He decided to pull back down the alley and regroup. He could hear the sounds of the two APCs arriving, so he thought it would be wise to move back to them before deciding whether or not to go after the gunmen.

Perkins pulled his men back under the cover of a further burst from Blakeman's machine gun. He met up with the APCs and decided it was too risky to advance back up the alley in the dark to engage the gunmen who had probably withdrawn anyway. He mounted his men in the APCs and directed Williams to inform Battalion Headquarters that Day was wounded and to take him and his section to the airfield so they could submit reports on the contact and have Day patched up. Fortunately, Day had a 'through and through' wound—a bullet had passed cleanly through the fleshy part of his shoulder, entering and exiting neatly. He was treated that night and

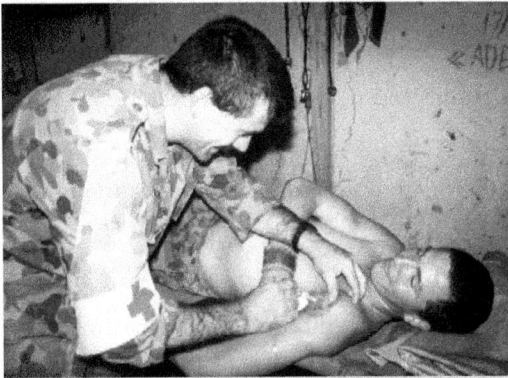

Sergeant Shane Marles, 1 RAR Regimental Aid Post Desk Sergeant treats Private Chris Day's shoulder wound. The wound was inflicted during a shoot out in Baidoa in February 1993
Image: G Gittoes

was evacuated to the Swedish Mobile Army Surgical Hospital (MASH) in Mogadishu the next day for further treatment.[148]

Counterintelligence investigations of Bill Perkins' contact revealed that the patrol had surprised a group of guards employed by the Somali Islamic Relief Organisation who were about to loot a warehouse.[149] One report suggested that guards employed by CONCERN were also there.[150] A further report suggested that the guards had decided to fire on the Australians because they were angry about the previous confiscation of arms. They feared that all NGO guards in Baidoa would eventually be disarmed and lose their jobs as the Australians took over more and more NGO compounds.[151]

The next day, Lieutenant Anthony Swinsburg was ordered to follow up blood trails from Hodges' fire fight at the pumphouse. He was talking to his APC Crew Commander, Lance Corporal Andrew Bradbrook, near the World Food Program Compound when a water truck came speeding down the road.[152] A girl about six years old was playing in a puddle beside the road when the truck hit her. She was tossed several metres, suffering massive head injuries. Private Graeme Brown, a trained medic, applied First Aid but she died.[153] In the meantime several diggers had rushed forward and stopped the truck. The driver complained bitterly about being stopped and wanted to continue his journey as if nothing had happened.

The Australians were angered by this little girl's death. Swinsburg contacted the Duty Officer at Battalion Headquarters and asked what he should do with the driver. BHQ indicated 'to give him a stern talking to and let him go.' After confirming this message I lost a lot of faith in our tasking. Here we were detaining persons for theft of grain but not manslaughter. This driver was dealt with appropriately.[154]

The Somali truck driver became very upset when the diggers vigorously chastised him for reckless driving. He felt that he had 'lost face' in front of a crowd that had gathered. It was discovered later that the driver made a number of threats, declaring that he would avenge his treatment and kill some Australians.[155] Both Hodges and Perkins became celebrities after their sections' contacts. They were interviewed by journalists and received a personal signal from Major General Blake, congratulating them on their leadership and coolness under fire.[156] On 20 February, Colonel Bill Mellor visited Baidoa and spoke with Hodges and Perkins, asking them for detailed descriptions of their encounters and subsequently congratulating them on their efforts.

After 24 hours of notoriety to break the routine of urban patrolling, Corporal Bill Perkins was rostered for another night patrol in the early hours of 21 February. Just before 2 am he directed his scout, Jason Flatley, up the same alley that they had been in during the contact two nights before. Suddenly, several Somali gunmen ambushed them from the windows of a building with long bursts of automatic fire.

Perkins and his diggers returned fire and went to ground. In the face of sustained automatic fire, Perkins decided to turn the tables on his assailants by ordering Private Mike Meehan to fire his 66 mm anti tank rocket at the muzzle flashes.[157] The rocket hit the building with a thunderous crash and the firing stopped. Perkins pulled back and fired a green flare, a prearranged signal for when a patrol wanted to be reinforced.

In a street a few hundred metres away Corporal Tom Aitken and his section heard the bursts of automatic fire and suspected that Perkins' patrol was in contact. He and his men rushed to the scene. At the same time Perkins' platoon commander, Lieutenant Ian MacGregor, arrived with a section of APCs, commanded by Corporal Darren 'Moose' Ferriday, and his third infantry section.[158] There was no sign of life from the building that had been hit by the 66 mm rocket.

MacGregor ordered Aitken and his men to search the building. Covered by the remainder of the platoon, Aitken rushed into the building and discovered four concussed and disoriented Somalis lying in the rubble and white-painted rendering that had blown in on them from the 66 mm

Corporal Bill Perkins whose section was involved in two firefights with NGO guards in February.
Image: ADF Electronic Media Unit.

blast. Aitken's concern about whether he would be fired on as he entered the building turned into uncontrollable laughter at the comical scene of the four Somalis looking like bedraggled white ghosts.[159] Three AK 47 rifles were recovered and the Somalis, all NGO guards working for the Somali Islamic Relief Organisation, were taken into custody.

Later that morning Bill Perkins was brought back to the site of his contact with a large group of journalists, and several video crews and photographers. Accompanied by Bill Mellor and David Hurley, he conducted a guided tour and recounted again what had happened the night before. Coincidentally, there was a higher than usual media presence in Baidoa that day to cover a visit by the veteran American actor, Charlton Heston.

Heston, a patron of CARE International, was on an African publicity tour. His visit had been eagerly awaited by many of the Australians in Baidoa who wanted a chance to see this internationally-acclaimed movie star. Lieutenant Troy Webb, Commander of the Direct Fire Support Weapons Platoon, and his men were given the task of providing security for Heston's visit.[160] He was flown by helicopter to Hobisole, CARE Australia's model village, and treated to a display of affection from Somali children and several Somali songs sung by a well-rehearsed children's choir. He then went onto the GOAL orphanage and the IMC Hospital.

At the orphanage a small donkey had been harnessed to a cart. Sending-up his famous scene in the film, 'Ben Hur', the good-natured

Corporal Tom Aitken, 8 Platoon, C
Company, 1 RAR Group in Baidoia,
talking about the responsibilities of
leading patrols in Baidoa.
Image: ADF Electronic Media Unit.

Heston sat on the cart and coaxed the donkey around the courtyard area of the orphanage, much to the delight of the Somali orphans and the media contingent. Heston was then a guest for lunch at the 1 RAR Mess at the airfield. He gladly agreed to be photographed with six members of the 1 RAR Group who had been selected by ballot to have their photograph taken with him. The photographs were developed quickly and autographed by him before he left. Those who had been photographed with him were each given a copy.

Major Mark Harnwell was pleased to be able to offer one of the photo opportunities with Charlton Heston to a member of his unit. Harnwell offered 21 year old Lance Corporal Matthew Holden the opportunity to be photographed with Charlton Heston because it was his birthday.[161] Holden queried Harnwell's offer by asking, 'Who's Charlton Heston?' and declined the offer. Forty year-old Warrant Officer Colin Young, who had seen most of Heston's films, volunteered and took Holden's place.[162]

Anticipating a reaction to the recent contacts and concerned that the town might become unsettled, David Hurley had decided to divide Baidoa down the middle and sent Major 'Ant' Blumer's Delta Company into town to occupy an area of operations adjacent to Major Mick Moon's Charlie Company. With two companies in town occupying compounds, patrolling 24 hours a day and maintaining a network of observation posts, Hurley hoped to deter criminal activity and politically-inspired violence.

Blumer favoured retaining the initiative against local urban-based criminals and visiting bandits by conducting house and building searches to confiscate weapons and keeping them guessing about when and where the Australians would appear next. The CI Teams appreciated Blumer because he was always ready to react to their information. Like Blumer, they felt that the Australians should continually search for and confiscate weapons, and detain suspicious individuals for questioning. Blumer assessed that there was no point in waiting and reacting to incidents. He instructed his platoon commanders to maintain the pressure on any suspicious individuals and groups, and continue a vigorous approach to searching buildings.

Hurley's deployment of two companies, each supported by a section of APCs,[163] into town was timely. He anticipated that the beginning of the Muslim period of fasting known as Ramadan might be used to stir up trouble among the crowds that would gather. The rival political factions in town were spreading disinformation among the populace about the impending arrival of troops sent by competing warlords to retake Baidoa. Blumer mused in his diary:

> ... Loud crowds and bands are moving around town. We are sitting here waiting for something to happen. There's a crowd of two to 3 000 in C Company's Area of Operations across the creek from us. We've been told that a 90 vehicle convoy is moving from Kismaayo towards us and a group of 2 000 [troops] are in Awdinle in a fairly riotous state with 800 of them moving east and south towards us. Glamour excitement and adventure ! It will be interesting to see how the evening develops.[164]

Hurley and his staff had received information from HQ 10th Mountain Division that 2 000 Somali National Front (SNF) troops loyal to General Omar Jess, a supporter of the deposed dictator, Barre, were approaching Baidoa. Hurley sent Thomson and Douglas into Baidoa on the evening of 22 February to verify this information. They interviewed a number of Somalis employed by the aid agencies and found that the common denominator was those reporting the advance of the SNF forces were associated with the political faction supporting General Aideed. Thomson and Douglas had been out on patrol with the Reconnaissance Platoon in the preceding week and had heard nothing from locals about an advance by 2 000 SNF troops mounted in 90 vehicles. Due to the clan and political affiliation of the locals, Thomson and Douglas knew they would have reported any impending advance.

Aideed's supporters appeared to be trying to embarrass the Australians by inciting panic among the populace that would result in an exodus from Baidoa. Hurley remained concerned because of the renewed rioting and looting in the port of Kismaayo where Belgian troops had been caught by surprise and had taken casualties. At the same time, warlords were making claims and counterclaims about movements of their troops. Responding to these statements, HQ UNITAF had put all units on alert.

Faced with conflicting reports and following the directions of his superior headquarters, Hurley decided to take tactical precautions. He put the 1 RAR Group on alert and gave orders for the deployment of two companies mounted in APCs and other vehicles, to block the approach of advancing troops. He decided to deploy the Reconnaissance Platoon on a wide front to find the location of the advancing SNF forces and provide early warning for the remainder of the 1 RAR Group. McKaskill sent Lieutenant Mick Toohey and Corporals Harry Smelt and Justin Hepburn in their APCs to conduct a reconnaissance around blocking positions.[165]

That afternoon Steer issued his orders and divided Reconnaissance Platoon into two groups to patrol south and set up observation posts 40 kilometres out from Baidoa. Each group was allocated a signals vehicle and a 6 x 6 Land Rover. His Platoon Sergeant, Wayne Thomson was not happy:

> I questioned Lieutenant Steer about fire support, air assets and [the need for] APCs as it seemed suicidal to be restricted to the roads by moving in vehicles with tyres (you could not move off-road without puncturing tyres as the camel thorn was everywhere, including on the ground). The thought of having 2 000 Somalis between us and base [Baidoa] did not gel, and if we were detected it would have been untidy to say the least.

> He went back and questioned BHQ while I chased up rations, stores etc. When he returned he told me we were by ourselves and that all the APCs had been pulled in to protect the airfield which made the platoon feel real good![166]

The Reconnaissance Platoon vehicles deployed south as ordered, accompanied by a detachment of signallers to ensure communications. Each vehicle was carefully hidden off the two main roads south of Baidoa and tracks cut through the camel thorn from their 'hides' so they could move back onto the road and withdraw quickly without puncturing their tyres. Wayne Thompson recalled:

> It was a prick of a day—the hottest I remember on the trip. We were laying under the Rover as there was no shade. The boys reported two small antelope and nothing else. We were ordered to return at 0600 hrs on 24 February—it ended up being a false alarm—thank God![167]

The false alarm did not enhance Battalion Headquarters' reputation among members of the 1 RAR Group. Hurley's precautions appeared to be an over-reaction. However, there were beneficial consequences. He ordered Thomson and Douglas to 'convey his displeasure to representatives from Aideed's political faction in town.'[168] Wayne Douglas confronted the leader of the faction:

> ... in a manner that rendered interpretation redundant and [that] could be heard hundreds of metres away, told him that LTCOL Hurley had just about had enough of the disinformation coming from Aideed's supporters. The dressing down had the desired effect. Shortly after, the faction leader and a number of key personnel departed for Mogadishu and were not seen [in Baidoa] again. The back of Aideed's faction in Baidoa was effectively broken.[169]

With the fear of attack from 2 000 Somali militia over, security operations returned to normal in Baidoa. On 23 February, Blumer was tasked to conduct a raid on a restaurant near the Mosque, reported to contain a cache of arms and ammunition. Steve McDonald had received this information from an English-speaking Somali teacher hired by Dick Stanhope to conduct Somali language classes for the Australians.[170] Blumer tasked two sections of Jan Van Der Klooster's 12 Platoon to conduct the raid and accompanied them with his signaller.

After an hour of frustrating indecision about which restaurant had been identified by the Somali teacher to be raided, Blumer called off the operation.[171] However, the Somali teacher then insisted that the Islamic Children's Orphanage and Mosque compound should be searched because there was a cache of arms and ammunition hidden there. Sergeant Alan Barrie, a CI Team member, who was travelling with Blumer, assessed that a search would turn up a cache. Earlier that day, Blumer had been told by Battalion Headquarters by radio that he was not to enter the orphanage unless he had specific information.[172] He took Barrie's advice and that of the Somali teacher. After receiving permission to conduct a search from Battalion Headquarters, Blumer directed Van Der Klooster to enter the orphanage.[173] After a comical episode of finding the main gate locked and trying to scale the wall only to find another gate open, Blumer followed the platoon into the orphanage. He wrote:

> Politely I explained that I was there to check their weapons with the paperwork. They initially agreed and asked me to wait for the boss man. After a while it was clear he would not appear so I decided to search the place. We had a dog as well. The Somalis—30 to 40 males, squillions of kids and lots of women—became uncooperative and resentful and then openly abusive and they caused quite a kafuffle inside the compound. Poor old Mohammed, the interpreter, became hot under the collar when they started giving him death threats. Eventually we found six weapons, a grenade, magazines and other paraphernalia. We left another annoyed and militant enemy there. However they had little to complain of as we found the weapons.

Blumer then decided to continue to search places that might yield weapons:

Back to Bar DEXXO and went through that without finding anything more than developing a little more ill-feeling when a bloke refused to come outside and was forcibly ejected. Sent a platoon into ICRC Compound area and they found five weapons.[174] Later that night we raided the Hotel Al Timmaan. It didn't provide much other than a few bayonets and a magazine. Hussein Barre Warsame from the SLA [Aideed's faction] appeared claiming all sorts of bullshit. He served only to piss me off and intimidate the interpreter. We have made a few enemies out of today I think![175]

The next day Blumer continued to achieve success with his surprise raids. After a chat with the newly-appointed Police Chief, Colonel Aden Nuur, who was responsible for setting up a new Somali Auxiliary Security Force in Baidoa, Blumer decided to raid the Hotel Al Timmaan again and search surrounding houses, including a brothel. This time Van Der Klooster's diggers turned up five assault rifles, two 20 mm Anti Aircraft guns, one heavy machine gun, numerous spare barrels and boxes of ammunition, as well as numerous documents. The raid on the brothel yielded an AK 47 assault rifle, a .45 calibre pistol and ammunition. There were also a number of uniforms, videos and documents. Blumer retired at midnight after a busy day.[176]

Despite these successes, Blumer and the battalion Headquarters staff were in conflict. Blumer wanted the freedom to act on information gathered in his Area of Operations and to conduct random searches for weapons and ammunition when he saw fit. He did not care whether he aggravated Aideed's supporters who were a minority faction implicated in criminal activity and political violence. He also felt that he should follow up leads in C Company's Area of Operations if he decided that immediate action was required. Moon agreed to Blumer's initial requests, but then refused any more because of the proximity of his forces and the frequency of the requests.[177] Though Blumer felt that C and D Company were 'very close and aligned for the vast majority of the deployment', there was some competitive tension in town between the two companies as they operated alongside each other.[178]

Major 'Ant' Blumer,
OC D Company, 1RAR
Image: ADF Electronic Media Unit.

For their part, Hurley and his staff insisted that Blumer obtain approval from them before conducting operations in his area of town. They also wanted regular reports after he had conducted local operations. Blumer acknowledged the requirement to provide regular reports but felt that his higher headquarter staff were demanding too much information that 'was filed, not processed or interpreted'. He continued, 'All the information was passed, but they wanted moment by moment descriptions of what was occurring, a lot of which was random "house searching".'[179] Blumer was becoming increasingly frustrated:

> [I am] frustrated, disappointed and [have] quite a feeling of loss and pointlessness. It really got me down and the more I see of OA [Battalion Headquarters], the more I feel a gulf developing between us in the field and those [in the] HQ.[180]

The gulf between Blumer and his Headquarters was not the only gap developing between himself and others in 1 RAR. He continued:

> Went for a drive around town looking for places for weapons and transmitters but many of them are outside our area of operations and in C Company's. OC C Company [Major Mick Moon] [got] stroppy about our movement in his area. Can't blame him but [we are] hardly getting into many of the places that we would like to.[181]

Blumer was frustrated for two reasons. He felt the boundaries of his Area of Operations were too small for a company and that his headquarters took too long to respond to his requests to authorise his operations and to allow his patrols to respond to situations that would require them to go outside his boundaries into Moon's area. He believed that, 'My relationship with Mick was generally very good'.[182]

Blumer felt that the battalion should model its operations on his style of random, aggressive house and building searches and vigorously looking for opportunities to close down criminal activities by confiscating weapons and ammunition. He questioned whether Hurley and his staff had an overall plan. In absence of a plan, he wanted to continue his approach to 'creating a secure environment' in Baidoa without interference from his Headquarters. 'I wanted a 'directive control' approach. A clear task and resources and left to do it. I did send in written reports as well as briefing IO, OPSO and CO.'[183]

On 23 February, John Caligari called a meeting between himself, Blumer, Moon and Jim Simpson, whose company was securing the airfield. He felt that the meeting went well and that everyone was open and frank about how to improve the coordination of operations between Battalion Headquarters and the rifle companies. Caligari stated at the meeting that he was happy to discuss any further concerns at any time. He also decided to increase his vigilance over communications between his Headquarters staff and the rifle companies.

Hurley returned from a visit to Mogadishu on 27 February to find the tempo of operations had increased significantly in Baidoa. Moon raided the derelict International Ice Cream Factory at 11 am and confiscated two M1 Carbines, two rocket propelled grenade launchers, a pistol and four and a half a million Somali shillings. Later that day, an enraged delegation of Aideed's supporters came to his headquarters and issued numerous death threats and accused Moon and his men of being thieves.[184]

Meanwhile, Blumer's men hit the Islamic Children's Orphanage again and retrieved an M 16 rifle and a grenade. Later they discovered a 2 inch mortar and a radio in a nearby building. That afternoon, acting on CI information, Blumer sent Lieutenant Don Tulley's 11 Platoon down to search several houses accommodating Aideed's supporters behind the creek and Police Station towards the Main Gate of the airport. This series of house busts resulted in the confiscation of numerous weapons, and $US 10 000 in a safe. It was now Blumer's turn to be accused of stealing. He and his company also received a number of formal complaints about damage to doors and property as a result of using sledgehammers to gain quick entry into houses.[185]

After what he regarded as a successful day of keeping Aideed's supporters off-balance and confiscating arms, ammunition and cash, Blumer attended David Hurley's evening conference at the battalion Headquarters. When Hurley called him aside for a chat, Blumer explained that he was doing his job but was not receiving the acknowledgment he felt was due to his diggers 'who were working their butts off', confiscating weapons and ammunition as well as putting the criminal elements on the back foot. Blumer reported that Hurley 'listened before talking and came around very quickly'.[186]

Hurley seemed to have been convinced by Blumer's arguments for the time being. Blumer's peers were not happy.[187] Caligari vowed never to put two companies in town at the same time again.[188] Mick Moon was annoyed that Blumer did not want to discuss his concerns about D Company patrols entering his area in town.[189]

Blumer continued to be proactive and innovative. He received reports from CI Teams that the same group of bandits had been extorting, robbing and raping in the Bay Project area over the past few nights. Coincidentally, he had four snipers in his area looking for work. A series of unrelated events had brought these four men together. Several days before, Corporal Darren Mortimer[190] and another sniper had been withdrawn from a surveillance position after being compromised. A Somali gunman had fired at them to show that they had been discovered. Since then Mortimer and his partner had been with Delta Company awaiting further assignments. Later that afternoon they had been visited by two other snipers. Mortimer and his fellow snipers were bored and keen to get out on a night patrol.

Blumer had agreed to Mortimer's request to give him and the three other snipers an opportunity to go out and apprehend the bandits who had been terrorising the women drawing water at the Bay Project. He saw this as an opportunity 'to reduce the patrol burden on the sections. They were not sent out as a sniper patrol but as a 'Bric' patrol—infantry'.[191]

At 10:40 pm that night Mortimer's patrol made up of Lance Corporal Darren Loftus, Private Rob Maitland and Private Steve Seath[192] was moving in the vicinity of the water points at the Bay Project. They prided themselves on their silent stealthy patrolling techniques. After moving into the area without being detected, Loftus observed a Somali male through his night vision goggles carrying a rifle in the company of two other Somali men. He wrote later:

> I advanced towards him, trying not to alert the man to my presence but at the same time trying to alert my own patrol of the situation. I became close enough to the man to challenge him to put down his weapon. The man turned, ran to his left raising his weapon to his shoulder. I initiated with two automatic bursts, killing the man.[193]

Loftus had fired a total of 13 rounds at the Somali hitting him 7 times which, as Blumer wrote later, 'was fairly conclusive'.[194] Simultaneously, the other two Somalis also turned and opened fire. Maitland, who was alongside Loftus in the patrol's box formation, was not wearing night vision goggles. He fired bursts in the direction of the muzzle flashes but the bandits slipped away.[195]

Blumer, who was only two hundred metres away at his headquarters in the Police Station, rushed to the scene with four APCs, commanded by Lieutenant Mick Toohey, and his QRF of two sections and a platoon headquarters. He wrote:

> I interviewed the diggers who all appeared rather calm about the whole deal. The Somali was a real mess! Took the body to the IMC hospital and got onto OA [Battalion Headquarters] to make the report [in person].[196] The CO thought too many rounds were fired and too few [soldiers] to use as a patrol. I objected on both points. Lengthy written report and back to bed.[197]

While Blumer was reporting back on the contact, Mortimer and his men continued their patrol until 1 am and, after an hour's rest, set out again at 2 am on another patrol which concluded at 5 am.[198]

Charlie Company was having a busy time in their area of Baidoa that night as well. At 1 am Hassam Hirsi Mohammed, known as Lesto, and a group of recently-sacked MSF guards decided to make a withdrawal of funds from their former employer at gun point. After bailing up the expatriate Dutch staff for several hours they took $US 25 000 and a quantity of Somali shillings valued at $US 10 000. This robbery had probably been precipitated by the expatriate Director of MSF sacking Lesto and 19 guards that morning. Before leaving with the money, Lesto promised the Director, as he was bailed up against the wall, that he and his Dutch compatriots would be killed unless they left Somalia in the next few days.

Moon was very annoyed about this robbery because it occurred less than 500 metres from his headquarters. In his opinion, the MSF staff had brought all of this on themselves by hiring Lesto and his henchmen in the first place and seeming to ignore Australian warnings about their criminal activities. He felt that the staff had not taken appropriate precautions after they finally decided to sack Lesto and his companions. The MSF compound and other MSF facilities in town were now not only defenceless but they were also possible targets for disgruntled ex-employees. Hurley ordered Moon to occupy the MSF accommodation compound to protect the Dutch staff. All of the Charlie Company sections were now spread out over the company AO occupying NGO compounds.[199]

February had been an action-packed month for those securing Baidoa. In light of the MSF robbery, the increasing fear among NGO guards that they were going to lose their jobs to the Australians and the anger of criminal elements supporting General Aideed to more assertive Australian operations, it remained to be seen whether the action would continue. It was difficult to determine whether the Australians had attained just the right level of deterrence needed to maintain a secure environment for humanitarian operations or had stirred up a hornet's nest of antipathy that would result in hit-and-run, payback attacks from Somali gunmen. There was little doubt from the contacts the Australians had been involved in during February that criminals and bandits knew that the Australians were not easy marks and would return fire accurately and decisively when engaged or threatened.

On 1 March, the 1 RAR Group celebrated the Australian Army's Birthday. For the first time in six weeks some of the diggers were able to have a beer. Two hundred cartons of beer and sufficient hamburger steak to feed all of the Australian military personnel in Baidoa had arrived at 2:30 pm on the New Zealand Andover flight. The beer had been donated by the Australian Returned and Services' League (RSL). Two platoons from D Company and B Company and half of B Squadron were the only combat troops available to enjoy a BBQ dinner and a few beers that evening. The remainder of the rifle companies, B Squadron and specialist platoons were on duty. Beer was kept aside for them to consume the next time they returned to Baidoa and were able to take time off.

The diggers who were non-drinkers had to pay for cans of soft drink while their mates drank beer for free. Blumer, McKaskill and Hurcum were very annoyed about this inequity and paid for soft drinks for their soldiers out of their own pockets.[200]

On 2 March the 1 RAR Group came under the operational control of HQ AFS in Mogadishu because the 10th Mountain Division was pulling out of Somalia. General Arnold, the Commanding General, attended a short ceremony at HQ 1 RAR Group and handed over command responsibilities to Colonel Mellor. Later that day David Hurley and John Caligari, had a long discussion about the challenges that faced them for the remaining 10 weeks of the operation. Caligari wrote:

> We talked for about an hour on various matters. I felt that he enjoyed talking to someone who at least understood the situation to a similar degree to him. He gives the impression on occasion s that he is not completely happy with the lonely life of command. We all have peers

Major Mick Moon, OC C Company
Image: ADF Electronic Media
Unit.

to laugh and talk frankly to but he has no one. I think he appreciates
being able to speak honestly to someone now and then. By the time
he left I had a better grasp of the complicated politics [pertaining to
our mission].[201]

On 3 March, the companies of the 1 RAR Group rotated around the four
operational tasks for the sixth time. Fraser's A Company returned from
Buurhakaba and took over airfield security, Simpson's B Company took over
eight NGO compounds in town and began a 24 hour patrolling program,
Moon's C Company took over escort duties for convoys and medical visits,
and Blumer deployed into the field in search of bandits. Dodd's Mortar
Platoon and Webb's Direct Fire Support platoons with a section of Assault
Pioneers deployed to Buurhakaba as a company minus. McKaskill also rotated
his personnel for the sixth time. Typically, his APCs remained in the outlying
patrol areas and crews were rotated. He could not afford to have his crews drive
for 6-8 hours in their vehicles to replenish, refurbish, catch up on some sleep in
Baidoa and then drive back into the field. Only vehicles in Baidoa and in the
vicinity of the airfield rotated.

Meanwhile, the pressures of command were being felt on three of the
battalion's platoon commanders. One was relieved of command[202] and another
was warned that he had to improve his performance or he too would be relieved
of command. His section commanders had approached their company
commander stating that they had lost confidence in their platoon commander.

The company commander counselled him that he had a lot of work to do to restore his platoon's confidence in him. A third platoon commander was similarly counselled in another company.[203]

Blumer was also concerned that his men were losing their concentration and motivation. 'With seven weeks down and 10 or 11 ahead, the glamour and panache of the overseas deployment has worn off somewhat. We have a long haul to keep motivation going and troops interested'.[204] Operation *Solace* was much longer than annual large scale field exercises in Australia. The living conditions at the airfield were endurable for a couple of weeks when the adrenalin was high and everything was new and exciting. After six weeks, the close living, the lack of privacy, sleep deprivation and the hot, dry dusty conditions began to wear people down.

Meanwhile, bandits were still at work in town. On 6 March, Brendon Thomson, Wayne Douglas and a visiting US Marine intelligence specialist, Corporal Doyle, went into town on a routine visit to chat with various community and political representatives. They were accompanied by their interpreter, Abdul, and a protection group travelling a discrete distance away led by Steve McDonald, who, with WO2 Rick Vine, and a signaller, Private Andrew Burns wanted to break the monotony of life at the airfield by conducting a patrol in town.[205]

At the northern end of the market area, a visibly-distressed woman approached Abdul and told him that two armed bandits were about to shoot storekeepers further south in the markets. McDonald assumed command of all those present, and deployed Vine, Thomson, Douglas, Burns and Doyle into patrol formation and headed south. Douglas was assigned to be the forward scout.[206]

Tension and expectation among the crowd increased as the Australians advanced; 'Somali women began a high pitched ululating'.[207] Suddenly there was a short burst of gunfire around a corner 10 metres ahead of Douglas. He turned to McDonald for permission to go ahead. McDonald called out, 'Go'! The nearby crowd scattered in all directions as the Australians and their American visitor swung into action, quickly positioning themselves to aim weapons around the corner where the shots had come from and to cover each other as Douglas moved forward. By this time Abdul had disappeared into the crowd, leaving McDonald's patrol without the means to communicate with the Somalis in the area. Around the corner were two Somali shopkeepers, one shot in the hand and the other, writhing on the ground, in the leg.

A young man beckoned to Douglas to follow him and, encouraged by other Somalis pointing further down the alley, McDonald ordered Douglas to advance. Before leaving, Vine insisted that everyone introduce themselves to the other members of the group by name so they could call to each other if they became involved in a firefight with the bandits. In less than a minute, a routine counter-intelligence visit was transformed into a fighting patrol.

A few metres further on Douglas saw two bandits scurrying through the crowd, one armed with an AK 47 assault rifle and the other with a G3 rifle. He quickened his step and called the remainder of the patrol to give chase. Soon the patrol were in a frantic foot race. Douglas, Doyle, McDonald and Burns were hard on the heels of the two bandits.

Vine and Thomson assessed which way the bandits were going and took a parallel route aiming to come at them from a flank and cut them off. The Australians began leaping fences and walls, charging through houses, backyards and shops. As they reached the end of the market area, the bandits separated, and both turned and raised their weapons, ready to shoot the pursuing Australians. Douglas, who was still leading the chase, hurtled towards one bandit who was aiming his G3. Fortunately, Douglas spotted the bandit and had time to react. With civilians still milling around, he did not risk a snap shot at the bandit and fired a warning burst above his head. Doyle, who was following closely behind Douglas, saw the second bandit with the AK 47 'take a bead on' Douglas. He fired warning shots instinctively. Simultaneously, this bandit fired off a hurried burst of rounds at Douglas before turning and running away. Possibly fearing that Douglas and Doyle would fire at him next, the bandit with the G3 dropped his weapon and fled.

With the bandits now separated, Douglas now set off after the unarmed bandit 'like a hound after a rabbit', with Doyle, McDonald and Burns following.[208] Douglas was 100 metres in front of them, determined to catch the bandit after seeing his handy work with the shopkeepers. By this time Somalis were screaming and scattering in all directions. Prior to the exchange of shots, they had been cheering the Australians 'like a crowd at a football game'.[209] Vine and Thomson took off after the second bandit.

Realising that the running bandit was increasing his lead against the Australians, who were wearing patrol webbing and flak jackets, Douglas decided to close the distance with a 5.56 mm bullet. Thoroughly exhausted, he propped and took aim. However, because the bandit was unarmed and running away, the ROE did not allow Douglas to shoot. He lowered his rifle and watched as the bandit made his escape.

Vine and Thomson also gave up their chase after a short distance, fearing that they would lose complete contact with the rest of the patrol. The Australians had chased the two bandits, who were initially only carrying rifles, for over three kilometres in 40C degree heat:

> All of the soldiers were drenched with sweat but still exhilarated from the excitement of the chase. As they made their way back to the vehicles, Abdul, the interpreter, materialised from a side street and asked where they had gotten to 'I have been looking all over for you.'[210]

Aside from the presence of bandits in town, the Australians were becoming fed up with the behaviour and apparent immunity from prosecution of the NGO guards. Blumer wrote:

> Hard decisions should be made now about whether we want to have long term effect on this country. The NGO guards are a classic example. At the moment they are all henchmen, outsiders who are hired gun slingers who have no roots or connections in Baidoa at all. They are the ones causing trouble after dark with their weapons and vehicles, who have no concern for anyone other than their own gross pay cheques and what they can steal from anyone. The leaders have

offered to provide locals for the job and would take responsibility for their actions. There would be obvious difficulties sacking the present guards and that is where we could bear the brunt for the interim before the system is up and running strong enough to support itself, along with the ASF [Auxiliary Support Force].[211]

Hurley continued to focus operations in Baidoa on protecting expatriate NGO staff. He wrote:

> The security of NGOs at night is still our major concern. To deploy two companies in town, with sections in every compound, would severely limit our activities elsewhere, particularly on the roads at night. It is important to retain the confidence and desire to continue among the NGOs—should they lose their will and leave, we will have failed in our mission. The direct personal threat to them is our most important challenge.[212]

By the end of the first week in March paybacks for the shootings, confiscations and house busts in February had not occurred and the level of reported violence in town had gone down significantly. The Australians appeared to have the initiative and were receiving sufficient information from the locals to know that they had friendly 'eyes and ears' throughout the metropolitan area. The turnaround in four weeks since the unlucky first week in February was very gratifying for Hurley, his staff and his commanders who had worked very hard to come to terms with an unfamiliar and dangerous urban environment.

The turnaround of the security situation in Baidoa did not go unnoticed or unappreciated. The expatriate NGO managers and their staff were grateful for the 1 RAR Group's efforts. In an unprecedented gesture, they all signed a letter of appreciation on 10 March addressed to General Johnston, and later distributed widely to other aid agencies and senior UNOSOM staff in Mogadishu. The NGO managers in Baidoa praised Hurley's pragmatic approach. They not only highlighted the protection they were receiving through foot patrols, escorts, the confiscation of weapons, a cash deposit and delivery system and the presence of troops in their compounds, but also the establishment of excellent communications with the NGOs and local elders. Hurley had correctly assessed that opening up and maintaining communications through the CMOT and CI Teams was making a major contribution to deterring criminal and violent political elements from operating freely.[213]

On 12 March, Alpha Company moved into town. They received a 'warm welcome' when a gunman fired at a patrol led by Corporal Terry Conner at the Bay Project Water Point on the first night.[214] There was no return of fire. The gunman was only keen on firing a burst and escaping quickly. Lieutenant Bob Worswick, accompanied by one of his sections, was sent into town to investigate this incident. He received information through local sources that an NGO guard had fired at Connor's men. The next day Worswick received permission to send Corporal Dion Jobson's section to search the NGO compound from where the burst of fire had come from. Their surprise visit was rewarded with a haul of 17 weapons, many more than were required by the guards working there.[215]

Unlike Blumer, Fraser preferred to maintain law and order in Baidoa through deterrence rather than conducting an active program involving searches and weapons confiscations, and the harassment of known criminals. Fraser was not as keen as Blumer to respond to CI information about where weapons might be and what houses and buildings should be 'busted'. Fraser tried to entice and motivate the newly-established Auxiliary Security Force (ASF) to increase their activities in Baidoa. He saw that the future for law and order in Baidoa lay with the ASF. He and his patrol commanders did what they could to give these reluctant, cautious Somali constables status in the eyes of the locals.

An incident on 15 March illustrated the dilemma being faced by the Australians who wanted to keep out of local political disputes and let the ASF sort out local law and order problems while they concentrated on their core business of deterring armed interference with humanitarian activities and protecting expatriate NGO staff. There was a confrontation in town between groups of women aligned to Aideed's USC faction and the dominant SDM faction. Stuart Dodds, happened to be in town with six mortarmen in two Land Rovers at the time. He had just exchanged one Landrover with A Company. The Land Rovers approached the melee cautiously. Two groups of women were facing each other angrily with a larger crowd watching—all the ingredients for a riot. Dodds could see over 20 rocks in the air at any one time as the women shouted insults at each other.[216]

Dodds directed the Land Rovers to push in between the two groups.[217] The rock throwing stopped. Dodds sent a radio message through Battalion Headquarters for the ASF to send some constables down to keep an eye on things so that the rock throwing would not start up again. Assured that the ASF had been alerted and would be arriving soon, Dodds and his mortarmen left the scene.

The Somali constables did not turn up. Soon the women went at it again—rocks in the air and scuffles on the ground. Several women sustained injuries from being hit by rocks, scratched and punched. This time Doug Fraser's headquarters was alerted. He sent a section down to disperse the warring women and the crowd that had gathered to watch. The Australians were the reluctant policemen again.

By the middle of March the focus of operations began turning from securing Baidoa to making the roads in the Baidoa Sector safe from ambush. Though difficult to prove conclusively, the outbreak of ambushing on the roads may have been a consequence of the urban operations that had deterred nomadic bandits from coming into Baidoa and other major villages to ply their violent trade. They were now having to make a living by ambushing vehicles on the roads. The contest in town focused on deterring NGO guards from causing trouble.

On 25 March, Delta Company took over from A Company in Baidoa. Fraser's tour of Baidoa had been quiet. Aside from Corporal Terry Connor's shooting incident two weeks earlier, Corporal Shaun Moore had apprehended a gunman who had fired shots at his patrol on 21 March and Corporal Chris Townson's section had captured four bandits who had been reported by the locals on 23 March.[218] Fraser was pleased with the effort and restraint shown by his soldiers:

> My soldiers worked extremely hard to achieve the aim of the operation, and whilst they would have liked to have had a fight, in every situation where this could have occurred, they instead disarmed individuals and defused situations, despite incredible peer pressure being placed upon them.[219]

Delta Company occupied 11 NGO compounds and began patrolling. The MSF compound and the nearby Bay Project continued to be a particular source of concern. Despite the Sniper Section's decisive intervention in February, bandits continued their villainy there. Local informants reported that MSF guards were suspected to have joined bandits in extorting money from those using the wells. There were also regular reports of break and enters in MSF accommodation and storage compounds and there had been a number of suspicious shootings in the vicinity. Despite the sacking of Lesto and his cohorts their replacements were suspected of being no better.

On the morning of 26 March, Blumer received information that bandits were going to be in the vicinity of the Bay Project that night and would use the MSF compound as a hiding place.[220] He decided to target the area and increase the pressure on the MSF guards. He tasked several patrols to include the MSF Compound and the nearby wells in their routes, and to question and search anyone behaving suspiciously.

At 10:10 pm Corporal Phillip Martin led a patrol of six men into the area of the wells and discovered two suspicious groups of Somali men, one of 16 and the other of six, loitering.[221] It was unusual to find groups of men near the wells at night; collecting water was women's work. Martin had been in the vicinity of the wells earlier in the day and had received reports from locals that a man in a red hat had been extorting money and had gone into the MSF compound.

Martin decided to split his patrol into two groups of three to conduct a visual search of the two groups and 'have a chat'. The group of six Somalis were suspected of being bandits but they were not carrying arms at the time. After some inconclusive conversation, Martin decided that further contact with the Somalis would not achieve anything. Hopefully, the presence of the Australians would deter them from intimidating the women queuing for water. The pattern had been that when the Australian patrols were in the vicinity women drew water. However, they quickly dispersed after the Australians left.

Martin's Gun Group, led by Lance Corporal Steve Chinner decided to use the pause in proceedings to look into the MSF compound to see if anything suspicious was going on among the guards.[222] The patrol had been briefed before leaving that the MSF compound was suspected to be a hiding place for bandits who operated around the wells. Like most NGO compounds, the MSF compound comprised a group of buildings surrounded by a brick wall about three metres high. Private Ashley Franklin, a machine gunner, was the first to jump up and scale the wall of the compound and look in. He noticed two people lying on beds in the courtyard.[223]

Soon after, Chinner did the same and looked over the wall. He shone his torch on the two people Franklin had noticed—a man sitting on the end of a bed and a person lying on a bed nearby. Exposed by torch beam, the man reached for a rifle under the bed sheet and brought it up in Chinner's direction. Chinner called out in Somali for the man to stop moving. Despite the warning, the Somali continued to raise his weapon in Chinner's direction. Franklin came off the wall at this point. Chinner decided that the man was about to fire at him so he fired a burst at him from his Steyr rifle and jumped back off the wall.

On the sound of the Steyr firing, Private Alex Kowarik-Berger, the No 2 rifleman accompanying Franklin and Chinner ran to the wall and looked over.[224] He saw a Somali man moving towards the wall with a gun raised in his direction. Kowarik-Berger fired at him and jumped back off the wall.

Both Chinner and Kowarik-Berger felt the impact of rounds hitting the other side of the wall where they had dropped off. As far as they were concerned they were now in contact with at least two armed Somalis, one of whom had not responded to verbal warnings and continued to raise his rifle in an aggressive manner, and the other who had pointed his rifle at Kowarik-Berger in a threatening manner.

Chinner, while clearing a stoppage in his rifle, ordered Kowarik-Berger to put suppressive fire into the compound to keep the Somalis heads down. Rather than risk putting his head over the top of the wall again, Chinner threw a grenade over the wall in the direction from where he assessed the return fire was coming from.

By this time Blumer had begun moving to the scene and had warned out his Quick Reaction Force, commanded by Lieutenant Don Tully, to deploy to the MSF Compound. He wrote:

> We got up to CRS house [Catholic Relief Service] when all shit broke loose to the SE [South East], heavy repeated fire, answered by our lighter Steyr staccato and automatic fire and grenades going off. ... The contact continued as we moved up the road and then we had another one to the NE [North East] around 10 Platoon location. I reacted the QRF and got a SITREP [situation report] enroute. As we moved past the MSF compound I heard a weapon cock and came very close to moving that one step closer to engaging. It was OC 11 Platoon [LT Don Tulley] waiting with the QRF for the APCs. Annoying and very dangerous. Anyway.

While Blumer was moving towards the MSF compound and the Quick Reaction Force sections were getting into their two APCs, commanded by Lance Corporals Peter Larsen and Glen 'Bear' Gough, Corporal Martin, moved with his two scouts, Privates Ron Dann and Craig Russell to the front gate of the compound.[225] Kowarik-Berger continued to fire bursts into compound, hitting the opposite wall. He heard a number of people escaping over a sheet of corrugated iron at the rear of the compound.

Martin decided to go after the Somali gunmen in the compound by jumping over the wall near the front gate. After some discussion with Chinner over the squad radio, Martin ordered him to throw another grenade and for Kowarik-Berger to provide covering fire. After the grenade exploded, Martin planned to scale the wall with Dann and Russell and jump into the compound. About 8 minutes had elapsed since Chinner had fired his burst at the Somali shouldering his rifle on the end of the bed.

After Chinner's second grenade exploded Martin, Dann and Russell scaled the wall and jumped into the compound. On landing, Dann fired three shots into the compound. Martin and Russell landed near him. There was no return fire. Leaving Russell to open and cover the front gate, Martin and Dann began searching the compound for the gunmen. This was a very dangerous and tense time, moving in the dark, in an unfamiliar area, looking for two gunmen who might already have them in their sights, or were hiding in wait. After searching the compound, Martin and Dann assumed that the gunmen had escaped. Kowarik-Berger reported that he had heard the sound of people running over a sheet of corrugated iron. Martin and Dann had found a sheet of iron on a possible escape route.

Martin, Dann and Russell left the compound to investigate the suspected escape route beyond. Russell was left at the front gate while Martin and Dann went a short distance across the road. While this was going on, Chinner ordered Franklin and Kowarik-Berger to set the machine gun up at the front gate and cover the area in front of the MSF compound.

Martin and Dann began searching a nearby house occupied by a man and a group of women who had been woken by the firing and grenade explosions. Martin wanted to be sure that no one was hiding there who had been in the compound. At the same time, Chinner decided to take Russell back into the compound with him to conduct a more thorough search of the main building. Chinner also gave the D Company QRF, mounted in APCs about 100 metres away, the 'go ahead' to move to the front gate of the compound but to remain mounted. This message was not heard by the QRF commander, Don Tulley, who decided to dismount against the advice of his APC crew commander, and have his men approach the compound on foot.[226] Franklin and Kowarik-Berger stayed at the front gate, covering the approaches to the compound.

Murphy's Law now began to operate. While members of the QRF moved in, Chinner and Russell rounded a corner of the main building in the compound. Suddenly they were confronted by an armed Somali who moved out from an open doorway into Chinner's torch beam. The Somali turned towards them with his weapon raised. There was no time to decide whether he would fire at them or not. Martin and Russell assumed that this was one of the men who had fired back earlier. As the Somali turned, Russell instinctively fired three shots at him from the shoulder and Chinner went down on one knee and fired an aimed burst which killed the man instantly.

Just as Chinner and Russell shot the gunman in the compound, another Somali gunman fired a burst of 7 to 10 rounds at the approaching members of the QRF. One of the diggers fired two illumination rounds from a 40 mm M79 Grenade Launcher over the compound. Private Phillip Denby yelled out, 'He's behind the wall. Can I throw a grenade?'[227] Don Tulley asked Denby if he was clear to do so. Denby called back, 'Yes'. Tulley then ordered him to throw a grenade into the compound. Denby did so and called out 'Grenade'![228] At this time Chinner was inspecting the dead Somali gunman and Russell was in the vicinity.

Hearing the order for a grenade to be thrown into the compound, Martin raced towards the front gate from the nearby house shouting, 'Ceasefire, friendlies'. He was too late. The grenade landed near the front gate and exploded sending a few small shards into Private Franklin's elbow causing a minor wound and spraying debris on Private Russell who was slightly injured.

Blumer arrived soon after the grenade exploded:

> When the QRF closed up with 42B [Martin's patrol], it was on again.
> Another grenade goes off and then, 'Cease fire, friendlies.' An awful

feeling hearing that on the radio. I shouted, 'Cease fire', as well. Then, 1 KIA [killed in action] and 1 WIA [wounded in action]', came over the [radio] net. Thankfully we soon established that the KIA was Somali but the WIA was ours, from our own grenade.

Blumer took charge of the contact area immediately. Martin administered First Aid to Franklin. Russell cleaned himself up. After Martin had briefed Blumer and Tulley on the contact, Tulley conducted a search of the buildings in the compound and discovered a blood trail inside a rear door and a Somali woman cowering in the veranda area of the main building. The woman was treated for shock and apprehended.

Russell returned to duty and Franklin was taken back to the airfield for further observation and treatment. Blumer also moved to the airfield to make his report on what was an untidy incident which was difficult to piece together. The woman from the compound was also taken back to the airfield and spent the rest of the night resting at the medical facility monitored by two female nurses.

Blumer wrote:

> I began making my report when CI told me that the woman we brought in from the compound claimed that the Australians took the dead guard out and shot him/murdered him. Really threw a different light on things and that caused quite a bit of reservation on the issue. I sat and listened to the woman and she had holes all through her story, so basically I did not believe her. It really put a big downer on the whole story though. I was quite bummed out.[229]

The claim that Australian soldiers had murdered a Somali guard created quite a furore at Battalion Headquarters. John Caligari directed Thomson and Douglas to speak with the woman and either confirm or deny her claims. After only a few questions, Thomson and Douglas identified, 'large discrepancies in her story and she subsequently retracted her claims. Further questioning revealed that the two Somali men involved in the contact were MSF guards who had reacted when Chinner's torch had shone on them. She claimed that she and the guards thought they were being attacked by a group of bandits.[230]

On 27 March a thorough investigation began into Corporal Martin's contact and the subsequent contact with Don Tulley's QRF. Later in the morning, after discussions with Thomson and Douglas, MSF officials brought in the second MSF guard who had escaped. He was armed with a G3 rifle which was the weapon used to return fire at Chinner and Kowarik-Berger, and subsequently to fire at Don Tulley's men. Questioning by Thomson and Douglas left little doubt that this man used his weapon in the contact.[231]

As March drew to a close Hurley learned that Warsame, one of Baidoa's two major criminal gang leaders, who had been arrested a few days before and sent to Mogadishu Prison after threatening to kill Australians, had been released after Gutaale, the other criminal gang leader in Baidoa who had strong affiliations with General Aideed's political faction, had gone to Mogadishu and bribed a judge and a jailer. Gutaale had returned to Baidoa. Hurley did not want to give Gutaale the impression that his actions had gone unnoticed, or that his previous crimes would be ignored. He directed Blumer to take Gutaale into custody and search his office and houses for further evidence of his illegal activities.

Blumer's diggers searched Gutaale's office at the Hotel 'Favourite' and the adjacent brothel where weapons and documents had been discovered in earlier raids on 29 March. Hurley also decided to reinforce Blumer's company with Dodds' Mortar Platoon. He was concerned that Gutaale and Warsame would try and organise some 'pay back' on the Australians who were not only putting pressure on their criminal operations but also gathering evidence to put them both in gaol. The CI Teams were now protected by an infantry section during their visits into town because they assessed that Gutaale and Warsame had probably deduced that they had been duped into divulging incriminating information about themselves.[232]

Gutaale's luck eventually ran out. The pleasure of capturing him was to belong to Thomson and Douglas. On the morning of 29 March they were driving along NGO Road in Land Rover with an infantry section in the back, when they recognised Gutaale's car approaching from the opposite direction.

The Australians stopped and waited. Just as Gutaale's vehicle came up to the Land Rover it had to stop for a donkey cart, right alongside. Thomson and Douglas got out immediately, dragged Gutaale out from behind the wheel and bundled him in amongst the diggers in the back of their vehicle. They drove off leaving Gutaale's car with its engine running, driver's door open and a group of shocked passengers with mouths agape in the back seat.[233]

Arresting Gutaale was a major step in keeping criminal elements in Baidoa on the back foot. Blumer's diggers had found sufficient arms and ammunition during their search of his offices to justify laying charges. This physical evidence combined with the affidavits from locals about Gutaale's murders and other criminal activity would be sufficient to bring a case against him in the re-established local court.

Gutaale was flown to Mogadishu and put into detention while Major Mick Kelly, the AFS Legal Officer, and Sergeant Peter Watson, of the AFS Military Police detachment, continued to gather evidence.[234] In the following days many death threats were reported to have been made against the Australians and Gutaale's gang were reported to be planning to avenge their leader's capture. However, 'amid fears that they would be the next to disappear into Australian hands', Gutaale's gang soon dispersed, with most packing up and leaving Baidoa.[235]

The arrest of Gutaale and the departure of his gang was the third major security coup for the Australians in Baidoa. The first had been the shoot out at the pump house with Gaaney and his gang that left Gaaney with a head wound. The second was the departure of key figures in General Aideed's political faction after they had spread disinformation about an advance on Baidoa by Omar Jesse's troops and received an ultimatum from Hurley, delivered by Wayne Douglas. Aside from these three events, the contacts in February and the MSF compound shootout in late March probably reinforced in the minds of criminals, bandits and NGO guards that the Australians meant business. By the end of March the Australians appeared to be well on top of the security situation in Baidoa.

The beginning of April marked a turning point as well as a high point in Australian-Somali relations in Baidoa. Security operations had been successful in creating a secure environment and mutual respect. There was renewed confidence in the community about the future. Local elders were more confident about asserting their authority. This new confidence in the community was highlighted by a community clean up day organised by Dick Stanhope. Personnel and vehicles from Administration Company were tasked to provide support for this community effort to change their municipal environment.

Another example of bridge building between the Australians and the local Somalis was a soccer match planned between an Australian team and a local Somali student group, called the Young Intellectuals, at the main soccer field on 2 April. The RSM, WO1 Greg Chamberlain organised a demonstration game of touch football as a curtain raiser. Planning soccer matches between Australian soldiers and the local Somalis would have been unthinkable eight weeks before. A lot had happened in just over two months in Baidoa.

The community clean up began on 1 April under the direction of Administration Company's 2IC, Captain Steve Grace.[236] All spare trucks and personnel were deployed to load rubbish, old car bodies and other wreckage from the civil war, and take it to the airfield tip. There was an upsurge of community spirit as Somali men women and children assisted in loading the trucks. Local leaders were put in charge and directed activities.[237]

On 2 April, a crowd of 3 000 Somalis attended the soccer match on a red dirt field in town. Several expatriate NGO staff played for the local team and received enthusiastic support whenever they kicked the ball. The joy of the crowd when the local team scored and maintained a 1-nil lead for the first half was followed by disappointment when the Australians scored three times early in the second half. A late goal, assisted into the net by several spectators, took the edge off the defeat and emphasised the informality of the game. The crowd took special delight when the Captain of the Australian team, Captain Tony Annetts tripped on a rock and had to be carried off' the field on a stretcher with a sprained ankle.[238]

About an hour after the soccer game finished and night had fallen, Lance Corporal Shannon McAliney directed his scout to open up the gate of the CONCERN compound and move out to begin another routine night patrol.[239] The scout cocked his weapon and, because he was the first man in the patrol, left the safety catch off so he could fire instantly. As he made his way out the gate he realised that he had left his bush hat in his small 'bum pack'. He turned to McAliney and asked him to hold his rifle while he took out his bush hat. Holding the rifle by the pistol grip he extended it forward to McAliney who reached out and took it. At point blank range the rifle discharged. A bullet ripped through McAliney's flak jacket and into his stomach.

Corporal Paul Nunan and Corporal Graeme 'Bud' Wehmeier heard the shot and a moment later a soldier from McAliney's patrol rushed up to them asking for their help:[240]

> ... Worm [Wehmeier] and I started giving him First Aid and patching him up. It was the worst thing I have ever seen or had to do. There was blood everywhere and, as I said, it was the worst thing I have ever experienced. He was still conscious whilst we were trying to fix him up and some of the words he was saying were horrifying. He kept saying, 'Help me please' and 'Oh God, I'm going to die'. He knew he was going to die but I kept telling him that he would be OK.[241]

The call for Shannon McAliney's urgent evacuation went out immediately at 6:50 pm. Lieutenant Mick Toohey picked up Blumer, his CSM, Ken Hudson, the D Company medic, Lance Corporal Ross 'Ossie' Osborne, and the Company Clerk, Corporal Kevin Brown, in his APC and was at the scene within a few minutes.[242] He used his APC spot light to illuminate the area while McAliney was being treat [243] Ossie Osborne took over from Nunan and Weiemer and bound up McAliney's wounds in preparation for his evacuation to the airfield in Toohey's APC. Shannon McAliney remained conscious during his ordeal. Thinking that he was going to die, he looked up at the circle of concerned faces around him and said, 'Tell Mum and Dad that I love'em. Swany, you're my best mate'. Toohey recalled:

> We drove back to the airfield at best speed. Smithy was flashing his high beam and low beam and I had the spotlight on flashing it from side to side in an attempt to get the locals out of the way[244] The cargo hatch was open and I could distinctly hear the OC [Blumer] yelling, 'You're not going to die' and 'You'll be alright', fairly regularly. At one stage we had the elements against us, a storm front blew in and, although it wasn't raining, it was like driving through a dust storm. Smithy did a top job. A young Somali was nearly run over when he was stunned by the white lights.[245]

Near the front gate, Toohey stopped his APC just long enough to lower the back ramp and pick up Captain Darren Keating, the Regimental Medical Officer.[246] Keating assessed that the chances of McAliney surviving were slim. At that range the bullet had caused massive internal wounds despite the protection of his flak jacket. Notwithstanding this grim prognosis, everything was being done to give McAliney his best chance of survival. The Americans of the 159th Aero-Medical Evacuation Company put a helicopter into the air while he was being evacuated.

On arrival at the Treatment Section, Darren Keating was joined by Darrell Duncan. All that could be done was to administer oxygen, stabilise the wound and comfort McAliney in his dying minutes. At three minutes past eight Shannon McAliney, aged 21 years, died. After spending 15 more minutes trying to revive him, the two doctors finally stepped away, exhausted and defeated.

Toohey had waited outside the treatment section drinking a cup of coffee with his driver. Blumer came out and told him that McAliney hadn't made it. Toohey recalled, 'Those few words made me feel "numb". To have been a part of the incident, doing my job and seeing the casualty alive and still fighting when he left my APC, it was a shock to find out he died ... The things I will always remember ... seeing the blood-soaked shirt being removed, the OC [Blumer] yelling encouragement, the blood in the back of my APC and the numb feeling, but not being upset when told he'd died'.[247]

Ant Blumer reflected on this tragedy in his diary:

> I stayed with him talking—about anything until he died. ...The medical people worked hard to keep him going but it wasn't to be, too much damage and he just went. Such a stupid and awful waste. I went back to tell his section and platoon who were all quite taken aback and stunned. Poor _ must be feeling like no one could understand— Poor, poor young digger ...I'm tired and deeply saddened by this affair. Such a disgusting waste ... CPL McSwan came to see me [the next day].[248] He was really cut up and upset and asked what had actually happened. He wanted to know himself for sure and not what rumours he had heard. I went through everything with him as I saw it and what I thought of it which seemed to help him a bit. He let it out for a minute and my heart goes out to him. They were flat mates and quite close.[249]

Lieutenant Mick Toohey, B Squadron, 3rd/4th Cavalry Regiment.
Image: 1 RAR Unit photographer.

Shannon McAliney's death from this tragic accident was the single most depressing event for the vast majority of Australians in Baidoa, including those who did not know him.[250] It had a unifying as well as a universal effect. Everyone openly shared their sadness and took time to console each other and to commiserate with each other on their feelings about the tragic accidental death of such a popular young man. His memorial service on 4 April unleashed a wellspring of emotion and deep sadness. Many of those who attended were moved to tears during Don Tulley's eulogy which concluded:

> The men of Delta Company knew 'Mac' was the sort of man that would always bring a smile to your face, he was a leader, a worker and, importantly, a mate. He gave his all to the section and platoon. He loved his job. We will remember you 'Mac', you'll always be a part of the team.[251]

On 5 April Blumer wrote:

> Back to company location and write a letter to LCPL McAliney's mother. It's something I've only ever briefly thought about and never really thought that it would be a letter I would have to write. It is certainly an uncheerful and saddening thing to do. I spent a long time thinking about it.[252]

Meanwhile the Reaper was still about. On 3 April, Charlie Company took over security duties in town. That afternoon an APC, commanded by Corporal Luke Entink, broke its Universal joint and the prop shaft speared up into Trooper Craig King's driver's compartment floor, fortunately without injuring him.[253] The APC careened out of control and came to rest against a wall, hitting an elderly Somali man and crushing him to death.[254] That night a single shot was fired into one of the compounds Moon's troops were occupying. The dangerous work wasn't over yet.

At 1.30 am on 5 April Lance Corporal Gary Lively led a patrol down an alleyway nicknamed 'Sniper's Lane.'^[255] The patrol had stopped to listen. Lively heard the sounds of at least two Somalis arguing at the top of their voices. One telling the other to get down and the other screaming. He went

Lance Corporal Shannon McAliney, D Company, 1 RAR died on active service in Baidoa, Somalia on 2 April 1993
Image: Gary Ramage.

forward to investigate. Suddenly a single shot was fired in his direction and Lively returned a single shot. This was followed by a burst of automatic fire that impacted a few metres away.

Patrol members then fired bursts in the direction of where the fire had come from. Lively and his men saw several Somalis scattering to the left and right away from the entrance to the alleyway. Concerned that the Somalis might come up on him from a flank or get around behind his men, Lively ordered his men to cover each other and withdraw down the alley. Nothing more happened. Later conversations with locals revealed that a Somali had been murdered near where Lively had exchanged fire and a group of bandits associated with the MSF guards had fled from the scene. The bandits had fired in panic when they noticed Lively approaching them up the alley.[256]

The two contacts on two successive nights put Moon's diggers on full alert for a busy tour in town. Like Blumer, Moon believed that security operations in Baidoa should be proactive and keep the pressure up on local criminals and visiting bandits. On 6 April he sent Corporal Tom Aitken's section on a search of NGO compounds and warehouses. They discovered several rifles and a store of ammunition. More importantly they uncovered a large quantity of medical supplies including 120 Intravenous bags and 6 000 syringes that had been stolen from UNICEF. This cache of syringes enabled a stalled inoculation program to go ahead as planned.[257]

With Hurley's concurrence, Moon mounted an operation to 'shake down' the NGO compounds in NGO Road. Charlie Company's searches resulted in illegal weapons being found at the UNOSOM Compound, the World Vision Compound, the CRS Compound, the CONCERN Feeding Centre and the GOAL Compound. Moon felt that these confiscations proved that the major security problem faced by the Australians was the illegal activities of NGO guards. It was also possible that the NGOs were getting in weapons for their own security after the Australians left in May. Many experienced NGO managers may have been planning for the time after UNITAF forces left and had possibly sought to conceal additional weapons from the zealous Australians.[258]

One of the most pleasing aspects of security operations in April from many members of the 1 RAR group was the trust that had developed between themselves and many Baidoans. The locals were more confident about giving the Australians information they needed to catch bandits 'red handed.' One example of the rapport between the Australians and the locals was played out on the morning of 7 April. After receiving information that a group of 15 bandits who had entered Daynuunay earlier that morning and looted a storeroom were still in the area, Mick Moon despatched Captain Mark Hankinson, his 2IC, with a Company Headquarters group in two Land Rovers and two sections of Lieutenant Todd Everett's 9 Platoon, his QRF, in APCs to a vehicle checkpoint on the outskirts of town.[259]

After arriving at the checkpoint, Hankinson interviewed local elders who informed him that the bandits had robbed the store in Daynuunay at 5 am and had left for Baidoa carrying their booty. They estimated that the bandits would not have reached the town on foot yet with the loads they were carrying. Hankinson reported back to Moon who directed him to follow up the bandits and try to cut them off. Hankinson directed Everett to follow the bandit's route across country in APCs while his group went by road in Land Rovers to Junet, a small village on the route into Baidoa.

While driving back through Gasarta and towards Junet, local camel herders signalled to Hankinson's vehicles to stop. They reported that the bandits had been through and had stolen milk before continuing on to Baidoa along a track to the south west. Closer to Junet, another camel herder reported that the bandits had headed up a track only a few minutes before. Hankinson dismounted with Sergeant Steve Boye and Corporal Darryl Heaslip and, accompanied by the original informer and the camel herder, patrolled brusquely forward on foot for about 600 metres.[260] Along the way they found a World Vision ID Card for Hassan Guudow Hassan. Hankinson recalled, 'Since we had positive ID of one of the bandits I called off the search and started home, we could now pick up this guy and his mates anytime. As we came out of the creek line just south of Junet we saw 9 Platoon in their carriers going hell for leather towards a creek across our front.[261]

While Hankinson had been following up the bandits along the track, Everett had been approaching Junet from a different direction in APCs. He was also stopped and told by locals that the bandits were in a creek line up ahead. Everett ordered Corporal Dean Caple's section to shake his men out into an extended line and move forward on a broad front into the creek line, weapons cocked and ready.[262] The two locals who had informed Everett about where the bandits were holed up acted as guides. As they went into the creek line, Caple's men observed a group of Somalis gathering water. The guides pointed two men out as being bandits. One bandit was grabbed immediately. He was carrying a sack which contained stolen goods from the Daynuunay raid. Caple chased the other bandit who ignored calls to 'Stop'. Caple fired two warning shots over the fleeing bandit's head and two more into the creek bank ahead of him. In his panic to escape the bandit ran straight into a camel thorn bush and entangled himself.

While Caple was detaining this bandit, two of his riflemen and Sergeant Michael Morrissey[263] were chasing two other bandits who had begun to run after the first shots were fired and locals drew the Australians' attention to their flight. Morrissey called out to the bandits to stop and, when they did not, he fired two warning shots. Two APCs, commanded by Corporal Ray Moore and Lance Corporal Peter Soppitt[264], joined in and fired bursts of .30 calibre machine gun rounds ahead of the bandits who were ducking and weaving through clumps of camel thorn. Morrissey continued to run after the bandits when he was suddenly confronted by a bandit breaking cover in front of him. The bandit had not seen Morrissey and ran close to him.

Just as surprised as the bandit, Morrissey instinctively lashed out with his rifle and butt stroked the bandit in the head, knocking him down. A rifleman near Morrissey detained another fleeing bandit, pointed out to him by the locals, by tackling him rugby-style and bringing him down.

Leaving the bandit dazed and guarded by the locals, the rifleman leapt to his feet and chased after two other bandits who by now were 50 metres away. He fired two more warning shots. At this time Hankinson, who was on the left flank, fired at the two fleeing bandits from a range of 150 to 200 metres, hitting one bandit who was carrying the AK 47 in the leg and wounding the other who 'stumbling to the ground, fell over and dragged himself off.'[265] Hankinson's firing resulted in the rifleman chasing the two bandits going to ground until he was able to ascertain where the firing had come from. When he realised it was Hankinson and his group, he went up to the bandit who had been shot in the leg and administered First Aid. The other wounded bandit made his escape.

In a separate chase, two other bandits fired at their pursuers and there was a brief exchange of shots before the bandits were able to make their escape. In all, Everett's men detained four bandits from the group of eight that had been surprised and captured one rifle. A subsequent search through the area uncovered a significant quantity of stolen items and a large amount of Somali cash. Later questioning of the detainees revealed that two members of the bandit gang worked as World Vision guards. Locals from the Daynuunay area confirmed that one of the bandits had murdered several people in the local area over the previous months.

Chasing and detaining bandits was the 'order of the day' for Charlie Company on 7 April. In town, locals pointed out a group of seven bandits to Corporal Bill Perkins who ordered his section to rush them immediately because they had become suspicious and were leaving the area. A chase ensued but because of the danger to other people in the area no warning shots were fired. Four bandits were detained and three others made their escape.[266] Moon was very satisfied with his men's work. Charlie Company had captured eight bandits, confiscated caches of weapons and ammunition in NGO compounds, returned a large quantity of medical stores to UNICEF so that an inoculation program could go ahead and maintained pressure on NGO guards suspected of engaging in criminal activity. In all cases information from the local people had led to success. In his opinion, the tide of public sentiment had turned in favour of the Australians.

The only sour note from Moon's point of view was the refusal of the ASF to keep bandits locked up in their new detention centre for fear of reprisals from other members of their gangs. The Australians still had no authority to detain bandits for longer than 24 hours in Australian facilities at the airfield. Typically, bandits passed onto the ASF were released within 24 hours without facing trial. The Australians countered this by sending bandits charged with serious crimes, such as murder or rape, to Mogadishu Prison for detention and trial. This also proved to be ineffectual after several were released from there after prison officials were bribed by members of their gang or relatives.[267]

The indecisiveness and weakness of the ASF was a constant source of frustration for the Australians in town. This was brought to a head when another notorious bandit named Gaardub was released twice, the first time after being captured by the Reconnaissance Platoon during a patrol to the west of town and the second time after being detained by one of Moon's patrols. During an operation to recapture him, there was a scuffle and Private Allen was injured.[268]

Another source of frustration for the Australians was the persistence of UDs. Blumer was particularly angry when one of his soldiers was involved again on 12 April:

> __ has a UD with a 9 mm pistol. He fired a shot through the floor of an NGO vehicle ... I was just so angry that we can have negligent weapon handling, especially when we have just had a friendly death [McAliney] from exactly the same thing ... Again I have to go to the CO and talk to him about another UD. I die a thousand deaths when that happens.[269]

Charlie Company's efforts were rewarded at the end of their tour in Baidoa when the town recorded several consecutive quiet nights. Moon handed over to Jim Simpson's B Company on 15 April. However, the problems with some NGO guards persisted. Andrew Pritchard wrote on 22 April:

> We have moved from the World Food Program to the local GOAL compound. This is the original site we occupied when we arrived in Baidoa town proper for our first urban tour.
>
> The only change is the locals guarding the compound. They appear very suspicious. I have upgraded weapon readiness to 'Action', whilst inside the compound. For my money, they are given too much responsibility after reports of treachery from other NGO guards. They are armed to the teeth and carry weapons as if they are play things ... I can't help but question putting our forces in NGO compounds already crawling with local guards. Instead of resting, most of the time inside the compounds is spent wondering what the local guards are up to. All day long they gibber away to themselves, point to us and laugh. I had to stop _a few times from taking violent action with the guards when they became particularly cheeky. Although maybe once would do the trick. They seem to have lost any fear they had originally of our weapons or ourselves.[270]

Fortunately, the quiet period in Baidoa coincided with a visit by a delegation of Australian Federal Parliamentarians made up of Bob Halverson, David MacGibbon OBE, Roger Price and Neville Newell. The visitors were escorted by Colonel Bill Mellor and they were impressed with the professionalism of the Australians and the praise from NGO representatives about the success the Australians were having in creating a secure environment for their operations.

On 25 April, ANZAC Day was celebrated in the traditional manner at Baidoa and at ceremonies in Mogadishu. Major General Peter Arnison , Commander of the 1st Division, was a guest.[271] ANZAC Day celebrations created a great boost to the morale of the troops in Baidoa. There was a Casino games night on ANZAC Eve organised by Lieutenant Shaun Voss and his troop from B Squadron.[272] Somali money was used as 'Monopoly' currency with each player beginning with 500 shillings. On ANZAC afternoon, B Squadron also organised donkey races using livestock they hired from in town.

That night it was business as usual in Baidoa. Corporal Wayne Prosser, led an eight-man patrol down NGO Road at 3 am.[273] He and his men were randomly searching Somalis congregated near the entrances of buildings. Many were young Somali men spending another night chewing *khaat* and listening to music. Prosser's forward scout, Private Bramwell Connolly, observed two men moving away from the patrol in a suspicious manner and called out to them in Somali 'Joog So'! [Stop].[274] They continued walking. Connolly moved up to one of the men and gripped him on the shoulder. He turned around and brought up a L2A3 Sterling Sub Machine Gun into Connolly's stomach and fired the action. Fortunately the round in the Sterling's chamber was 'hard struck' and did not go off.[275] Connolly grabbed him and the weapon and wrestled the man to the ground. Prosser stepped in and helped Connolly subdue him. The second man was detained by other members of the patrol as he tried to escape into a hotel across the street.

Prosser saw two MSF vehicles outside the hotel and a number of people milling suspiciously outside the entrance. He asked for and received permission by radio from his Company Commander, Jim Simpson, to search the hotel. While the search was being conducted two Somali men approached from the rear of the patrol and then quickly changed direction when they saw the last members of Prosser's patrol, Privates Matthew Greene and David Hawkins.[276] Greene and Hawkins called out for them to stop and, when they didn't respond, began to follow them.

Greene apprehended one of the men and began searching him. During the search, the other man, who was sitting sullenly on the other side of the alleyway, reached behind him and pointed what looked like a pistol at Greene.[277] Hawkins saw the movement from a few metres away, brought up his Steyr and fired a three round burst instinctively at the man, hitting him in the stomach. After the burst, Prosser supervised the administration of First Aid by the B Company Medic, Corporal Peter Vigar.[278] After he was stabilised, the stricken Somali was evacuated in a Land Rover to the IMC hospital. He had sustained a fatal wound and died a few minutes later.[279]

On 27 April justice was finally meted out to the notorious criminal, Gutaale. He was tried by Somali judges under the provisions of Somali Law drafted in the year of Somalia's independence in 1962. He was found guilty of a number of murders and sentenced to 20 years in prison. His appeal against this sentence was unsuccessful, and in an unusual turnaround by Western judicial standards, he was sentenced to death by the judges and summarily executed. This execution was another violent milestone in creating a secure environment in Baidoa. A clear message was sent to local criminals that they could be brought to justice for murder. For many of the families of loved ones who had been killed by Gutaale there would have been some comfort in the knowledge that he was no longer a threat to them or to others.

Events in Mogadishu were the focus of interest in the early days of May. An Army legal team led by Colonel Les Young had arrived from Australia with the Prosecuting and Defending Officers, Majors Mick Griffiths and Roy Abbott, to conduct a Court Martial for two members of the 1 RAR Group who had gone Absent Without Leave in Mombassa several weeks before and for the scout who had been charged with dangerous and prejudicial behaviour over the accidental death of Shannon McAliney.

The case of the two soldiers who had overstayed their period of leave in

Mombassa had received wide publicity in Australia because of speculation that they might be found guilty of the serious offence of desertion, because they were on active service. Members of the AFS were concerned that the foolish actions of these two soldiers would tarnish the excellent reputation they had established through their work in Somalia. In the end, both were found guilty of being AWOL only and received periods of detention and fines. By trying them in Mogadishu, the Army avoided their Court Martial being sensationalised by the media in Australia. McAliney's forward scout was found not guilty of all charges because there was insufficient evidence to prove any wrongdoing or neglect on his part.

In the days leading towards the end of the Australian tour of duty, Baidoa remained quiet and everyone hoped that there would be no incidents to complicate the redeployment back to Australia. There were several quiet, private farewells with those in the community that were recipients of Australian kindness. Two weeks before Sergeant Andy McCarthy, Corporal Tom Britton, and Privates Michael Dale and Andrew Pattullo had collected surplus stationary items, such as pencils, pens, paper and writing pads, and driven to the Bartamaha school in Baidoa, one of the four schools the Irish NGO, CONCERN had reestablished.[280]

Mr Ali Mahummud, the Headmaster, had greatly appreciated the Australian generosity. The school was established in 1958 but had been closed after its buildings were bombed during the civil war in 1990. It was only after the Australians had provided a secure environment in Baidoa that CONCERN had decided to open the school and re-employ teachers. Now that the Australians were leaving, Ali Mahummud, said that there was a strong fear that the teachers would be murdered when the clans renewed their violent political struggle. 'Our group left with mixed feelings of satisfaction and regret, of having made a difference to these people yet leaving the job unfinished.'[281]

It fell to Delta Company to cover the Australian withdrawal and hand over to French paratroopers. On 12 May, Lieutenant Jan Van Der Klooster's platoon was assigned to man observation posts and the Main Gate to the airfield. Blumer had told them to be especially vigilant because the criminals in Baidoa who knew the Australians were withdrawing in a few days time might give the troops covering the withdrawal a 'farewell gift'.[282] Two incidents during the withdrawal constituted what Blumer described facetiously as 'farewell gifts': one to remind the Australians that Somali society was still rooted in ancient customs of violent retribution and the other to show that after their departure bandits were already poised to close in on Baidoa again like prowling hyenas.

Van Der Klooster's diggers had not been in position for very long when he received reports that a crowd was gathering in the warehouse area near the Main Gate. He wrote later:

> Over the next half hour the crowd increased, then all of a sudden there was heaps of yelling and screaming and the crowd became rather excited. A call from the warehouse informed me that it looked like the crowd were stoning somebody. We had been told to stay out of Somali politics and religion—a stoning is usually a result of both. However, my men were not going to stand by and watch someone stoned to death, and neither was I.

> I gathered four soldiers and we ran down the road towards the crowd. The rest of the platoon was manning OPs [observation posts] or escorting visitors inside the base. Upon seeing our approach most of the Somalis ran off but a few decided to stick around and throw rocks at us, so we replied in kind and they too backed off.

> When the crowd cleared two 16 year old girls were left, both or whom were the victims of the stoning. They had large lumps swelling up on various parts of their bodies and one girl was spitting blood and broken teeth. The crowd now stood off, watching and were angry but strangely silent which made it very eerie. As there was five of us and a few hundred of them, we didn't feel very secure, so I decided we would grab the two girls and withdraw to the front gate. Pointing our weapons at anybody and everybody we got back to the front gate with the crowd following 150 metres behind.[283]

Van Der Klooster radioed Battalion Headquarters and, after describing the situation, asked what he should do with the girls. He was told that it was his decision. 'I was rather surprised by this answer which left me wondering what the hell to do with two teenage girls who had a couple of hundred people chasing them, trying to stone them to death'.

He called for the company interpreter, Mohammed, to find out why the crowd were so angry with the girls. He wrote:

> We discovered that the girls had been seen inside the warehouse with members of C Company just prior to the company departing Baidoa. As they had left the building they had been chased by some of the boys who were selling cigarettes and Pepsi outside the base. The boys were calling these girls 'harlots' and soon everybody who was in earshot got into the act. Before the girls could make their escape, they were apprehended and after some accusations had been levelled at them, the crowd decided to stone them there and then.

Van Der Klooster faced an acute moral dilemma. He remembered watching coverage of a similar incident on the news in January before the 1 RAR group had deployed. A Somali woman was accused of sleeping with French soldiers in Mogadishu and was being stoned by a crowd before some French soldiers stepped in and saved her from an horrific death. The French were forced to release her and she was tried by a Muslim court, found guilty and summarily executed. 'I didn't want these young women to suffer the same fate.' Delta Company was due to leave Baidoa the next day so he decided that all he could do was guarantee them a safe trip to the Baidoa Police Station where they could be protected from the crowd until newly-established judicial authorities decided what to do with them:

> We bundled them into the back of a Land Rover and covered them with blankets ... When we attempted to drive down the road, we discovered that the Somalis had pulled large rocks and poles across the road as barriers. I sent the vehicle back and myself and some of my soldiers clear the barriers while the rest of the platoon covered us. Although I was worried about the two women, I was more worried

that the roadblock was constructed deliberately to set us up for a sniper's bullet. Needless to say we were happy chappies when the Rover got through, dropped the girls off at the Police Station and we returned safely to the airfield. That was the last we saw or heard of them.

'Thinning out' forces during a withdrawal is a risky business. The thinning out of Australian forces around the airfield meant that the number of observation posts were reduced progressively, creating gaps in the airfield defences. Local Somalis realised that the Australians were leaving. Crowds gathered on the airfield perimeter looking for opportunities to get in and see what was worth stealing. Bandits moved among the crowds, assessing the situation, looking for any opportunities and possibly relishing the prospects of returning to business as usual now that the Australians were leaving.

At 10 am on 13 May Lance Corporal 'Jake' Jason Verschelden, Private Jonas Hollingsworth and Private Michael Lange were manning an observation post at the NGO warehouse complex a few hundred metres from the main gate to the airfield.[284] Until that morning a section of six to eight men would have occupied this key position. Corporal Bud Wehemeir had taken the rest of his section back to main gate area at 8 am as part of the thinning out process for the Australian withdrawal. A group of French officers and soldiers had visited the area the day before. A French squad was due to take over the observation post in two hours time.

With two hours to go, Verschelden and Lange were discussing the prospects of beer and female companionship back in Australia. Hollingsworth had not been feeling well during the night and was resting. A group of Somalis had gathered on the other side of the rolls of concertina barbed wire, selling cans of Pepsi, trinkets and cigarettes. There were more Somalis gathered near the wire, spending the day watching for opportunities to steal.

Two older Somalis interrupted Verschelden's and Lange's conversation by pointing to the end of the warehouse building that could not be observed from the observation post and calling out the word 'thief' in Somali. That end of the warehouse was normally kept under surveillance from another observation post but the post had been withdrawn that morning in anticipation of the French takeover. It had not taken the Somalis long to realise that they could use this newly-created blind spot in the Australian perimeter defence to sneak in and steal grain, timber and other commodities stored in the warehouse.

Verschelden and Lange moved out to the end of the building to investigate, leaving Hollingsworth to man the observation post. On arrival they found that Somalis had breached the wire barrier by dropping rocks and heavy poles across the concertina rolls. There was no one on their side of the wire so they slung their rifles on their shoulders and began to move the poles and rocks off the wire to let the rolls spring back into shape.

Soon after they began this work, the crowd near the observation post moved down to the area and had gathered to watch. Members in the crowd began to gesture towards some scrap timber that was on Verschelden's and Lange's side of the wire barrier. Verschelden assessed that the crowd would breach the wire barrier again and take this timber as soon as they returned to the observation post. Verschelden directed Lange to throw the few pieces of scrap timber back over the wire to the crowd in the hope that this would satisfy them for the time being.

Unbeknown to Verschelden and Lange a group of bandits had observed the crowd gathering and the outbreak of scuffles and arguments as individuals fought to get pieces of timber being thrown to them. Like a pack of African hyenas scouting the edges of animal herds, they moved closer to investigate. Fortunately for the bandits, an Australian section standing patrol that had been positioned forward of the line of observation posts covering the warehouses and main gate areas had been withdrawn a few days before. They approached the two Australians unobserved.

After four months on operations Michael Lange had a sixth sense about danger. He observed the small groups of people moving from the scrub and buildings behind the main crowd while picking up and throwing pieces of timber over the wire. He saw two men standing slightly off to the side 70 metres away from the main crowd. There were other men gathering in the same area. Inner alarm bells began to ring and a squirt of adrenalin entered Lange's stomach as he caught a glimpse of a cut down AK 47 and the familiar banana shape of an AK 47 magazine being held by one of the men down his side, under his robes. Bandits also have a sixth sense about danger. Just the cock of Lange's head was enough to alert the bandit with the AK 47 that he had been observed. It was now a matter of whether Lange or the bandit was quicker 'on the draw'.

This deadly contest, at a range of 70 metres, was between the Steyr rifle with its telescopic sight in the hands of a trained Australian infantryman and an AK 47 rifle with a shortened butt in the hands of a Somali bandit. Lange wrote later:

> The next thing I knew I was raising my weapon, cocking it and yelling to Jake, 'He's got a gat [gun]'. At the sound of me cocking the weapon, Jake looked up, the crowd started bolting in all directions and the Somali started raising his AK. 'Shoot him', Jake yelled. I took a quick sight picture and released a snap shot at him. The next thing my sight was blocked by some of the main group running in front of me ... I caught another glimpse of him running off with the others, but carrying his arm limp beside him.[285]

Verschelden was told later that the bandit eventually turned up at the IMC Hospital in town, having lost a lot of blood and suffering from severe shock. He died later from complications related to his wound.[286]

Single shots were the book ends of the Australian operations in Baidoa. A Somali gunman had begun the contest in town four months before by taking a potshot at an Alpha Company patrol on the first night the Australians patrolled into town, scrambling away unscathed. Within two hours of their withdrawal, a single shot from a digger, eyeball to eyeball with a Somali bandit, ended Australian participation in the contest as well as a Somali's life.

These two shots, separated by four months, typified the shoot-outs in Baidoa. Most were chance encounters followed by split second decisions to fire. The Somalis and Australians were rarely in a position to take well-aimed, carefully-considered shots under reasonable conditions. Shots were almost always exchanged at night, amidst confusion and under pressure. The Australians won all of these fleeting firefights but not without some good luck. Some might argue that superior military training and professionalism creates its own luck. True enough, but a hard-struck round in Sterling sub-machine gun pointed into a man's stomach, a bullet passing through night vision goggles dangling in front of a digger's throat and one passing through a shoulder a few centimetres from a beating heart count as simple good luck.

LESSONS

The measurements of success in securing Baidoa begin with the uninterrupted conduct of humanitarian operations by aid agencies and maintaining the personal safety of expatriate NGO managers and staff for four months. These two successes were hard won in face of a persistent threat posed by some NGO guards and the illegal activities of urban criminals, visiting bandits and local hooligans. Occasionally the inexperience of a few expatriate NGO staff put them in mortal danger when they should have exercised more discretion, especially when hiring and managing their Somali guards.

The added bonus for the citizens of Baidoa was a decline in the level of violence because of the pressure the Australians put on local criminals, bandits and hooligans. In this climate, local elders were able to reassert their authority, a police force was recruited and a judicial system was tried out. Ian Harris, the National Director of CARE Australia assessed in June 1993, two weeks after the Australians had left, that, 'Today Somalis living in and around Baidoa are no longer hungry, the children are healthy and the people have enduring memories of the good work done by the Australian Army in their city.'[287]

The political situation in Baidoa changed significantly over the four months. Supporters of General Aideed's political faction, who mixed their politics with violence and crime, were hamstrung by Australian house and building searches as well as confiscations of weapons, munitions, supplies and cash. By hitting this violent political faction hard, the Australians created a better environment for political reconciliation and nation building by more moderate and popular local political groups. However, had the Australians stayed longer and kept the pressure up on Aideed's supporters, they may have attracted retaliation, like the Pakistanis did in Mogadishu in June and the Americans did in October.[288]

Whether the Australians should have been drawn into influencing local politics, closing down criminal activities, such as robbery, rape and extortion, and restoring law and order were important issues. All of these activities could be justified within the parameters of the mission of creating a secure environment, but the harassment of criminals, bandits and NGO guards increased the possibility of pay back. The key appeared to be to apply sufficient pressure for deterrence, but not to escalate the level of violence. Arguably, criminals and bandits were likely to fight back to save their lives if they were being hunted down, cornered and killed. They would be less likely to pick a fight for the opportunity to loot humanitarian aid or to kill an uncooperative expatriate NGO manager if they knew that they were putting their lives on the line against heavily armed, capable Australian infantrymen.

In the end pressure was not applied uniformly anyway. Different levels of pressure were exerted by companies and platoons guided by the different attitudes of their commanders to creating a secure environment in Baidoa. Some commanders decided to harass hostile groups in town through aggressive patrolling, house and building searches and vigorous questioning of anyone behaving suspiciously in order to force them to curtail their activities. Other commanders, deciding that maintaining a presence through patrolling and offering a friendly, 'Hello' in Somali to those they met along their way would diffuse tension in town, won support among the populace and provided sufficient deterrence to hostile groups who knew that if they confronted the Australians they would lose. There is insufficient evidence to validate either approach. The combination appeared to work, neither the proactive or reactive companies took significant casualties, and Baidoa settled down over time.

In the final analysis, the Australians achieved the same outcome as the Marines, but in a different way because they were there longer and approached the mission differently. The Marines had protected the distribution of humanitarian aid quite successfully during the five weeks they operated in Baidoa but had shown little interest in improving the personal safety of citizens or NGO staff, aside from confiscating weapons they found during searches and conducting vehicle and foot patrols through town. Hurley took a more proactive approach. From the beginning he and his CMOT staff fostered close working relations with local Somali authorities and NGO staff. Though he held back from closing down the arms trade, drug trafficking and vice, he decided to devote resources to protecting citizens from being robbed, assaulted or being forced to pay money for water. He also decided to occupy NGO compounds to deter the illegal activities of NGO guards and provide close physical protection for expatriate managers and staff. This approach worked. The Australians went on to receive the added bonus of a formal letter of thanks from the NGOs of Baidoa, not a formal letter of complaint that had been signed by the same NGOs when the 1 RAR Group arrived.

The 1 RAR Group completed a dangerous and difficult mission in Baidoa with the loss of only one Australian life. Only a handful of soldiers had sustained light wounds and minor injuries. The Australians made their presence felt during the last two weeks of January, won the shoot outs and earned respect in February, dominated in March and controlled in April and May. Competitive professionalism between companies, platoons and sections, loyalty to mates and fears for personal safety rather than job satisfaction kept most members of the Group motivated and alert. Very few

Australians felt that their security operations would make a lasting impact on the situation in Baidoa. The majority were satisfied that their conduct had set memorable example of a 'Firm, Fair and Friendly' approach to human relations. From their point of view, the citizens of Baidoa could take it or leave it.

The story of how the 1 RAR Group secured Baidoa is replete with lessons and insights that should inform the thinking of those who have the complex task of creating a secure environment in modern Dark Age cities and towns in the future. The groups that threatened life and property in Baidoa in 1993 will be similar to the groups that will threaten life and property in any city, town or village that is caught in the midst of unresolved civil war, tribal conflict or violent politics. Security forces will be up against local gangsters organising crime, drug runners and gun runners protecting their illicit trades, bandit groups attracted from the countryside into urban areas for richer pickings and foolhardy, angry youths with time on their hands and testosterone to burn. Security and politics in the broken-back states of the Third world are inseparable.

The activities of hostile groups will be intertwined with the agendas of Western-style political factions and traditional tribal elders vying for power. Political factions will often be supported by trained military forces, guerrilla bands, semi-trained militia or just local louts. An intervening security force will find it impossible to keep out of local politics while it goes about the business of protecting life and property.

The first thing of note in the story of the 1 RAR Group's securing of Baidoa was the level of media interest during the settling in period. Within 24 hours of arrival in Baidoa, the 1 RAR Group had to take over from the Marines, occupy and secure a base camp as well as begin security operations in unfamiliar urban environment where the populace had been stirred up recently by vigorous building and house searches and some manhandling. All of this had to done under the scrutiny of journalists, video camera crews and photographers. Commanders at all levels stated that they felt intense pressure to perform well immediately so that images and stories broadcast back in Australia would be positive, assuring their fellow Australians, loved ones and friends that all was well and under control.

Close media scrutiny of the arrival and initial forays by a force deployed overseas has been and will continue to be the norm. The challenge for commanders is to ensure that their troops understand the importance of media coverage and accept it as part of their mission, and not as an intrusion. The 1 RAR Group took some risks allowing journalists to accompany the first patrols into town when the security situation there was volatile. In 1965 the 1 RAR Group only allowed journalists to accompany routine clearing patrols close to Bien Hoa Airbase in the first 24 hours while they settled in. The images broadcast back in Australia were of diggers doing their job professionally. The lesson would appear to be that troops who are depicted should be given tasks that do not put them or accompanying media representatives in danger or under pressure. This may involve setting up action sequences and contriving tactical scenarios for voice overs and storylines. However, this deception is preferable to taking the risk that troops and media representatives might find out the hard way that they have entered a dangerous area. The author was interested to read recently that much of the footage shot of Allied troops going into battle in World War II was contrived before or after the battle started to ensure that troops, journalists and camera crews were not endangered. This footage was played over and over again accompanied by voice overs that suggested that viewers were actually following the troops into real battle. The Public Relations officers supporting the US Commander in the South West Pacific, General Douglas MacArthur, were particularly adept at the technique of creating 'genuine historical moments' from set up situations or recreating the historical moments of battle being joined or landings being made after the battle had already begun or beachheads had already been established.

Identifying why the 1 RAR Group was so successful in Baidoa for four months in 1993 is important. Answers to this question may save lives in the future if Australian forces are tasked to secure urban areas like Baidoa. The 1 RAR Group appeared to have created the right level of deterrence. If so, how did they do it?

The Australians arrived with seven significant cutting edge advantages over their opponents in Baidoa; their individual training, teamwork, physical and mental toughness, superior weapons, flak jackets, low-level communications and night vision technology.

The demanding standards of individual training maintained by the Australian Army gave each member of the 1 RAR Group high levels of personal and professional confidence in themselves, their weapons, their commanders and those around them. This situation not only applied to infantrymen on patrols but also to the cavalrymen, engineers, signallers, medics, drivers, administration staff and others who supported operations in town. The high standard of teamwork came from rigorous collective training and the Australian Army's investment in the career development of junior NCOs; sometimes taken for granted in the Australian Army but not done as well in other armies. The physical toughness came not only from traditional daily physical fitness programs, battle conditioning activities and hard training regimens, but also from competitive inter and intra unit sport. Mental toughness came from the physical and mental challenges of demanding training as well as from a strong individual and collective ethos that the Australian Army in general and their unit and subunit in particular meant business. They believed that they had the guts to get the job done and looked forward to the contest with anyone who would oppose them.

The attitude of Somalis to flak jackets were an unexpected bonus for the Australians. Despite the reality that flak jackets worn by members of the 1 RAR Group only protected the torso from fragments, flying debris and possibly spent bullets coming in at long ranges, Somalis believed that they would stop bullets fired at closer ranges. Consequently Australians were not only hard targets because of their individual skills, weaponry and teamwork but also because their torsos were perceived by the Somalis to be bulletproof.

The competitive edge offered by low level communications and night vision technology did not come without some hard bargaining and advocacy. The Army had invested in modest numbers of Motorola handheld radios and night vision devices to maintain the state of the art in selected specialist units, but could not afford to fund comprehensive low level communications and night vision capabilities for front line ODF units. Major Bannister, the DQ of 3 Brigade, insisted that the 1 RAR Group be given a communications and night vision competitive edge in Somalia at the expense of specialist training and loss of some night vision capability in Australia. Extra Motorola radios were bought especially for Operation *Solace*. However, without his advocacy, the support of his Brigade Commander, Brigadier Peter Abigail, and the loan of a small quantity the next generation of NVGs being used on trials in Australia by the Defence Science and Technology Organisation, the 1 RAR Group would have arrived in Baidoa unable to equip each night patrol with NVG. These goggles gave Australian patrols additional confidence and a competitive edge when they needed it—right from the start of operations. Subsequent shooting incidents in Baidoa vindicated Bannister's advocacy and probably saved Australian lives.

Somali lives were also saved because Australian forward scouts wearing NVG held their fire more often than they might have if they had to depend on moonlight to decide the intentions of Somalis ahead and around the patrol. The goggles also acted as a deterrent. Diggers wearing NVG caused comment among Somalis in town. Once they deduced that the goggles enabled the wearer to see at night, Somalis contemplating a hit and run attacks on Australian patrols may have decided that it was not worth the risk. CI reports verified that Somalis were 'spooked' by the knowledge that the Australians could see at night.

Australian and Somali lives were also probably saved by NVG and thermal imagers 'spooking' Somali infiltrators at Baidoa Airfield. Armed Somalis tested the airfield's defences from the first night the Australians occupied their positions and before the perimeter was better protected by a barbed wire fence. Sentries were able to observe Somalis infiltrating into the airfield area and use Motorola radios to guide troops moving out to intercept them. Infiltrating Somalis soon realised that they could not just prop behind a bush under the cover of darkness and avoid discovery, arrest and an uncomfortable day in the Australian detention area. The word soon got around that the Australians could see at night at the airfield. This knowledge not only appeared to be a deterrent to thieves but also to groups of armed Somalis who moved at night like trained infantry. These groups were seen in the first couple of weeks moving around the airfield and the outskirts of town but disappeared soon after. These marauding straggler groups from warlord armies could have been a threat to life and property if they thought the pickings were easy and the Australians were not vigilant at night.

The key to success at the cutting edge of urban operations was not only the seven advantages the Australians brought with them but their individual and collective attitudes, and responses to the hostile environment in town. The patrol commanders and their diggers had been given plenty of guidance on the conditions under which they were authorised to use deadly force and all of the dire consequences of not obeying the ROE and OFOF. In short, they knew when they were allowed to kill, when they were allowed to protect themselves and the penalties for killing or wounding if they had not judged the situation correctly. No one had clarified for them how they should respond to being stared at malevolently, having rocks thrown at them, taunted or spat on. The implications from their training were that these acts should be ignored. They were cautioned not to be drawn into reacting physically to provocation or they would be in trouble for exceeding the guidelines for using minimum force.

Platoon commanders and section commanders realised quickly that ignoring provocation was only encouraging further provocative acts, especially from local youths with time on their hands and displaying more bravado than common sense. During several research interviews and many off-the record conversations with the author, NCOs asked the rhetorical question of whether the police back in Australia would tolerate these acts without presenting the perpetrators with consequences. Since the Australians were not getting any respect initially from local Somali youths and were left in no doubt about the dislike of Somali males for their presence in Baidoa, patrols began 'adjusting Somali attitudes' and ensuring that anyone provoking members of an Australian patrol by spitting, taunting or throwing rocks could expect a swift, controlled and uncomfortable response. Over time the Australians won the grudging respect of local youths and Somali males who maintained their dignity by turning away and ignoring patrols as they moved about the town. For the vast majority of patrol commanders, this was an extremely satisfying outcome. Many of them noticed respectful acknowledgment and occasion appreciation in the eyes of many ordinary Somali citizens after the battle for respect had been won over local louts and hostile Somalis who resented the presence of foreign troops.

The decisive, controlled and aggressive response of Australians when they were under fire also contributed significantly to their success in Baidoa. They avoided firing at anything suspicious automatically, especially at night, but returned high volumes of accurate fire when engaged or split seconds before they were about to be engaged. Poor fire discipline would have stirred up and unsettled the town rather than diffusing tension and uncertainty. The levels of reported violence and numbers of reports of weapons being fired in town dropped significantly after the Aitkins, Hodges, Mortimer and Perkins fire fights in February and continued to drop in the following weeks. By April the town could almost be described as quiet on most nights.

Overall, the key to the Australian success appeared to be the controlled, measured responses they brought to breaches of security by Somalis in town and at the airfield. Somalis knew what barbarism and excessive violence looked like, they had experienced both intermittently since the late 1970s when the dictator, Barre, and his forces had enforced his rule at the point of a gun. They grew to know that the Australians were tough but not brutal. Even the controversial removal of the ISO container from town in early February demonstrated that even under intense physical pressure, Australian infantrymen would prefer to put their bodies on the line and, as the platoon commander stated, 'face the music' rather than fire into an unarmed stone-throwing crowd. Other UNITAF forces in Somalia did not achieve this level of fire discipline on all occasions.[290]

Though it is difficult to prove, it is logical to surmise that the word got around among local Somalis that Australians could be expected to respond firmly but fairly to infiltration and stealing at the airfield and threats to life and property in town. In other words, ordinary citizens, minding their own business and getting on with life, could feel safe, but those up to no good could expect to face consequences aligned to the seriousness of their wrongdoing.

It is easy to write about when the Australians opened fire but more difficult to find information about when they didn't. This chapter focussed on the contest between the Australians and hostile groups for control in Baidoa by describing the unfolding drama and peak confrontations. However, it is clear that the many times Australian patrols decided not to fire, and showed restraint, which contributed to their success in Baidoa. The tactics of presence, friendliness and communication help explain why many ordinary Somalis came forward to pass on information through interpreters to Australian CI teams and commanders on operations.

A prevailing view among the members of the 1 RAR Group, supported by the opinions of Somali interpreters and NGO representatives in Baidoa at the time was that many ordinary Somalis grew to trust, respect and like the Australians because most of them were firm, fair, good humoured, friendly and compassionate. They found that aside from a small minority of racists whose contempt for Somalis was obvious in their body language, eye contact and tone of voice, the vast majority of Australians gave ordinary Somalis a fair go. Language and cultural barriers may have precluded genuine friendship, but the body language and messages sent through Australian and Somali eyes showed that each affirmed the other's humanity and empathy with the other's circumstances.

With hindsight, 'white light' ambushes aimed at curtailing the movement of suspicious vehicles and groups of armed Somalis as well as rumours that permission had been granted to 'take out' armed straggler groups from warlord armies, brought the Australians close to breaching their ROE. This situation arose soon after arrival in January partly because the Australians had expected to confront groups of militia and fight them conventionally, partly because of poor judgement, but mostly because of an understandable desire to deter infiltration into the airfield area that might result in death or injury to personnel as well as loss of property. To the credit of David Hurley and his senior commanders, everyone in the 1 RAR Group was quickly and firmly reminded that putting in a road block or positioning troops on a likely approach route for infiltrators did not constitute permission for a 'white light ambush'. They emphasised that there had been no change to the ROE and that anyone in breach would be charged and dealt with severely. No more 'white light' ambushes were mounted after the end of January in the Baidoa area.

It is debatable how much the confiscation of weapons contributed to the success of security operations in Baidoa. Defining a weapons confiscation policy will be a challenge for security forces operating in modern Dark Age urban areas because most households and businesses will have weapons on hand to deter intruders and individuals over the age of about 14 years will consider carrying or having easy access to a weapon as a necessity for survival. The arguments for and against a policy of confiscating all weapons are similar to those made by pro and anti gun control lobbies in Western nations. Pro-gun control lobbyists argue that registering all guns and banning certain types of weapons will reduce the loss of life through gunplay. Anti-gun control lobbyists argue that gun control will deprive ordinary citizens the right to defend themselves from criminals who will have access to guns whatever controls are put in place.

Those advocating the confiscation of all weapons might have a valid argument if many of the shots fired in the vicinity of Australian patrols at night were from frightened householders who mistook them for armed bandit groups. Frightened householders firing warning shots over the heads of approaching groups at night does seem to make sense. If there was a significant danger from this type of contact, then confiscating all weapons found in houses also made sense. One could also argue that the distinctive silhouettes of the Australian bush hats and the different sounds of the foot fall of bare-footed bandits compared to combat-booted Australians prevented mistaken identity. Consequently, leaving householders and businesses with guns and ammunition to protect themselves also made sense. With hindsight a 'one gun—one magazine—one box of ammunition' rule for each household and business premises may have been an effective compromise. An armed population protecting themselves against robbery, rape and violence may be better than a disarmed disgruntled population attracting crime because they were made defenceless by the security forces deployed to protect them.

The 1 RAR Group's experience of finding and confiscating weapons throughout their tour of duty in Baidoa suggests that there were enough guns and ammunition to go around for everyone anyway. However, the confiscations of caches of automatic weapons, heavy machine guns and mortars from political factions and criminals suggests that the Australians did reduce their capability to pose a more serious military threat to them while they were there. The registration of the weapons carried by NGO guards was well intended but ineffectual. The guards had access to as many weapons as they wanted as Charlie Company's raids on NGO compounds in March seemed to prove.

The policy of banning the open carriage of weapons, unless employed as a security guard, appeared to be sound. This meant that the Australians knew that anyone carrying a weapon openly or concealing a weapon in their clothing was committing an offence and could be challenged and arrested. If everyone was carrying guns, bandits and criminals would be able to move about without detection This policy also gave Australians reason to speak with locals and search them for weapons. Consequently, there was a lot of communication and interaction with Somalis. It would not have taken long for the word to be passed around that the Australians would search randomly as well as focus on persons behaving suspiciously in a firm, professional way. Consequently, those with nothing to hide grew to respect the thoroughness of the Australians and those with something to hide knew that the risk of being arrested had increased significantly.

The contribution of CI teams and their interpreters to the success of security operations in town was significant. Initially, Hurley's staff were cautious about involving outsiders in their intelligence gathering operations and were unfamiliar with the benefits of providing interpreters down to company and, in some cases down to platoon level when platoons were operating independently. It should also be said that members of CI Teams had to learn on the job as well because none had received specific training in how to operate in a town like Baidoa. The adapted their general Intelligence Corps training to good effect and 'had a go'.

In the end CI teams provided 90 per cent of the exploitable intelligence for urban operations and contributed to all of the security coups achieved in Baidoa. They infiltrated and undermined the efforts of General Aideed's supporters to gain influence in Baidoa through intimidation of the locals and discrediting Australian efforts to restore law and order. They provided information that resulted in caches of weapons and munitions, as well as stolen humanitarian supplies, incriminating documents and cash being found and confiscated. They gathered much of the information that was used to indict local gangsters such as Lesto, Warsame and Gutaale. They were also the major links between local informants and citizens who wanted Australian security operations to succeed against the violence of criminals, bandits and the political factions, and the 1 RAR Group.

With hindsight, the living conditions for the combat elements of the 1 RAR Group at the airfield were not conducive to efficient security operations in Baidoa. The heat combined with the distances from the airfield into town and having to wear flak jackets as well as webbing quickly exhausted those

Seventeen weeks in hot, dusty, cramped and noisy conditions added to the fatigue and stress of the combat elements of 1 RAR Group. Pictured are the living conditions for the Reconnaissance Platoon, 1 RAR. Lance Corporal Brendon Anderson (L) Private Shannon Brown (R foreground), Lance Corporal James Wilson (R on stretcher) and Corporal Mark Retallick behind the group playing cards.
Image: Wayne Thomson.

involved in urban patrolling. The living conditions at the airfield did not help them recover mentally or physically. An earlier and more thorough reconnaissance may have foreseen the requirement to set up a tented camp with some amenities that would have offset some of the privations the 'Poor bloody infantry' had to endure once again. The 1 RAR Group that had deployed to Bien Hoa Airbase in 1965 faced a similar situation and had to beg, borrow and steal from the American logistic units in their vicinity to make up their shortfall in construction materials, tentage and other camp stores. Fortunately, in 1993, the 1 RAR Group were collocated with US Army and Navy engineer units that agreed to help establish the Australian camp without resorting to the dubious tradition of stealing from allies.

Many members of the 1 RAR Group tolerated their cramped living conditions because austere living conditions, interrupted sleep and no recreation facilities were the norm for field exercises in Australia. The difference was that those exercises lasted for two to three weeks. After four weeks in Baidoa, diaries kept by members of the 1 RAR Group and interviews conducted in May by Army psychologists debriefing troops before return to Australia suggested that close living conditions and the paucity of amenities for the combat elements contributed to a build up of frustration, resentment and fatigue and probably contributed to the risk of poor weapon handling leading to Unauthorised Discharges.

The combat elements of 1 RAR Group were not accommodated as well as their counterparts in American, Canadian, French and Italian combat units whose camps were prepared by construction engineers and included recreation facilities where personnel could relax. The lesson would appear to be that those costing and planning to send forces overseas for longer than the period of a typical field exercise in Australia should allocate resources to set up a tented camp with essential services and recreational facilities, such as Messes, or lease comparable accommodation and facilities.

4 SECURING THE COUNTRYSIDE

Most Somali men seem to have three wives following behind them. The women carry the load—the man has an elegant stick. I think part of the problem here is that women do most of the work—leaving the men free to squabble and fight. ... Saw no modern structures. Nothing modern has come here, except guns, in thousands of years.

George Gittoes,
10 March 1993

Securing the countryside around Baidoa for the distribution of humanitarian aid was another operational challenge facing the Australians. The Baidoa Sector comprised 17 000 square kilometres of flat featureless terrain, occasionally studded with rock formations, populated by an estimated 380 000 people. The Australian patrolling strength was about 350 personnel. This equated to a ratio of one Australian soldier for every 50 square kilometres of terrain and one for every 1000 Somalis reported to be living in the Sector.

The focus for UNITAF was not on securing each Sector in its entirety, but on ensuring there was a secure environment for the

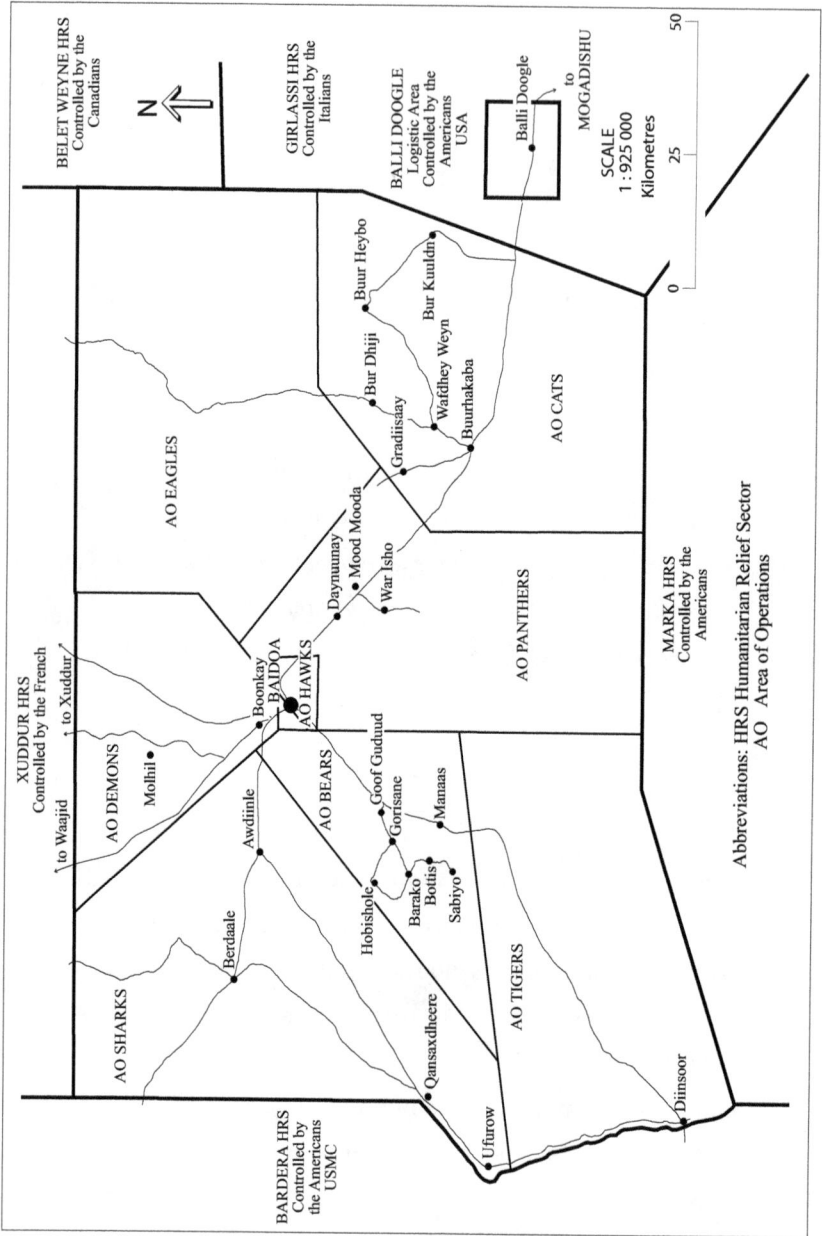

Map 6: Baidoa Humanitarian Relief Sector

distribution of humanitarian aid. From the beginning it was unlikely that convoys escorted by heavily-armed UNITAF troops would be attacked. For the time being the warlords had agreed to a ceasefire. They knew that UNITAF had the fire power and mobility to hunt their forces down and punish them severely if they went back to interfering with the distribution of humanitarian aid by ambushing convoys with technicals and heavy weapons.

The contest for the protection of humanitarian aid was to be between the Australians and groups of nomadic bandits. Bandits preyed on the rural populace, raiding isolated villages, ambushing vehicles on the main roads, sexually assaulting women, and extorting humanitarian aid and cash. In addition to these core criminal activities, bandits often stole humanitarian aid from villagers and refugees soon after it was distributed, sometimes demanding a percentage of the quantity delivered from village chiefs who controlled the distribution to the locals.

Back in Australia Hurley and his staff did not think it would be sufficient to just accompany food and other humanitarian relief convoys and vehicles to and from their destinations in the countryside. He wanted to control the roads of the Baidoa Sector with vehicle check points, conduct cordon and search operations of villages to clear out bandits and confiscate illegal weapons, as well as react forcefully to sightings of armed bandit groups.[1] The infantrymen and cavalrymen of the 1 RAR Group were very enthusiastic about the prospect of finding and apprehending bandits. From the beginning the capture of bandits was regarded by every infantry company and platoon as a tangible measurement of success. They wanted to hunt bandits down and bring them to justice. If the bandits chose to fight, then so be—bandits dead or alive would be their trophies.[2]

The first security task in the countryside fell to Delta Company at short notice. Ant Blumer arrived in Baidoa on 22 January after his diggers had spent several days on the docks at Mogadishu unloading vehicles, equipment and stores from *Tobruk* to be told that the next day his mission would be drive 70 kilometres back down the Mogadishu-Baidoa Road and occupy the town of Buurhakaba There were two reasons for quickly asserting Australian authority in the Buurhakaba area. The first was that the town was astride the main Baidoa-Mogadishu road. If a militia force occupied the town, the main supply route to Baidoa and the northern sectors would be cut. This situation would isolate some UNITAF units, threaten the Baledogle logistics area and close down the movement of humanitarian relief supplies by road from Mogadishu to the northern regions.

The second reason was that Buurhakaba, according to the Americans, was, 'one very bad town' dominated by 'thieves and the families of thieves' who comprised 'almost 50 percent of the population.' The American intelligence staff told their Australian counterparts that Buurhakaba was controlled by pro-Aideed forces who coordinated extensive criminal activities in the region. Every house concealed a weapon and there were caches of weapons and ammunition buried in the area. The dominant clan leader was the local commissioner, Abdala Ibrahim, whose clan allegedly controlled NGO food and humanitarian aid distribution. This control resulted in riots at food distributions points when rival political groups and clans missed out.[3]

After setting up camp off the road near Buurhakaba in the evening of 23 January, Blumer prepared to conduct a cordon and search of Buurhakaba and surrounding villages, called OP *Rosella*, the next morning. Much to Blumer's chagrin he was told that earlier that day US psychological operations helicopters had dropped pamphlets and announced over loud speakers that the Australians would be securing the town within the next 24 hours. Shortly before Delta Company assembled closer to Buurhakaba area in their APCs, an American vehicle drove through the town with loudspeakers announcing that the Australians had arrived in the area. These actions angered Blumer and his diggers who had been trained to achieve surprise for cordon and search operations. They looked forward to catching bandits off guard and capturing weapons and other munitions, not arriving to find that bandits had been given plenty of time to bury their weapons and flee. From their point of view the Americans had compromised the operation despite advice from their accompanying CI Team's advice that, 'should Australian forces enter the town from one side, thieves will withdraw through the other ... the provision of prior notice should be avoided.'[4]

Pressing ahead despite the loss of surprise, Blumer deployed two of his platoons as blocking forces around the town after dark that night. He then sent Jan Van De Klooster's platoon, mounted in APCs, commanded by Lieutenant Mick Hanna, accompanied by the CI Team, Brendon Thomson, Wayne Douglas and their Somali interpreter, Mohamed, into town at 10.30 pm.[5] Simultaneously, a sniper team and members of the 1 RAR Intelligence Section led by Sergeant Dave McKay, equipped with a Thermal Imager, climbed up the large rock formation adjacent to the town carrying rations and water to set up an observation post for two to three days.[6]

A large crowd of local Somalis were delighted to see the Australians and greeted them warmly, lining the streets and shouting encouragement. After driving along the main street, Van Der Klooster directed Hanna to drive his APCs south and ordered his diggers to get out and patrol on foot through the town. Thomson wrote in a report later, 'The diggers were keyed up, patrolling through a new and reputedly bandit-ridden town at night.'[7] Thomson and Douglas, accompanied by their Somali interpreter, advanced with the infantry speaking with locals to ascertain if any bandits were in the vicinity. The inhabitants told their Somali interpreter that all of the bandits had made their escape some hours before, prompted by the American warning of the Australian arrival. The Commissioner, his family and clan cohorts had also fled. Locals who kept weapons for their own protection had buried them on the outskirts of town to avoid confiscation.[8]

After patrolling through town, Van Der Klooster's platoon remounted in Hanna's APCs and proceeded to conduct 'snap' cordon and searches of hamlets north of the main town area. The procedure was the same for each hamlet; two APCs would manoeuvre behind the hamlet while the remaining section dismounted and patrolled into the hamlet from the opposite direction. Patrolling through Buurhakaba and the surrounding hamlets was not a stroll through the streets, akin to police walking their beats. Soldiers covered each other as they rushed forward from fire position to fire position; running, getting down 'on their guts', crawling and taking aim. Their weapons were raised and pointed down alleyways, between huts, and at doorways and windows where bandits might appear and take a shot. This was gruelling, exhausting work wearing webbing and flak jackets and squinting in the dark to detect any movement that would reveal the location of waiting or fleeing bandits.

The Australians worked hard all night to secure the area and apprehend bandits, but there were no arms to confiscate and no bandits to capture. Hanna's APC and his two other Troop Headquarters vehicles, commanded by Lance Corporals Mick Jollife and Tony Miller, were put through their paces in the dark. Fortunately, there were no break downs and they kept up with the tempo of operations set by Van Der Klooster. There was some consternation early the next morning when an APC drove across a disused well that had been covered over. The sides caved in, resulting in the APC becoming perched precariously on its 'belly' on top of the well walls, unable to move back or forward for fear of falling into the well. While the APC sat on its brick perch, there was a 'less than conventional debussing by the infantry and the carrier crew'.[9]

Dawn revealed Buurhakaba to be a filthy town with streets covered in human and animal faeces and rotting rubbish. During the night the Australians had noticed an over-whelming stench but had got on with the job at hand. They now realised from their own smell and the dark brown smears on their webbing and uniforms that they had stood, knelt, crawled and lain in this mess repeatedly during their night's patrolling.

For the next three days Blumer conducted a number of cordon and search operations of nearby villages. None of these achieved surprise. However, the Australians had established their presence and local criminals knew that they could not operate with impunity.[10] What concerned the Australians was the reluctance of the population to provide information on bandit activity or to even report sighting bandits when it was obvious that they had been seen. Near Buulo Nasib, a hamlet of 60 inhabitants a kilometre south of Buurhakaba, a group of Australians had chased a Somali armed with an AK 47 across the open fields. 'Despite most members of the village having been working in areas that allowed unimpeded observation of the chase, all claimed that they had not seen anyone run from the village, or seen a 12 tonne APC chasing him across an area that was devoid of vegetation and which, aside from a few creek lines was as flat as a billiard table.'[11]

Blumer was proud of the efforts of his diggers who had carried three days rations and ammunition in their webbing, only managed to snatch one or two hours sleep each day, patrolled at night and maintained their security by employing correct tactical techniques. All this was achieved in 40C heat, covered in 'dust and shit'. He was also proud of cavalrymen from B Squadron who had operated and maintained their vehicles around the clock to support his company.[12]

He also appreciated the bravery of Thomson and Douglas, accompanied by their terrified Somali interpreter, Abdul, who often went ahead of the advancing infantry to alert villagers not to open fire on following soldiers who might have been mistaken for approaching bandits. With a section of keyed up Australian soldiers behind them, a darkened village, possibly containing armed bandits, in front of them and the prospect of being caught in the middle of an ensuing fire fight, Thomson and Douglas felt that they were earning their $ 68 a day allowance for service in Somalia.[13]

Aside from losing the element of surprise during their four days in the Buurhakaba area, Blumer and his men were frustrated by the lack of support from local villagers who were afraid to identify bandits and their weapon caches. The locals were 'extremely evasive' about who were bandits and where their weapons were hidden. They assured the Australians that they had seen nothing, knew nothing but encouraged them to move onto other villages where 'bad men' were living. The Australians followed up reports of bandits being seen in neighbouring villages but had no success capturing bandits or siezing significant quantities of weapons and ammunition.

On 27 January Blumer handed over responsibility for the Buurhakaba area to Alpha Company. The next day Doug Fraser began OP *Rosella II*, a reconnaissance-in-force to find out the situation in the outlying areas of Buurhakaba region. Leaving Lieutenant Richard Adams' 1 Platoon to guard the base compound the company had occupied on the edge of town, Fraser led Bob Worswick's 2 Platoon and Gavin Butler's 3 Platoon and an Engineer group on a tour of local villages mounted in APCs.[14]

At this time Fraser's main concern was running into technicals from one of the militia armies. He had been given out-of-date intelligence reports that technicals were still operating in the area. Accordingly, he operated in a conventional manner keeping his APCs in formation with everyone watching their arcs either side of the roads, ahead and behind. As he approached the village of Surgle, Fraser ordered his force to encircle the village, block off the roads leading into it and approach from three different directions at the same time.

On the way in the Australians heard several warning shots being fired and observed a number of people running to and from houses. Leaving his forces astride the roads, Fraser dismounted and walked into town with a protection party and a group of engineers who were looking for signs that mines had been laid. On arrival Fraser noticed several recently-slaughtered camels lying in the village streets. Through his interpreter, Fraser questioned the local village chief and found out that they had interrupted a funeral procession. Fraser was told that 150 people from a nearby village had attacked and ransacked the village the previous night. The intruders were reported to have killed two villagers, raped several women and stolen household belongings and food. The village chief assured Fraser that his village planned to take retaliatory action against a neighbouring village that was responsible as soon as possible. Fraser informed him that this action would not be tolerated and that Australian troops would be in control of the area from now on. 'This was my first taste of inter-village conflict and I must say that I was naive. ... Two days later I found out that he [the village chief] was meeting with the chiefs of the other villages he claimed had attacked his.'[15]

Unaware of the complicity of the village chief at Surgle, Fraser decided to cordon and search the town of Shimbile where the intruders who had conducted the rape and pillage of Surgle had allegedly come from. He wrote later:

> As we were speeding over the fields to encircle the town, it started to pour with rain and almost immediately the APCs were slowed to a crawl.[16] The problem was compounded by the fact that the village was one kilometre long and the inhabitants lived at the far end. We finally rounded up the locals but immediately there appeared to be a discrepancy. The village constituted mainly old men and women. These people could not have attacked other villages. They told their story to me.
>
> Two years before the rampaging armies had gone through their town and rounded up all the inhabitants present and burnt them alive.
>
> They showed me where it [the burning] had occurred.[17]

Fraser was struck by the brutality of Somali society: clan on clan, village on village, the strong against the weak. However, he had still not properly understood how he and his diggers were being manipulated. Over the ensuing few days Alpha Company were led on a number of 'wild goose chases' by local Somalis reporting bandit sightings, nominating villages that needed to be searched for weapons and bandits, describing minefields that did not exist and generally confusing Fraser with inaccurate Somali versions of numbers, times and distances.

Fraser and his men became more and more sceptical about the value of reacting to information provided by local elders through his interpreter. Fraser's experiences highlighted two dilemmas for Australian security operations in the countryside. The first was the appropriate response by the Australians to information they received from local Somalis. Rival clans and villages were only too happy to provide information about each other so the Australians would search and confiscate weapons belonging to other clans and villages. Often Fraser had to trust his Somali interpreter to judge what information should be acted on. This put the interpreter in a powerful position as well as putting some law-abiding Somalis at risk from deceitful criminals and rival clan leaders.

The second dilemma was deciding what weapons to confiscate. Fraser did not like taking weapons from store owners and local householders who probably needed them for self protection. All Somalis feared losing their weapons for this reason. It was impossible to know for sure whether a weapon found in a house or store was there for criminal purposes or self defence. Fraser and his men started to use their judgement on what weapons to confiscate as they became more and more aware of the consequences of stripping a Somali household or business of its means of deterrence and protection.

On the day that Fraser entered Surgle, the Mortar Platoon led by Lieutenant Peter Connolly for this task was sent to Borakro, a village near Baidoa to retrieve two vehicles that had allegedly been stolen from NGO compounds the night before by 10 armed men. Connolly wrote later:

> Departed an hour late ... with the joyful attachments of Captain [John] Hill from CMOT and an Army media team including Mick [McKinnon] from the *Townsville Bulletin*.
>
> Because I knew bugger all about what I was getting into I dismounted three sections 300 metres short and patrolled in by foot leaving my fourth section in reserve with the trucks. At this stage I was expecting 14 bad dudes. I had an interpreter and two village elders with me. They were very vague but we found the place. There were two stolen [Land] Rovers (one GOAL and one ICRC) in a compound with several smouldering fires. We were quickly told that the bandits had fled to the west.
>
> ... Johno [18] saw two [bandits] hurrying through the sorghum to our west toward a copse of trees. When Fish [19] returned we went to investigate. 100 metres short of the copse we were shot at from another copse 50 metres to the right. Almost simultaneously

heavy rain began to pour down. As we hit the ground in the field, adrenalin pumping, the sound of the rain on the sorghum seemed very loud.

Target indication from Johno's section. Not much for it: platoon attack.[20] I got the section commanders in: 2 Section assault, 3 Section depth, spacing 15 metres between men and 1 Section fire support from the copse to my left. Cut in only if we are fired on and be sure to walk the fire off us on visual.[21] 'Fish, we will not move until you are firm.[22] 'Send the word and we'll be off. Questions? Lets go!' Admittedly a quick set of orders, but I was sure they [the bandits] would withdraw quickly.

We stalked, stooped over, through the sorghum, expecting to be shot at but we got there without incident. Everyone had a look on their faces which made them somehow different—fear, concentration, aggression—maybe all three. Voices in huts to our right. Swept through three different compounds. Found a child and an old lady close to death in one hut. No one understood that we were trying to protect them, and no one would tell us where the bad guys had gone.

The sorghum went on forever into the distance with a copse every 200—400 m. We could have played this cat and mouse game all day, and still would have only known where they were by one of the fellas getting shot. We'd been gone for an hour and a half and our mission was to recover the stolen vehicles by dark. I decided to return to town.

... We limped back to Baidoa, our two trophies being towed behind two trucks. We gave the sick woman a lift and dropped her off at the ICRC hospital. One hell of a day.[23]

By the end of January the Australians had made their presence felt in the Buurhakaba area. They were also reacting to reports of bandit sightings and criminal activity by cordoning and searching villages and following up on vehicle thefts. During these countryside operations the Australians searched for and confiscated weapons and ammunition they found in accordance with UNITAF policy requiring them to confiscate all weapons and register those carried by NGO guards.

During this time the potential for armed confrontation with militia groups was high. On 29 January Australian snipers and reconnaissance troops observed several small groups of well-armed militia, suspected to be stragglers or deserters from General Morgan's army in Kismaayo, moving through in the Baidoa area. They were on the prowl, surviving through banditry.[24] Battalion Headquarters staff wanted to deter them from remaining the area. Stuart Dodds was given the task of setting up a 'white light' ambush along a likely route for these groups. After a disorganised briefing and minimal time to plan and rehearse, Dodds deployed to the ambush site. He wrote later:

> We went forward with NVG and cleared some huts and a compound.[25] I set the APCs up about 80 metres apart and we placed out two Claymores and a Patrol Ambush Light.[26] These ran back to me with the 2 Section gun group with the left hand carrier [APC].[27] The moon set about 2400 hrs and we waited. All night we could hear the camel herd, a baby crying and dogs barking about 800 metres to the front of my killing ground.[28]

> The plan was to spot the enemy with the white light and call 'Jhoog So !' (Stop). If they fired or pointed their weapons, I would initiate with the claymores and the 50's, 30's and our Minimis would join in.[29] We were prepared to capture them.

> After a sleepless night no one moved along the track. I was not disappointed. My concern was controlling the fellas if civilians had moved down the track. Thank goodness they did not get tested. I had no qualms about my task but I was reasonably confident after all our other 'hot tips' that we would not see anyone. We moved back at 0600 hrs. As soon as we pulled up the claymores, a group of civvies moved along the track ![30]

As it turned out, the straggler groups seemed to make their own assessment of the situation in the Baidoa Sector and moved on after a week or so of sightings. It is possible that their observation of the Australians and information they gained from the locals about the Australians may have been sufficient for them to look for less secure areas to base themselves. Had they decided to stay, the Australians would have faced an additional challenge to creating a secure environment.

*Lieutenant Colonel David Hurley, CO
1 RAR Group, directing cordon and
search operations in the countryside
of the Baidoa Humanitarian Relief
Sector.*
Image: ADF Electronic Media Unit

For February Hurley and his staff planned to conduct a series of cordon and search operations in the countryside to clear bandits out and assert an Australian presence. Humanitarian aid was being distributed without interference, but he was determined to keep nomadic bandit gangs as well as town and village-based criminals off balance and unable to operate with impunity. These operations were also an opportunity for the Australians to practice manoeuvres and procedures that would stand them in good stead if they had to operate against militia units in the future.[31]

Hurley ordered Charlie Company, mounted in Hanna's APCs, to cordon and search the southern towns of Goof Guduud and Manaas on 31 January and 1 February respectively (OP *Kookaburra*). Goof Guduud was assessed as a town that would be neutral or friendly to the Australians despite containing a bandit group known as 'The Brothers', described by Australian counter-intelligence teams as persons who, 'regularly slaughtered animals, raped women and preyed on their own and other villages'.[32] Manaas was known as 'Bandit City'.[33] At least 25 bandits were reported to live there with their families. Many inhabitants were healthy and prosperous by Somali standards because of their ill-gotten gains. Moon was told that the majority of young men in town were criminals of some sort and his troops might encounter armed opposition.[34]

For the cordon and search of Goof Guduud, Moon decided to deploy two platoons on foot by night into blocking positions behind the village and then to make a noisy entrance with infantry in APCs from the opposite direction. His plan depended on Lieutenants Todd Everett and Ian MacGregor getting their platoons into the correct positions independently by night over unfamiliar terrain.[35]

All appeared to be going well when Moon's headquarters lost radio communications with Everett and MacGregor. He spent a very uncomfortable night and early morning worrying about how aero-medical evacuation would be organised if there were casualties in a platoon that was out of radio communications. By the time his headquarters regained communications the operation was well under way. Everett's platoon had apprehended two bandits carrying AK 47s trying to escape from advancing APCs. Thirteen rifles, five machine guns and a pistol were confiscated during the subsequent search of Goof Guduud by MacGregor's platoon.

At Manaas the next day Moon employed the mobility of his APCs to simultaneously cut off escape routes and sweep through the village. More rifles, machine guns and Rocket Propelled Grenade rounds and launchers were confiscated but no bandits were captured. Groups of bandits were observed escaping to the east of the village before the APCs could get into position to cut them off.

Hurley placed great importance on gathering intelligence before cordon and search operations were mounted. Before these operations he sent CI Teams into the villages around Baidoa. On 2 February Thomson, Douglas and Abdul, accompanied by a security group led by the Regimental Sergeant Major of 1 RAR, Greg Chamberlain, drove to Daynuunay, a notorious village reported to harbour over 75 bandits who conducted frequent raids on Baidoa.

During their time in the village Chamberlain was approached by a young child who had recently sustained severe hand and abdominal wounds from an exploding grenade or anti-personnel mine. After administering First Aid, Chamberlain's team flagged down one of the many vehicles passing by to take the child to the MSF medical facility in Baidoa. The occupants refused indignantly, irritated by the interruption to their journey. Occupants of the next vehicle that had room on the back seat to take the child demanded money in return for providing the service. Chamberlain then asked Thomson and Douglas to ask the village chief they were questioning through their interpreter to see if he could use his authority to have a vehicle take the child to hospital. The elder was completely uninterested in the child's fate. By that time an Australian convoy had been flagged down and took her to hospital. The CI Team found that they had both personal and professional reasons for recommending that Daynuunay be cordoned and searched. [36]

A week later Hurley led C and B companies on a cordon and search of Daynuunay (OP *Cassowary*). The Australians did not apprehend any bandits during the operation but they found 15 weapons, ammunition and a large quantity of stolen food. 'The village elders are no doubt corrupt as there were fat people in the same vicinity as starving refugees. Quantity of food aid was discovered in underground silos which the elders claimed belonged to someone else. They were warned that we would be back again if we heard of it [hoarded food] again.'[37] The elders of Daynuunay knew that the Australians had arrived and meant business.[38]

Hurley decided to keep the pressure up on the bandits at Goof Gudduud and Manaas by returning in force using UNITAF helicopters from Baledogle. On 9 February C Company conducted an airmobile operation around Goof Guduud as a show of force (OP *Kookaburra II*). Moon left a platoon in the area for several days after the operation to deter the return of bandits and to gather information from the locals.

Charlie Company's airmobile operation was followed on 14 February by a similar airmobile show of force by B Company around Manaas (OP *Plover I*). Hurley's final operation to demonstrate that the Australians could turn up at any time and at any place was OP *Lyrebird*. He ordered Delta Company, mounted in Lieutenant Shaun Voss's APCs, to conduct consecutive cordon and searches of the villages of Berdaale, Awdinle, Ufurow and Qansaxdheere in the south western region of the Baidoa Sector .

Two days earlier Hurley had dispatched the Reconnaissance Platoon, supported by Thomson, Douglas and their interpreter, to Berdaale for a seven day patrol to gather information before Delta Company arrived to conduct cordon and searches. The Reconnaissance Platoon assessed that the Berdaale area was affluent by Somali standards, surrounded by arable land that produced a surplus of sorghum, beans and maize. This surplus, including food distributed by aid agencies, was moved by road to Baidoa and Mogadishu for sale. There was little bandit activity in the area because the people of Berdaale were from the one clan group, well-armed and united in the protection of their prosperity. The Reconnaissance Platoon decided that the arms they noticed should be left with the people and the focus of cordon and search operations in the region should be on the nearby town of Ufurow where bandit activity was high.

Members of Reconnaissance Platoon patrolling in the countryside of the Baidoa Humanitarian Relief Sector.
Image: G Gittoes

This advice from Reconnaissance Platoon did not please Blumer. His mission was to sweep through the south western region, mounted in APCs to assert an Australian presence He assessed that the Reconnaissance Platoon and their accompanying CI team 'did not have a feel for the area' and that the Reconnaissance Platoon had denied him the opportunity to catch bandits by surprise by being in the area before he arrived. He moved out with his small armada of armoured vehicles and 'proceeded to conduct aggressive operations in the [Berdaale] area, his APCs roaring from one side of the area to the other, often through fields of crops, searching villages for weapons.'[39]

Both Delta Company and the Reconnaissance Platoon were doing their jobs. However, their differences of opinion about whether to leave the Berdaale area alone or not, illustrated that the policy of HQ UNITAF about the confiscation of weapons and deterring bandit activity may have needed more flexible interpretation when the local situation was better known. Hurley's mission, like all of the other Sector commanders, was to deter bandit activity. To do this, HQ UNITAF had ordered all Sector Commanders to confiscate any weapons they found and to treat all Somalis in their Sectors consistently to avoid accusations of political bias from competing warlords.

Members of Reconnaissance Platoon going through their morning routine in the countryside of the Baidoa Humanitarian Relief Sector, accompanied by APCs from B Squadron 3rd/4th Cavalry Regimemt.
Image: G Gittoes

Despite the initial excitement of operating in the countryside in APCs, helicopters and on foot in search of bandits, Moon, Blumer and Dodds became fed-up with these operations. Moon wrote after the Daynuunay cordon and search, 'Long morning with nothing to show.'[40] Blumer reported that, 'We didn't find anything and did not achieve surprise. I do not think the troops know what a cordon and search really is. Alas there is a lack of a sense of urgency at times and I find it frustrating.'[41] Dodds felt that the surprise necessary for successful cordon and search operations was not achieved.[42]

Unlike some of his subordinate commanders, Hurley was pleased with the results of these operations. By the end of February, he felt that the Australians had made their presence felt throughout the Baidoa Sector after a series of gruelling cordon and search and airmobile operations. He was more concerned with the show of force than the numbers of bandits captured. He was satisfied that the successful manoeuvre of sub-units by air, APC and on foot and the demonstration of mobility and firepower would deter bandit activity. [43]

February had been a particularly challenging month for Hurley and his company commanders, but had also been a frustrating time for some of their subordinates. Hurley did not want to provoke armed clashes and risk taking casualties. He placed the emphasis on 'search and clear' operations.[44] His objectives were to deter bandit activity and keep bandit groups guessing about where and when the Australians would turn up next. His company commanders shared his views but would have liked to have achieved more with these types of operations. They did not seek a fight with bandits but they were ready for one. [45]

Most of the NCOs and soldiers in the rifle companies and the troopers in B Squadron wanted to conduct 'search and destroy' operations. They sought the chance to surprise bandit groups and to take them on. Unlike their CO and company commanders, they were excited and hopeful about contact with bandits. When they realised that the element of surprise was being lost deliberately and they were involved in demonstrations of force rather than the use of manoeuvre and force to capture bandits, they became frustrated and critical of the way they were being employed. Disillusionment with 'cordon and search' operations in the countryside grew over time.[46]

Aside from bandits, the other dangers facing the Australians during security operations in the countryside were land mines and unexploded ordnance, such as grenades, artillery and mortar shells. Most villages were strewn with these lethal objects and some villages were protected by minefields that were not marked in accordance with international convention. The civil war had left a deadly legacy for local Somalis, especially children who moved about their villages playing and scavenging. However, there were some instances where the children were better mine detectors than the Australian soldiers. In Buurhakaba, Stuart Dodds remembered one incident,

As we returned from the ASF some children stopped us and indicated a 'Bomba' (bomb). I was driving so Fish and 1 Section investigated.[47] I saw them pick up a small white tin and then tell the kids 'No bomba'. The kids kept saying 'Bomba !'. Fish and his fellas identified it as a coil from a vehicle electrics.

I went up and had a look and told them to place it down and walk away. They had picked up an Egyptian hand grenade with the friction band removed. All it would take would be the ball bearing to move and it would go off. The Pioneers [48] were called in and said, 'Whatever you do, do not touch it.' They did not move it and blew it in the middle of town. The kids, who obviously identify bombs better than my soldiers, became our DMEO [mine] locators.[49] We paid them in coffee and rations. Everyday they would identify at least six different bombs and we would blow them.[50]

The dangerous task of finding and disposing of mines and unexploded ordnance was shared by the sappers from 17 Field Troop and 1 RAR's Assault Pioneers. The sappers had three explosive ordnance detection dogs accompanied by experienced handlers. These dogs were extremely successful in detecting weapons, explosives and ammunition during building searches in Baidoa and in outlying villages. The Somalis detested and were frightened of the dogs. Most patrol commanders were delighted to have them and their handlers accompany their patrols to detect mines and unexploded ordnance as well as to deter young Somali men from foolhardy provocation.

Alpha company had returned for a second tour of duty at Buurhakaba on 21 February. This time Fraser was a lot wiser about mounting futile cordon and search operations at the behest of rival village chiefs. He turned his attention to freeing the Mogadishu-Baidoa Road from vehicle ambushes.

Typically, bandits would wait on the edge of the road on a hill or near a bridge where vehicles would have to slow down. The vehicle selected for ambush would be sprayed with gunfire until it stopped. Occupants would then be lined up and robbed. Often young women would be taken a short distance away and raped by gang members. Sometimes passengers were killed or wounded in the initial spray of gunfire. Depending on the mood of the bandits and the clan affiliations of passengers, one or two passengers might be shot or stabbed to death on the spur of the moment. Bandits also liked to maim. Their favourite acts of intimidation were to shoot their victims in the hand, the foot or in the knee cap.

Fraser was not alone in his desire to rid the roads from the scourge of these brutal ambushes. He had full support from Hurley and his staff and his fellow company commanders. Technically, the Australians were keeping the roads clear for the passage of humanitarian aid and the protection of aid agency staff who used them This was a preventative measure because humanitarian aid convoys and vehicles driven by UN and NGO expatriate staff were not being targetted at the time.

A counter ambush patrol from the Defence Fire Support Weapons Platoon before heading out on the Mogadishu-Baidoa Road. The patrol was commanded by Corporal Chad Hughes. His driver, Private Jeremy Tauri is standing near the driver's door.
Image: Stuart Dodds

Underlying their enthusiasm for preventing vehicle ambushes was an enduring desire to confront and apprehend bandits. Morally and militarily, the Australians did not want to stand by and do nothing while bandits killed and maimed innocent Somalis in what was temporarily their 'turf.'

To deter bandits from ambushing vehicles Fraser wanted to modify some of his company vehicles and adopt tactics he had discussed with Lieutenant Troy Webb, the Direct Fire Support Weapons (DFSW) Platoon Commander. Two weeks before Webb had agreed to allow his Platoon Sergeant, Mark Thorp,[51] to adapt the two DFSW 6 x 6 Landrovers for night counter-ambush operations.

Thorp borrowed infra-red lights from the B Squadron Technical Support Troop and, with the help of members of the 1 RAR Technical Support Platoon, fitted the lights and an extra battery to power them. The vehicles were then fitted with tripod mounts behind the drivers cab to support ANTAS 6A Thermal Imagers and the floors were sandbagged to lessen the impact if the vehicle detonated a landmine. Three MAG 58 medium machine guns were then mounted; one over the passenger compartment hatch, and two centrally mounted left and right on the vehicle tray. These vehicles were then capable of being driven in the dark by a driver wearing Night Vision Goggles with all head lights turned off. They were also able to generate a substantial volume of automatic fire in any direction.[52]

On 9 February Webb took out the two modified vehicles on the first night-time counter-ambush task, clearing 20 kilometres of road on the Buurhakaba-Baidoa Road successfully. Webb also authorised the use of the platoon's Carl Gustav 84 mm Anti Tank Guns to fire illumination rounds over groups of suspicious Somalis moving in the scrub near the roads. The firing of these weapons and 40 mm illumination rounds from M 79 grenade launchers created fear among the Somali groups who did not understand that the Carl Gustav rounds were harmless, but sounded like artillery. By the time Fraser moved to Buurhakaba later in February the DFSW Platoon had been sleeping by day and deploying out on most nights deterring night ambushes.

Fraser was keen to have a go. He wrote later:

> As the ambushes occurred at night, I developed a plan whereby we would either kill or capture the perpetrators or at least scare them off. I had vehicles fitted with Infra Red headlights and placed a section [six to eight soldiers] in the back with a machine gun in the cupola.[53] This vehicle would follow Somali vehicles at a distance where it would not be detected. Should an ambush occur to its front, it [the machine gunner in the cupola wearing Night Vision Goggles] would engage the ambushers and destroy them ... By following vehicles the ambushers would never know whether there was an Australian force

behind a Somali vehicle or not. I hoped this would discourage them from carrying out their ambush.[54]

Members of Alpha Company enjoyed the aggressive nature of these innovative counter-ambush patrols that were dubbed 'Night Rider' operations. Speeding through the night in a blacked-out vehicle, looking for bandits through NVG and Thermal Imagers and awaiting the sound of a sprung ambush to initiate a counter-ambush drill to sweep through and engage bandits created 'quite a buzz'. In Fraser's opinion, the ROE permitted his troops to apply deadly force after ambushers had opened fire on a Somali vehicle because they would be responding to a life-threatening situation. He cautioned his men that they did not have authorisation to open fire on ambushers, even if they were in an ambush position. They would have to be verbally challenged and given the opportunity to throw down their arms in accordance with the ROE.[55]

During Alpha Company's tour of duty in Buurhakaba from 21 February until 2 March the night ambushes along the Mogadishu-Baidoa road stopped. Later Fraser found out from CI reports that his first Night Rider patrol had driven through an ambush which was not sprung because the bandits had heard the different engine sound of the Australian Land Rover following a Somali bus up a hill. This incident appeared to have deterred them from mounting more ambushes while Fraser's men were in Buurhakaba.

While Fraser's Night Rider patrols had been successful in the Buurhakaba area, there was a significant increase in ambushes on the roads in other areas of the Baidoa Sector. On 2 March a vehicle carrying 10 Somali civilians was ambushed at 8 pm on the Mogadishu-Baidoa Road near Baidoa. One occupant was murdered and four seriously wounded with gunshot wounds to their hands and legs.[56] Expatriate NGO staff and local elders implored Hurley to take more action to stop these murderous activities.

After consultations with his subordinate commanders, Hurley assigned Mortar Platoon as well as DFSW Platoon to deter vehicle ambushes using Webb's and Fraser's Night Rider techniques. [57] Both these platoons had their own vehicles and were made up of experienced soldiers. On 4 March the mortarmen became 'Night Riders'. Dodds recalled:

> At 1900 hrs 1 and 3 Section deployed in two 6 X 6 Landrovers on a counter-ambush patrol down to Buurhakaba. Both vehicles drove on blackout [58] with the driver on NVG. The lead vehicle had a Thermal Imager mounted and this showed up people, donkey carts and vehicles easily up to a kilometre away and further by their heat.
>
> We would drive along in complete darkness and search vehicles as we came across them. What we were looking for were people lying on the side of the road in ambush positions. ... At 0415 hrs we woke to leave on our return journey. I drove SM 5 [a Land Rover] wearing NVG. It was eerie. I could easily drive at 40-60 km and see everything clearly but on the occasions when we stopped and I took off the NVG it was pitch black with no moon and I could not see a thing. What we were doing would not have been dreamed of ten years ago. [59]

While Hurley stepped up the counter-ambush program, he planned further search and clear operations in the countryside where bandit activity persisted. On 12/13 March Charlie Company conducted a series of sweeps through the villages of Junet, Makina and Gaserta. Moon felt, however, that the capture of one weapon and one bandit after the commitment of 350 troops, including a troop of APCs and six hours of helicopter time was not productive. He was not happy about losing the element of surprise during these operations which allowed groups of bandits to escape hours before Australian troops arrived. [60] Surprise was being lost because APCs and helicopters were noisy, the Somali 'bush telegraph' appeared to work quickly after Australian forces left the airfield and there were those among the Australians who were suspicious of the Somali interpreters.

From the beginning of operations Australian intelligence personnel realised that Australian operations were probably being compromised by their American-based Somali interpreters brought to Somalia by the Marines. Thomson and Douglas had insisted that all interpreters should not be allowed to attend briefings before operations and were not to be told where they were going when troops departed from Baidoa. WO1 Bill Bowser summarised his views in a report stating that the US Marine interpreters had strong political affiliations. He pointed out that one interpreter was a relation of the Mogadishu-based warlord, Ali Mahdi and another interpreter had strong links to the other Mogadishu warlord, General Aideed. Bowser had observed these interpreters trying to subvert and recruit other interpreters. This situation raised a conflict of interest and, in his view, made the maintenance of security of Australian operations 'problematic'.[61] The affiliations of the interpreters also suggested that their advice and their reporting of conversations with other Somalis to the Australians were biased.

Initially, company commanders had been ambivalent about the use of interpreters. It was interesting for the CI Teams to observe the changing attitudes of the Australian infantrymen to using Somali interpreters over time. Thomson and Douglas had from the outset repeatedly recommended to Caligari and his staff that more interpreters should be hired and assigned to each company. This advice was dismissed initially on the basis that Australian patrols would not want Somali civilians to accompany them. As company commanders realised the importance of being able to communicate with the locals, they asked for interpreters to be assigned to them permanently. By March Hurley had authorised the employment of ten interpreters.

In contrast to Bowser's assessment, many officers found Somali interpreters to be brave and intelligent companions, very committed to making a contribution to their country's future. Blumer was one of them:

> Today for the fourth time Mohammed Tamea has had his life threatened. The success the Australians have enjoyed has largely been due to innovative and concerned methods and approaches to the tasks ahead. All of these, however, are for nought if we don't possess the ability to pass information to the local people or receive it. ... We spend millions of dollars on computers and radios yet really lack the resource of human 'interpreters'. Mohammed has been a blessing in disguise. Entertaining, keen, friendly, sensitive and always sporting an excellent sense of humour. He has been invaluable. He

is also protective of the soldiers who show a little compassion for him. His wife and son were killed in Mogadishu by Morgan's men, yet he bears remarkably little hatred and bitterness that we can see. The continued contribution of Mohammed and others like him has been the key to our success. We have floundered at times because of a lack of interpreters.

They also provide the human touch which we can sometimes forget with the authority of a weapon or an APC to show. They have been the key that has allowed us to unlock many of the problems that we have faced here. Their survival after our departure is probably not on too many people's minds ... we put them in very compromising situations that cause a backlash to their community acceptance. ... We owe it to them to be more than 'passingly' interested in their future well-being. ... They are the unsung heroes of OP *Solace*.[62]

Sometimes the element of surprise was not necessary to achieve results when searching for bandits, weapons and munitions. A key factor was the attitude of local Somalis. On 5 March Lieutenant Justin Haddock's 10 Platoon headed out with an EOD team and a CI Team to Habare, a village a short distance from Baidoa, in response to information that the villagers wanted to hand in weapons and ammunition.[63] After entering the village the Australians were received warmly and offered food and drink. There were extensive NGO-funded agricultural projects around the village growing a diverse range of crops such as mangos, carrots, onions, oranges and peanuts. During the civil war the local irrigation system and all the crops had been destroyed. Since then, the villagers, with the assistance of aid agencies had painstakingly restored the system and replanted crops.

The villagers told the Australians through their interpreter that they now felt secure and wanted all of the heavy weapons and grenades removed from in and around the village so their children could play safely—they had enough of war and its destructive ordnance. They presented 10 Platoon with a collection of over 200 hand grenades, Rocket Propelled Grenades, artillery rounds and anti-tank mines. Most of this lethal load was put into the back of an APC, commanded by Corporal Dean Commons, and driven by Trooper Jason 'Kiwi' Robinson back to Baidoa for destruction.[64] The Australians were impressed with the 'honesty and willingness of the villagers to help.'[65] However, the experience at Habare was a rarity.

Meanwhile, 'search and clear' operations continued in March. On 14 March with little time to rest from the previous two days of cordon and search operations, Hurley ordered Charlie Company to deploy two platoons to the southern area of the Baidoa Sector and move with one platoon to Ufurow in the south west. He sent Everett's platoon to Buurhakaba, Swinsburg's platoon to Goof Guduud and moved by road with his headquarters and MacGregor's platoon to Ufurow for a demonstration of force in the area. On the way there he was told to join up with the Reconnaissance Platoon for a special cordon and search operation.

Following the earlier patrol to Berdaale by the Reconnaissance Platoon supported by Thomson, Douglas and their interpreter, Hurley had authorised the platoon to patrol to the south west through the towns of Qansaxdheere, Ufurow and Tugerhosle. They had been in the area for three days before the Charlie Company group arrived. They had discovered that three bandit gangs operated in the area. The first was led by Abdullahi 'Dhegeweyne' who concentrated on ambushing vehicles in the Tugerhosle-Ufurow area. The second was led by Katin Issack Dolol who operated near Qansaxdheere, rustling camels and ambushing vehicles. The third gang was led by Hassan Mohamed Issack, known as 'Gaardhub', who normally operated around Manaas south of Baidoa but occasionally ventured into Dhegeweyne's area. Gaardhub was described as a professional bandit, with a formidable reputation among the Baidoa Sector's bandit gangs.

Dhegeweyne, whose name meant 'big ears' in Somali, was also a legend in the Baidoa region.[66] He had been a member of Barre's elite commandos in the 1970s but had deserted the Somali Army to join a notorious bandit gang operating over a wide area of southern Somalia. He became a gang leader in 1988 and led the local Luay clan against Barre's forces during the civil war.

He had loyal supporters among the Luay clan and had resumed his bandit activities in 1991.

The Ufurow police were as keen as the Australians to capture Dhegeweyne and the other bandit leaders. On the night of 9 March Dhegeweyne had ambushed a vehicle transporting *khaat* near Tugerhosle, wounding all of the occupants in a spray of gunfire. On 13 March the Ufurow ASF alerted Thomson and Douglas through their interpreter that Dhegeweyne was staying in a hut in Ufurow. The Charlie Company group were now available to back up Reconnaissance Platoon with a cordon around Ufurow while they attempted to capture Dhegeweyne who was expected to try and shoot his way out of any situation rather than be taken alive. After confirming with the Ufurow police that Dhegeweyne was still in one of a small group of huts on the outskirts of the town, the Australians planned to capture him on the afternoon of 14 March.

After maintaining the appearance of conducting a routine foot patrol around the streets of Ufurow, a combined CI/Reconnaissance Platoon patrol approached the huts reported to contain Dhegeweyne and some gang members. The Australians suddenly leapt over the surrounding camel thorn fences and charged into the huts, hauling out half a dozen shocked Somalis. Dhegeweyne was not among them. By this time Moon had deployed his troops in a cordon around this section of Ufurow. Their vehicles suffered two punctured tyres during this manoeuvre which had hampered their ability to close the cordon off quickly. After questioning by Thomson and Douglas through their interpreter, those apprehended informed them that Dhegeweyne had left several hours before they arrived.

The Australians' disappointment was short lived. They were told that Dhegeweyne's 'Number Three Man', Mohamed Moalim Mayow, was in the main market area. The patrol raced to the market area where the police chief pointed out a man standing to one side. Members of the patrol raced up to him and took him into custody. After questioning, he revealed that he was in town to pick up the proceeds from the earlier ambush of the vehicle transporting *khaat*. He told Thomson and Douglas that Dhegeweyne and the rest of the gang were in a nearby village.[67] In fading light, Moon assigned two sections mounted in APCs, accompanied by Thomson and Douglas, to act on this information. On arrival at the village, Thomson and Douglas discovered from the villagers that the noise of their approach had prompted Dhegeweyne and his gang to flee. The chase was called off.

Acting on information from locals keen to rid their area of bandits, Moon conducted a cordon and search of the nearby village of Miisra on 16 March. He had received reports that a 120 mm mortar and a technical armed with a 106 mm Recoilless Rifle (RCL) had been used to support bandit operations in the area. Subsequently the Charlie Company group found a large cache of 120 mm mortar and small arms ammunition, nine sophisticated Milan anti-tank rockets, Anti-tank mines and 106 mm RCL ammunition, probably belonging to bandits who operated in the Dinsoor region.

Meanwhile, across the other side of the Baidoa Sector in the Buurhakaba area, Everett's platoon was being kept busy apprehending bandits and reacting to local incidents. He wrote later:

> MSF officials reported that one of their vehicles had been ambushed by bandits. I dispatched 33B and uniform elements to the ambush site.[68] They found one wounded Somali who had been shot off the back of the vehicle. There had been two other Somalis shot, one had died on arrival at the local hospital, the other had been wounded in the leg. No bandits had been found at the ambush site.

> A local policeman from the ASF came with information on some thieves who were in a coffee shop only 300 metres away! Sent some men down to investigate. ... the two locals in the shop had stolen 12 million [Somali] shillings ... The two were detained and they will be sent to Baidoa for trial.

> ... a camel herder informed us that some armed bandits had stolen his camels. A section was sent out in APCs to find the bandits with the herder acting as a guide. They found two men, one armed with an SKK [assault rifle], who fled. Two warning shots were fired and the bandits stopped and were brought back to the base camp to be detained.

The bandits are being pushed out of the main towns and appear to be more desperate in trying to find some sources of income. The locals are giving us more and more information. It is good that they are starting to trust us.[69]

After the unsuccessful pursuit of Dhegeweyne, Reconnaissance Platoon returned to Ufurow. Reacting to further advice from the ASF about camel rustlers being sighted north of the town, the Platoon, accompanied by several police, drove north. On reaching the area where the camel rustlers had been reported there was a flurry of shots fired at the vehicles by a group of Somali males who began to run off. After a short chase two were captured and brought back to Ufurow. The Australians had been lucky. One of the them was Katin, the leader of the Qansaxdheere bandit group: one bandit leader down and two to go.

After patrolling around outlying villages for two days, the Reconnaissance Platoon returned to Ufurow on 18 March to find that three bandits had attacked the Red Cross storehouse and seriously wounded a woman. Thomson and Douglas were able to ascertain that the bandits were likely to leave town in a truck. The Australians set up a vehicle check point on the road out of town to try and cut the truck off before it left the area.

Within the hour this timely action was rewarded. Diggers manning the check point stopped a truck and two passengers attempted to run off. They were apprehended and found to be carrying pistols. One was Gaardhub, the bandit leader from the Manaas area. When the patrol returned to Ufurow with Gaardhub and his accomplice, the villagers were fearful and stared in disbelief that Gaardhub had been caught. For his part, Gaardhub exhibited no fear at all. He gazed malevolently at the crowd and saved his most fearsome looks for members of the ASF who appeared to be terrified of him. After several statements were taken that implicated Gaardhub in several murders, he was taken to Baidoa to face trial. Within 24 hours the Baidoa ASF had released him, possibly out of fear of retaliation if they proceeded with a public trial.

Unaware of Gaardhub's release, the Reconnaissance Platoon continued operations in search of Dhegeweyne, the last of the three bandit leaders operating in the area. The Platoon Sergeant, Wayne Thomson, lead half the platoon in the search. The CI Team, led by his brother Brendon, supported the patrol. Investigations revealed that Dhegeweyne frequented Tugerhosle village and was being protected by relatives and members of his own clan, whom he used to visit frequently. However, there were others in the same area who despised him because of his ruthless acts of violence during ambushes, and his killing of rival bandits and members of their families, as well as innocent people if they got in his way.

The information the CI Team received about Dhegeweyne's movements was very specific—times, routes and numbers in his group. Two locals said they would notify the CI Team when the bandit leader would be in town next, but warned that any attempt to capture him would involve gun play. After some discussion, the patrol decided that the best way to capture him and avoid casualties among civilians in the area was to intercept Dhegeweyne on his way in or out of the village. 'If he surrendered, good.' However if he tried to fight his way out, the Australians were confident that it would be 'a one-sided shoot out.'[70]

Later that day, three snipers, dressed as locals joined the crowds in town, keeping an eye on the house Dhegeweyne might visit that evening. The Platoon Commander, Lieutenant Graeme Steer ordered a 'stake out' and decided to go ahead with surveillance without waiting for the CI Team's informants to specify when the bandit leader was coming to town.[71] He ordered Wayne Thomson's half platoon patrol, accompanied by the CI Team, to deploy to Qansaxdheere that afternoon. The next day the CI Team returned to Baidoa and briefed Hurley personally on Dhegeweyne's movements and the methods being used by Steer to capture him.

Hurley cancelled the 'plain-clothes' surveillance operations and the 'stake out' by the Reconnaissance Platoon immediately. He assessed that they contravened the ROE as they could be construed as predatory ambushes.

Staff Sergeant Brendon Thomson,
Counter-intelligence Team No 1
(L) with his brother, Sergeant
Wayne Thomson, Platoon Sergeant,
Reconnaissance Platoon..
Image: Stuart Dodds

He preferred to use infantrymen from the rifle companies in APCs to cordon and search once it was verified that bandits, such as Dhegeweyne and his gang, were in the area. From the CI Team's point of view, this approach posed practical problems and dangers; the sound of the APCs would alert Dhegeweyne and his cohorts and Tugeresole was too large for a company to cordon effectively. If Dhegeweyne and his gang decided to fight their way out in the unlikely event that the cordon went in without them being able to escape, the subsequent fire fights might endanger civilians in the area.[72]

The difference of approach between Hurley and Steer was based on differences in the premise for security operations in the countryside. Guided by the ROE, Hurley's premise was deterrence through demonstrations of force, presence and manoeuvre. Steer's premise, which was shared by many junior officers, NCOs and soldiers in the 1 RAR Group, was that operations should create a secure environment by capturing bandits, not frightening them off. They sought tangible results, such as numbers of bandits captured or killed, as the outcomes of operations, not the less tangible result of deterring bandit activity. They also felt that innocent Somali lives would be saved and the community would be safer if bandits were captured and brought to justice.[73]

Members of the Reconnaissance Platoon and the CI Team took some satisfaction that their efforts had produced results when it was reported to them that Dhegeweyne no longer felt secure and was now very wary. The captures of Katin and Gaardhub, and the close calls with Dhegeweyne, had resulted in the gang leaders and their men knowing that they could no longer operate with impunity. However, the Australians were extremely frustrated when they learned that all the bandits they captured and sent to Baidoa were released by the ASF without trial.[74]

The Charlie Company group returned to Baidoa on 21 March. Hurley ordered the Reconnaissance Platoon to secure Dinsoor. He had decided to occupy Dinsoor permanently as a patrol base to cover the south western area of the Baidoa Sector. He also planned to provide support to newly-established ASF and a judicial system in the south western area by occupying Dinsoor and building a detention centre in Ufurow. This was part of an overall plan to set up and support ASF police in several regional towns. The local Chief Elder in Dinsoor was about 40-45 years old, a relatively young man compared to other Elders in the Baidoa Sector. He had the support of local radical youths and had allegedly struck a deal with Dhegeweyne and other local bandits to give them an outlet for the sale of their stolen goods as long they carried out their criminal activities elsewhere. As a result Dinsoor was a prosperous town with plenty of foreign goods and household items for sale. 'It amazed me that, while people from other clans were starving to death in camps on the outskirts of town, you could always purchase a can of ice cold Pepsi for $US 1 in this town in the middle of nowhere.'[75]

By establishing and backing up the ASF in the Baidoa Sector using Australian troops, HQ UNITAF, with the encouragement of UNOSOM officials, hoped to validate a model for similar efforts in the other

Humanitarian Relief Sectors of southern Somalia. These efforts were well-intended. However, the release of bandits from detention in Baidoa and Buurhakaba in February and March by the unarmed ASF without trial did not augur well for the future and created a significant source of frustration for the Australians trying to create a secure environment in the countryside. Stuart Dodds had written a critical report on the ineffectiveness of the ASF in Buurhakaba on 11 March:

> My report centred around the inability of the ASF to detain and hold prisoners, the contempt and lack of respect held by the public of them and their fear of bandits. I recommended that we should not give them uniforms or pay them and that we send in a training cadre to take control until a reasonable level of training had been achieved.[76]

Aside from the problem of bringing known bandits to justice after they had been captured and turned over to the ASF, the Australians operating in the countryside also faced the complicity of NGO guards and some expatriate NGO managers who hid weapons. While searching for Degeweyne, Wayne Thomson had taken his half platoon patrol into Qansaxdheere on 20 March where they visited the French MSF compound in town to gather information. Two French expatriate staff discussed the local security situation and assured Thomson that they only had two rifles in the compound, both belonging to their Somali guards. Thomson noticed a number of large sheds adjacent to the compound but decided not to search them to test the Frenchmen's word. He continued his patrol around the town and returned to the compound a few hours later to bid the Frenchmen farewell and to thank them for their assistance.

While in the compound Thomson noticed a .30 calibre round lying on the ground that he had not noticed on his previous visit. He quizzed the Frenchmen about where the round had come from. They became abusive, telling Thomson to mind his own business. Thomson then requested that the sheds be opened so they could be searched. The Frenchmen became more abusive and stated that they would comply. He then gave the Frenchmen the option of either opening the sheds or he would order his men to break into them. Shortly before, a section from Delta Company, passing through Qansaxdheere, had stopped to talk to members of Reconnaissance Platoon. Corporal Steve Chinner, the Section Commander, had been listening to the exchange between Thomson and the Frenchmen and moved behind Thomson to break into the sheds. Prompted by this move, one of the Frenchmen produced a key. Chinner moved in with his men and found a truck equipped with a pintle-mounted .30 calibre machine gun with a hundred round belt hanging off it and several hundred more rounds in boxes ready for use. He grabbed the machine gun, stating, 'beauty, another one for D Company', and he and his section jumped on their Land Rover and departed, leaving Thomson to deal with a torrent of verbal abuse from the Frenchmen and their Somali assistants. He decided that 'enough was enough' and departed to search the other MSF compound in town.

On arrival Thomson's patrol found the MSF compound gate locked. They were denied entry by Somali staff who claimed that there were no weapons in the compound. Thomson ordered members of his patrol to scale the walls and open the gate. During a subsequent search they found six rifles and a quantity of ammunition. None of the Somali staff could explain why the two guards employed at the compound needed six rifles. Thomson confiscated four rifles and left two for the guards. He directed the staff to have the two rifles registered with the CMOT in Baidoa or they would also be confiscated. Two days later Thomson returned to find that no effort had been made to register the weapons so, good to his word, they were confiscated.[77]

The confrontation with the French MSF expatriates in Dinsoor highlighted a gap in the Australian security arrangements for the distribution of humanitarian aid in some of the outlying areas of the Baidoa Sector and the longer term agendas of some aid agencies. The Dutch MSF, based in Baidoa, was accorded security for their humanitarian medical operations as a part of the Australian convoy escort and town security program. The French MSF, based in Mogadishu, had been in Somalia since the worst days of the civil war in 1991. They were distributing food and other humanitarian medical services over a wide area of southern Somalia that overlapped with the Baidoa Sector, but had not requested convoy escorts nor had they attended meetings of the Relief Committee or the NGO Committee in Baidoa. They had been, and were planning to be, in Somalia for the foreseeable future. Hence they were sceptical of the long term security arrangements offered by UNITAF or UNOSOM, preferring to hire reliable guards and arm themselves against marauding bandits. Their anger at the Australians taking their machine gun and other weapons can be understood within this context.[78]

The challenge of deterring bandit activity in the Buurhakaba area and keeping the Mogadishu-Baidoa Road free from vehicle ambushes continued in April. The Australian companies had become wiser and more canny. On 3 April Fraser returned with two platoons, a detachment of engineers and section of APCs to Buurhakaba for another 9-day tour.[79] The Night Rider patrols conducted by Support Company platoons had stopped vehicle ambushes at night but Fraser was briefed at Battalion Headquarters that bandits were now resorting to ambushing Somali vehicles by day.

Based on information collected from earlier operations in the Buurhakaba area by the other companies, Fraser identified two hot spots for ambushes on the Mogadishu-Baidoa Road: an iron bridge five kilometres east of town and a stretch of the main road 15-20 kilometres west of town. The CI Teams had ascertained that the bandits ambushing in the iron bridge area lived in the vicinity of the bridge and those ambushing the stretch of road to the west were from the BuurHeybo area.[80] Despite the best efforts of Bravo Company platoons, there had been 10 vehicle ambushes during their nine-day tour of duty at these two locations before Alpha Company arrived to take over.[81]

Fraser established a 24 hour vehicle check point on the iron bridge and radio communications between those on the bridge and what were dubbed 'Road Runner'[82] sections mounted in 6 x 6 Landrovers. By day these vehicles travelled two kilometres behind Somali vehicles, such as buses and trucks carrying humanitarian relief supplies, that might be targeted for ambush. [83]

Despite these new security arrangements, bandits still managed to ambush a bus load of civilians on 6 April. A Road Runner section found a bus swerving from side to side along the road driven by a wounded driver with several seriously wounded passengers on board. Fraser tightened controls over vehicle traffic, continued the permanent occupation of the iron bridge and put those occupying the bridge in charge of assigning Road Runner sections to likely ambush targets. The platoon commander on the bridge would assess vehicles as they approached and direct vehicle-mounted sections to follow likely ambush targets along the road until they were clear of the dangerous road section. As a result, Australian vehicles from Alpha Company and Mortar and DFSW Platoons were driving up and down the dangerous stretches of road randomly 24 hours a day following vehicles that would have been attractive targets for bandits.

After the ambush on 6 April, Fraser was determined to catch the perpetrators. He surmised that the ambushers would have another go along the same stretch of road. He assessed where the most likely ambush sites would be and sent two platoons to patrol through the area the next day. At 5 pm this pre-emptive patrolling was rewarded when Corporal Shaun Moore's section observed two armed Somalis moving towards the road.[84] He challenged them and they immediately fled into thick camel thorn. Moore and his men gave chase for two kilometres but the Somalis escaped; a Somali carrying a rifle dressed in T-shirt, trousers and sandals running through clumps of camel thorn has a decided advantage over an Australian soldier wearing webbing, a flak jacket, and combat boots. Fraser assessed that these men were the forward scouts of an ambush party.

Fraser also realised that he needed to stop the movement of bandits around Buurhakaba and villages in the vicinity of the Mogadishu-Baidoa Road if he was to deter ambushes and robberies. He decided to plan and rehearse his platoons in the conduct of white-light ambushes. He wrote,

It [The white-light ambush] was placed as a normal ambush, manned at 50percent to allow a rotation of personnel. The ambush would be sprung by switching on the PAL [Patrol Ambush Light]. The Platoon Commander would then be required to give a warning to anyone with weapons to place them down. If they fired, or pointed their weapons at the direction of the [Platoon Commander's] voice, they could be engaged with a single shot. However, if they ran they could not be engaged. It [this ambushing procedure] provided some very interesting ROE problems but was a necessary development if the gangs of armed men which had been reported [to be in the area] were to be stopped. Platoons taking part in this activity were required to rehearse all the possible reactions by persons caught in the ambush, as there could be no room for mistakes. As it was very difficult to discern between a stick [carried by many Somali males] and a rifle through night vision devices, a hand held Infra Red headlight running off 77 [radio] set batteries was developed to illuminate people entering the target area.

Fraser set white light ambushes on the tracks around Quardho, 'a particularly suspicious town' on the Mogadishu-Baidoa Road near the 'ambush stretch' that was close to Buur Heybo. None of Fraser's ambushes caught armed bandits but he assessed that they would deter bandit activity. He instructed his platoons to set up ambushes without being detected, but to make noisy exits from ambush sites after they had been in position for some time. He hoped the word would get around the villages that the Australians were lying in wait for bandits moving around at night.

Fraser also decided to deter bandit activity by cordoning and searching local villages. The change from his company's first tour of duty in the Buurhakaba area was that he did not mount operations until he had received reliable information. A productive example was when he cordoned and searched the town of Garanka based on timely information. He deployed a platoon mounted in APCs on the main routes in and out of town to prevent inhabitants from leaving. He recalled:

As we had an informant, we rounded up all the menfolk and he [the informant] went through picking out those he claimed to be bandits. We ended up with nine detainees, although it was hard to take it seriously as in Somalia everyone claims that everyone else is a bandit. ... I contacted Captain Mohamed, the Chief of Police, and asked him to get anyone who had been victims of robberies or been in ambushes to come and have a look at the detainees. To my surprise five of the nine were positively identified as bandits and two of them as ambushers by the people they had shot.[85] This was a big win as we now had witnesses and statements. Later Counter Intelligence Reports revealed that many of the residents at Quadho who had made a living off ambushing, packed up and moved north to Buur Heybo.[86]

During A Company's tour of duty in Buurhakaba in early April, there were no reported vehicle ambushes after the 6th of April incident, there was a drop in bandit activity and several bandits were detained and sent to Baidoa with sufficient written and oral evidence to put them on trial. The major factors contributing to Alpha Company's successes were Night Rider and Road Runner vehicle patrols, white light ambush techniques and the receipt of timely information from informants. [87] Fraser wrote, 'We left Buurhakaba on 15 April and, whilst it was a hot, dry, dusty, windy place, as a Company Commander it was the most enjoyable. ... Much to the work of my subordinates, I found that it was my favourite place in Somalia.' [88]

While Alpha Company were in Buurhakaba, Hurley deployed Bravo and Delta Companies to the outlying areas of the Baidoa Sector to create a secure environment for establishing the ASF. Two platoons from Bravo Company were protecting Australian sappers who were building detention centres in Ufurow and Qansaxedeere. Simpson based his headquarters in Dinsoor with his remaining platoon to bolster the local ASF who had suffered a set back when the Police Chief in Dinsoor was assaulted by a camel thief, leaving him with a fractured skull.[89] In the end, Bravo Company platoons were spread over a front of 120 kilometres.

During this period, platoons from Delta company were also in the countryside conducting cordon and search operations in the north western area of the Baidoa Sector around Berdaale. Haddock's platoon were conducting a building search in Berdaale on 13 April when a shot rang out near a shop in the market area. A digger had come out of a shop and, 'was confronted by a man who pulled a pistol on him. The member shot the man. An accessary to the bandit [sic] then retrieved his shot partner's weapon, and ran off down the street heading in North-East direction.'[90]

The Somali who had drawn the pistol was seriously wounded with a bullet wound in the stomach. Lance Corporal Darren Coleman[91] administered First Aid after borrowing a pair of surgical gloves. Corporals Greg Evans and Walter Davis[92] ordered their sections to pursue the Somali who had fled after the shooting. Haddock arrived soon after and ordered the wounded Somali to be loaded into a Land Rover and driven to his platoon headquarters for further medical attention.[93] At the end of the street near a Red Cross compound, the pursuers came to an open area on the edge of town. Haddock, who had caught up with the two sections, ordered Evans to cover

Major Jim Simpson, OC B Company, 1 RAR Group. His company was spread over 120 kilometres to establish 'presence' in the towns and villages of the Baidoa Humanitarian Relief Sector.
Image: ADF Electronic Media Unit

Davis's section as they moved out into the open. Suddenly two Somali gunmen opened fire on Davis's section from their left. Evan's section returned fire immediately as Davis's section dived for cover. Davis then ordered his section to crawl around so they were in position to face their assailants while Evans kept the gunmen's heads down with fire. Two children then broke cover and ran in front of Davis's section. Haddock called out, 'Cease Fire !'. The pause in firing gave the gunmen sufficient time to make their escape into thick camel thorn. Haddock formed his two sections into an extended line and swept in the direction the gunmen had fled for 300 metres, but they had disappeared. The Somali who had been shot died later from complications from his wound. This shooting was the only serious incident that occurred during Delta Company's tour of duty in the north west.

On 15 April Delta Company took over from Alpha Company in Buurhakaba. Blumer saw this last tour of duty in Buurhakaba as an opportunity to undertake a series of cordon and search operations and for his company to live and work in the field. He wrote, 'I don't want to occupy the compound like all the others have done. This is the last chance to be able to conduct patrols on operations as a company ... I am most keen to take advantage of it. Get out there and do what Australians do best—patrol and live in the bush.'[94]

Unlike Fraser, Blumer did not seek cooperative arrangements with the local police, whom he assessed were ineffective, but used informants to target villages suspected of harbouring bandits. He divided his area of operations up and sent his platoons out on foot through the countryside, living in the field and approaching villages at first light to conduct searches. Delta Company were successful in finding and confiscating weapons and ammunition in most villages but was not able to apprehend any bandits.

Meanwhile the Mortar Platoon had taken over from Bravo Company in Dinsoor and were sending out local clearing patrols mounted in APCs to maintain a presence in the area. On one of these patrols Corporal John Moore directed his section to continue to patrol towards a nearby village on foot after the APC they were travelling in broke down. He left the APC crew waiting for a forward recovery team to come from Baidoa and fix it.[95] Private David Woodrow recalled, 'It was about 1500 hrs when we walked into the next village. Doogs [96] yelled out, "He's got a rifle." The man ran away to our left. We dropped packs and gave chase with Reidy [97] and myself staying to look after the gear.'

While Woodrow and Reid waited with the patrol's packs, Reid recognised the man who had ran away walking back among a crowd that had gathered in the vicinity. Reid acted quickly and with Woodrow's help they apprehended the Somali and covered him with their rifles until Moore returned with the rest of the patrol.

As it was becoming dark, Moore sent Woodrow and Lance Corporal Darryl Lowrey to a position 100 metres beyond the village to guide in the APCs being sent by Stuart Dodds from Mortar Platoon Headquarters to their location to pick up the Somali they had detained. Just after nightfall they heard two people moving down the road. Woodrow and Lowrey went down onto one knee, pushed their safety catches off and aimed their rifles at two Somali men approaching. 'When they were close enough, Kiwi [Lowrey] put a torch onto them and at that point I saw a weapon pointed at us.'

Woodrow, looking down the sights of his Steyr rifle, had a split second to decide whether or not to shoot the Somali who was pointing a Russian-made RPD light machine gun at him and Lowrey. The ROE permitted Australian soldiers to use 'deadly force' against anyone carrying an automatic weapon like an RPD, whether they had raised it to fire or not. As instinctively as the Somali had raised the RPD when surprised by the torch light in his face, he lowered it and held it across his body. Woodrow assumed he had decided to do this because he was facing Australian soldiers who would not open fire unless threatened. Had he and Lowrey been armed bandits he had no doubt that the Somali would have opened fire immediately.

> I then stood up to take the weapon from him. He didn't want to give it up. The second man came around behind me and got me in a

headlock. Kiwi jumped up and butt stroked him, he fell to the ground dazed. The one with the weapon dropped it and ran off towards the village. We called out to the rest of the section, and realising that he was surrounded, he stopped and gave up. A few minutes later, the two detainees were confirmed as bandits after a torch-light search of the sides of the road where they had come from turned up several knives and a large quantity of Somali shillings.

On 30 April the bandits of the Baidoa Sector demonstrated that they were still able to terrorise the roads of the Baidoa Sector. At 1130 am a Forward Recovery Team, made up of Corporal Harold 'Huck' Berry, Craftsman Andrew Goss, in a Technical Support Vehicle[98] and Craftsman Scott Burdon and Private Gary Kirkpatrick in a protection vehicle was travelling 66 kilometres south west from Baidoa along the Ufurow-Dinsoor Road to repair one of the Mortar Platoon vehicles that had broken down near Dinsoor.[99] Goss wrote later,

> We were approaching a Somali vehicle that looked like it was broken down and on the other side of the road there were about 20-30 people sitting down. As the lead vehicle approached the people stood up and started jumping up and down and pointing to the broken-down vehicle. ... both Corporal Berry and I were watching the people who looked very desperate for us to stop. As we passed between the vehicle and the people a burst of automatic fire came from behind their vehicle. I wasn't sure whether to stop or keep going, I turned to Huck to ask him what I should do only to find him half out the window returning fire, there were empty cartridges ricocheting around the cabin. After Huck pulled himself back into our vehicle we checked ourselves for any injuries. After pulling up the lead vehicle we found out that they had been shot at [as well].[100]

As both vehicles were out of range of the ambush site, Berry sent Burdon 50 metres down the road towards the Somali vehicle to act as sentry while he contacted Battalion Headquarters by radio to report the incident. Fortunately the area was quite open so bandits would not have been able to follow up Berry's team without being observed from a distance. After a few minutes Burdon reported that the Somalis had got back into their vehicle and left the area in the

Corporal 'Huck' Berry, Technical Support Platoon(L), discussing a road ambush he encountered with (left to right) Craftsmam Shane Scott, vehicle mechanic, Warrant Officer David Lancaster ASM, Technical Support Platoon, and Warrant Officer Dale Sales RSM, 1 RAR Group.
Image: Bob Breen

opposite direction. Berry assumed that the bandits had made their escape rather than continue with their robbery. He continued with his task and repaired the Mortar Platoon vehicle before returning to Baidoa that same day.

This incident emphasised that technical support troops as well as infantrymen needed to be well trained in instinctive counter-ambush drills. Berry and his team were the envy of the Technical Support Platoon when they returned as some of the few members of the Australian Force Somalia who had come under fire. Berry and his team was just glad to have survived the incident unscathed.

On 1 May Bravo Company arrived for the last rotation in Buurhakaba before redeployment back to Australia. Support Company platoons occupied towns in the western region of the Baidoa Sector and the remainder of the 1 RAR Group remained in Baidoa and at the Airfield packing up and cleaning vehicles and equipment in preparation for their return home. There were no serious incidents reported in the country side during the last two weeks of

OP *Solace*. The bandits, like members of the 1 RAR Group, may have been counting the days off before the French arrived, and the time came for the Australians to depart.

Lessons

Australian operations to create a secure environment in the countryside of the Baidoa Sector for the distribution of humanitarian aid began with determination and enthusiasm in January. The Buurhakaba area was secured quickly and companies operated there in 9-day rotations from then on. Initially, the Australians operated in the Buurhakaba area as they had been trained—cordoning and searching villages, setting up vehicle check points and patrolling over wide areas in Landrovers and APCs. They also had expectations of capturing bandits and bringing them to justice. This desire to capture bandits was motivated partly by a desire create a secure environment, partly as a service to the long suffering inhabitants of the region, but mostly to satisfy expectations of active service and pass the warrior's test.

During February and March cordon and search operations were conducted systematically throughout the entire Baidoa Sector to establish an Australian presence and deter bandit activity. For many members of the 1 RAR Group their expectations of achieving the element of surprise, capturing bandits and confiscating significant quantities of weapons and munitions were not met. Without the element surprise, bandits had time to escape and there was also time for locals as well as bandits to hide their weapons and ammunition.

Losing the element of surprise was only one of the frustrations for the Australians in the countryside endured in the early weeks. Until they learned more about local Somali politics and culture they were sent on several wild goose chases, wasting time and energy reacting to unreliable information. At the same time local inhabitants, whom they thought would help them to find and apprehend bandits, did not provide information. Once the Australians established their presence, employed CI personnel and interpreters more effectively, thus gaining the trust of local Somalis, informants appeared and cordon and search operations began to get results.

Freeing the main roads from bandit ambushes was a major challenge. The Australians developed Night Rider and Road Runner techniques that did much to deter ambushes over time. Using night vision equipment, and radio communications, vehicle patrols shadowed likely ambush targets along the roads reducing ambush incidents significantly. Bandits were also deterred by white light ambushes that inhibited their movement at night.

When reliable information began to flow to the Australians about bandit activity, the ROE and a reluctance by senior officers in the 1 RAR Group to risk the chance of taking casualties, frustrated the efforts of some junior commanders to close with bandits and capture them. Many members felt that the Australians had a professional as well as a moral obligation to capture as many bandits as possible and bring them to justice. The senior officers operated from the premise that deterrence was the outcome, while their subordinates sought the more tangible outcomes of numbers of bandits, weapons and quantities of ammunition captured.

When they captured bandits, the Australians had expected that they would be brought to justice or at least prevented from returning to banditry. They were to be disappointed on both counts. Despite the encouragement of HQ UNITAF officers and UNOSOM officials, the provision of detention centres, training, uniforms, vehicles, equipment and pay, the newly-raised ASF were unable and, possibly unwilling, to detain bandits and present them with evidence of wrong doing for trial. Even with the protection afforded by the Australian presence, they operated ineffectively in an environment of intimidation and potential retaliation.

In the end, the Australians achieved their primary mission of creating a secure environment for the distribution of humanitarian aid in the countryside during the 17 weeks they had been responsible for the Baidoa Sector. They had also achieved a lot more by providing some respite to the Somali people in the Sector from the previously high incidence of physical and sexual violence associated with vehicle ambushes, robberies and village raids.

5   GETTING THE FOOD OUT AND DOING SOME NATION BUILDING

Today was good and it's the reason we came here—to help feed the people and give them perhaps a little bit of hope.

Major Anthony Blumer,
4 February 1993

While the 1 RAR Group was protecting aid agency employees and property in Baidoa and asserting their presence in the countryside, they were also providing close security for humanitarian aid convoys and vehicles transporting NGO medical staff visiting outlying villages and refugee camps. Coordinating the protection of the delivery of humanitarian aid brought the Australians into the frontline of crowd control, as well as community and aid agency politics. This political involvement and a desire by the Australians and HQ UNITAF to leave a legacy led to some nation building.

Flooding Somalia with food to lower its price and make humanitarian aid less attractive to looters began soon after UNITAF troops secured Mogadishu early in December 1992. By the time the Australians arrived in Baidoa in mid-January 1993 the US Marines had been providing security for food distribution in the Baidoa Sector since 22 December 1992. With the exception of small refugee groups living in remote areas or near villages where they had no clan allegiances, no one was starving. The Marine presence had also deterred nomadic bandits and local criminals from extorting money from aid agencies, or looting convoys *en route* or during food distribution.

The UNITAF Civil Affairs Branch had set up a Regional Relief Committee in Baidoa comprising a central committee of elders, NGO and security representatives and three sub-committees: food distribution, employment and contracts, and resettlement and construction. In addition they had established a Security Committee to co-ordinate security activities, an NGO Committee to co-ordinate NGO activities and a Council of Elders Committee to facilitate communications with the local populace. The Australians inherited these committees, and their membership and meeting arrangements.

David Hurley assigned Dick Stanhope and members of his liaison team to represent the 1 RAR Group at commitee meetings and to coordinate with all civilian agencies and individuals who had business with the Australians in the Baidoa Sector. Stanhope and his team had not trained for or had expected to be involved in this type of work. They anticipated liaison duties between the 1 RAR Group, its superior headquarters and neighbouring military units. HQ UNITAF did not require UNITAF battalions to have direct radio contact with each other or for liaison teams to be exchanged between units in neighbouring Humanitarian Relief Sectors. Consequently, Stanhope concentrated on establishing civil and military operations teams.[1]

Major Dick Stanhope, OC Civil and Military Operations Team, outside 1 RAR Group Headquarters early in 1993
Image: ADF Electronic Media Unit

In the beginning, Stanhope's Battery Commander's CMOT team remained at the airfield in charge of civil-military communications and operated as a point of contact for all civilian agencies. He deployed the other Forward Observer CMOT teams with food convoys and with whatever company was on operations in the countryside. Stanhope detached his two spare gunners, Gunners Peter Malone and Marc Peters,[2] to Support Company where they performed duties as Regimental Police.

Soon after the arrival of the 1 RAR Group in January, Captain Stu Bagnall[3] and his CMOT team were detached to HQ 10th Mountain Division in Mogadishu. His role was to attend briefings and report back on matters effecting the 1 RAR Group. This group quickly proved their worth by providing Hurley and his operations staff advanced warning of tasks and operational issues before they were communicated through normal channels.

Unlike the Marines, who Stanhope assessed 'were extremely hesitant to assist the NGOs', Hurley and his staff always assumed that they would work closely with UN aid agencies and NGOs to protect the distribution of humanitarian aid. However, Stanhope and his team found each NGO to be 'reasonably well organised individually, but not well coordinated [collectively].' Indeed, there was some rivalry among the NGOs that made Stanhope's efforts to forecast security requirements for their activities challenging.[4] He found that the Regional Relief Committee was 'not functioning well, as the NGOs would rather not attend meetings and the Committee has not been able to achieve results to date. 'The NGOs say that some of the members of the Committee are the people who organised and participated in the looting prior to military involvement.'[5]

Many of the expatriate NGO staff chose to avoid close associations with the 1 RAR Group. They did not want to compromise their neutrality and their ability to negotiate with all Somali political factions and elders. The experienced NGO managers realised that one day the Australians would leave and they would be responsible once again for their own security.

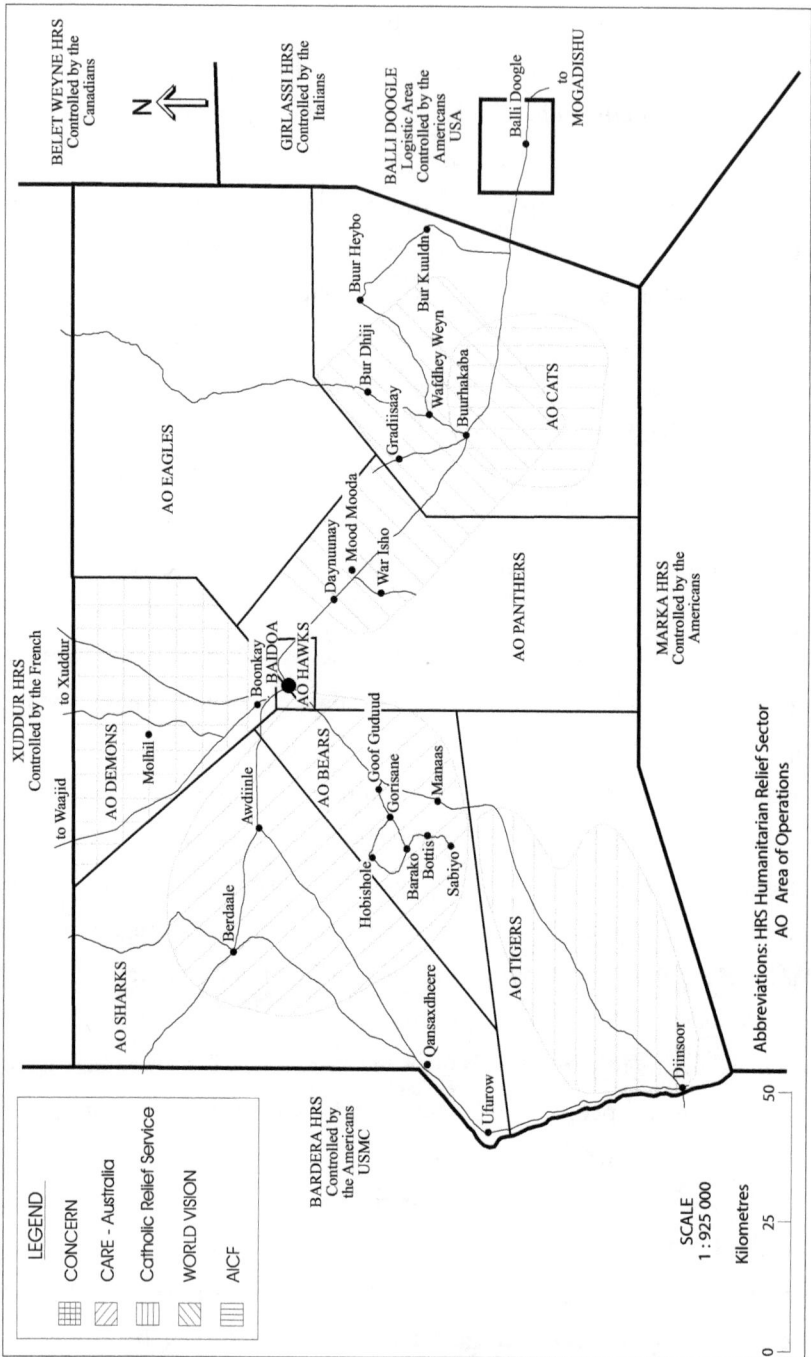

Map 7: Unofficial NGO Areas of Operarion for Food Distrubution.

Sergeant Joanne Cook, Medical Assistant,
Medical Section, Battalion Support Group,
assisting a starving youth during a clinic.
The Australians did much to supplememt
the humanitarian relief effort in the
Baidoa area.
Image: Gary Ramage

This attitude led to some misundertanding between the Australians and the NGOs in the beginning. From Stanhope's perspective some NGO staff appeared to be uncooperative and recalcitrant.[6] However, the majority of NGO managers were grateful for the physical security of food convoys, techical assistance and support for their humanitarian relief efforts.

The NGOs were getting a good deal from the Australians whose mission was confined officially to protecting the distribution of food and providing security for ex-patriate NGO managers. Major Darrell Duncan,[7] the Medical Officer with the Treatment Section at the BSG, and his staff began conducting daily clinics at the IMC Hospital in Baidoa and Duncan performed many surgical procedures. The Australian medical staff also began assisting World Vision, UNICEF, CARE Australia and GOAL with clinics and a large scale immunization program for villagers in outlying areas. The NGOs were also able to call on the Australians for minor construction tasks, technical repair of communications equipment, generators and refrigerators, movement of containers and recovery of stolen vehicles.

Notwithstanding the initial challenges of coordination, the system for the physical distribution of food was well established by the time the Australians arrived. The UN World Food Program arranged for food to be transported to Baidoa Airfield by air or by road from Mogadishu. An average of twenty C 130 transport aircraft landed at Baidoa each day and were unloaded by Somali contract labour. The NGOs would bid to World Food Program managers for sufficient food to distribute to villages in their unofficial areas of operation in the countryside. These tonnages would be loaded onto trucks by Somali workers hired by the World Food Program. Expatriate NGO staff were contracted for onward distribution of the food. This involved making arrangements with village elders, leading convoys to their destinations and supervising the distribution of food from temporary food distribution points or directly to feeding centres near villages.

The system for protecting food distribution had also been established. However, the Marines preferred to dump rather than distribute food to save time and minimise security risks to their troops. Unfortunately, this dumping policy allowed bandits to steal large quantities of food before villagers and refugees could get their share. Somali drivers were also happy to dump rather than wait while food was distributed. Stuart Dodds wrote:

> Each day a line of trucks, often overloaded with people who acted as loaders and porters who the driver had to pay out of his money, would appear at the front gate of the Airfield in the early hours of the morning, waiting for first light so they could be searched by the Marines. The trucks would then move one at a time [into the Airfield to be loaded].

> The NGOs hired the trucks at $US280 per load. Consequently the drivers were keen to get to the front so they could get out with an early convoy, unload and return to the Airfield for another load. On a good day, a truck driver could make three round trips and earn almost $US850 a day. '8

*Somali trucks with plenty of passengers
and sacks of grain lining up to join a
convoy in Baidoa Airfield in early 1993*
Image: ADF Electronic Media Unit

The Australians had not protected the distribution of food before but they had been trained in crowd control and convoy protection procedures before embarkation. While they were taking over security responsibilities in Baidoa and at the airfield, several officers and NCOs accompanied the Marines on convoy protection tasks to observe their techniques. Todd Everett wrote:

> When the Marines escorted convoys to food distribution points, there was an impressive display of firepower on the ground and in the air. Helicopters prowled the skies and a company of about 120 men rode in vehicles, ready to shoot. The orders to commanders were very specific and restrictive. Time out, time in, routes to follow etc. When anything out of the ordinary happened, or they began to run out of time, the food was either dropped quickly near its intended location or it all came back. The NGOs found the Marines strict adherence to timings to be very frustrating. Then again, they [the NGOs] did not have their act fully together either.[9]

The performance of NGOs in organising the distribution of food depended on the experience of the expatriate manager in charge and the comprehensiveness of prior arrangements made with village elders. Some expatriates were inexperienced and their methods for identifying those who was entitled to receive food were slow and confusing. For example, one expatriate in charge of food distribution used to call out names of people and invite them to come forward to receive food. Every time a name was called out several people raised their hand identifying themselves as that person. It did not take long for fights to break out between Somalis claiming to be the same person.[10]

Frequently bandits and village elders colluded to steal food as well as to deny food to anyone except members of their own clan. The town of Daynuunay was one place where this type of collusion occurred:

> They [the bandits] come into town to steal as much food as they can cart away, this being accomplished with eight to ten donkey carts. The gunmen reportedly fill up the carts and make numerous trips to town. No sources told of any murders or shootings, but they did say the gunmen hit people with their rifle butts. The food that is left is distributed by the Council of Elders to people in the township proper. By township proper, it is meant the people of the same clan who reside in a strict geographic boundary. A village of 50 refugee families located at the northern end of Daynuunay receives no free food…. An old man in the main part of Daynuunay, who acted like he was an authority figure, was asked about the living conditions of the refugee village north of town. The old man stated that he only takes care of his own clan.
>
> Mr Jouko Ala-Outinen, the supervisor of the food drops and an employee of the World Food Program [WFP], stated that his organisation was only responsible for the delivery of the food, and that CARE International supervises the distribution of the food. He further stated that WFP delivers enough food each time to feed the town for a few months, however, the food only lasts a few days.
>
> Mr Martin Morris, Chief of CARE in Baidoa was questioned about the distribution problems.[11] He stated that he was aware of the problems in this locality, but that his organisation operates by strict guidelines to distribute the food through the local Council of Elders. Mr Morris further stated the the UN is trying to saturate the various hard-hit areas with enough food to cause the market price to fall. This action would take away the incentive to steal or hoard the food. He then noted that despite the problems with distribution, compared to the conditions just this past August, the people of Daynuunay were positively fat. (Note: One wonders about the fate of the refugees [north of town] under this UN policy.) [12]

In January Stanhope's CMOT Command Post co-ordinated all the bids for security forces to escort food convoys. The NGOs distributing food at the time were CARE Australia, Irish CONCERN, World Vision and the US Catholic Relief Service. For the initial convoy protection tasks, David Hurley used his experienced Support Company Platoons, who had their own vehicles, and Charlie Company. The Mortar Platoon escorted the first convoy on 19 January to distribute food to a series of CARE Australia client villages south-west of Baidoa. The platoon was accompanied by Phoebe Fraser, a contingent of CARE Australia public relations personnel, a group of Australian and British journalists and several camera crews.

Lieutenant Peter Connolly wrote at the time:

> The day was hectic and confusing. Too many 'hangers-on'. The obligatory press (some of whom should be shot!) led by a bloody LTCOL (?!?!) who insisted on letting them get in the way.

> It was funny watching one reporter making his report to camera. He looked like a peacock in the mating season, redoing his spiel 10 to 15 times to get a camel in the background.[13]

Stuart Dodds was there the same day:

> The first village was fine ... the men [Somali workers] off-loaded the trucks singing and dancing, racing each other to be the first to empty their trucks. The team that came first had a dance to celebrate. ... We set up wire [barriers] to keep the people back but it wasn't necessary and they were well behaved. The last village was a model village [Hobishole] and had an Australian windmill pumping water. We did not have to secure the trucks to unload there. I flew my kite and the people came out and performed a dance while singing to welcome us. Quite a buzz.[14]

The next day the tempo of convoy escorts increased significantly as Charlie Company provided support to CONCERN, World Vision and the Catholic Relief Service. Mick Moon could only afford to send a platoon and sometimes a section of seven to eight men to protect some convoys because of the number of convoy protection tasks required of his company.

Despite learning a great deal from observing the Marines, the Australians were to learn the hard way how things could get out of hand. Moon wrote:

> Met Paul Enright (Irish CONCERN), set out with three 6 x 6 and one FFR.[15] Got away at 0745 hrs. Long trip with a few stops. All OK for about 2 hrs. A fight broke out with Somali males swinging machetes. Crowd get restless. These people who are causing the trouble are not from the village receiving food.
>
> Elders then lost control. An axe fight broke out at the rear of the village. Troops deployed to quell this problem. Took over from NGOs. Started [distributing food] again after 25 minutes. Fighting started again. Others not entitled to food turn up. The crowd has swelled to 3 000. There are many youths in the crowd. Security situation now critical.
>
> Paul decides to pack up and go. We had a tense extraction while we pulled in the [barbed] wire. Somali workers hard to control. Just as crowd [was] closing in we had a breakdown. ... Fixed with spanner and we managed to get out ...
>
> Men are shattered. Todd [Everett] did very well. I tried to get the food out right up to the end, but the NGO wanted out. Got back to airfield at 1715 hrs.'[16]

The next day Charlie Company were involved in another difficult crowd control situation. Moon took Ian MacGregor's platoon to a village near Buurhakaba. He wrote:

Somali women lining up for food distribution, under the watchful eye of village men with 'elegant sticks'.
Image: G Gittoes

As 8 Pl [platoon] set up their perimeter, people started streaming from Buurhakaba. All quiet for 2 hours then trouble. Elders could not control people from outside the village. Soldiers working hard to keep crowds off the perimeter. ... People started to run. Couldn't keep them out of the compound. A big rush meant World Vision guy decided to make a point and take the last truck back [full of food]. Willesee [17] turned up at food site and then departed very quickly. Fairly tense as we withdrew (Food truck first). A quick trip home.[18]

Journalists accompanying Charlie Company's first food distributions reported that the diggers had been involved in controlling dangerous food riots. There were several dramatic photographs in the Australian press of soldiers surrounded by distressed and struggling Somali women. Moon wrote, 'Read front page news of "Food Riot" at Buurhakaba. Complete fabrication. Quite pissed off'.[19] He hoped that when the media lost interest in the Australian operations in Somalia he would be able to get on with the job without fear of similar sorts of sensationalism.[20] Left unsaid at the time was that the Australians did not sit back protecting trucks while crowds battled for food. They immediately took responsibility for the orderly distribution of the food to the people in addition to protecting vehicles and NGO staff. A benign form of mission creep had begun.

Not all of Charlie Company's food escort operations on 21 January involved controlling volatile crowd situations. Anthony Swinsburg's platoon's distribution had gone well and his men had been entertained by Somali 'thank you' dances and their morale was boosted by cheering,

Lieutenant Todd Everett with Paul Enright from CONCERN before food distribution in January 1993 Image: ADF Electronic Media Unit

welcoming crowds. Todd Everett's platoon had also been welcomed and cheered at Qansaxedeere.

The command and control arrangements for protecting food convoys had some teething problems. CMOT teams accompanied the first food convoys and insisted that they be the contact point for liaison with expatriate and Somali NGO staff. In effect, they were trying to be intermediaries between the Australian commander protecting the convoy and the expatriate manager responsible for distributing the food.

Though well-intended, there was no need to send CMOT teams on food convoys. Infantry convoy commanders proved to be quite capable of working directly with expatriate NGO managers. Stanhope found that he and his gunners were able to do their liaison job much more effectively back in Baidoa and leave convoy commanders to deal directly with NGO staff on the ground. After some clarification, the arrangements were that Captain John Hill,[21] Stanhope's Battery Captain, received requests for convoy protection from NGOs and passed them to Captain Peter Short,[22] who was appointed Convoy Master for all humanitarian convoys in the Baidoa Sector. These arrangements worked well. Short wrote:

> The largest factor in improving the coordination of convoys was the decision to force the NGOs into planning their distributions in advance. Priority of support was given to those who could plan ahead. World Vision were the best for this. Convoys were decided 24 hours in advance. This allowed for thorough staff checks and briefings. Flexibility was still exercised which allowed for last minute changes ... .I was able to tell an NGO well in advance when support was, or was not, available. It meant that NGOs could reschedule their own distributions which helped planning further.[23]

Not all food distributions were conducted by road. When the risk of land mines was perceived to be high, the Australians asked HQ 10th Mountain Division for helicopter support to deliver food. Todd Everett was involved in the first of these heliborne food distributions:

Paul Enright, the CONCERN representative, and myself conducted an aerial reconnaissance of the village [Molill] on the day before the distribution was to occur. ... We were using GPS to locate the village. I indicated to Paul that we were over Molill but he said it was the wrong place. We then flew over what he said was Molill and an appropriate landing zone was selected.

... Our [start] time [the next day] was put back due to a large media interest in the distribution ... The insertion [of troops] went very well. The platoon secured the landing zone and we waited for the food to come in. Paul and his translator went off to find the Chief Elder of the village. I detected quite a bit of embarrassment when Paul came back and told me that we were in the wrong place ... We were unable to fly to the original village as the Americans were locked into the plan and the food was inbound. The media naturally wanted to talk to us about this little problem.

... Paul found a man from Molill and sent him to bring a truck from Molill to the LZ, [presumably] over the same mines that we had been afraid of.[24] The truck arrived after the second of three loads of food came into the LZ [Landing Zone]. Time was running out because we were due to fly out after the final load of food came ... The Somalis [from Molill] worked like wild men to load the truck and drive it back and forward to Molill.

There was still 100 bags of grain left over when it came time to fly out and a large crowd from the village we were in had gathered. The small group of elders [from Molill] who were accompanying the truck attempted to hold back the crowd as we flew out. They couldn't hold them, the [ensuing] fight looked like an ant's nest gone crazy ... We were disappointed that the Somalis fought so savagely for the remaining food instead of sharing it out. This was our first glimpse into how the Somali people treat each other.[25]

Charlie Company were keen to deter bandits from entering villages to steal food after it was distributed and convoys had left. The Americans had reported that there were distribution problems at Danuunay. Charlie Company was keen to confront the bandits on their own 'turf'. Moon decided to follow American counter intelligence team advice, 'that a positive way to discourage gangsters stealing food in this area is to engage them as they come into town to steal it.'[26]

On 25 January Swinsburg's platoon deployed on the first Australian-protected food distribution to Daynuunay. Moon wrote:

> 7 Pl [platoon] food drop off main road to south of Daynuunay. First drops go OK then locals report that there are bandits outside town with weapons who will steal food once the Australians have left. It took 1 1/2 hours to get permission from OA [Battalion Headquarters] to stay and wait for bandits.[27] Travel 2 km down the road, wait 20 mins and then double back and catch the bandits red handed. Some drop weapons and run off, others run with their weapons through fields. 5 rifles captured. P [platoon] very happy with outcome. No shots fired.'[28]

Efforts to deter bandits from stealing food after it had been distributed was another form of well intended mission creep but it did increase the chances of shots being exchanged between Australian troops and bandits. There appeared to be three motives.[29] The first was to give local Somalis a 'fair go'. The second was to demonstrate that the Australians were willing to protect the people against the bandits and the third was to confront bandits as a part of creating a secure environment. For most of the infantrymen and cavalrymen of the 1 RAR Group, if deterring bandits from stealing food involved gun play, then 'so be it'. Hurley's staff at Battalion Headquarters had the difficult task of arbitrating on whether stalking bandits who were stealing food, or were suspected of moving in to steal food, fell within the parameters of the UNITAF mission of 'creating a secure environment for the distribution of humanitarian aid'.

Distributing food near Buurhakaba continued to be a challenge after Moon handed over convoy protection duties to Delta Company. On 4 February Delta Company encountered a similar situation that had faced Charlie Company two weeks before. However, Blumer had learned from Moon's experience:

> Third distribution was too close to Buurhakaba and people from the town came out to have a second go at the food. The elders had done

the right thing and had sat everyone down, so we threw the wire around them quickly to isolate them inside the perimeter and adjusted the seating area so the crowds wouldn't get at them before they had broken down the food sacks and tied them up and headed off.

Those not getting food became increasingly annoyed but realised there was little they could do as all the vehicles etc [and those entitled to food] were inside our wire ....

The troops worked very well and it was really hot out there with helmet, flak jacket and working in 40C plus heat. Its remarkable how fast you become used to it and the weight of our webbing, by comparison to when we first arrived. Today was good and its the reason we came here—to help feed the people and give them perhaps a little bit of hope.[30]

Most of the crowd control problems the Australians encountered occurred in the first three weeks after their arrival in January. Moon briefed Blumer, and Blumer briefed Fraser, who took over convoy protection duties after him, about their experiences. Moon made recommendations on how to assess the NGO preparations for each food distribution and how to anticipate trouble. Each company improved their techniques as time went on. Controlling crowds at food distribution points as well as protecting vehicles and NGO staff became routine by the middle of February. Peter Short wrote:

The people responded favourably towards our soldiers and that is understandable as they were being brought food. Different companies had different attitudes [to the Somalis] which tended to reflect upon their commanders. Some companies adopted the firm and friendly approach whilst others opted for the more aggressive stance. I believe the aggressive approach was unnecessary, however, no company stood out from the others as being better or worse at conducting such operations.[31]

Relations between the Australians and Somali truck drivers were an ongoing problem. The drivers were held in contempt because they were perceived by the diggers as uncaring, money-hungry thieves.[32] Warrant Officer Paul 'Gus' Angus,[3] Charlie Company's Company Sergeant Major, had an altercation with a driver after he ignored Angus's direction not to back out of a food distribution area because he would run over groups of women who were assembled behind his truck. Angus punched the driver through the open window of the driver's compartment several times after he had ignored Angus's directions, started his motor and was driving off with Angus standing on the running board.[34]

These tensions resulted in a more serious altercation with Somali drivers and Australian infantrymen some weeks later when Lieutenant Jan Van Der Klooster's platoon was protecting a food distribution point near Awdinlee, 30 kilometres west of Baidoa. The distribution had begun routinely. The sections had secured an area around the trucks, erected a concertina barbed wire barrier around the intended distribution area, formed an outer perimeter of sentries to keep crowds off the wire barrier and positioned machine gunners on truck cupolas to maintain surveillance of the approaches to the food distribution area.

After six hours in the hot sun, the Australians were looking forward to packing up and leaving. The Somali drivers were also very keen to leave because they had been ordered to stay within the barbed wire perimeter until all trucks were unloaded. Van Der Klooster wrote later:

> The normal procedure was to let the Somali trucks go as soon as they were empty. On this occasion, the trucks could not get out until we had finished, unless they drove over the seed mats and all the people on them.
>
> Anyway, I was standing to the rear of the first truck when I noticed the truck's driver climb into the cab and try to start the engine. I moved over to him and told him he could not leave ... He became fairly angry and started arguing, but hopped out of the cab and moved over to one of the utes [utility trucks].
>
> Thinking the situation was over, I moved back to watch the distribution.
>
> All of a sudden he rushed over and jumped back into the driver's seat. He started up his truck and slammed it into reverse and proceeded to drive into the women who were kneeling down collecting seed from behind his truck.'[35]

Private Chris Keyte[36] who was standing nearby saw Van Der Klooster rush to stop the driver reversing his truck into a group of women by jumping up on the running board of the truck and reaching in to take out the truck's ignition keys. As Van Klooster was looking down towards the dash board, grabbing for the keys, the driver reached behind him and brought out a long handled metal torch. 'I saw what was going to happen, so I pushed past the boss [Van Der Klooster] and pulled the troublemaker down and began adjusting his attitude. I was all over the poor bugger when I got hit from the side by another Somalian.'[37]

Van Der Klooster took on the other Somali driver who had attacked Keyte and 'a fairly well aimed punch flattened the driver who swung a punch'.[38] A few more drivers then joined in the fracas. Van Der Klooster's men came running in a manner reminiscent of football brawls in Australia and a melee ensued. Private Jason Watson joined in:[39]

> With no hesitation I decided to join the boss and Keyte in their re-education [of the Somali drivers]. I jumped the Somalian that Keyte was wrestling with, I grabbed him in a choker hold, and got more than I bargained for. He was as strong as an ox. I thought to myself for someone who is supposed to be hungry and undernourished, he was doing all right ... to my left one of his mates was going off his nut 'Please Australia, let my friend go'. 'Tell your mate if he calms down I will let him go, shit, the bastard has just head butted me.' ... Anyway, after calming him down, I released him. At this time everyone was pretty hyped up.'[40]

Almost as soon as it had begun, the brawl was over and four Somali drivers were 'on the deck.' However, one more driver decided against better judgement to advance on Watson with a truck tyre lever. As he approached Watson, who raised his weapon, Lance Corporal Adam Pavlic[41] ran over and grabbed him in a strangle hold and he soon dropped his iron bar.

> The next thing I know I had been bowled over by [Lincoln] Carson[42] who must have done a zoom dash in about two seconds to get into this action. Then I ended up on top of him, still in a strangle hold. Then 'Kit' threw two punches—both connected with me and not the Somali.'[43]

This untidy incident ended with the Somali driver who had attacked Watson being detained and taken back to Baidoa to face charges of assault. The other drivers dusted themselves off and drove back in convoy to Baidoa—sullen and obedient.

Despite their fees for each load, Somali drivers often pilfered food.[44] The Australians took special delight in ensuring drivers did not get away with this practice. Blumer wrote,

> The truck drivers slash bags for spillage and get really stroppy[45] when we sweep the backs of the trucks out and give it away. We will also let it fall to the ground rather than letting them get it ... We followed up one of the trucks that dropped off a sack untorn and promptly confiscated it.[46]

The corruption evident in the distribution of food also angered most Australians who had come to Somalia to feed the needy, not increase the wealth of village elders and bandits. Moon commented after a food drop at Ufurow, 'Grain was taken away and immediately put on sale in the town. I can see many Somalis in this area wearing gold jewellery. All the houses on the main road have corrugated iron roofs. Bandit city.'[47] After watching food distributed on 16 February, Fraser commented, 'it is very evident that corruption and theft abounded.'[48]

By early February the Australian CI Teams had discovered how well organised the syphoning off of humanitarian relief supplies into the local economy had become:

> When food arrives (2000 bags) the elders give 100 bags to bandits from outlying villages and direct 150 bags to the Buurhakaba markets. ... The remaining aid goes into store and is sold to businessmen from Baidoa and Buurhakaba. It takes about seven days for all the food from one drop to be sold off or paid off to bandits.
>
> In excess of 75 percent of all bulk aid deliveries are redirected to Buurhakaba/Baidoa markets at elders' direction. The bandits use wells as staging areas for operations and the wells are important as focal points for transient activity. They extort money at these wells

which forms a steady source of income. Refugees, who are most in need of succour, bear the greatest burden of banditry.'[49]

The CI team assessed that the World Vision guards, who have been hired by the elders, are used to secure the aid from refugees and rightful recipients, and that the information supplied on the bandits [to the Australians] was an attempt by the elders to use the Australians to eliminate this source of loss from their profitable operation.[50]

The Australian reaction to the black market in humanitarian relief supplies and theft by bandits after deliveries were made was to establish vehicle check points and mount several cordon and search operations in the Buurhakaba and Daynuunay areas [See Chapter 4].

However there were insufficient troops to continually return to villages or maintain surveillance of villages after food had been distributed. Generally it was timely reactions by Australian convoy commanders to local information about the presence of bandits that seemed to deter bandit activity. Aside from verbal warnings to village elders that hoarding food for distribution to the black market was not equitable and denying refugees their share was unfair, the Australians had no authorisation to interfer with the elders' control of the distribution of food.

The other form of corruption related to the distribution of humanitarian aid was the extortion of NGOs by village elders who would not permit children to be immunised or other medical services to be provided unless they were paid in grain or cash. GeorgeGittoes, an Australian artist on assignment with the 1 RAR Group in Baidoa, wrote:

> There was considerable delay while Suzanne, [a World Vision expatriate manager], argued with a group of elders. She came and explained to us that the elders were saying, 'If you want to give aid to our region, you must pay for the privilege.' In other words the elders intend to exploit the misery of their people for gain, aware of the keenness of the aid organisations to get on with their work. The greedy elders want to be paid in grain or whatever, to give the aid workers permission to proceed. Suzanne made a deal ...[51]

Set to this background of redirecting humanitarian aid for commercial gain and the extortion of NGOs by village elders, HQ UNITAF with encouragement from UNOSOM, became directly and indirectly involved in nation building. Nation building was a term used to describe efforts to rebuild southern Somalia's infrastructure, such as roads, water supplies and municipal buildings, and local government services, such as police, judiciary, administration, education and sanitation.

Southern Somalia benifitted from the presence of UNITAF. The Texas-based American contractors, Brown and Root, and the US Army engineers and US Navy Sea Bees repaired the runway and taxiways at Baidoa Airfield, drilled and established water wells, and repaired and upgraded main roads within the Baidoa Sector. Most of these construction tasks had short term military and humanitarian motives, but also served to upgrade Somalia's regional and national assets.

For UNOSOM a central pillar of nation building was the encouragement of moderate Somali political factions and facilitating the processes of political reconciliation. UNITAF commanders, whose mission was to create a secure environment for the distribution of humanitarian aid, approached involvement in Somali politics cautiously and were concerned about mission creep.[52] However, their military jurisdiction and protection of the distribution of humanitarian aid brought them into close contact with local Somali authorities and competing political factions. Hence, political involvement was very difficult to avoid.

In the Baidoa Sector David Hurley's involvement in Somali politics began with having to establish a working relationship with the Governor of the Bai Region, Dr Mohamed Ibrahim Hussein, who was also Chairman of a pro-Aideed faction of the Somali Democratic Movement (SDM), a mostly moderate political movement made up of four factions in the Baidoa Sector. The SDM was based on the Rahaweyn Clan that constituted 90 percent of the population in the Baidoa Sector. The agenda of moderate SDM factions was to reconcile the pro-Aideed and pro-Ali Mahdi factions and form a united front against the violent agenda of both warlords. This was made more difficult when the Governor was pro-Aideed and the Australians had to acknowledge his status in front of members of the pro-Mahdi SDM faction.

Aideed's and Ali Mahdi's supporters were the main political protagonists in Baidoa, both having the support of factions of the SDM. General Aideed's faction made its presence felt soon after their arrival when Mohamed Jumale Osobleh proclaimed himself the Chief of Police and told the Australians he had a force of 300 loyal policemen who were ready to restore law and order. Brendon Thomson assessed Jumale as being, 'a sly, self serving individual', looking for recognition from the Australians so that his men could be declared as the official police in Baidoa and achieve political advantage over their opponents.'[53]

Police General Muhman Ibrahim Isaak, one of Aideed's strongmen, had travelled from Mogadishu to install Jumale as the Police Chief, establish an office for the Somali National Alliance (SNA) and raise a unit of the Somali Liberation Army (SLA). Aideed's USC faction stood little chance of acceptance as the legitimate representatives of the people of Baidoa, so the SNA was founded with the declared purpose of representing all the clans of Somalia. In reality the SNA was a new title for Aideed's old USC faction and the SLA was to be its military arm. The SLA opened an office on NGO Road and began recruiting on 1 February. Colonel Abdullahi Aden Hussan, known as 'Ciro', the newly-appointed CO of the SLA, met with Thomson and Douglas and advised them that the SLA's mission was to recruit men who would support Australian operations in Baidoa and had no political ambitions.

Thomson and Douglas assessed that Ciro was a weak, mild-mannered individual, easily manipulated by others in the SLA. One of his senior advisers happened to be the notorious local criminal and Aideed supporter, Gutaale.[54] It was a small world in local Baidoan politics. The Australians soon realised that they were in the middle of a complex power struggle by Aideed's supporters with the moderate SDM factions. Any recognition they gave to the SNA, Jumale's police force or Ciro's SLA would have significant political consequences.

Hurley was suspicious of the motives and capabilities of the SLA, who claimed that they had a force of 4 000 men. He tasked Thomson and Douglas to find out more about them. They soon achieved cordial relations with SLA officers, to the extent that Douglas reviewed a number of SLA parades that often numbered over 450 soldiers and addressed meetings of the SLA cadre through an interpreter. As the days went by, Thomson and Douglas were able to provide information on SLA personnel that needed to be removed from office because they were criminals. This information also helped to defuse disinformation and deter planned political violence originating from the SLA.[55]

In addition to efforts to reconcile the two Mogadishu-based warlords, the SDM factions were running a more traditional agenda. From the beginning the Governor, representing the pro-Aideed faction, made it clear that his objectives were to collect taxes and restore local government institutions, such as civil administration, hospitals, schools, sanitation, a police force and a judicial system. For some observers, the Governor's desire to restore a tax collection system was motived by self-interest rather than civic duty. He had allegedly been involved in many corrupt and violent activities before the Marines and the Australians had arrived.[56]

The other players in local politics were the members of the Council of Elders. This group of 12 men represented the 41 traditional clans in the Baidoa region and were not formally aligned to any of the Western-style political factions competing for power, although they appeared to be united in their opposition to Aideed's SNA and SLA. From their point of view Aideed belonged to another clan that had no business asserting his authority in the Baidoa area.

The Council of Elders were the links to the village chiefs and the local populace, maintaining their status and power because Hurley and the aid agencies had no other option but to work through them to co-ordinate the delivery of humanitarian aid as well as having them authorise construction projects and the establishment of wells. The Council of Elders were also consulted about the provision of medical services, employment and resettlement. By Western democratic standards, the Council of Elders was a corrupt group of individuals whose role was to promote the interests of their clan at the expense of other clans. The major difference between the Council of Elders and a Western local government council was accountability. There were no institutions, such as a free press, judicial authorities or community organisations, that had sufficient influence or authority to monitor the activities of Council, to question decisions, or to take action to curb corrupt practices.[57]

Aside from trying to establish working relations with the NGOs, Stanhope found his role of working with the competing Somali political factions and the Council of Elders a significant challenge.

> It is extremely difficult to determine the agenda of the local people.
> ... Our long term aim with the community is to assist them to re-
> establish the area's infrastructure. Our first step is to determine 'who
> is who' amongst the leaders of all interested groups and identify
> 'which clan is which'. So far the community has not done much to
> help itself.[58]

During January and February Hurley was drawn increasingly into the local politics of food distribution, the management of national building activities and the struggle among the Council of Elders, Aideed's SNA/SLA group and SDM factions for power. Once they found out that Hurley was the CO of the 1 RAR Group, the Council of Elders and SDM factions would not accept Stanhope or members of CMOT as having sufficient status to negotiate with them. They insisted that they had to speak with Hurley, the Australian 'Chief Elder'. The processes of negotiation and consultation in Somalia were slow and meticulous. Hurley found it increasingly difficult to give as much time as he would have liked to the supervision of 1 RAR Group operations and visiting his troops because of the requirement to attend meetings with elders and political figures in Baidoa as the *de facto* military governor: He recalled.

I was very much a novice as military governor. It's not something that rested very comfortably with me because there were quite a number of responsibilities that I was given that I had very little preparation, and certainly no training to be able to achieve—particularly in my relationship with emerging political organisations and with the eldership in the area which was a very important and powerful community organisation.[59]

Despite the military power Hurley had in the Baidoa Sector, he had no authority to take formal action against members of the local Council of Elders or the elders in villages who were syphoning off humanitarian aid for commercial gain. The best he was able to do when these practices became obvious was to lecture the Elders on their responsibilities for the equitable distribution of humanitarian aid and occasionally deliver a firm warning to village elders when their activities were exposed during Australian operations.[60]

While Hurley was adjusting his politico-military role in the Baidoa Sector, the UN agenda for nation building in southern Somalia began to unfold. In his report to the UN in August 1992 Peter Hanson had recommended the establishment of local Somali police forces, recruited on objective criteria, and then trained, equipped and monitored by UN agencies. From the beginning of the UNOSOM military intervention in September 1992 Dr Boutros-Ghali had advocated UN-sponsored national reconciliation and reconstruction in Somalia. His vision was for UNOSOM to broker a political solution to Somalia's civil war and then facilitate nation building using UN and NGO agencies. An important part of nation building was to be the establishment of a judicial and penal system supported by a Somali police force.

Staff at HQ UNITAF, aware of the political and security challenges facing UNITAF commanders in their Humanitarian Relief Sectors, supported the restoration of law and order through the establishment of a police force, and a judicial and penal system. From the UNITAF perspective these nation building activities would facilitate a transition from UNITAF peace enforcement operations to UNOSOM peace keeping operations. Within HQ AFS the strongest advocate for Australian involvement in setting up a police force and restoring judicial and penal systems was Major Mick Kelly, the AFS Legal Officer. Kelly took a personal and professional interest in contributing to the development of policy in HQ UNITAF's Judge Advocate's Branch. He cultivated a close working relationship with Colonel Spitari, the Judge Advocate, discussing and promoting his ideas enthusiastically. Kelly was able to convince Spitari that the Baidoa Sector was the place to start setting up the Auxiliary Security Force (ASF) and reestablishing a penal and judicial system as a precursor and model for the re-establishment of these systems in the other Humanitarian Relief Sectors of southern Somalia. As a consequence of Kelly's advocacy, the 1 RAR Group received an unforeseen additional task on top of their security operations. The Baidoa Sector was to become the test-bed for Kelly's model for the establishment of the ASF, and a judicial and penal system.

On 4 February David Hurley presented his plan for the establishment of the ASF in the Baidoa Sector. He gave Dick Stanhope the job of being the liaison officer between the 1 RAR Group, UNOSOM and the ASF. The process began with the distribution of leaflets in Baidoa and a briefing to the Council of Elders advising them that anyone with at least two years previous police service, no criminal record, and the approval of the Elders could be a member of the ASF.[61]

During this time Kelly visited Baidoa regularly and facilitated the appointment of two judges, Mohammed Isaac Ibrahim and Abdullahi Aden Isaak. The establishment of the ASF in Baidoa was to be the first stage of the introduction of 'a national police force.' Kelly continued his visits to Baidoa in February with American legal officers to monitor progress in the establishment of the ASF and to begin the process of recruiting more judges and judicial staff to operate a new judicial system based on the 1962 Somali Criminal Code.

Meanwhile, after attending several meetings with the Council of Elders, the Governor and the Relief and Security Committees, Hurley and Stanhope recognised that there was substantial local support for the moderate SDM agenda of non-violent reconciliation and a united front against the warlords in Mogadishu. Opposing these factions and claiming to be legitimate representatives of the people were Aideed's SNA and SLA, whose members were involved in criminal activity and colluded with NGO guards and bandits to loot relief supplies and conduct robberies.

Though it was not a formal part of their mission to support the fortunes of any political group, Hurley and Stanhope saw it as in the best interests of maintaining a secure environment and promoting national reconciliation to support the moderates in the SDM and distance the 1 RAR Group from the SNA/SLA. In effect this meant denying SLA requests to work with the Australians on security operations. Concurrently, Hurley and Stanhope worked hard to establish the ASF as a neutral Somali security force and facilitate ASF involvement with Australian operations.

While Hurley and Stanhope became increasingly involved in the politics of nation building, humanitarian aid operations continued and services to the NGOs increased. After the MSF compound was robbed by ex-MSF guards, Stanhope and his team had become a cash deposit bank for the NGOs. They held several hundreds of thousands US dollars in a safe in their office at Battalion Headquarters. They had also become the de facto air controllers at Baidoa Airfield when UNITAF withdrew air control specialists. The NGOs were also able to call on the Australians for minor construction tasks, technical repair of communications equipment, generators and refrigerators, movement of containers and recovery of stolen vehicles. For their part, the Australians supported the GOAL Orphanage by building a playground, donating stationary and refurbishing kitchen and toilet facilities.

Concerned about the syphoning off of food into the local markets by village elders and the denial of food to refugees, the Australians developed

Lieutenant Colonel David Hurley (centre) chairing a meeting of the Council of Elders at Baidoa Airfield. On his right is Patrick Veracammen, UNOSOM representative, and in the lower right, a the end of the table, Major Dick Stanhope keeps the minutes of the meeting.
Image: G Gittoes

proposals to curb these activities on behalf of the local populace. At Council of Elders meetings Hurley put proposals forward to confiscate all trucks and control their movement. He also proposed that his troops would detain drivers who were stopped a vehicle check points and found to have food supplies on board that should have been distributed to villages. None of these proposals were adopted by the Council of Elders.

Stanhope also put proposals to the Relief and NGO Committees that the exorbitant wages being paid to guards and drivers could be reduced substantially if there were uniform wage rates among all of the NGOs. He also proposed that the movement of all NGO vehicles should be regulated, no weapons should be carried openly, except in compounds, and that all weapons should be registered or they would be confiscated. None of these proposals were supported. Like the Council of Elders, the NGOs wanted the protection of the Australians and their material support with construction and other nation building activities, but they did not want to support tighter controls over vehicle movement and small arms.[62]

While the Australians negotiated with the Council of Elders and the NGOs for control measures and facilitated the establishment of the ASF and a judicial and penal system, discontent was building among NGOs about the lack of leadership and effort by UNOSOM to get on with the business of nation building and ensuring a secure environment after UNITAF forces were withdrawn. By March a steady build up of tension between the NGOs in Baidoa and UNOSOM representatives from Mogadishu came to a head in the form of a letter to Dr Boutros Boutros-Ghali. The letter signed by representatives from the NGOs in Baidoa stated that:

> The UN lacks the leadership, the organisation, the will and the respect of Somalis to provide relief agencies a secure environment in which to work. ... Since its inception, UNOSOM has made no real achievements. Progress on political reconciliation has been non-existent. ... Total failure to establish any stabilising and nation building infrastructure, be it in security, health, education, water etc, exposes profound operational problems. Lack of support for its own field staff based in Baidoa either reveals indifference or misplaced priorities.[63]

Stanhope and his team did not involve themselves directly in these issues, but empathised with the NGOs and the UNOSOM field representative in Baidoa, Patrick Vercammen, who had been sent by HQ UNOSOM to set up nation building efforts in Baidoa. He was not given any resources or personnel. The NGOs were fearful that once the Australians left Baidoa in May, UNOSOM would have done nothing to ensure their security.

In the absence of effective effort by UNOSOM to encourage reconciliation among Somali political factions or to co-ordinate nation building efforts, Hurley, Stanhope and the CMOT teams filled the gap. This role drew Hurley away from his duties as a battalion commander:

> ... a lot of my time was torn between trying to be the battalion Commander in the field, which I did in the initial stages of the operation, but later ... a vast amount of my time was taken up either talking with local officials, NGOs, or in meetings with the eldership trying to work out how we were going to rebuild the Baidoa area or assist them to rebuild the area. So I certainly found that I was drawn away from the normal day-to-day life I expected as a battalion commander.[64]

Not being able to spend time with his soldiers was an intense source of frustration for Hurley. Not only was he being forced to spend more time as the *de facto* military governor and UNOSOM political representative, but he did not have the means to visit and encourage his troops in the field or to monitor operations closely by going to the field in person. This was partly a consequence of having no intergral aviation support in the 1 RAR Group:

> I think it [absence of aviation support] was quite a serious deficiency. If I had helicopters I could have reached the far end of the HRS in 20 minutes. With only vehicles it would take me an hour to three hours to reach most locations. ... Without a command and liaison aircraft I was reduced to road travel. If I wanted to visit a company, it would take me the entire day. ... We were used to air mobility, that's what the ODF is all about. So, overall, I found it quite limiting.[65]

While there was pessimism among the NGO community in early March about the effectiveness of UNOSOM, the restoration of the judicial and penal systems in Baidoa proceeded on schedule. On 5/6 March CMOT supported a meeting of American legal officers from HQ UNITAF and Somali lawyers in Baidoa. Mick Kelly contributed his ideas on how the new legal system should work based on the legislation that had been drawn up when Somalia achieved its independence in 1962. The 1 RAR Group was asked to provide tents and other resources to set up a court house in town and construct a new detention centre.

On 7 March seventy-five uniforms for the ASF arrived. By the middle of March the ASF were also issued handcuffs, night sticks and blue berets. However, Stanhope observed that there were significant differences of opinion between UNITAF legal officers, and the Governor and the SDM on the role of the ASF. Kelly and the American legal officers envisaged the ASF performing the normal duties of Western police forces, namely, prevent and detect crime, arrest and detain suspects, protect lives and property of citizens and foreigners, and maintain law and order while treating all citizens equitably. The SDM envisaged the ASF comprising SDM supporters who would collect taxes and work against the ambitions of the SLA to become a recognised security force. There appeared to be little chance of creating an impartial police force.

The best intentions of Kelly and the American legal officers had come up against the enduring political agenda of the Rahaweyn Clan in the Baidoa Sector. In effect, the the moderate factions in the SDM wanted the ASF to become the security arm of the SDM, controlling the community and collecting taxes to secure sufficient revenue for the SDM to pay salaries and set up SDM-managed local government services. In short, the ASF would be tied inextricably into the political agendas of the SDM factions and the anti-Aideed agenda of UNOSOM and UNITAF. The process of politicising the ASF had begun in February when an SDM supporter, Colonel Aden Nuur Sheik Mohammad, was selected to be the Chief of Police in Baidoa and the first 30 names put forward for approval to become members of the ASF were all SDM supporters.[66]

As the weeks went by, the Australians began to pick up more nation building and community support responsibilities. In mid March the wells and the two larger water sources serviced by pumps and generators, known as Bay Projects One and Two, began to dry up. In absence of any effective co-ordinated action by UNOSOM or the NGOs to respond to this pending disaster, Hurley gave CMOT the tasks of co-ordinating the establishment new water points and improving existing ones. Within a few weeks, Stanhope managed to have three bores dug and four pumps repaired or installed by combining the resources of the US Navy Sea Bees, Brown and Root and several NGOs.

Meanwhile, at the national level, UNOSOM was trying to facilitate national reconciliation. In early March over 750 Somali delegates had been flown to Addis Ababa by UNOSOM for two conferences. The first conference was on improving the arrangements for 'Humanitarian Assistance for Somalia' and the second, planned for later in the March, was on how to achieve 'National Reconciliation'. Sergeant Rick Abraham, an Air Force member of Greg Jackson's UNOSOM Movement Control Unit, was assigned to coordinate customs clearances and accommodation for the delegates in 11 local hotels:

> On 13 March the first conference was completed and the time had come to send the delegates [who were not invited to the second conference] home.

> This was a nightmare job as the Somalis did not want to leave. They wanted to have a paid holiday. There was actual rioting at the Ghion Hotel. Because of the rioting, the Swedish Embassy advised us that they would pick up the hotel bills for all those who wanted to stay for the second conference. From the Movement Control Unit's point of view this was disastrous as delegates knew they could dictate the terms.'[67]

At the regional level UNOSOM was also trying to facilitate political reconciliation and nation building. Over 300 delegates gathered for an SDM conference in Baidoa on 23 February. The conference was protected from interference by Aideed's militiamen in the SLA by Delta Company mounted in a troop of B Squadron's APCs. The SDM delegates recommended that UNOSOM pay the salaries of a scores of SDM chairmen who would have jurisdiction over groups of villages in the Baidoa Sector. These SDM appointees would be supported by the ASF and local administrators who would collect and administer taxes. These taxes would come from local businesses and from citizens who would pass through ASF-manned toll gates on the main roads. The first week of April was set as the time that the ASF would begin tax collection.

While the program of nation building continued, the Australians maintained their escorts of food convoys. On 18 March Jan Van Der Klooster's platoon was escorting two World Vision vehicles and a convoy of Somali trucks delivering food and seed to four villages south of Buur Heybo. After the trucks were emptied they returned to Baidoa. He had been 'stuffed around' for most of the day with conflicting reports about bandits being in Buur Heybo. In the first village the numbers of bandits in Buur Heybo were reported to be about sixty. At the next village 100 bandits were reported to be there. One of the World Vision drivers who had been in Buur Heybo recently reported that there were 'about 20 men with guns there two days ago.'[68] Van Der Klooster recalled:

> I contacted Battalion Headquarters and informed them that there was a possibility of there being bandits in the next location and that we were going to proceed with caution. I gathered my section commanders in and informed them of the situation and gave quick orders. The convoy now consisted of my Land Rover, three Unimogs (one section each), three Somali trucks full of seed and tools, plus two World Vision utes.
>
> The first part of the plan was to drive slowly towards Buur Heybo to keep the dust down in a vain attempt to cover our approach and avoid an ambush. My Rover was first, followed by the two World Vision utes, followed by a Unimog at the front of the three Somali trucks, one after the first truck and one behind the last Somali truck. This practice was essential in order to contain the Somali drivers as they would drive anywhere and everywhere except where you wanted them to go. They would throw bags of grain off the trucks into the bush to retrieve on their homeward trip or inform a friend in the next village who would shoot off and retrieve them. They needed to be driven like sheep and watched as if they were foxes.[69]

Van Der Klooster planned to stop two kilometres short of the town and dismount two of his sections who would patrol forward and surprise any bandits in Buur Heybo. Before he reached the point where he planned to dismount his troops, Van Der Klooster's convoy was met by 30 Somalis and an elder from Buur Heybo. After questioning, the elder said that there had been bandits in Buur Heybo but they had seen the dust from the convoy and had 'run off into the bush.' He also informed Van Der Klooster that he was a lot closer to Buur Heybo than his map suggested. 'By now the whole area was aware of our presence so surprising anybody with a sweep on foot was not such a viable option.'

> I decided that the possibility of gunmen being in the village was minimal so I chose to revert back to our normal practice for food distribution. My vehicle would lead into town with the NGO utes and pick a good position to lay out the distribution point. The rest of

the platoon vehicles would stay back just out of town so they could not be swamped by rioting Somalis.

They would then be called forward by my Platoon Sergeant who was responsible for placing each truck inside the position.[70]

Van Der Klooster's vehicle and the World Vision utilities travelled 200 metres ahead of the remainder of the convoy. Buur Heybo turned out to be two kilometres down the road. After rounding a corner, Van Der Klooster's vehicle entered the town area. He observed a large crowd of men and women and a herd of about 80 camels in the market place. He ordered his driver to move the vehicle up to the crowd slowly. The curious crowd started to move towards them. Suddenly, all those in the Landrover saw a Somali male rushing away with the butt of an AK 47 rifle protruding from under his arm pit. In a split second Van Der Klooster, his Platoon Sergeant Bob Elmy[71] and their signaller, Private Jason Williams,[72] decided to get out of the vehicle and capture the bandit. As they did, the crowd, donkey carts and camels began to scatter. Van Der Klooster wrote later:

> All except the driver immediately debussed and rushed after the gunman. He began to run and the crowd parted revealing another seven gunmen who also began running away. The three of us separated as the bandits split up. I had covered about 60 metres when rounds started to come in from the right. I then fired a warning shot over the bandit's head to my front and dropped to my knee. Shots were then fired at me from the left and I saw Private Williams return fire. The gunman to my front continued running up a small mound about 60 metres away and jumped behind a small shrub. He then began to turn around and point his weapon at me, so I shot him and he dropped into the dead ground [down the slope beyond the shrub]. I then had to take cover as there were more rounds coming from the left.[73]

Meanwhile Private Jason Williams was in the thick of things:

> I caught a glimpse of the man at the same time as the boss and Sarge exited the vehicle. I threw my HF handset to the driver and joined the chase. ... At this stage I had drawn level with the boss and had my weapon pointed at a bandit whilst running. We burst out onto

an open plain where locals were gathered in groups. The bandit kept going and was suddenly joined by seven more locals with rifles and three or four with machetes, all doing the runner.

I broke off to the left and had four or five bandits in front of me. The boss went up the guts chasing the initial bandit and Sarge broke off to the right chasing the rest. I had by this time covered about 80 metres, when I heard shots going over my head, the boss returned fire and was answered very quickly with more enemy fire.

I gave fire to my front, possibly downing one bloke, he fell behind a mound into a gully about 100 metres from me. At this time the other bandits were out of range and I had drawn level with the boss. It was then I noticed the bandit the boss had shot. He was sitting up with blood pissing out of his leg. His rifle was five metres to his side and he was praying to Allah. I went looking for the Sarge and found him in a fire position, with five or six bandits about 200 metres away walking, occasionally glancing back but not daring to raise their weapons.

Sarge and myself moved back to the boss's position, the locals started screaming and shouting and were moving towards us.[74]

Van Der Klooster wrote:

By this time the shooting had stopped and myself and Private Williams moved over to where the bandit I had shot dropped. I discovered him, still alive, but bleeding very badly from a gun shot wound to the upper thigh. I pulled him out of the camel thorn bush after removing his weapon which had been lying a few feet from him ... I could see by the amount of blood that his femoral artery was cut. The smell was unforgettable, as he had also dropped his bowels.[75]

Van Der Klooster was now in an awkward situation:

There were still groups of armed bandits moving in the bush out to our front and the soldiers were keen to go after them. However our task was to provide protection for the convoy, not chase bandits and we now had an angry crowd situation on our hands. So I decided to cancel the distribution and pull out before the crowd got too violent.

... All this time the Somali men were yelling, and women were wailing and crying.[76]

Van Der Klooster ordered Private Bill Lovis and Lance Corporal Jason Verschelden,[77] who were the first soldiers to arrive from one of the following sections, to administer First Aid to the wounded bandit under Sergeant Robert Elmy's supervision. He then directed Williams to go back and tell the section commanders to secure the vehicles in an suitable area. Once this was done, he went to his headquarter vehicle and requested an aero-medical evacuation helicopter to fly in and evacuate the wounded Somali, and send a contact report back to Battalion Headquarters.

> Williams moved back through the crowd, 'scared shitless, but adrenalin flowing. I reached the vehicles but the sections were already deployed and keen to follow up.'[78]

> Williams then passed on Van Der Klooster's orders. Corporal Graeme Wehmeier's [79] section deployed to assist with the administration of First Aid and prepare for the evacuation of the wounded Somali, Corporal Paul Nunan's[80] section secured the area where the bandit had fallen and Corporal Andrew Parnaby's [81] section secured the vehicles. By now, the Somali NGO staff were very concerned that the situation would get out of hand. Members of the crowd were threatening to kill them.

Tactically, the platoon was in a vulnerable position. The crowd was swelling as villagers came out to lend their support to the locals who were remonstrating with Van Der Klooster's soldiers and the NGO staff. Parnaby had already been forced to manhandle an irate Somali male who was related to the bandit who had been shot and have him tied up. The area was also overlooked by a large rock formation that could be used by the escaped bandits to fire on the Australians as they regrouped and evacuated the wounded Somali.

Once the wounded Somali was stabilised and loaded onto a vehicle, Van Der Klooster ordered his men to mount up and drive two kilometres away from town to get out of range of anyone firing from the rock formation and away from the crowd. The wounded Somali died during this journey and the aero-medical evacuation was cancelled. The platoon drove back to Buurhakaba and the bandit's body was handed over to the ASF.

By the last weeks of March the Australians were involving the ASF as much as possible in their operations. Since the ASF had been raised in Baidoa and issued with uniforms, night sticks and their own vehicles in early March, ASF constables had been recruited in Dinsoor, Buurhakaba, Ufurow and Qansaxedheere. They accompanied Australian sub units on security operations and humanitarian aid distributions. Mick Kelly had been able to secure significant quantities of building materials from UNOSOM and several NGOs. These materials were used by Australian sappers from 17 Field Troop to build detention centres in each of these locations.

On 18 March the first trial under the new judicial arrangements was conducted in the recently-renovated Police Station in Baidoa. A man was sentenced to five years jail for armed robbery. By this time Stanhope had one of his CMOT teams permanently located at the Police Station to monitor detention and legal proceedings, and to co-ordinate the provision of ASF constables to accompany both Australian patrols in Baidoa and humanitarian aid convoys.

Despite the best intentions of HQ UNITAF, the newly-established penal and the judicial systems struggled to detain criminals and bring them to justice. From the beginning of their operations the Australians were frustrated because any criminals or bandits they detained could only be held in Australian facilities for 24 hours before being transferred to Somali jurisdiction. There was no police force in Baidoa with the power or facilities to hold offenders. The Australians hoped that once the ASF was established with detention centres, legal jurisdiction and a supporting judicial system, this problem would be solved. These expectations were not met. Urban criminals and bandits captured during security operations in Baidoa and the countryside were handed over to the ASF and the majority were released within 24 hours.

The judicial system was also proving to be ineffective. Two bandits were arrested and, on the basis of several sworn eye-witness statements, charged with murder late in March. They were held for a short time in the detention centre and then released by Somali judges on 28 March.

The explanation given to Stanhope was that the newly-appointed Public Prosecutor had not given the paperwork to the judges in time and the bandits were released before charges could be formally laid. Stanhope found this to be highly suspicious.[82]

The use of the ASF constables during food distributions was also proving to be ineffective. Stuart Dodds wrote:

> When we arrived there was an APC and half of 3 Section (about 5 men) trying to control a crowd of over 300 people and it wasn't going well. I called up for two other sections to assist and spoke to Belly [83] who was the section 2IC in charge. The people would not line up and kept rushing the entrance to the barbed wire enclosure we had set up. I drew a line on the ground and had the interpreter tell them that if they did not line up they would not get any grain. This worked to a certain extent. We also pushed and shoved people away from the sides of the wire enclosure. The majority would still not line up and we kept telling them through the interpreter that they would get no grain. But they kept surging. ... It was a great game for them to see if they could sneak into the queue without being seen by an Australian soldier. The ASF were as usual useless. Eventually the distribution was sorted out after more troops arrived.'[84]

Orderly food distribution in early May 1993; grain trucks in the background and cans of cooking oil in the foreground. Lieutenant Gavin Butler, 3 Platoon, A Company, is escorting a French paratroop officer around the area (centre of photograph) to the rear of the trucks
Image: Bob Breen

As time went by many members of the 1 RAR Group grew more pessimistic about the future of the Somalis now that famine conditions no longer applied. They were concerned about the continuing dependence of the villages on the distribution of international aid and the lack of equity evident in the distribution. On 18 March Blumer wrote:

> I'm sitting on a 6 x 6 at Tugar about 3 hrs drive SW of Baidoa. It is a CARE Australia clothing, blanket and soap distribution and its got great potential to screw up. They went to a different village in between two [villages] that had been done before. There must be well over 2 000 people here and the distribution is planned for 800. ... Other villages are streaming in.
>
> I wonder too with all this clothing [whether] we are building their dependence on Western things. They're a real hotch potch of colour and look like the bargain basement of St Vincent de Paul. What, I wonder, did they do before us. Probably got on with life pretty well but probably now are so far down the track there is now no choice [but to be dependent].
>
> We've started the distribution now. Its for the kids at present. They get 2-3 pieces each and are being kept inside [the barbed wire perimeter] because they [CARE Australia staff] think all the others outside will mob them. Funny thing is that they will be mobbed anyway regardless of when we let them loose. At least if we are here we can stop it. ... Its an interesting feeling sitting here detached, wondering what I would be like with my family in this situation. It's terrible ...
>
> Have finished the distribution now and there are 1 500 people plus who have to go without. They're getting restless. We are going to do a bolt.[85]

Accompanying this concern about the dependence of the local populace on international aid, was a feeling that Somali society was flawed by inequities between men and women, and an exploitation of the weak by the strong. The most typical comment was that Somalis were dominated by a 'dog eat dog' approach to human relations. Many members of the 1 RAR Group felt that Somali society was too dysfunctional to cooperate for the equitable distribution of resources or to achieve the political reconciliation required to avoid a return to civil war. They were pessimistic about the nation building efforts begun by UNITAF and dismissive of UNOSOM's nation building agenda.

Despite the substantial surpluses of food that were being distributed throughout the Baidoa Sector, there were still problems when people from nearby villages who had received a food drop but travelled to neighbouring villages to get more food. There was growing concern among the Australians that the locals were becoming too dependent and too greedy. On 6 April Mortar Platoon supervised a food distribution at the village of War Have about 40 kilometres north of Baidoa. Stuart Dodds wrote:

> We went with 2 Unimogs and [Land Rover] SM5 with CONCERN and 12 trucks. When we arrived we let them park their trucks side by side in two rows backed into each other. We then surrounded them in wire. One area was set aside where all the women were to sit and wait. The men and children had to wait outside the compound. Eventually with a bit of pushing and using the wire we got all of the women back and sitting down. We then started to take one row at a time. They filed through and their cards were checked then each received a tin of cooking oil, about 10 kg of beans and a 50 kg bag of grain. Porters carried the grain from the truck to the end of the enclosure for them.

Somali woman with child and her allocation of foodstuffs. Behind her walks her husband, carrying nothing.
Image: ADF Electronic Media Unit

The women would then strap the sack of grain onto their backs,[86] stoop, pick up the tin of oil and sack of beans, sling a baby onto their hip and step off. Ahead of them would be their husband, who would stride forward proudly with a stick, and carry nothing.

Many of the women were pushy and wanted to jump the queue, so we kept changing the order and sending them to the back of the line. We had 1560 bags of grain. There were about 3-4000 people there. I felt little compassion for the 50 or so women who missed out at the end of the day as they had been the most pushy and no one that day was in real need. As the CONCERN workers said, they should stop food distribution altogether soon and force them back to planting seed and growing crops.[87]

At the national level, there appeared to be little hope for creating a stable political environment for Somalis to return to planting seed and growing crops. The UN convened a conference for national reconciliation in the last week of March. Earlier in the month a conference trying to co-ordinate and ensure the safe distribution of humanitarian aid had ended in embarrassment as Somali delegates had refused to leave their hotels return to Somalia.

The conference on national reconciliation ended on 28 March with a final document signed by 16 faction leaders agreeing to an interim government, total ceasefire, total disarmament, the establishment of local police forces and the return of all illegally-gained assets to their rightful owners. However, UNOSOM was again faced with the embarrassment that no delegates wanted to return to their homeland. They wished to remain in their hotels and continue to 'run up the tab.'

After negotiations between UNOSOM and the 16 faction leaders failed to set a date for their return to Somalia, the centuries-old antipathy between Ethiopians and Somalis surfaced to break the impasse. The President of Ethiopia summoned the faction leaders to his palace. He told them that they had two days to get out of his country or face dire consequences. The Somali delegates were all gone by 31 March, each privately vowing to continue to their fight to be the President of Somalia.[88]

The criminal, Gutaale, executed by small arms fire at Baidoa, on 27 April 1993, under the provisions of Somali Law.

The culmination of the Australian efforts to introduce a viable police force and judicial system was the execution of the criminal Gutaale on 27 April. Mick Kelly, Captain Matt Carrodus, CMOT Team leader attached to the ASF, and WO1 Bill Bowser, head of CI, attended the trial and were not pleased when Gutaale appealed against a 20 year sentence for murdering numerous Somali citizens. Subsequently, Gutaale not only lost his appeal but was sentenced to death. Under the provisions of the 1962 Somali Law he was executed by small arms fire minutes after his appeal failed and his death sentence was pronounced. However, many felt that without the presence of the Australians during the proceedings, this outcome might not have been achieved.[89]

While there were some doubts about the longevity and effectiveness of their nation building efforts on behalf of HQ UNITAF and UNOSOM after they left Baidoa, by the time the 1 RAR Group handed over to a French battalion in mid-May they could rightfully claim to have achieved their primary mission of protecting the delivery of humanitarian aid. Over 8 300 metric tonnes of food items were delivered by 402 convoys in a 17 week period. Convoys had also delivered cooking sets, agricultural tools, seed, blankets, clothing and medical supplies.

LESSONS

The 1 RAR Group started out in January as capable novices in convoy protection, crowd control and working with UN agencies and NGOs. The Australians quickly mastered these unfamiliar tasks and ensured that humanitarian aid and services, such as medical clinics, were delivered safely to towns, village and feeding centres in the Baidoa Sector. Within the confines of their UNITAF mission, this was 'a job well done'. Anne Morris, CARE Australia's acting team leader in Baidoa wrote in April 1993, 'Without the army to escort aid convoys into the villages, the anarchy that we witnessed here from last August would still be on, and the people would still be dying in their thousands.'[90]

The close physical protection of humanitarian operations did not prove to be a difficult task for a well-trained and well-equipped military force like the 1 RAR Group. They quickly sorted out the competing priorities of UN aid agencies and the NGOs, established procedures, including a reliable communications system, and allocated sufficient troops and transport to cover all humanitarian activities that might be threatened.

The major lessons to be learned from protecting humanitarian operations in a war-torn developing country were political and moral, not administrative. The safe, efficient delivery of foodstuffs to villages did not always result in its equitable distribution. The Americans had reported this soon after their arrival in Baidoa in January:

> The biggest problem, however, is the clan factionalism that prevents refugee communities from receiving any of the food aid. It is a problem of distribution and one of centuries-old hatreds—neither of which can be solved by the military.[91]

This assessment was born out during the 1 RAR Group's tour of duty. Refugees living on the outskirts of villages struggling for survival were excluded from the food distribution by many village elders, or had their food taken from them by bandits or local louts once convoys had left. [92] After satisfying their needs and those of their families and clan, village elders and bandits redirected surplus humanitarian aid into the black market for sale, not to their displaced, impoverished and starving compatriots

After observing this situation themselves, the Australians had two options: either ignore it or fix it. Hurley and his staff decided to intervene out of a sense of fairness, born of a moral commitment to give everyone a 'fair go' and an anger about those who were exploiting the system. Hurley delivered warnings to the Council of Elders and, either personally or through his commanders, to village chiefs not to deny refugees food and to stop syphoning off humanitarian aid into the black market. Hurley and Stanhope put proposals to the Council of Elders and the NGO Committee for tighter control of vehicle movement in the Baidoa Sector to deter the illegal movement of stolen humanitarian aid. They also proposed penalties for truck drivers carrying stolen food and those caught stealing it or selling it. Troops at distribution points handed sacks of grain and other foodstuffs to refugees when it appeared that they would miss out. On many occasions widows and the elderly were escorted to their homes, if they were nearby, to prevent sacks of grain and food stuffs being taken from them by groups of Somali males standing about looking for an opportunity to steal from the weak. Drivers and porters were warned about pilfering food and watched closely.

None of the Australian control measures were supported. Representatives from the World Food Program and the NGOs did not appear to be surprised by redirection of surplus humanitarian aid into the Somali black market. There appeared to be an assumption among some of them that surpluses would stimulate economic activity. The Council of Elders showed no interest in curtailing a lucrative trade in humanitarian aid, aside from encouraging the Australians to keep the pressure up on bandits who were taking their cut of the trade at gun point. With the bandits out of the picture, the elders and village chiefs would maximise their profits. From the Australian point of view, nations who donated the humanitarian aid expected it to be distributed equitably to the needy, especially starving refugees, and also that surpluses created deliberately, unintentionally, or by theft should not syphoned off into a black market for the profit of a few unscrupulous individuals. In the end, after hours of negotiation and advocacy, nothing was done. The Australians were left standing alone on the moral high ground.

Forces deployed in the future to protect the distribution of humanitarian aid in the developing world will face the same moral dilemma as the 1 RAR Group did in 1993 The question to be answered will be, 'How far should military forces protecting the distribution of humanitarian aid be responsible for its equitable distribution'? For the US Marines and the French paratroopers the answer was simple, 'None.' On the evidence available, some NGOs also felt no responsibility either. They left it to village elders to decide on who was entitled to aid and who would receive it after delivery. In effect the World Food Program and NGOs were good at the processes of distributing humanitarian aid and the delegation of responsibility to local managers and authorities, but they were not in a position to enforce an equitable outcome for refugees in the Baidoa Sector. Military forces have more power than the NGOs to enforce a more equitable distribution but the challenge will be to deter corrupt village elders, criminals and bandits from excluding groups outside their families, villages and clans, and prevent them syphoning off stolen or surplus aid into the black market. For some military forces the risk and the extra work is not worth it.

The vast majority of Australians wanted to make a useful contribution to the welfare of the Somalis living in the Baidoa Sector during the four months they were there. They wished to use their time and resources productively. Most had been touched by the images of starving Somalis in the Australian media before they deployed. At the unofficial level the Australians gave expression to their compassion in a number of ways. The diggers gave away rations, water, stationary, and other items as they

*Corporal Daryl Heaslip, C Company, 1RAR
cradles a starving Somali child. Many members
of the AFS were motivated by the plight of
the starving children to show kindness and
compassion in Somalia.*
Image: Gary Ramage

went about their business. These gestures of friendship by individual
soldiers and small groups made the Australians feel that they were making
a direct, personal contribution to the welfare of Somalis and showing
them kindness. On the evidence available, this widespread positive,
friendly interaction with local Somalis probably helped to create a secure
environment by reducing the tension many citizens would have felt having
unfamiliar heavily-armed foreign troops in their midst.

At the official level, Dick Stanhope and CMOT responded patiently
and fairly to requests from elders and NGOs for 1 RAR Group resources.
The impact of the 1 RAR Group's philanthropic nation building efforts in
general and CMOT in particular was summarised well by Peter Kieseker,
Manager of CARE Australia, in Baidoa later:

> According to the UNOSOM directives, the Australians were not
> required to rebuild warehouses or schools or jails, or help with
> the town's water supply. The forklift drivers did not have to move
> containers all over town so that NGOs could have secure stores.
> The diggers did not have to build playground equipment for the
> orphanage. The CMOT personnel did not have to listen to endless
> elders and try to arbitrate on domestic issues. They did have to let the

elders come in close to them—they could have kept them at arm's length as the French did. But they did do all of these things and more. The Australians were there to 'rebuild a nation' and to do that you need to the nation's people to take the initiative.

... the Australian forces were the only army to receive a letter of commendation from the NGO community in Baidoa—not to mention being the focus of a thanksgiving song by the community of elders of Baidoa.[93]

The 'Firm, Fair and Friendly' approach and the numerous acts of kindness by members of the 1 RAR Group when dealing with local citizens is a major lesson for future military forces conducting security operations in a developing country. While this may appear to be obvious, not all units serving with UNITAF were kind to ordinary Somalis. On the evidence available, the most serious consequences of not adopting this approach at the tactical level is that citizens who have no respect or trust in foreign troops will not provide information to protect them or themselves from hostile groups. At the operational level, there can be significant repercussions if international units are percieved to condone racism and brutality. The national and international repercussions of the alleged treatment of Somalis by Canadian troops during Operation *Restore Hope* are a case in point.

While the members of the 1 RAR Group were comfortable with the idea of helping out needy Somalis when the opportunity, time and spare resources allowed, there were differences of opinion about the outcomes of the efforts that were made to establish the ASF, appoint judges, and rebuild detention and judicial facilities. Many members of the 1 RAR Group pointed out that the ASF were never in a position to establish law and order while they were a partisan group selected by one political faction, and unarmed, under-trained and poorly administered by UNOSOM. They and the Somali judges appointed by the Judge Advocate Branch of HQ UNITAF operated under the shadow of retaliation from criminals and bandits, as well as the retribution of the clans. From the perspective of many members of the 1 RAR Group, the ASF and the penal and judicial systems never performed well and were artificial creations, propped up and protected while the Australians secured the Baidoa Sector but likely to disintegrate as soon as the Australians left.

Bill Mellor disagreed with the assessment of many of his compatriots in Baidoa:

> It would be overly simplistic to say that the re-introduction of the police was an unqualified success, but despite the set backs of some organisational and personnel problems, and a great deal of difficulty in getting the UN to fund the project, the re-appearance of the police on the streets of Baidoa and other surrounding villages contributed significantly to OP *Solace*.
>
> ... [The setting up of the judicial system] was the initiative of the Staff Judge Advocate, HQ UNITAF, but in this instance the energy and determination of the Legal Officer, HQ AFS ,[Major Mick Kelly] ensured that the proposal became a reality in the Baidoa HRS. The Court House was rebuilt, copies of the Somali Penal Code located, and judges and lawyers encouraged to participate. ... the efforts required to re-establish a functioning judiciary cannot be overestimated. ... Western values and logic do not necessarily hold sway, with frustration often accompanying effort. Nonethless, the judicial system did re-emerge and began functioning in Baidoa in April 1993.[94]

A number of issues emerge from the 1 RAR Group experience with nation building. The first was the relevance of nation building activities to UNITAF's mission of creating a secure environment. The military rationale for establishing the ASF and a judicial and penal system was that bringing criminals and bandits to justice under local arrangements would decrease the threat to humanitarian operations and UNITAF personnel and property. The humanitarian rationale was that the ASF and the judicial and penal systems would eventually restore law and order and reduce the suffering being endured by ordinary Somalis at the hands of criminals, bandits and violent political factions.

Neither of these rationales were vindicated over time. The ASF were ineffectual in investigating and detaining persons who posed a threat to humanitarian operations, UNITAF personnel and property and the lives and property of ordinary Somalis. If the SDM factions had been given their way, the focus of ASF enforcement duties would have been on tax and toll collection, and possibly the elimination of political opponents. From the 1 RAR Group's point of view, the ASF may have been able to re-establish law and order if its members were well-selected, well-trained, well-armed, well-equipped and well-administered. During OP *Restore Hope* the will, the time and resources were not available to achieve this level of external support.

With hindsight, if the UN and intervening international military forces wish to help create a secure environment by bringing criminals and bandits to justice, then combat units should be supported by Civil Affairs units running military police, judicial and penal systems. Military Police, CI Teams and combat units could assist each other to apprehend criminals and bandits. Civil Affairs units could then try them under international laws and conventions, and incarcerate those found guilty from posing a threat to life and property for the duration of the intervention. In this setting, locals could be selected, trained, armed and administered to work with international personnel on the job.

The second issue was the additional pressure that participation in local politics and nation building put on David Hurley. As the weeks unfolded he spent more and more time as a de facto military governor and less as the tactical commander of the 1 RAR Group. His rationale for his participation was that effective liaison with local authorities, UN aid agencies and the NGOs would enhance the effectiveness of security operations. He was vindicated because Australian security operations were not only very effective but also appeared to be safer than if there was minimal liaison and no good will. However, the issue will be to assess how much time tactical commanders should spend on local liaison, negotiation and nation building when the leadership of their troops is their primary responsibility.

Hurley argued that if he had been given a command and liaison helicopter he would have been able to combine his political and military responsibilities more easily. The evidence suggests that his troops did expect to see him in the field more often. Hurley's point about being able to move by helicopter to visit and keep in touch with his troops in the field, make his presence felt around the Sector by meeting with village chiefs in remote villages and attending meetings in Baidoa is valid. Tactical commanders will have to become involved in civil affairs during future humanitarian operations in the developing world. There is an obligation to give them the resources to meet their commitments.

In the final analysis, it was the philanthropic nation building conducted in conjunction with some of the NGOs that gave members of the 1 RAR Group most satisfaction. They enjoyed knowing and being able to observe their efforts touching the lives of needy Somalis. They had little interest in higher level nation building. The Australian engineers took particular satisfaction that their water points would quench thirsts, and that their refurbishment of schools and orphanages would improve the living conditions and prospects for thousands of children.. The Australian medicos from the BSG Treatment Section, Dental Section, Health Section and the 1 RAR Regiment Aid Post felt that they had enhanced the lives of thousands of Somalis through numerous clinics, surgical procedures and vaccinations as well as mosquito eradication programs and community sanitation programs in Baidoa and outlying areas.

6 COMMANDING, RESUPPLYING AND GETTING BACK

You are to report to me immediately on receipt of any order, directive
instruction from any headquarters under which you have been
placed for operational control which, in your judgement, jeopardises
the safety or does not allow for the prudent employment of your force.
You may consult with me at any time as to what constitutes prudent
employment of your force.

Colonel W.J.A. Mellor AM
Directive to Lieutenant Colonel D.J. Hurley,
10 January 1993

Within the Baidoa Humanitarian Relief Sector David Hurley was the tactical day-to-day commander of the 1 RAR Group. The resupply of his troops was the responsibility of Major Rod McLeod, his Quartermaster, and Major Mark Harnwell, OC Battalion Support Group (BSG). Beyond the dusty, dangerous world of the Baidoa Sector there was a chain of ADF headquarters and units, including ships and aircraft, that played an important part the story of OP *Solace*.

Colonel Bill Mellor, Lieutenant Colonel Graeme Woolnough, Mellor's logistics staff officer and Chief of Staff, and their staff in Mogadishu constituted the next level of command, co-ordination and resupply for the 1 RAR Group.[1] Mellor's headquarters was the 1 RAR Group's link back to Australia, responsible for liaison with senior UN, US and NGO headquarters staff in Mogadishu, and co-ordinating resupply from Australia, UNITAF and through local purchase, as well as visits to the 1 RAR Group and media coverage of Australian operations. Though Mellor did not have operational control over *Tobruk* which was assigned by the Navy to support the AFS offshore, the ship's commanding officer, Commander Kevin Taylor, was responsive to Bill Mellor's requirements. In reality, *Tobruk* was commanded from Maritime Headquarters in Sydney. Taylor had to report back daily about *Tobruk's* operations and his future intentions. Neither he nor Mellor had much discretion about the employment of his ship or the Sea King helicopter that had been deployed with it.

Mellor and Woolnough respectively reported back to Colonels Phil McNamara and Brian Vale at LandHQ in Sydney. McNamara was Major General Murray Blake's Principal Staff Officer (PSO) responsible for operational aspects of OP *Solace* and Brian Vale was Blake's PSO responsible for resupply, administration, movement and personnel aspects. McNamara and Vale were not commanders. Their roles were to co-ordinate, communicate and advise on behalf of Blake who was the Lead Joint Force Commander for OP *Solace* at the operational level.[2]

Beyond LandHQ, McNamara and his staff reported to Captain Russ Swinnerton, RAN, and his staff at HQ ADF in Canberra on operational matters. Vale and his staff reported to Brigadier Bill Traynor and his staff at HQ ADF on resupply, administrative, movement and personnel matters. Like McNamara and Vale, Swinnerton and Traynor were PSOs at the strategic level of command and resupply. They were responsible for co-ordinating, communicating and advising on behalf of the Chief of the Defence Force, General Peter Gration, who had delegated the command of OP *Solace* to General Blake. Ultimately, however, Gration was the commander responsible to the Defence Minister, Senator Robert Ray, for the success of OP *Solace*.

In addition to reporting upwards along the chain of command, Brian Vale and his staff were responsible for co-ordinating resupply of the AFS with Colonel Tim Winter[3] and his staff at HQ Logistics Command in Melbourne. Logistic Command depots at Moorebank on the outskirts of Sydney and at Bandiana, located on the outskirts of the inland city of Albury, were the key depots for the resupply of the AFS.[4] The Moorebank Logistic Group (MLG), commanded by Colonel Greg Thomas, was the principal depot assigned to resupply the AFS.[5] The demands for resupply from the AFS that could not be satisfied from the warehouses at Moorebank, or by purchase by MLG Fleet Managers, were to be directed back to Tim Winter's staff who would direct other depots, such as the Bandiana Logistic Group, commanded by Colonel Jim Campbell, to satisfy the demands.[6] All resupply items were to be sent to the Moorebank Freight Terminal for onward movement to Somalia.

Major Michael King on Tim Winter's staff at HQ Logistic Command was responsible for the movement of items of resupply from the Moorebank Freight Terminal to Somalia. He acted as HQ Movement Control for OP *Solace*. His role was to co-ordinate with Wing Commander Ian Jamieson,[7] the senior movements staff officer at AirHQ, and his staff, located near Glenbrook on the outskirts of Sydney, to fly stores to Somalia, either on Air Force aircraft or by commercial air freight. King also had the option of sending less urgent items by sea.

This chapter is the story of the experiences of the officers and staff who constituted the chain of command and the resupply system back to Australia. It was their job to support the operations of the 1 RAR Group in the Baidoa Sector and get the AFS back home safely.

The appointment of Colonel Bill Mellor as national commander of the AFS in December 1992 had several precedents. Like his predecessors in Zimbabwe, Uganda, Namibia and Iraq, he was made responsible for the operational success of an Australian Service Contingent serving under foreign operational control. Other Western nations did the same for the American-led OP *Restore Hope*. For example, Colonel Serge Labbe, who, like Bill Mellor, was the Chief of Staff of his Army's 1st Division, was appointed as the national commander of a group of Canadian Defence Force units, including a ship, Her Majesty's Canadian Ship *Preserver*, assigned to participate in OP *Restore Hope*.

While Bill Mellor's appointment followed convention, raising and deploying his headquarters was *ad hoc* and haphazard. In the days after the Government's announcement on 15 December, Mellor was too busy until he departed with the National Liaison Team to Mogadishu on 21 December to raise his headquarters personally. Despite receiving information that he would be appointed as the national commander of the AFS earlier in December there was little he could do about raising his headquarters until the official announcement was made on 15 December. By that time he and his family were on holiday at Caloundra, a beach-side holiday resort north of Brisbane. He returned to work in Brisbane immediately after the announcement on the 15th and flew to Townsville on 16 December for a press conference. For the remainder of the week he attended meetings and briefings in Canberra, Brisbane and Sydney and commuting to work from Caloundra when he was in Brisbane. After packing and family farewells on Sunday 20 December, he flew to Mogadishu on Monday 21 December.

Initially, Lieutenant Colonel Graeme Woolnough had been in charge of 'standing up' HQ AFS. Ably assisted by Warrant Officer Kevin Yeo,[8] Woolnough drew up a vehicle and stores list that was largely satisfied from the Q Store of 1st Signals Regiment, a neighbouring unit to HQ 1st Division in Enoggera Barracks, Brisbane.[9] However, just as arrangements were coming together, Woolnough was directed by LandHQ to deploy with Mellor's National Liaison Team to Mogadishu. Major David Creagh, Woolnough's logistic staff officer, was left to make final preparations and then deploy personnel, vehicles and stores from Brisbane to Townsville in time to join Hurley's advance party that was due to depart on 8 January.

In a small Army, personal relationships formed during initial training courses and enhanced by serving together in units, often in isolated locations, can assist in binding groups that have been assembled at short notice. This was not the case during the raising of HQ AFS. The officers from HQ 1st Division in Brisbane, and the officers from HQ 3rd Brigade and 1 RAR in Townsville did not know each other well and had not served together. In the absence of prior personal relationships and some negative feelings among Townsville-based officers to the importation of headquarters staff from Brisbane, the initial reaction by most officers in Townsville to Creagh and his group was cool and uncooperative.[10]

Creagh and his group were irrelevant to the success of the preparatory phase of OP *Solace*. They participated in training and administrative activities after arrival in Townsville early in January, but remained isolated. Without a senior officer like Woolnough accompanying them, they were not in position to influence what was going on around them. LandHQ was mounting the operation and HQ 3rd Brigade was responsible for dispatching the AFS from Townsville. Consequently HQ 1st Division personnel in Townsville were unable to earn the respect of members of the 1 RAR Group or the young officers from HQ 3rd Brigade who had been seconded at short notice to serve with them.[11]

In some ways Mellor was similarly isolated during the preparatory period from 15 December until the arrival of Hurley and his advance party in Mogadishu on 10 January. He and Hurley did not know each other well, had not served together and were from different Corps.[12] Their relationship was formal and cordial, but not based on a mutual understanding developed through shared experience. Both Mellor and Hurley were very busy during the Christmas-New Year period with their individual responsibilities. Their geographic separation prevented the development of a closer relationship.[13]

Despite capping the numbers for the Mellor's headquarters at 10 personnel in the proposal approved by Senator Ray in December, HQ AFS almost quadrupled its size to 39 personnel during the preparatory phase of the operation.[14] Because of Blake's overall cap on the AFS was set at 927 personnel, every person added to Mellor's headquarters meant that Hurley had to cancel the deployment of a person from the 1 RAR Group.[15] Though this policy had nothing to do with Mellor and his staff, dropping personnel from the 1 RAR Group to increase the capabilities of the National Headquarters did not help develop closer relationships.[16]

The arrival of Majors Steve McDonald, Greg Hurcum and Dick Stanhope as an *ad hoc* 1 RAR Group reconnaissance/liaison team two weeks after Mellor's National Liaison Team deployed on 21 December further estranged Mellor and his staff from the 1 RAR Group. McDonald arrived on 2 January and told Mellor that he had been deployed to co-ordinate the reception of the 1 RAR Group.[17] By the time Hurcum and Stanhope arrived on 4 January, differences of opinion had developed between McDonald and Mellor over McDonald's authority, the reception arrangements and subsequent activities of the 1 RAR Group. Hurcum and Stanhope bore the brunt of Mellor's frustrations at not being able to influence the preparations of the 1 RAR Group in Townsville. Mellor made it clear that wanted the 1 RAR Group to adhere to his plans after their arrival and not plans being made by David Hurley and Paul Retter, General Blake's principal planning officer for OP *Solace* at LandHQ.[18]

Mellor had been contacting Hurley since 23 December on Hurley's home telephone. Hurley briefed him on the preparations of the 1 RAR Group but expected Paul Retter, who spoke to Mellor twice a day on secure satellite communications equipment to keep Mellor up to date with the decisions being made about the deployment of the 1 RAR Group.[19] It became clear after his deployment to Mogadishu that Mellor expected to be consulted more frequently and be involved in decision-making, not just informed of decisions after they had been made.[20]

Retter and Hurley felt that Mellor was receiving sufficient information. They were also concerned that he was asserting authority he did not have, or that he may not be in a position to exercise from Mogadishu over preparations being made for the 1 RAR Group deployment. From Mellor's point of view, he was the national commander on the ground in Mogadishu, had met key UNITAF officers and had visited Baidoa. Furthermore, given Blake's personal advice at Sydney Airport before he left in December that he would be held responsible for the success of OP *Solace*, he felt that he was entitled to influence the 1 RAR Group's preparations because he would have to deal with the consequences after the 1 RAR Group arrived in Somalia.[21]

After an argumentative meeting with Mellor and Woolnough on 4 January, McDonald, Hurcum and Stanhope adopted a guarded approach in their dealings with Mellor and his staff and subsequently advised their peers to do likewise.[22] As they prepared for the arrival of Hurley and his advance party, they focussed on establishing relationships with headquarters staff from the 3rd/9th Marine Battalion and American Civil-Military operations unit in Baidoa. Stanhope reported to Mellor once a day by telephone and requested that information he and Hurcum had obtained in Baidoa be passed onto Hurley.

On 7 January after being awakened by tanks firing a kilometre from the Embassy Compound and watching General Johnson's Cobras in action pounding General Aideed's warehouses, McDonald, 'Rang CO [Hurley] this morning and passed lots of info on Baidoa. Mellor upset with length of time I spent on the phone (10-15 mins). I'm sure the CO is 'hanging out' for info so I pushed it a little and kept talking!'[23] This testy incident underlined the difficulty Mellor and his staff were experiencing trying to have input into the deployment of the 1 RAR Group and the attitude of the first group of 1 RAR Group officers to arrive in Somalia. Hurcum, Stanhope and McDonald had first loyalties to Hurley. They were concerned that Mellor and his staff would attempt to interfere with Hurley's tactical control of the 1 RAR Group.[24]

Aside from personal impressions, the root of the differences of opinion between Mellor and the 1 RAR Group officers, as well as with Hurley and Retter, lay in the absence of a clear definition of what 'national command' entailed. Mellor was not given a definition in his informal or formal directives from LandHQ.[25] In absence of a shared definition, Hurley, Retter, Hurcum, McDonald and Stanhope, interpreted Mellor's role as being that of a senior liaison officer, handling sensitive higher policy issues, VIP visits, logistic support and media coverage to enable the 1 RAR Group to get on with the job at hand in Baidoa under American command. Hurcum's, McDonald's and Stanhope's first impressions were that Mellor would exceed his liaison and monitoring role and assume the role of a *de facto* Brigade commander, directing Hurley on how he was to conduct operations in Baidoa.

On the day Hurley and his advance party arrived in Mogadishu on 10 January Mellor spent a day with him bringing him up to date with the situation in Somalia and introducing him to Lieutenant General Robert Johnston, Commanding General of UNITAF, Major General Steven Arnold, Commander of the 10th Mountain Division, who would command the 1 RAR Group and Brigadier General Anthony Zini, Johnston's senior operations officer. He also gave Hurley two formal directives. The first reminded him that, 'From arrival in Somalia you come under command Headquarters, Australian Force Somalia. You may be placed under the operational control of other headquarters from time to time as the situation requires it. Such changes to your status of command will be issued in writing by CAFS.'[26] The second advised Hurley that, 'You are placed under the operational control of ARFOR (HQ 10 MTN DIV) from 110001C Jan93. You remain under command HQ AFS.'[27]

The relationship between HQ AFS and the 1 RAR Group had moved a long way from that intended by General Peter Gration and Major General Murray Blake during their meeting on 16 December in Canberra. Neither had envisaged Mellor's Headquarters commanding the 1 RAR Group, but both had left it to Mellor to decide on command and control arrangements once he arrived in Mogadishu. Mellor had discussed his directives with Blake and obtained his concurrence before issuing them to Hurley:

General Blake made it clear to me at his briefing [at Sydney Airport] that I was to make command and control decisions based on my assessment of what I found on the ground [in Mogadishu], and that is what I did.

Mellor's directives paralleled the command relationship Colonel Serge Labbe had with the Canadian units assigned to OP *Restore Hope*. However, unlike Mellor and his staff who had been sidelined during the preparatory phase of OP *Solace*, Labbe had been in charge from the beginning of the equivalent Canadian operation called OP *Deliverance*. HQ 1st Canadian Division had mounted and dispatched the Canadian Joint Task Force to Somalia. As the Chief of Staff, Labbe had raised his national headquarters from his own headquarters and immediately set them to work concentrating units prior to embarkation, planning their deployment and accompanying them during deployment. The Canadian national headquarters was in control from the beginning. Subsequently, Labbe continued as the Commander of all Canadian forces assigned to support OP *Restore Hope* after their arrival in Somalia. However, once the Canadian units were on the ground in the Belet Weyne Humanitarian Relief Sector, he placed them under the operational control of HQ 10th Mountain Division, sometimes referred to as ARFOR (Army Force).

Aside from the advantage of being involved in concentrating and deploying Canadian forces, Labbe was in a better position to exercise national command than Mellor because he had a larger, better-resourced headquarter group. Canadian and American headquarters personnel, including civilian staff working for the US State Department, were housed in pre-fabricated, air-conditioned accommodation and worked in containerised operations rooms and office areas. They deployed with the resources to 'set up shop' independently. Contracted or general duties military personnel maintained living and working areas. When there were buildings available, they were refurbished by tradespeople, painted and then modern office equipment and furniture was installed. Some of the other donor nations to OP *Restore Hope* set up their national headquarters staff in adapted ISO shipping containers with windows, doors and openings for air-conditioning units cut into the sides.

From the beginning Mellor and his staff did not enjoy the same living and working conditions as their Canadian, American and some other donor nation counterparts. The National Liaison Team had to depend on the goodwill of HQ UNITAF to accommodate them. They were assigned a derelict building in the American Embassy Compound that had been used as a library. The building was gutted, leaking and rat-infested. Even after the arrival of the remainder of his staff and stores on 10 January with Hurley's Advance Party, there was little that could be done to improve the cramped living and working conditions. Everyone had to 'make do' and improvise. Fortunately Graeme Woolnough was able to have the headquarters area cleaned out and electrical wiring fitted. The Chief Clerk, Kevin Yeo, spent part of each day for four weeks after deployment 'scrounging' items from neighbouring units and their Q Stores to make living and working conditions easier for HQ AFS staff.

Mellor and his staff accepted their circumstances and 'made do'. Mellor did not seek additional personnel and resources from Australia to refurbish his headquarters area and improve living and working conditions. Everyone was expected to pitch-in and help out with domestic duties and security rosters. The surprise deployment of a Navy Clearance Diving Detachment to Mogadishu in late January helped HQ AFS with their facilities problem. They arrived unannounced at Mellor's headquarters seeking accommodation, transport and rations. Maritime Headquarters had despatched them to Mogadishu to support OP *Restore Hope* without the knowledge or concurrence of General Blake. Eventually, they were returned to Australia without conducting any operations. While they waited for their status to be sorted out, they built a shower facility and other amenities for Mellor's staff.[28]

After the arrival of the 1 RAR Group Advance Party, John Caligari concentrated on responding to 1 RAR's American headquarters while

Mark Harnwell concentrated on establishing relationships with UNITAF logistics staff and supply units. He assessed that he would only need to communicate with Woolnough at HQ AFS if he struck any problems setting up resupply arrangements with the Americans, or to follow up service and staff demands for stores from Australia. He envisaged Woolnough monitoring the resupply situation affecting the 1 RAR Group by attending daily briefings in Mogadishu and keeping in touch with Brian Vale's staff in Sydney.

On the eve of the start of urban patrolling by Doug Fraser's A Company in Baidoa on 17 January, Caligari and his staff were focussed on their operational chain of command to HQ 10th Mountain Division in Mogadishu, not to their Australian National Headquarters. Like Harnwell's assessment of his reporting requirements to Woolnough on resupply matters, Caligari assessed that Mellor and his staff would receive sufficient information in Mogadishu to monitor 1 RAR's operations. His staff would only report to HQ AFS when 'serious incidents' occurred, and on a daily basis, 'at 1700 hrs to be received by 1800 hrs', in accordance with Mellor's directives.[29]

On the night of 17 January an Alpha Company patrol was fired on with one single shot at about 10 45 pm.[30] The patrol went to ground, but did not return fire because they did not know exactly where the shot had come from, and there were Somali civilians in the area. They 'froze' and listened for 15 minutes before continuing on their way. This type of occurrence was expected. It was handled routinely by the patrol and by the Duty Officer, who reported it to Caligari as well as to the Duty Officer at HQ 10th Mountain Division.[31] Caligari assessed that the incident was a 'one off' and not significant enough to warrant waking Hurley. The incident was noted in the 1 RAR Operations Log Book, recorded in a patrol report and Caligari anticipated that it would come up as an item in the next morning's HQ 10th Mountain Division briefing in Mogadishu.

On 18 January Steve McDonald and Glen Crosland flew from Baidoa to Mogadishu on the daily New Zealand Andover air shuttle to supervise the movement of stores and equipment to Baidoa. They visited Mellor's headquarters after arrival. In reply to a series of questions from Mellor about how things were going in Baidoa, McDonald mentioned the shot fired at the patrol the night before. Mellor forcefully made the point that he and his headquarters should have been notified immediately after this incident occurred. He did not want to wait for daily briefings from American staff officers to find out that Australians had been shot at in Baidoa. He was also conscious that several journalists were in Baidoa who might report the incident in Australian newspapers before he had been able to inform LandHQ in Australia. He wrote later:[32]

> The import of this incident was not its routine nature, but rather that this was the first time that a shooting involving the Battalion Group had occurred. This coupled by the presence of more than twenty Australian journalists in Baidoa provided it with significance beyond the usual. To expect HQ AFS to wait for the information to filter through HQ 10th Mountain Division and HQ UNITAF, while it was potentially news back home would have placed me in an unacceptable situation with LandHQ, who were expecting to get news of happenings in Somalia from HQ AFS, not the media.[33]

Mellor had a personal policy of hitting subordinates hard early in his relationship with them when he was not happy with their performance.[34] After his conversation with McDonald, he rang Hurley and clarified his requirements for information in a direct and forthright manner. He emphasised that all information deemed notifiable to HQ 10th Mountain Division was to be sent to HQ AFS at the same time.

This incident brought previous tensions between the two Australian headquarters staffs to a head. From Mellor's point of view, as the National Commander and senior Australian officer in Somalia, he felt entitled to be given information when he wanted it, and in as much detail as he wanted it.[35] Hurley kept his own counsel on what he thought of Mellor's requirement for double reporting. He had dual responsibilities to be loyal to Mellor as his national commander, and to lead and motivate his subordinates.

Hurley directed Caligari and his staff to comply, pointing out that Mellor had the authority to demand information so he could fulfil his responsibilities as national commander.

Hurley, however, concurred with Caligari's understanding of the role of HQ AFS. He recalled later:

> HQ AFS played a very important role as the national command element for the Australian force. Their main jobs were to look after the prudent use of our force [1 RAR Group], secondly, to ensure that the logistic support was maintained and we were receiving what we required on a timely basis. Colonel Mellor and his staff were very much the bridge for me back to Australia and the means by which my force's particular needs could be expressed to in-country American headquarters.[36]

Privately, Caligari and his staff resented the double reporting system and the growing number of enquiries from Mellor's staff seeking further clarification and elaboration of information sent to them. He and most other 1 RAR officers assessed that Mellor was directing his staff to keep pressure on the 1 RAR Headquarters to provide more and more information as a way of keeping his five Majors and five Captains busy.[37] For his part Mellor pointed out that, 'My motive in seeking information from the battalion group was to get an accurate picture of what was occurring, as reporting, at least early on, was scant, sometimes misleading, and occasionally contradictory.[38]

By the end of the January HQ AFS had formalised the 1 RAR Group's reporting requirements. There were to be eight routine daily reports, one weekly report and eight immediate reports required as certain types of incidents occurred. Two of these reports were service and staff demands for stores. Justification for all reports were referenced to 3 Brigade Standing Operating Procedures, Australian Defence Instructions and directives from the staff of HQ UNITAF on their reporting requirements from subordinate units.[39]

While formalising their information requirements from the 1 RAR Group, Mellor and his staff were also busy carrying out their other duties as the Australian National Headquarters during the last weeks of January; attending meetings and briefings in Mogadishu, supervising media coverage of Australian operations and co-ordinating security arrangements for a visit by John Kerin, Minister for Trade and Overseas Development, to Somalia planned for 25/26 January.[40]

Mellor arranged two important meetings during this time, one with Robert Oakley on 18 January and another with Brigadier Imtiaz Shaheen, the UNOSOM Commander, the day after. He wanted to ascertain the timings and nature of the transition arrangements between UNITAF and UNOSOM II.[41] From the beginning, the aim of OP *Restore Hope* was to stabilise the security situation in Somalia to make way for the resumption of UN peace keeping and nation building efforts. The US Government did not want to be drawn into a longer term security commitment in Somalia.

Oakley assessed that the transition to UNOSOM II would occur in early February and the ROE for peace enforcement would most likely continue for some time until the security situation improved. Shaheen was much more cautious about predicting when UNOSOM II would take over from UNITAF and whether the ROE for peace enforcement or peace keeping would apply thereafter.[42] The transition was 'far less pressing' from the UNOSOM perspective. The UN agenda may have been to keep US troops and resources committed in Somalia for some time. The lack of urgency by the UN to make transition arrangements was underscored by the absence of Mr Kittani, the Political Head of UNOSOM, who was on leave until the end of January.[43]

It was at his meeting with Shaheen that Mellor became aware that, 'The command and control position would seem to make HQ AFS redundant post transition'. He proposed to General Blake that 'some members [of HQ AFS] may be made available to serve on HQ UNOSOM II until the completion of the tour' or that Blake might 'place HQ AFS in the operational line of control, although this may seem to be adding an unnecessary step.' He concluded that, 'It may also be possible, given the extant situation and the

Lisa Keen, Commander of the Media Support Unit in Somalia. Lisa and her unit produced a documentary, numerous photographs and hours of videotape to record the work of the AFS.
Image: ADF Electronic Media Unit. Terry Dex

time limited nature of our commitment, to seek an accommodation from the UN for HQ AFS to continue to exercise national command until the withdrawal of the force [AFS].'[44]

During this time Mellor was clarifying his authority over the media coverage of Australian operations. Early in January he had asked Captain Nigel Catchlove, the Public Relations officer who accompanied the National Liaison Team, to specify his responsibilities before the arrival of Lisa Keen's Media Support Unit (MSU). Catchlove reported back that Mellor would be responsible for the administrative support of the MSU and would be invited to participate in determining the guidelines for what constituted 'Operational Security.'[45]

Catchlove's findings meant that Mellor had no command relationship

Major Mark Harnwell, OC Battalion Support Group, at Baidoa Airfield.
Image: ADF Electronic Media Unit.

with the MSU and thus no say on how the AFS was to be protected from disclosures by the media that might have a negative impact on the 1 RAR Group's operations.[46] In effect, Mellor was responsible for the co-ordination of media coverage of Australian operations but did not have control over the MSU. This situation paralleled his relationship with the 1 RAR Group where he was responsible for the operational performance of the 1 RAR Group but had to strongly assert his authority to receive information from its headquarters. He now faced the challenge of asserting what he assessed to be appropriate control over the MSU.

Fortunately, Lisa Keen was a canny operator. After she arrived with the MSU and 23 journalists on 12 January, she ensured that Mellor was the principle spokesperson for the AFS and was kept well informed on media matters. She worked well with Major David Tyler, Mellor's senior Public Relations officer based in Baidoa, to give journalists equitable access to Australian activities in the critical two weeks after arrival in Somalia when the Australian public would be most interested in the fate of the 1 RAR Group.[47] Most of the journalists who had deployed with the MSU flew back to Australia on 27 January after covering John Kerin's visit.[48] Aside from exaggerated reporting of crowd control problems during initial food distributions, members of the AFS were happy with the media coverage they

The rear of HQ 1 RAR Group, Baidoa Airfield. The roof was provided by the US Navy Sea Bees and the tents in the foreground were left by the US Marines. The tent second from the left was VIP Accommodation for visiting senior officers.
Image: G Gittoes

received in Australia. The average digger had been portrayed as being both tough enough as well as compassionate enough to meet the challenges of operating in Somalia.

While the command and control arrangements were settling in during the last two weeks of January, the resupply arrangements for the 1 RAR Group were being finalised. By the time John Kerin flew to Nairobi on 26 January after a successful visit, Mark Harnwell, building on the valuable ground work done by Woolnough and his logistic staff, Major David Creagh and Captain Mick Fulham, had negotiated mutual support arrangements with Lieutenant Colonel Quick, Commander of the UNITAF Composite Logistics Battalion. Quick agreed to provide water, fuel and rations under the cross-servicing agreement that had been negotiated by Brigadier Bill Traynor in Australia earlier that month. Harnwell took responsibility for providing local medical, dental and maintenance support and running a staging area at the airfield for up to 80 UNITAF vehicles each night. He also agreed that his unit would be the distribution point for all packaged water and rations to UNITAF units in the Baidoa Sector. He worked hard at building up a rapport with American logistic staff whom he would have to work with for the remainder of the operation. [49]

The first local logistic challenge for Woolnough and Harnwell was how to overcome the unforeseen requirement to establish a tented camp with essential services, such as electricity, waste disposal and ablutions, for 1 RAR. From the beginning of OP *Solace* Harnwell had expected to set up his unit in a serviced, tented camp. As a second line logistic unit, the BSG was designed and equipped to be located in a secure base area. At the Baidoa Airfield his personnel were accommodated in a large aircraft hanger and had plenty of tentage, lighting and refrigeration to create comfortable living and working conditions.

Two hundred metres away 1 RAR were accommodated in a few roofless, derelict buildings and abandoned US Marine Corps tents amongst camel thorn bushes. The infantrymen and the cavalrymen were 'hurting for power generation, lighting, flooring and tarpaulins'.[50] They needed significant engineering support to upgrade their camp area with roofing for derelict buildings, floor boards and drainage for the tent lines, shower and toilet facilities, roads and power. The task of upgrading the camp area was well beyond the resources of the Australian sappers and 1 RAR's Assault Pioneer Platoon.

Harnwell's first piece of luck was finding that the neighbouring 600-strong 43rd US Engineer Combat Battalion (Heavy) was commanded by Lieutenant Colonel Robert L. Davis, who had been on exchange at the Australian Army School of Military Engineering at Casula, Sydney, from 1986 until 1988. He wrote later:

> I had the benefit of knowing Australian equipment, capabilities and organisations gleaned from my exchange experience. This allowed me to better understand how I could support 1 RAR with the capabilities of my own unit. Equally beneficial was the immediate rapport established as a result of my respect for Aussie soldiers.
>
> While I did not personally know any of the members of 1 RAR, we had several friends in common and that just aided in getting off on the right foot.[51]

Davis's positive attitude to helping the 1 RAR Group out with its predicament was matched by a similar show of goodwill by a company of 50 US Navy Sea Bees also located at Baidoa Airfield.[52] Graeme Woolnough worked hard in Mogadishu to have US Engineer Task Lists changed to include the upgrading of the Australian camp. He was sympathetic to the 1 RAR Group's circumstances, having been posted to the ODF twice in the 1980s and been the DQ of 3rd Brigade in 1990/91.

Davis's combat engineers and the Navy Sea Bees pitched in to put roofing on buildings and to build other camp facilities. Thousands of dollars of building materials were used and many man, vehicle and equipment hours were allocated to develop the Australian camp over the next few weeks.

In addition to the good fortune of being co-located with co-operative and friendly American Army and Navy engineers, the departing Marines were generous to the 1 RAR Group. The Marines not only left sufficient tentage to house the 1 RAR Group but also substantial quantities of water, rations, insecticide and other consumable items. These stocks were put on the 'tick' and billed later under the Cross Servicing agreement.

Aside from consumables, such as water, rations, fuel, detergents and common camp stores, Harnwell was not able to plug into the American logistic system in Somalia to meet all resupply requirements. The Australians had deployed with vehicles, weapons and equipment that could only be maintained from Australian supply depots. By Western military standards, B Squadron's 36 APCs were not state-of-the-art. The Americans had not brought M113 APCs to Somalia. They and other Western armies, such as the French and the Canadians, preferred either wheeled armoured vehicles or more sophisticated armoured fighting vehicles. Consequently, the UNITAF supply system had no engines, transmissions, differentials or other spare parts for the Australian APCs. Similarly, the AFS's fleet of Mack and Unimog trucks, Land Rover four-wheel drive vehicles and forklifts were mostly modern, robust vehicles but, like the APCs, were not in the US military inventory. Consequently, there were no spare parts available for them from UNITAF.

Like its vehicle fleet, the AFS's communications equipment was either no longer in the American inventory or was unique to the ADF. The resupply of radio spare parts was important because Australian radio equipment was proving to be unreliable and required frequent repair and maintenance. The ageing AN/PRC 77 radio set, carried by platoons and patrols, 'broke down continually and the range achieved was poor.'[53] The Australians had to rely on this equipment because there was a severe shortage of the Australian-designed and developed RAVEN High Frequency radios. This equipment also 'suffered a high incidence of breakdown' and its antennas 'performed poorly.'[54] The old American 524 radios in the APCs also 'suffered frequent breakdown'. Maintaining the RAVEN radios was further complicated because malfunctioning or unserviceable equipment had to be sent back to Australia for repair because the 1 RAR Group had not deployed with a field repair facility for these radio sets.[55]

The AFS's computer equipment had been brought 'off-the-shelf' in Australia, mostly stand-alone Toshiba lap-tops and bubble-jet printers. This equipment had not been modified to cope with the harsh operating conditions in Somalia and were not compatible with American military computer systems. The Americans had to provide the computer and communications equipment to link the Australians to the HQ UNITAF local area network, logistic data bases and communications system.

The 1 RAR Group's power generators were also old and unreliable. Some broke down immediately or were incorrectly used by untrained operators and required spare parts from Australia. This problem was exacerbated when Harnwell realised that many generators had not been checked and serviced before departure. As a consequence, some were unserviceable on arrival.[56]

Many of the spare parts for the 1 RAR Group's vehicles and equipment had either arrived too late for loading on *Tobruk* and *Jervis Bay* in December, accumulating to the 50 pallets of stores that had been left on the wharf in Townsville. These stores were returned to their depots in Australia and not forwarded to Somalia. Thus the 1 RAR Group arrived in Somalia with a deficiency in spare parts.[57]

The resupply system for OP *Solace* had been defined by both LandHQ and HQ Logistic Command by the end of January. Brian Vale issued an Administrative Order on 19 January and Tim Winter issued the resupply plan for Operatio *Solace* on 26 January, based on Vale's Order. [58] Logistic Command accepted responsibility for satisfying all service and staff demands from the 1 RAR Group that Harnwell could not satisfy through the UNITAF resupply system or that Woolnough could not satisfy through his local purchase system.[59] HQ Logistic Command responsibilities included co-ordinating the delivery of stores to Mogadishu from Australia. The Air Force appointed a liaison officer to work with Michael King on a part-time basis at HQ Logistic Command, to co-ordinate the delivery of stores from the Moorebank Freight Terminal to Mogadishu in the timeframe specified in demands.[60]

Winter, through his senior operations officer, Lieutenant Colonel Tony Ayerbe,[61] had been handling the resupply of Australian Service Contingents serving with UN forces overseas for several months. However, OP Solace was the first major logistic challenge they had faced since they had received the responsibility for planning and conducting logistic operations in support of land operations from Logistic Division in HQ ADF several months before. Logistic Division had the corporate experience of resupplying the Navy's ships during the Gulf War two years before and there were advantages in being located close to the senior logistics staffs from all three Services in Canberra.[62] It remained to be seen how the Army's logisticians in Sydney and Melbourne would handle resupplying a land operation in Africa. Ayerbe's staff had not taken leave over the Christmas-New Year period and had been involved in stocking up the 1 RAR Group, guided by Brian Vale and his staff at LandHQ before deployment.

Despite officially deploying with stocks to last 60 days, the 1 RAR Group needed urgent resupply from Australia within 10 days of starting operations. Mark Harnwell and David Creagh were able to alert Lieutenant Colonel Peter Neuhaus,[63] the acting Operations Officer at LandHQ, in time for the RAAF C 130 aircraft flying to Mogadishu from Australia to pick up the journalists who had deployed with the MSU to be loaded with high priority items. However, Neuhaus emphasised that this was a 'one off' quick fix' and LandHQ 'was not keen to move the remaining stores listed ... unless absolutely necessary. ... The remaining stores were to be forwarded to Mogadishu on an 'as required basis' as a result of service demand action.'[64]

Compounding the 1 RAR Group's deficiencies in spare parts for vehicles and equipment, caused in part by the redirection of the 50 pallets of stores back to issuing depots in Australia, was the toll the Somali countryside and Hurley's tempo of operations was taking on the Group's vehicle fleet. By the end of January it was clear to Hurley, McKaskill, McLeod and Harnwell that the actual usage rates of vehicle spare parts were well above the anticipated consumption rates.

In particular, the requirement for urgent resupply of track pads and track link for the APCs caught everyone by surprise. No one, including McKaskill, had anticipated that constant driving over rough roads and tracks in a hot, dusty environment would take such a significant toll on the APCs.[65] Every APC had been refurbished meticulously, had had new parts installed and had been serviced before departure. However, after two weeks in Somalia, McKaskill knew that his vehicles were consuming spare parts, track pads and track link at unexpected rates and that he would have to replace some engines, transmissions and differentials to keep up with the tempo of operations. [66]

Hurley and McKaskill worked out that a resupply of 7 000 track pads and just under 2 000 track link were needed for the duration of the operation, if the current tempo was maintained. The first consignment had to reach Baidoa by 23 February or all APC operations would cease. Further consignments would be required periodically thereafter, using a 'trickle system' to maintain the APC fleet in the field for the remainder of the operation. [67]

On 4 February Harnwell detected that all was not well with the resupply system from Australia:

> At 1400 hrs LTCOL Woolnough arrived for a 24 hour stay. He spent the next three hours describing the reasons for delay and the way the system of resupply would work to myself, Majors Crosland, McDonald and McLeod. We [the BSG] are yet to receive any formal guidance on the system of resupply from Australia and no one seems to know whether the C-130 will ever arrive, or if it does, how frequent air resupply will be.[68]

McLeod, McDonald and Harnwell were expecting an Air Force C 130 aircraft flight every two weeks in accordance with Major Gary Banister's HQ 3rd Brigade Administrative Instruction written on 19 December. Banister had been guided by an ODF contingency plan for the deployment of a battalion group overseas at short notice. In theory, this contingency plan would have also been used by higher headquarters to develop their own Administrative Instructions for OP *Solace*. None of the subsequent administrative instructions developed by higher headquarters promised fortnightly air resupply but all had confirmed that service demands would be satisfied on time. From Woolnough's point of view, 'At this meeting I was seen as an ogre because I told them the truth. They were naturally uptight and angry that the system advised to them by Gary Banister would not occur.'[69]

During his visit on 4 February Woolnough stated that he was not happy with the number of Priority One service demands that were coming to his staff from the 1 RAR Group. He felt that McLeod and McDonald had not adjusted to the fact that they were not on a field exercise near Townsville where deficiencies in stores could be made up by sending a truck 'down the road' to a supply depot.[70] He cautioned Crosland, McLeod, McDonald and Harnwell not to use Priority One demands that had to be satisfied in seven days unless they were operationally essential.[71]

> The point I tried to make was that the system would clog up if every demand was Priority One ... I also know that for Log Comd to satisfy a Priority One demand within the timeframe of 7 days would take exceptional support and a fast and efficient resupply system.[72]

For their part, the 1 RAR Group officers knew that the Group had deployed with insufficient spare parts. They wanted to stimulate the resupply system to get these parts in stock in Baidoa before the vehicle fleet ground to a halt. McLeod took the view that OP *Solace* should be given top priority because the 1 RAR Group was the only Army unit overseas on active service. He wanted the Group stocked up in case the warlords went back on the warpath. After two tours of duty in Vietnam, including the Battles of Fire Support Bases *Coral* and *Balmoral*, McLeod was of the opinion that the level of threat could change quickly. He continued to submit Priority One service demands for spare parts. [73]

On 8 February, Harnwell discovered that Woolnough's staff had been returning his service demand vouchers and not forwarding them to Australia. Twice these urgent Priority One demands, including a further request for

APC track pads and link, were returned by Woolnough's staff because they were not in signal format. Initially Harnwell's staff had written demands as vouchers used routinely in Australia. After rewriting all the vouchers in signal format and sending them to Mogadishu, the demands were sent back once again by Woolnough to Baidoa with directions to rewrite them in a new abbreviated signal format developed by his staff.

Exasperated by the delay, Harnwell directed that one of his clerks take the rewritten demands to Mogadishu that day on the New Zealand Air Force daily shuttle that operated between Baidoa and Mogadishu, and every day thereafter to ensure the format for every demand met Woolnough's requirements. The clerk was instructed to rewrite the demands on the spot in Mogadishu in any format specified by Woolnough's staff to get them accepted so that there would be no further delay in sending them to Australia. Woolnough asked that this clerk remain in Mogadishu permanently to help identify items of stock in the American logistic computer system. Woolnough had found that the stock numbers being used by the Australians were not the same as the numbers in the American system. This incompatibility meant that someone had to manually search the American computer system to identify the stock numbers the Australians should use when requesting items from the Americans.[74]

Woolnough knew there was tension and misunderstanding between his staff and Harnwell's staff. He also knew from his visit on 4 February that Battalion administrative staff were critical of the time it was taking to respond to their urgent service demands, HQ AFS is neither a Control Office at 3 BASB[75] or 2 Field Supply Battalion[76] or a Brigade Headquarters.[77] Our structure is somewhere in between, mainly focussing on staff work and future intentions. ... The LOG Staff [at HQ AFS] are working very hard and are having minor concerns. Until we get the BSG [Harnwell's staff] in the right frame of mind and thinking more about what is going on around them and we educate them in priorities, then we will continue to have some misunderstandings. ... I am not sure the 1 RAR Group understands the situation but I am continuing to push my point about demanding using Priority One. I will not budge from my position. '[78]

Late on 8 February Harnwell received a message to call Woolnough immediately. He did so and Woolnough told him that Vale's staff and Winter's staff were questioning the quantity and timeframe for delivery of track pads and track link. He demanded further justification because the quantity and priority of seven days Harnwell had requested would cost close to one million dollars to air freight to Nairobi from Australia. Woolnough emphasised that the resupply should be spaced over the duration of the operation which was 17 weeks, and should not be a Priority One demand for delivery in seven days. Somewhere between Harnwell and Woolnough there had been a misunderstanding. McKaskill had requested a 'trickle flow' system of resupply, not a bulk delivery.[79]

Harnwell found it galling to have logistics officers back in Australia question his judgement, and that of Hurley and McKaskill who had extensive experience with APC operations.[80] Like Crosland, McLeod and McKaskill he had expectations that OP *Solace* was a sufficiently important operation to warrant high priority for resupply. Hurley, McKaskill and Harnwell met again and worked for several hours on a detailed explanation of why 7 000 track pads and 2 000 track link were required to keep McKaskill's APCs in the field. Several days later the quantities and 'trickle flow' delivery were agreed to by logistic staff officers back in Australia.

Tensions between Harnwell and Woolnough's staff in the first week of February were exacerbated on 7 February when a quantity of long-handled torches, purchased by Woolnough's Local Purchase Officer in Mombassa, Captain Mark McKeon, arrived in Baidoa. This brought to a head differences of opinion about the efficiency of Woolnough's local purchase arrangements.[81] Hurley had requested long-handed, sealed-beam 'MAGLITE' torches be purchased urgently after operations began on 17 January. Despite having pairs of Night Vision Goggles for two members of each night patrol and small hand-held field torches, he assessed that his soldiers needed to be able to shine torch light from a distance onto buildings, into doorways and windows, and at any suspicious persons they found in the streets of Baidoa. The torches would also be useful inside unlit buildings during searches.

Hurley envisaged soldiers holding the torch handle along the side of his rifle so that the rifle barrel and the torch beam pointed in the same direction. This would enable them to fire a burst along the beam if a person was discovered reaching for a weapon or taking aim. This use of light along the barrel also had the advantage of temporarily blinding a suspicious person so that the initiative remained with the Australian soldier.

The tactical advantage of having these torches on night patrols and during building searches was not identified during the reconnaissance phase of the operation in December and early January so they were not purchased in Australia before the departure of the main body of troops. Two weeks after submitting Priority One demand, a quantity of torches arrived on 7 February. In Harnwell's words they were, 'Woolies Specials,[82] Lion Brand, Made in China, so weak that the metal cases flexed when pressed.'[83] These torches did not have a strong penetrating beam and their casing was not strong enough for the length of the torch to be held along the barrel of a rifle without creating a bend that would eventually make the torch unserviceable.[84]

Woolnough had assessed that 'anything was better than nothing' and had directed McKeon to buy the Lion Brand torches as an interim issue. He had also obtained six 'MAGLITE' torches from US military sources in Mogadishu which he sent to Baidoa. Several weeks later, more 'MAGLITE' torches arrived from Australia.[85] It was not the outcome expected by Harnwell and many members of the 1 RAR Group, but it was Woolnough's and McKeon's best effort. Despite these best efforts, the interim issue of the 'Woolies Specials' and the delay in the arrival of the remainder of the MAGLITES became a major source of ill-feeling between the two Australian Headquarters.[86]

Others were also having problems with local purchase arrangements. McKaskill requested a resupply of 3 inch screwdrivers. Subsequently, he was asked to clarify the requirement. After clarifying the requirement for the third time, McKaskill's Squadron Quarter Master Sergeant, Warrant Officer Simon Bingham, produced an exact scaled drawing of a 3 inch screw driver. Four weeks later a quantity of 7 inch screwdrivers arrived from Nairobi.[87] While the Australian supply staffs in Baidoa and Mogadishu had their differences of opinion in the first week of February, there were also differences of opinion among commanders and their operations staff.

On 6 February a group of Australian troops in Baidoa were stoned by an angry crowd and had a grenade thrown at them [See Chapter 3]. These incidents were precipitated by John Caligari's decision, on the advice of Dick Stanhope, to retrieve an ISO shipping container from Baidoa for use by the Australians back at the airfield.[88]

Mellor was critical of the decision to persist with the retrieval of the ISO container when it was obvious that a local crowd opposed its removal,

> The fact is that the soldiers at the incident site persisted in attempting to remove the supposedly abandoned container despite the clear indication that the local population was not happy with their actions. I had difficulty seeing how their actions were in accordance with our mission, was concerned that safety and community relations had been jeopardised, and advised David [Hurley] accordingly.[89]

For their part, Hurley and Caligari obtained a written report from Captain Amon Reid from Major Dick Stanhope's Civil and Military OP Team verifying that the container had been abandoned. Setting aside the merits or otherwise of the Platoon Commander's handling of the incident and the original decision to retrieve the container, Caligari perceived that Mellor and his staff were interfering in matters that should be sorted out within the 1 RAR Group,[90]

> The CO was summoned by CAFS [Mellor] and I was spoken to by SO2 Ops HQ AFS [Major Geoff Petersen]. Both seem to have little else to do but micro-manage the CO. They do not appreciate

the information the CO has put together [about the ISO container incident]. [There are] Some things a CO must be able to assess and act on himself without having to explain in minute detail what he has done and why.[91]

This incident brought to a head the issue of Mellor's authority as national commander to monitor 1 RAR Group operations and react forcefully when he was not happy with what had gone on.

Privately, Mellor believed that the temperament of some members of the 1 RAR Group and their expectations of combat were not aligned to the ROE.[92] Hurley was also aware of his troops eagerness for combat if the opportunity arose. He had the dual responsibilities of ensuring that his troops would fight instinctively when required under the ROE but that they also maintained a firm and fair attitude to Somalis, especially provocative Somali males. Mellor's directives from Blake and Hurley's position as Commanding Officer entitled them both to intervene strongly whenever they were concerned about the performance of Australians on operations.

By the middle of February logistic staff officers in Australia were questioning why the AFS needed resupply so soon after deployment. They expected the AFS to sustain itself for 60 days with stocks brought on *Tobruk* and *Jervis Bay*. On 13 February Captain Mick Fulham from Woolnough's logistic staff penned a signal for Woolnough's signature explaining what he called the 'incongruous' service demands from the 1 RAR Group for 180 line items within one month of beginning operations.[93] He pointed out that 85 percent of these demands were for spare parts to maintain APCs, vehicles and weapons that were not available from the American supply system in Somalia, or through local purchase in Mombassa.

Fulham went on to state that the AFS had not deployed with sufficient spare parts and were using existing stocks at unexpected rates. He mentioned the pallets of spare parts had been left on the wharf in Townsville and the significant quantities of spare parts had not arrived in Townsville by the time *Jervis Bay* and *Tobruk* had sailed. Fulham then went one step further and pointed out that funds for the purchase of spare parts for the ODF had been reduced in the years leading up to the deployment. Consequently there were long term deficiencies in stock holdings of spare parts in Townsville. He concluded by stating that the resupply system for spare parts that applied to the ODF in Townsville was not suited to remote operations in Africa, several thousand kilometres away from Australian and overseas suppliers.

Fulham's point about the policy on resupply of spare parts to the ODF was significant. In Townsville only limited supplies of spare parts were 'held on the shelf'. They were sought from local Australian or overseas suppliers only after vehicles, weapons and equipment had been inspected and put in for repair and maintenance. Now that the 1 RAR Group was operating 11 000 kilometres away from military supply depots and commercial suppliers in eastern Australia, it was difficult to resupply spare parts in 7 to 28 day timeframes.

While the customer-end of the resupply pipeline justified its requests, re-wrote paperwork and sent increasing numbers of service demands in late January and early February to Australia, the supplier-end of the pipeline broke down. The problem was not the speed of handling the paperwork, but monitoring and satisfying over two hundred Priority One service demands from Somalia using the peacetime Army logistic system. Paperwork moved swiftly, but action came slowly.

Service demands from Somalia were being sent quickly by Brian Vale's staff at LandHQ to Tim Winter's logistic operations staff at HQ Logistic Command. The demands were then sent to supply depots. By early February scores of containers full of stores addressed to the AFS in Somalia were accumulating at the Moorebank Freight Terminal, apparently unnoticed by the logistic staff at LandHQ and HQ Logistic Command. [94] No one appeared to be tracking service demands from the AFS through the Army's logistic system. The clerical staff and warehouse managers working in the system processed the paperwork, satisfied what demands could be satisfied from existing stocks, packed the stores up and forwarded them to Moorebank. The stores arrived at the Moorebank Freight Terminal and nothing happened. The air resupply system specified in Vale's and Winter's administrative orders for OP *Solace* had not 'kicked in'.

The challenge of providing air resupply to the AFS had been considered during the early days of planning for OP *Solace* in December 1992 Lieutenant Colonel Bill Nagy, the Logistic Operations Officer on Bill Traynor's staff, and Wing Commander Barry Newham, a senior air movements officer, at HQ ADF in Canberra had envisaged QANTAS 747 aircraft being contracted to resupply the AFS for the duration of OP *Solace*. Newham, who had been a supply officer and air movements specialist since 1977, realised that, though there were plenty of air hours[95] set aside for the ADF to fly national tasks in financial year 1992/1993, the Air Force did not have air hours allocated to support OP *Solace*. He also realised that it cost twice as much to send a C 130 aircraft on a round trip to Somalia compared to contracting a 747 jet. A 747 also had a significantly higher payload than either a Air Force C 130 transport aircraft or a 707 jet configured to carry freight. [96]

Senior ADF staff in Canberra had accepted Nagy's and Newham's recommendations for commercial air freight resupply of the AFS and allocated $ 2.4 million to the Army's financial management staff at Army Office in Canberra to pay for a commercial air freight link to Nairobi, the closest secure commercial airport to Mogadishu. Nagy and Newham intended that these funds would be transferred to AirHQ from Army Office because AirHQ was responsible for co-ordinatingthe provision of both Air Force and commercial aircraft in support of ADF operations.[97]

During the first days of February Corporal Jim Rickard,[98] who was responsible for co-ordinating the arrival of stores at the Moorebank Freight Terminal, had been noticing that more and more Priority One stores, including increasing numbers of pallets of track link and track pads, were being delivered to a transhipment area set aside for OP *Solace* in Warehouse 90. No one appeared to be taking an interest in moving them to Somalia.

Many of the containers of stores marked as 'Priority One' were days and, in some cases, two weeks overdue.

On Friday 5 February Rickard passed his concerns onto Sergeant Wayne Schafferius, [99] who co-ordinated the movement of all stores from the Freight Terminal to customer units in Australia and overseas. Schafferius had no information on what to do with the stores addressed to the AFS in Somalia once they arrived. He contacted Major Michael King, on Ayerbe's staff in Melbourne. [100]

On Saturday 6 February Captain John Yeaman, the Operations Branch Duty Officer at LandHQ, started to receive telephone calls from Duty Officers at AirHQ, HQ Logistic Command and the Moorebank Logistic Group about the accumulation of Priority One stores for Somalia at Warehouse 90. No one appeared to have the authority or resources to move the stores to Somalia. All Yeaman could do was record their concerns in his Duty Officer's Log. [101]

The job of sorting things out fell to Major Jim Kirkham, who had replaced David Bucholtz as the Major co-ordinating logistic operations on Brian Vale's staff at LandHQ, on Monday 8 February. By Thursday 11 February, Kirkham had held discussions over the telephone with Jamieson's movements staff at AirHQ and King at HQ Logistics Command. Jamieson's staff pointed out that Vale's Administrative Order had made Winter's staff responsible for selecting the means for the movement of stores to Mogadishu, and that AirHQ only responsible for flying the stores to Mogadishu, the Point of Entry for OP *Solace*. From there, HQ AFS were responsible for transporting stores to Baidoa. Jameison's staff then stated that no one from HQ Logistics Command had formally requested them to move stores to Somalia from Moorebank.

At 4 10 pm on 11 February, just under four weeks after the 1 RAR Group began operations in Baidoa, the first signal requesting air resupply of the AFS was sent to Jamieson's staff. King asked them to move 16 pallets of track link and track pads, and two pallets of general stores by commercial air freight to Nairobi urgently. [102] As an alternative, King recommended that the Air Force allocate 150 air hours for the stores at Warehouse 90 to be flown to Mogadishu by C 130 aircraft. A staff officer at AirHQ annotated on King's signal that this option would cost the Air Force $A 800 000.[103]

King's signal stimulated Jamieson's staff to task the Air Lift Group to allocate a C 130 aircraft to take a load of stores to Mogadishu on 18 February and for Barry Newham at HQ ADF to seek diplomatic clearances for the aircraft to fly into Mogadishu. King's preferred commercial air freight option of flying stores into Nairobi was not acted on. On the evidence available, this C 130 seems to have been deemed by movements staff at the various headquarters to be a 'one off' air resupply of urgently needed stores. It is likely that there was an *ad hoc* allocation of air hours by HQ ADF or AirHQ for the return trip from Richmond to Mogadishu. At this time, no one at the various higher headquarters appeared to know the magnitude of the stock pile building up at Moorebank. [104]

This *ad hoc*, 'quick fix' did not solve the problem of what to do with the remaining Priority One stores at Warehouse 90 or the increasing quantities of inbound stores. On Thursday 14 February John Yeaman penned a pithy minute to Jim Kirkham pointing out that HQ Logistic Command and the other agencies designated to manage air resupply to Somalia 'had little or no idea of what was happening.'[105] Yeaman had spoken at length with Wayne Schaffarius by telephone to come to a first-hand understanding of what was going on. Schaffarius reported that the problem was that two more C 130 loads of overdue Priority One stores were on the floor of Warehouse 90 but no aircraft had been tasked to fly them to Somalia. More overdue Priority One stores were arriving daily and 24 pallets of APC track link and pads were due to arrive in a few days. [106]

Schafferius also pointed out that the stores were arriving in Multi-pack cartons and packing boxes with no attached information on what was in them except their priority, their weight and their size. Consequently, he had a warehouse floor piling up with 'Priority One' cartons, boxes, crates, cylinders and envelopes. He needed some guidance on what should be the priorities among the Priority One stores. He was ensuring that 'today's Priority One stores were segregated from tomorrow's Priority One stores', but had no idea of what urgency to assign any container because he did not know what was in it and how essential the items might be for operations in Somalia. [107]

Yeaman made several telephone calls after speaking with Schafferius trying to discover why the 'log jam' of Priority One stores discussed on 11 February had not been unclogged by 14 February. He found that Jamieson and Ayerbe believed that three C 130 aircraft were due to fly to Somalia, whereas staff at the Air Lift Group, who co-ordinated the use of C 130 aircraft at 86 Wing, verified that there was only one C 130 sortie on 18 February. Despite Rickard's, Schaffarius's and Yeaman's advice about the accumulation of stores at Warehouse 90, no one at higher headquarters appeared to have sorted out which of the overdue Priority One stores would go next and how the rapidly accumulating quantity of stores would get to Somalia from now on. Significantly, no one appeared to be talking about the $2.4 million that had been transferred by HQ ADF to Army Office for AirHQ to support OP *Solace* with an air freight link to Nairobi.[108]

By the middle of February, four weeks after the 1 RAR Group had commenced operations and two weeks after Priority One stores had begun arriving at Warehouse 90, the key question remaining unanswered was, 'Who had the authority and resources to move stores to Somalia and who was going to co-ordinate that movement? According to administrative instructions issued by Vale and Winter, Michael King at HQ Movement Control located at HQ Logistic Command had the authority to choose how to move stores to Somalia. According to the intentions of Nagy and Newham at HQ ADF, Jamieson at AirHQ had been given sufficient funds to pay for a commercial air freight link to Nairobi. It appeared that King had authority but no resources, and Jamieson had resources but no authority. For the time being there was a stalemate between Army and Air Force staffs while the situation was being sorted out: Priority One stores continued to accumulate at Moorebank.

While the air resupply system from Australia yielded only one C 130 sortie on 18 February, the system to supply canteen stores to the 1 RAR Group was being set up. After the 1 RAR Group had set up camp in Baidoa, canteen stores, such as soft drinks, confectionary, potato chips, and cigarettes, were not available except through a visiting American Tactical Field Exchange, or by mail from Australia. Woolnough's staff had made arrangements for the issue of toiletries and personal hygiene items through the American supply system.[109]

Four weeks after arriving in Somalia those doing the physically demanding work on the operation could not get a cold can of soft drink or cold water after a hard days patrolling, or buy a packet of cigarettes except from local Somali vendors or from a visiting American Post-Exchange. There was no refrigeration at company level for soldiers to cool containers of water or cans of soft drink brought from Somalis in town. The only refrigerators were at the BSG and at HQ AFS in Mogadishu.[110]

On field exercises in Australia Glenn Crosland was responsible for the provision of canteen stores, purchasing stocks and sell them to members of 1 RAR as a commercial transaction through the unit's Regimental Funds account. He had not brought funds from this account to Somalia so he sought public funds from the AFS financial adviser in Mogadishu, Major Richard Clarke, to buy an initial quantity of stores to start up a canteen. [111]

Woolnough informed Harnwell on 14 February that Clarke had not supported loaning public funds to purchase initial stocks of canteen stores. The staff at HQ AFS thought this was 'a ludicrous situation but we were hamstrung by the rigid regulations.' [112] Woolnough recommended a $ 15 levy of all soldiers in Baidoa to start up the canteen. If this was not acceptable, he directed Harnwell to put a request for funds via his headquarters to LandHQ in Australia. [113]

Both Crosland and Harnwell expected a full range of canteen supplies to be provided through the establishment of a permanent American Field Exchange at the end of February.[114] Woolnough had been assured by

American staff officers on 14 February that a Tactical Field Exchange (TFX) was to be set up in Baidoa, manned and supplied by the Americans in about a weeks time. He wrote at the time:

> The battalion has not agreed to my proposal [for a $ 15 levy on each soldier] and seems to think that the Commonwealth should fill the bill initially. This can be done but requires an on-cost of 15 percent for administrative costs. I think the system needs reviewing and I will make a case when I write the post-exercise report. The TFX will take seven days to set up but will be a good thing once established.' [115]

Crosland decided not to wait for either the Australian or the American systems to provide canteen stores. On 15 February Michael McKinnon, a journalist with the *Townsville Bulletin* who had accompanied the Australians to Baidoa, wrote an article deploring the delay in finding public funds to set up a canteen for the diggers. McKinnon emphasised that Australian lives could be endangered if they were forced to buy cigarettes and soft drink from the Somali vendors in Baidoa. He quoted an unnamed source at the battalion Headquarters saying that a packet of cigarettes could be booby-trapped. [116]

McKinnon's article produced results. Within 48 hours Lieutenant Colonel Andrew Barton from Vale's staff had briefed General Blake and obtained approval to ask Army Office to transfer $10 000 to Richard Clarke's AFS account in Mogadishu to purchase canteen stores. [117] On 22 February Blake's Chief of Staff, Brigadier Ian Macinnis, had to answer Ministerial enquiries about the provision of canteen stores, probably prompted by McKinnon's article. He affirmed that $10 000 had been transferred and, 'Acquisition of supplies occurred last week and distribution commenced immediately.' [118]

Harnwell had also decided not to wait for the system to work. He arranged with Kevin Taylor to buy $4 500 worth of canteen supplies on credit from Mombassa when *Tobruk* docked there next. [119] Harnwell planned to open his unit's canteen on 19 February. He envisaged the profits from his enterprise going to purchase special OP *Solace* T-shirts, plaques and a range of videos for members of his unit.[120]

Though *Tobruk's* direct support to the troops in Baidoa during the first weeks of February was minimal, Kevin Taylor had made good use of *Tobruk's* time and resources since unloading AFS personnel and stores in January. After re-joining the ships of Multi-National Naval Force on 19 January about five kilometres off Green Beach where the Marines had landed six weeks before, his crew had conducted training in bridge procedures, seamanship, flying operations with the Sea King helicopter and communications. He was able to negotiate a series of personnel exchanges with other ships and *Tobruk* participated in training exercises with American and Canadian ships during the next week.[121]

He sought and received permission from Maritime Headquarters to carry cargo from Mombassa to Mogadishu in support of UNITAF and NGOs. On 14 February *Tobruk* was returning from Mombassa with a second load of cement, timber, fuel and general stores for UNITAF. By this time Taylor had transported almost 1 000 tonnes and was seeking further opportunities to assist as a 'coastal freighter' for OP *Restore Hope*. Concurrently, he maintained 24 hour contact with Mellor's headquarters in Mogadishu in case there was a requirement to provide direct support to Australian operations.

Meanwhile, Colonel Bill Mellor visited Baidoa on 15 February, refreshed after arriving back from two days leave in Nairobi a few days before. He had also taken the opportunity to call in on the Australian High Commissioner. He wanted to clarify his command status over the 1 RAR Group once and for all in anticipation of a change of operational command arrangements. The 1 RAR Group's operational headquarters, HQ 10th Mountain Division, was due to leave Somalia on 2 March. The two options for new operational command arrangements were for the 1 RAR Group to report directly to HQ UNITAF or to HQ AFS who would then report to HQ UNITAF.

Mellor carefully considering these options, deciding that his headquarters should assume operational command of the 1 RAR Group. He assessed that HQ UNITAF (HQ 1st Marine Expeditionary Force), was too high a level headquarters to become involved in directing operations in a Humanitarian Relief Sector. HQ AFS was already placed under operational command of HQ UNITAF. To place the battalion group separately under operational command of HQ UNITAF would have thoroughly confused the Americans, and served no useful purpose.[122]

Soon after it was announced that HQ 10th Mountain Division was leaving, Mellor put his proposals to Blake who agreed to Mellor's new appointment as the operational commander of the Baidoa Humanitarian Relief Sector.

The Canadian Defence Force had done the same thing with Colonel Serge Labbe and his national headquarters. Labbe was to be appointed as the Commander of the Canadian-controlled Belet Weyne Humanitarian Relief Sector after 7 March. He would command the Canadian Airborne Regiment Battle Group, a unit similar in size to the 1 RAR Group, an engineer squadron, HMCS *Preserver*, a medium air transport detachment and a helicopter flight. The Canadian Government had financed the deployment of a capable force with independent aviation and engineer assets to participate in OP *Restore Hope*. Like Mellor, Labbe planned to exercise command from a headquarters in Mogadishu close to HQ UNITAF. Unlike Mellor, he had a helicopter to commute between Mogadishu and Belet Weyne whenever he wished. [123]

Mellor's quickest link between Mogadishu and Baidoa was the New Zealand Air Force air shuttle. The Maritime Commander- Australia, Rear Admiral Robert Walls, had decided to closely manage *Tobruk's* Sea King helicopter during OP *Solace*. There were also concerns at Maritime Headquarters about the dusty conditions in Somalia adversely affecting the operations of the helicopter. Thus, Mellor was not permitted to use the Sea King for routine command and liaison missions over land or for medical evacuation. [124]

Lieutenant Colonel Richard Greville, SO1 Operations, Land Headquarters, during Operation Solace.
Image: Richard Greville.

Lieutenant Colonel Gary Martin, CO 1st Ground Liaison Group, during Operation Solace.
Image: Gary Martin.

Mellor directed Hurley to convene a meeting of all the Majors in Baidoa at 7 30 pm on 15 February after the evening Intelligence briefing so he could discuss the new command and control arrangements. Mellor began the meeting by asking the majors what was the Australian mission in Somalia and what was his role. After some hesitation among those assembled about his role in Somalia, he stated firmly that he was the commander of the AFS and after 2 March he and his staff should be regarded as a tactical Brigade Headquarters. [125] He then went on to speak for just over two hours about transition arrangements, the requirement to augment HQ AFS with additional staff and a range of matters that he felt needed to be resolved face-to-face between himself and the majors.

The reaction among the majors was mixed. Several questioned whether Mellor could be an effective operational commander of the Baidoa Sector with himself and his headquarters based in Mogadishu. For Mick Moon, his discussion with Mellor about this concern after the meeting, 'Incited an active, serious debate which did not end.' [126] From Mellor's point of view, 'I considered redeploying HQ AFS to Baidoa, but could see nothing to be gained by such a move. It would only serve to crowd David and his HQ and considerably complicate the necessary liaison and staff work that my headquarters conducted with HQ UNITAF. [127]

After 3 March Mellor would have the authority to command the Baidoa Sector. However he faced a major challenge trying to raise his command profile with the 1 RAR Group. Aside from the differences of opinion about whether he could command operations from a headquarters 240 kilometres away with no dedicated aircraft for commuting back and forward, most of the majors did not assess that Mellor and his staff had the experience to command a battalion group on complex military and civil operations. They emphasised that without an infantry officer co-ordinating operations at HQ AFS this situation was unlikely to change.[128] From Mellor's point of view:

> ... from arrival to 2 Mar 93, I had no operational responsibility, that having been delegated to HQ 10th Mountain Division. After 2 March the nature and tempo of operations were well established and required little or no guidance from myself or the AFS Headquarters. The main source of irritation, in my opinion, between the battalion group staff, and myself and my staff, was my insistence on being kept fully informed of what was occurring and our questioning of what we regarded as incomplete or confusing messages. ... I doubt that the presence of an Infantry officer on my staff would have prevented it.[129]

After speaking with the majors, Mellor was invited by David Hurley to accompany a patrol into Baidoa that night. Patrol commanders were not happy having lightly-armed senior officers or, for that matter, any outsiders accompanying them on patrol. They were wary of being held responsible for the safety of someone who was not part of their team and who might not be sufficiently well-trained in close-quarter combat to take care of themselves, and shoot quickly and accurately if the patrol was in contact. Hurley assessed that it was important that Mellor obtained a first hand understanding of operational conditions. Mellor was not in a position to refuse;

> I was very conscious that I would be a burden to the patrol commander, but on the other hand, had word got around that I had refused the CO's invitation to accompany a patrol, I am sure that would not have helped either.[130]

While new command and control arrangements for the 1 RAR Group were being clarified, logistics and movements staff officers in Australia were trying to unclog the air resupply system to the AFS. Unbeknown to Nagy and Newham at HQ ADF, who assumed that HQ Logistics Command staff would task Jamieson's movements staff to establish a commercial air freight resupply link to Nairobi using the $ 2.4 million supplementation to meet air freight costs, the intended use for the additional funds had not been well communicated. [131] Apparently ignorant that funds had been allocated for this purpose, staff officers from LandHQ, HQ Logistic Command and AirHQ spent almost two weeks in February trying to find a solution to moving the Priority One containers of stores from the Moorebank Freight Terminal to Mogadishu.

Eventually, after meetings, numerous telephone calls and facsimiles and a personal visit to Moorebank, King wrote a signal to all of the Service headquarters and units involved in the resupply of the AFS on 26 February specifying that the Moorebank Logistic Group, through Captain George Kosciuszko[132] at 2nd Movement Unit, would advise movements staff at AirHQ when stores for Somalia constituted an aircraft load ready for onward movement, or when high priority stores arrived that needed urgent dispatch. Jamieson's staff would then decide whether commercial or Air Force aircraft would be used to deliver the stores. [133] Kosciuszko was responsible for delivering the stores to the 'air point of departure.'

Having been told unequivocally by Jamieson and his staff that the Air Force had no additional funds or air hours to support OP *Solace*, Lieutenant Colonel Richard Greville, the newly-arrived Operations Officer at LandHQ, was asked to provide air hours from the Army's annual allocation to fly stores into Mogadishu by C 130 aircraft. [134] Greville and Lieutenant Colonel Gary Martin, [135] the newly-arrived Commanding Officer of the Army's Ground Liaison Group located to AirHQ, negotiated with Wing Commander John Ward, the air operations staff officer and Greville's counterpart at AirHQ, to co-ordinate resupply flights. Martin was forced to find sufficient air hours from Army's annual allocation from HQ ADF to help fund the next resupply sorties to Somalia. [136] As a result of Greville's and Martin's efforts, another C 130 aircraft flew to Mogadishu in late February.

Meanwhile the delay in the delivery of vehicle spare parts to Somalia was having an adverse effect on operations in the Baidoa Sector:

> The vehicles are falling apart around our ears after only one month. This is mainly due to the way they are being driven and the conditions, dust, camel thorn and shocking roads. It is not helped by the fact that there appears to be no parts for even the most simple repairs, (brakes, shock absorbers, batteries etc).[137]

The arrival of the first two C 130s in Mogadishu in February exposed further problems with the resupply system. On 25 February Harnwell wrote,

> I have advised the Chief of Staff [Woolnough] about my dissatisfaction with HQ Logistic Command and their consigning of stores to us. We have been given five days notice about the arrival of each C 130. We are not advised what stores will be on the plane and until we have unpacked and identified [them], we cannot marry up the stores and vouchers.[138] The receipt of stores is therefore slow. Many stores are Priority 3 which have been ordered in the past week. Priority 1 items, meanwhile, ordered two weeks ago have not been on the aircraft.[139]

Harnwell had to send convoys to Mogadishu to pick up the stores arriving by C 130 aircraft at short notice. Unlike the C 130 aircraft supporting the NGOs and the UN World Food Program in Baidoa, the Australian Air Force C 130s would not land at Baidoa Airfield. Harnwell was unable to find a reason for this policy.[140]

Unbeknown to Harnwell air movements staff at AirHQ had assessed that there was insufficient handling equipment and technical personnel to unload and load stores at Baidoa. They did not want to deploy Air Force personnel and equipment with incoming C 130 aircraft because space would have been taken up room that should have be used for urgently-needed stores. Indeed the AFS were fortunate that Air Force movements staff had decided to off-load stores only 240 kilometres from Baidoa at Mogadishu. Serious consideration had been given to using airports at Mombassa or Nairobi because airport and accommodation facilities were superior to those at Mogadishu and security was less of a problem. [141]

Graeme Woolnough passed on Harnwell's concerns about not knowing what stores were on inbound aircraft to LandHQ:

> Major concern is visibility, in particular the [service] demands to be satisfied from AS [Australian] resources. Lack of information is making it difficult for HQ AFS to plan and disseminate information to battalion group. Battalion group is in turn frustrated by lack of information.

Mutual frustration over delays in resupply and the 'invisibility' of inbound stores contributed significantly to tension between the two Australian headquarters in Somalia. Woolnough and his staff were responsible for financial management and local purchase arrangements, but aside from sending strongly-worded signals, they were unable to influence resupply from Australia. Brian Vale, Andrew Barton and their staff at LandHQ who reported to General Blake were in a better position to fix problems with the resupply of the AFS. However, there were perceptions in the 1 RAR Group that HQ AFS staff may have been contributing to resupply problems, and not solving them.[142]

Woolnough saw himself as being between a 'rock and a hard place'. He was aware that the 1 RAR Group were frustrated and concerned that he was not doing his best to solve the resupply problems. He was trying to be as loyal as possible to his higher headquarters but he felt that his many telephone calls to Vale's staff and follow up signals were 'falling on deaf ears' and that no one seemed interested enough to fix the problems.[143]

On 1 March the 1 RAR Group celebrated the Australian Army's Birthday and their sixth week on operations in Somalia. By this time Harnwell had set up his unit's canteen and, prompted by the repercussions of McKinnon's article on 15 February, Mark McKeon had arrived a week before with several hundred cartons of soft drinks from Nairobi that had been bought under local purchase arrangements.

The next day a handover ceremony was held outside Hurley's headquarters. General Arnold praised the work of the Australians and said farewell. He handed over command of the Baidoa Sector to Bill Mellor. The communications arrangements for operational command were already in place as a result of a double reporting system instituted by Mellor several weeks before. John Caligari and his staff just discontinued communications to the 10th Mountain Division. On the logistic side Mark Harnwell wrote:

> We have now reverted to under command CAFS and this was made evident when, at 1600 hrs, I received a signal from the Chief of Staff [Woolnough] directing that all conversations between Colonel Harper [US Logistics Commander in Somalia] and the battalion be staffed through HQ AFS.[144]

Later that day Harnwell received news that the American Tactical Field Exchange was not going to be set up to provide a canteen for the troops in Baidoa because over the next few weeks American units would be withdrawing and there would be insufficient demand from American personnel to justify the establishment of canteen facilities. Hurley directed Crosland to establish a 1 RAR Group canteen as quickly as possible using the funds allocated by General Blake.[145]

Aside from canteen stores, the Australians were going to have to become self-reliant to other areas of resupply and support now that American units were leaving. The withdrawal of American logistic units, such as the 62nd Supply Company and the 43rd Combat Engineer Battalion, would leave the Australians with increased responsibilities for maintaining their camp, and receipting and distributing supplies.

During the previous weeks Mark Harnwell had been reminding Woolnough that many urgent service demands were well overdue. He sought permission to contact depots in Australia so he could speak with logistic staff he knew personally who might be able to follow up demands more vigorously if operational conditions were described to them first hand. Woolnough did not want Harnwell to by-pass him and deal directly with staff at supply depots and logistic headquarters in Australia. He signed a signal to Harnwell drafted by Mick Fulham on 1 March directing Harnwell to only 'initiate hastening action' with HQ AFS when stores were overdue.[146]

On the same day Woolnough signalled the Moorebank Logistic Group that despite the arrival of two C 130s in February and the earlier flight in January, there were still many unsatisfied service demands, mostly spare parts needed to maintain the AFS vehicle fleet. He pointed out that 'Many outstanding demands are affecting our operational capability and your early response is appreciated.' Andrew Barton gave Jim Kirkham this signal and he requested Major Peter Tweedie from Tim Winter's staff at HQ Logistic Command to provide an update on the status of the demands by 3 March.

Woolnough's signal on 1 March formalised that the inefficient resupply system was now adversely effecting 1 RAR Group's operations. Kirkham waited in vain for Tweedie's update on the status of service demands. There was no automated system for tracking service demands from Somalia through the Army's logistic system.[147] In addition, no effective action was being taken to satisfy many of the demands. Every time the Moorebank Logistic Group staff found that they could not satisfy a service demand from stocks on-hand or by commercial purchase, Corporal Oakley, a supply clerk, sent a signal to Winter's staff at HQ Logistic Command advising them that Moorebank had 'Nil Stocks. Other depots to provide.'[148]

Members of B Squadron changing the engine of an M113 APC at Baidoa Airfield in March 1993
Image: G Gittoes.

When Peter Tweedie checked on the situation on 3 March he found fifty six service demands were caught in a paper 'log jam' at HQ Logistic Command, waiting for Fleet Managers[149] to take action. Tweedie, assisted by Major Benny Woodward, an American exchange officer responsible for materiel management, then began the time-consuming task of following up each service demand from Somalia by telephoning Fleet Managers at Logistic Command, Moorebank and Bandiana as well as staff at other supply agencies to see if service demands from Somalia were in their 'In-Trays' and when they expected them to be satisfied.[150]

Another resupply crisis that had been brewing for several weeks came to a head in the first week of March. By 2 March 1 RAR Group drivers were taking inner tubes out of Landrover trailer tyres and putting them on their Landrovers to keep them operational. Camel bush thorns, many up to 10 centimetres long, had punctured over 100 inner tubes beyond repair, there had been over 350 reported punctures. The BSG's supplies were exhausted.[151] The high usage of inner tubes had been identified early and initial requests for resupply had been forwarded to Woolnough's staff for local purchase in late January. A request for 100 inner tubes and heavier ply tyres for the Landrovers had also been included in the first service demand raised by the 1 RAR Group in early February and accepted in the new abbreviated format insisted on by Woolnough's staff on 6 February.[152]

The frequent puncturing of tyres and damage to inner tubes had a significant impact on operations. Punctures occurred during off-road movement and left a vehicle and its passengers stationary for the 20 to 30 minutes it took to repair the inner tube and get going again. In some situations soldiers felt like they were 'sitting ducks' for bandits while the punctures were being fixed. Mick Moon commented that it was difficult to be an efficient 'pit stop crew', dressed in webbing and a flak jacket.[153] From David Hurley's point of view, running vehicles on operations with an average of 15 repaired punctures per inner tube 'was probably beyond a safe limit.'[154]

Rod McLeod waited for four weeks before he decided to 'stimulate the system.' He rang HQ ADF in Canberra about the problems of resupply, using the inner tubes as an example.[155] Having spoken to 'Canberra' direct, he went on to telephone Woolnough's staff in Mogadishu to express his disappointment at the delay in supplying these critical items that were available through commercial purchase in Mombassa. He wrote in his diary at the time:

> I expected to find 100 inner tubes for our Landrovers which I was promised. However, only five showed up. I then rang Mogadishu [HQ AFS] and got right up them but I don't know what good it will do ... Boy, am I disappointed in the system. They might as well just send me home as I cannot do my job over here ...

Two days later he wrote,

> Had lunch and on return, guess what, there were 95 inner tubes there. You could have knocked me over with a feather. Amazing what can happen when you spit the dummy. One had to ask why did it take me all that shouting and whatever. They could have released them 30 days ago. ... My hair is getting grey now. All they have to do is get the demand system [for stores] sorted out and all should be OK.[156]

The resupply of a field kitchen and APC engines and spare parts became the next major resupply issue by mid-March. On 24 February Woolnough had advised Barton that the American field kitchens which the Australians had depended on after two of the 1 RAR Group's field kitchens had burned down in the first week of February, were to be withdrawn by the Americans when they left in early March. He also pointed out several days later that 'Situation with M113 engines and transmissions is critical' and that B Squadron had only one spare engine and transmission left.[157]

On 4 March Woolnough told Harnwell that another C 130 aircraft that had been planned to arrive in Mogadishu on 13 March bringing an Australian field kitchen and trailer, cooking equipment, APC engines, transmissions and a range of spare parts had been cancelled. Michael King advised that he would move the back log of stores at Warehouse 90 by the freighter 'SABRINA' (sic) and commercial air freight to Nairobi.[158] Woolnough's staff now had the challenge of co-ordinating the reception of stores from Australia at Nairobi Airport and at the port of Mombassa, and to arrange their movement from those two locations to Baidoa. Once again the Australians would have to depend on the goodwill of the Canadians who were flying air transport shuttles to and from Mogadishu and Nairobi. The Canadian aircraft were flying in support of their own forces in Mogadishu, Belet Weyne and Bale Dogle, not units from other donor nations. Consequently, first priority would go to Canadian freight and personnel.

After he received Woolnough's news that the anticipated C 130 resupply sortie had been cancelled, Harnwell assessed that there was little chance that the field kitchen would arrive via Nairobi Airport before the Americans withdrew kitchen facilities from Baidoa in four days time on 8 March.[159] He asked Woolnough to take up the case with Andrew Barton at LandHQ. Later on 4 March Woolnough contacted Barton and impressed on him that urgent resupply of the kitchen equipment and APC engines and spare parts was needed or there would be embarrassing operational consequences.

At this time McKaskill was faced with the dilemma of continuing APC operations under increasingly dangerous conditions or grounding vehicles until track link and pads arrived. He had pointed out that a bulk delivery:

> ... was useless for track pads and link, as a 'trickle system' was required. This had been explained *ad nauseam*. At this stage [I] decided to keep running the APCs—although dangerous. [I directed that] Crews were to monitor tracks and stop if gaps became dangerous.
>
> He also instructed his mechanics to inspect and adjust tracks on each vehicle daily in an effort to maintain support to Hurley's operations for as long as possible.[160]

On 5 March it was decided to fly the field kitchen equipment and APC spare parts to Mogadishu by Air Force C 130 as soon as possible and move remaining stores by ship to Mombassa and by commercial air freight to Nairobi. Later that day, Richard Greville sent a message to AirHQ requesting an urgent air resupply. Sufficient hours from existing Army allocations and from Air Force training hours were found to authorise this flight.[161] On 10 March the field kitchen and other urgently required general stores were loaded on a C 130. Fortunately, the Americans had agreed to leave their kitchen facilities in place until the Australian kitchen arrived. The C 130 left Richmond on 10 March but, due to confusion about whether certain consignments were going by commercial air freight to Nairobi or on the C 130 to Mogadishu, the aircraft departed without any APC engines, transmissions or spare parts aboard.

The air freight charter to Nairobi, co-ordinated by movements staff at HQ Logistic Command and AirHQ, turned out to be an administrative nightmare for Woolnough and his staff. Initially, they were not sent airway bill numbers and, after requesting that the numbers be sent to them urgently, they were given wrong ones. Mick Fulham spent several days trying to locate and clear the Australian stores through an inefficient Kenyan customs system after they eventually arrived three days late. He then had to wait until there was space on Canadian transport aircraft to fly the stores from Nairobi to Mogadishu.

From Woolnough's point of view, he had neither the staff or the patience to persevere with a commercial air freight link from Australia to Nairobi. 'My view is that it is the responsibility of LOG COMD to get stores and equipment into the country, not mine. Our lack of air transport precludes us from having the self-sufficiency we need to react quickly.'[162] On 11 March, Michael King sent a signal to Major Mike Prain at Army Office asking for $ 200 000 to pay for the recent urgent air charter to Nairobi.[163] Meanwhile, MV SABRIAN, carrying containers of track link and track pads and other stores, was reported to Woolnough's staff as being due to dock in Mombassa on 24 March.

On 11 March Bill Mellor sent a personally signed signal to Phil McNamara and Brian Vale about the resupply situation,

> I am most concerned about the timeframe it is taking for service demands to be received in Somalia. There are Priority 1 demands that have yet to be satisfied 30 days after their satisfaction date.[164] The situation in regard to Class 9 [Spare parts] and MUAS [Major Unit Assemblies] is of particular concern. For example, M113 engines have not arrived in the requested timeframe and as a result are now critical items. AFS has one spare M113 engine in theatre;'[165]

On 12 March Jim Kirkham formally acknowledged to Woolnough in a signal that the Message Demand system was in trouble,

> Problem of the service demand visibility is acknowledged. Until recently there has been no system in place to track service demands through demand/supply/movement systems. Log Comd has initiated action to solve problem. Please assess service demand situation after C 130 is unpacked on 12 March. ... There is a 2nd Line to 4th Line 'weakness'.[166]

On 14 March, after the fourth Air Force C 130 aircraft to fly a resupply sortie in support of OP *Solace* arrived, Mark Harnwell wrote,

> The stores from the C-130 will be delayed one further day. I have no idea what is on the plane. The list that was finally sent to us from Log Comd merely mentions boxes and packages with various weights.

> The first indication of the stores sent from Australia to us will be known when the storeman opens the box and receipts the stores. I had hoped that the supply system could provide a better advance notification method to us, perhaps a signal or even a fax ![167]

The next day the convoy sent to Mogadishu to pick up the stores from the C 130 arrived back at 7 30 pm. The stores had to be unloaded from the trucks by hand. The Australian warehouse forklift had broken down within days of the Australian arrival at Baidoa in January and needed a small seal replaced. Harnwell wrote, 'Unfortunately the forklift seal is still to be received. The seal, no larger than a bottle top, has now been outstanding for eight weeks.'[168] The forklift operator could not understand why the Army resupply system could not provide a simple, small seal that was easily purchased in Australia for a few dollars. He wrote to a friend in Australia who purchased the seal and sent it to him through the mail. The forklift was operational ten days later.[169] Similarly, Warrant Officer Michael Robinson, the Artificer Sergeant Major of B Squadron, had spare parts for a generator sent through the international mail system by a friend after they failed to arrive through the resupply system.[170]

Since the discovery of the 'freight log jam' at Moorebank in mid-February, and the 'paper log jam' at HQ Logistic Command on 3 March, the resupply system had not been fixed. On 17 March Woolnough had advised Barton that, of 369 service demands submitted by the AFS since 30 January, 184 had been satisfied and 185 were overdue. Almost none had been satisfied in the timeframe requested. Most of the unsatisfied service demands were for spare parts for the 1 RAR Group's vehicle fleet, generators and communications equipment.

The military and civilian Fleet Managers at HQ Logistic Command, Moorebank Logistic Group and Bandiana Logistic Group were unable to satisfy all the service demands from Somalia within the timeframes specified by McLeod in Baidoa. All of the service demands from the AFS ended up in their In-trays where they joined the service demands from other units in Australia.[171] A significant number stayed there and the priorities of some of them were adjusted by hand so that they would not appear to have exceeded their original timeframes. [172]

Time would tick by while depot and headquarter Fleet Managers sought out the items requested so they could be forwarded to Moorebank for onward movement to Mogadishu. During this time the status of each service demand in general, and the individual items constituting the demand in particular, was unknown. No one on Winter's staff could provide updates back to Greville, Barton or to Woolnough about when items would arrive at Moorebank, let alone when they would arrive at Mogadishu.

Once items for Somalia were ready for consignment to Moorebank for onward movement, depot staff would pack them together in Multi-pack containers. Subsequently, items arrived in containers without accompanying descriptions of their contents or the service demand numbers Rod McLeod had given them originally. No one knew what was in the containers or what service demand the items were fully or partially satisfying. In effect, after Woolnough sent McLeod's service demands to Barton, they were 'invisible' until containers of stores turned up at Mogadishu and Harnwell's staff unpacked them in what amounted to a logistic 'lucky dip'.

By 15 March, eight weeks after Australian operations had begun in Somalia, Tweedie and Woodward, were following each service demand from Somalia through the Message Demand system manually, following up each demand with Fleet Managers—one-by-one, day-by-day. To ease this workload Woolnough was asked on 23 March to 'sub-categorise' outstanding Priority One service demands into 'operationally essential demands' and demands that had less priority. He refused on the basis that he had monitored the priorities and all were essential items. This put him in the invidious position of being seen by both the 1 RAR Group and higher headquarters as being inflexible.

To speed up the flow of spare parts to Somalia, staff at Moorebank and Bandiana were authorised to purchase APC, vehicle and weapon spare parts for OP *Solace* without reference to Logistic Command Fleet Managers. Woolnough was allowed to speak directly with Colonel Greg Thomas's staff at Moorebank for General Stores, including vehicle and weapon spare parts, and to Colonel Jim Campbell's staff at Bandiana on APC engines and spare parts.[173]

Meanwhile, on 15 March the next round of negotiations for air resupply to Somalia began. George Kosciuszko notified movements staff at AirHQ and Richard Greville at LandHQ, that a load of APC track pads and link, APC engines, transmissions, and other spare parts and general stores was ready for movement to Mogadishu.[174] Once again Richard Greville and Gary Martin found the air hours to authorise the flight.

The air resupply to Somalia now had hand-to-mouth, emergency-like characteristics; each C 130 load was 'urgent' and notice was short. Each flight became a 'one-off' negotiation between Greville, Martin, Ward and Jamieson. The bargaining chips were the annual allocation of air hours the Army had each year authorising Air Force aircraft to support Army operations, training and other activities.[175] In absence of any air hours being allocated by HQ ADF for OP *Solace*, Greville's agenda was to have the Air Force use their own training air hours to support OP *Solace* as a means of training air crew in long range air transportation tasks. Training C 130 air crew on operational missions to and from Mogadishu was very appealing to Wing Commander Greg Hartig, Staff Officer Operations at 86 Wing, and his air crew: better to train on international operational flights than on routine flights around Australia.

Despite the enthusiasm of air crew at 86 Wing to use Air Force air hours allocated for training for operational flights to and from Mogadishu and Martin's astute staff work to identify air hours that could be 'found' from the Army's allocation, there was pressure from more senior Air Force staff officers to conserve Air Force air hours and use Army's allocation of air hours to resupply the AFS. One AirHQ brief assessed that the Army had under-estimated the initial resupply and support requirements for the AFS and was requesting the Air Force to make special arrangements to overcome these planning deficiencies. The brief's author deplored that resupply was being organised 'very much over the phone or through personal manipulation or guilt inducing argument such that Army should support its troops "in need".'[176]

In reality there were two air support systems working at the same time in support of OP *Solace*. The first was the logistically-correct system involving Michael King as HQ Movement Control, and Jamieson at AirHQ. The second was an operationally-expedient system involving Greville, Martin and Ward. The first system involved King requesting Jamieson to move stores to Somalia by air.

Jamieson would ask King to identify air hours to cover the use of C 130 aircraft. Since King had no air hours allocated to him and was not aware that Jamieson had $ 2.4 million dollars to pay for air freight, he could not move stores unless he found air hours or funds from Army sources. He did this to send stores by commercial air to Nairobi earlier in March and by sea on MV *Sabrian*.

The operationally-expedient system of air resupply had been operating since late January when Peter Neuhaus had used his contacts to have stores loaded onto an empty C 130 flying to Mogadishu to pick up Australian journalists. Subsequently when King realised that he would not be given air hours or funds to move stores, he referred the problem to Andrew Barton. Like King, Barton had no air hours or funds to satisfy Jamieson's 'user pays' requirements. Consequently, King by-passed Barton and referred the problem onto Mc Namara and Greville as an operations problem.

McNamara and Greville did not have an allocation of air hours or funds to support the resupply of the AFS but they did have a 'can do' attitude and personal contacts in the military operations world. In absence of any formal means of solving the problem, Greville asked two Duntroon classmates to help him out.[177] Consequently, Colonel Jim Wallace, Commander, Special Forces, underwrote part of the resupply of the AFS by offering up unused air hours allocated to Special Forces training to AirHQ.[178] Gary Martin 'found savings' from air hours that had been set aside for Army field exercises and training activities. He also assessed that HQ ADF would reallocate air hours to Army that had not been used on national tasks before the end of the financial year in June to cover his underwriting of air resupply to Somalia.[179]

After receiving a transfer of air hours from Special Forces training and borrowing air hours from Army training activities scheduled for later in the year, Martin would advise John Ward that he was providing sufficient air hours for another resupply sortie. Once the air hours were officially transferred to AirHQ, Ward would advise Jamieson that 'the user had paid' and a C 130 from 86 Wing would be tasked to fly.

From the point of view of the staff at HQ AFS , Michael King's use of commercial air freight to move stores to Nairobi was not as convenient as having C 130 aircraft fly direct to Mogadishu. On 19 March David Creagh pointed out to logistic staff at LandHQ and HQ Logistic Command that the commercial air freight link to Nairobi had its problems: 'Difficulty to date, apart for obvious lack dedicated air support [between Nairobi and Mogadishu] has been Kenyan time (this has to be seen to be believed) and the Kenyan customs process.' Subsequently, Creagh spent some time developing guidelines for working with the Australian High Commission and Kenyan contractors to unload and clear military cargo more efficiently.

King's use of sea freight to move stores also had its problems. Unbeknown to Creagh in Mogadishu and King in Melbourne, the MV SABRIAN carrying a container of track link, track pads and APC spare parts had changed course to pick up cargo in Durban, South Africa, before sailing onto Mombassa. This would delay delivery of containers to Mombassa by at least a week.

During this period *Tobruk* continued in its role as a coastal freighter for UNITAF between Mogadishu and Mombassa. In the first weeks of March Taylor not only transported almost 400 tonnes of stores for UNITAF but had also used his initiative to help several NGO organisations move tonnes of humanitarian relief supplies and food from Mombassa to Mogadishu.

Tobruk was also being used as a communications link for several weeks while repairs were made to the Army's High Frequency MEDPORT communications equipment supporting HQ AFS Mellor and his staff were keen to see how well *Tobruk* could transmit 'hardcopy traffic' and send it via Navy communications facilities in Darwin to destinations in Australia. The MEDPORT equipment with its large, cumbersome antenna array had been 'performing extremely poorly' and had resulted in Mellor's staff having to rely on expensive INMARSAT communications for voice and data links back to Australia for the first six weeks of the operation. MEDPORT was down for 53 percent of the time due to power generation problems. When it was operational, MEDPORT could only send and receive messages for four hours in any 24 hour period because of 'HF frequencies issued to the force.' [180]

While *Tobruk* provided rear link communications from 3-23 March Mellor's staff had 24 hour rear link communications with Australia. Satisfied with Navy's rear link capabilities, Mellor recommended to McNamara that the MEDPORT equipment be packed up and returned to Australia with its supporting Army signallers.[181] Lieutenant Colonel Gary Allen, [182] who was responsible for operational communications at LandHQ, assured McNamara that he and his staff could fix the strategic communications link with Somalia without Navy assistance.[183]

Subsequently Mellor was told to cease using *Tobruk* as his rear link communications facility to Australia. Allan's staff issued new technical instructions and 12 additional HF frequencies for the MEDPORT which had been sought by Mellor some weeks before. Overnight, Mellor's strategic communications link went from being available 95 percent of the time to 55-60 percent availability. This was deemed 'to be sufficient to handle the traffic demand.'[184] Thus, Mellor had INMARSAT equipment for immediate communications to Australia and MEDPORT for routine communications.

On 20 March Lieutenant Colonel Trevor Jones [185] arrived to take up an appointment as a staff officer at HQ UNOSOM and to be the national commander of the Movement Control Unit. He flew in with Major Bruce Scott [186] who had been appointed the Operations Officer for HQ AFS Mellor had requested an experienced Infantry Corps operations officer some weeks before. Scott was delighted to be given the unexpected opportunity to serve overseas on active service. Within a few days he understood that he would have to be a careful communicator to ensure that he gained the trust and respect of his Commander as well as his infantry colleagues in Baidoa. [187]

By the time Scott arrived both Australian headquarters were working together formally, but not fraternally. The tempo of operations had slowed down to the point where most communications were routine. Caligari and his staff had more time to send as much information as staff at HQ AFS wanted on current operations and future intentions. Administrative staff in Baidoa had resigned themselves to the inefficiency of the resupply system and routinely reported on how many service demands remained unsatisfied without expectation that the situation would change quickly.

On 22 March Tony Ayerbe, convened a meeting at Logistic Command to 'further improve Logistic Command support to OP *Solace*'. He briefed Major General David McLachlan, the General Officer Commanding Logistic Command, that the meeting had decided to delegate responsibility for tracking all demands for OP *Solace* to the Moorebank Logistic Group (MLG). [188] Responsibilities that had belonged to HQ Logistic Command since the beginning of OP *Solace* were now being passed down. On 23 and 24 March Winter, spoke with Thomas and followed up with two signals informing him that;

> ... it is incumbent on MLG to track all consignments from their being raised [in Baidoa] to dispatch from Australia '... [and] 'as the major supporting logistic depot for OP *Solace* [MLG] is to track and maintain visibility for all service demand serials submitted by HQ AFS.' [189]

Thomas and his Senior Logistic Officer, Lieutenant Colonel Bruce Cook, initially told the Fleet Management Cell at Moorebank to raise registers to track service demands from Somalia and to discover the status of demands already in the system. Within a few days the Fleet Management Cell, like their counterparts at Logistic Command, demonstrated that they could not cope with intensively managing service demands for Somalia as well as their other work for units in Australia.

Thomas and Cook then decided to form a Special Projects Cell to manage the service demands from Somalia.[190] Fortunately, they had a spare subaltern to assign to this task. Lieutenant Chris Stevenson, arrived back from his Regimental Officers Basic course on the morning of 1 April.[191] By mid-afternoon he was in charge of a Special Projects Cell comprising himself, Sergeant Graham Weston and Corporal Debbie Fisher.[192] Stevenson recalled that Thomas was not happy to have to make special arrangements for OP *Solace* because Fleet Managers, and the administrative system they operated in, should have been able to handle operational requirements as well as their commitments to units in Australia. However, General McLaclan had made it clear that he had to take decisive action to get items moving to Somalia, so he supported Bruce Cook's recommendation that *ad hoc* arrangements needed to be made.[193]

Thomas told Stevenson that service demands from Somalia would no longer be sent to Barton's staff at LandHQ or to Ayerbe's staff at HQ Logistic Command. His cell would receive them instead and would be responsible for distributing them to supply depots and tracking them from then on. Stevenson was then handed a list of all the service demands from the AFS that Woodward and Tweedie had been keeping track of at HQ Logistic Command. He was told to establish the status of each supply item on each service demand and take 'hastening action' to have them satisfied if they were overdue.

Stevenson was given the telephone and facsimile numbers for HQ AFS and told that he could deal with Woolnough's staff direct and had authority to direct logistic staff to purchase any items HQ AFS deemed operationally essential without having to refer to Fleet Managers at HQ Logistic

Command. He was told to maintain a register of purchases and at the end of the operation to tally the costs and submit them to Cook who would then send them on for payment. Cook then told him to report to him daily on the progress he was making. Thomas concluded the meeting by telling Stevenson that he had his authority to deal directly with supply depots and Fleet Managers to ensure that the resupply system to Somalia improved its performance. [194]

The establishment of the Special Projects Team at Moorebank on 1 April marked the beginning of the next phase of the effort by higher headquarters in Australia to manage service demands from the AFS. Coincidentally, there had been a summit meeting to improve the management of the movement of stores from Moorebank to Somalia a week before.

Representatives from Winter's staff, Vale's staff, McNamara's staff, Ward's staff and Jamieson's staff and Gary Martin met at AirHQ on 24 March to discuss the procedures being used to manage the air resupply to the AFS. Senior Air Force staff appeared to have had enough of HQ Logistic Command's ineffectual involvement in air resupply, *ad hoc* negotiations with Greville and Martin over air hours and informal, late-notice finalisation of aircraft loads by Kosciuszko over the telephone. [195]

The results of this meeting took Winter's staff and Vale's staff out of the air resupply business. In future, Kosciuszko would advise Jamieson and Greville by signal when stores were ready to move from the Moorebank Freight Terminal. Subsequently Jamieson would co-ordinate the provision of aircraft, and Greville would decide when the C 130s would fly. Greville was told that he would have to find the air hours from Army's allocation to pay for resupply sorties. Martin was told to stop negotiating with Greg Hartig at 86 Wing for Air Force training hours to fly sorties to and from Mogadishu. [196]

The month of April began with renewed hope that the resupply system to the AFS had been fixed. Chris Stevenson's Special Projects Cell at Moorebank became Woolnough's 4th Line support co-ordination office.

Greville, in conjunction with Jamieson's movements staff, had the job of ensuring that stores moved quickly to Somalia after Kosciuszko had got them to the Moorebank Freight Terminal.

Meanwhile the new command and control arrangements for the AFS had settled in. Mellor's exercise of command had strong inspectional characteristics. He insisted that Hurley's headquarters continue to provide detailed information about operations and future intentions to his staff. He scrutinised operational reports sent to his headquarters and also directed his staff to present detailed briefings to him each evening on 1 RAR Group operations and intentions. He questioned them closely about the information they had provided and frequently told them to obtain further information if he was not satisfied that he had the full story. These evening sessions began to dominate the lives of Mellor's Majors and Captains. As a consequence, Caligari's staff in Baidoa spent significant amounts of their time sending information to Mellor's staff and elaborating on information previously sent. [197]

Mellor also spoke with Hurley daily by telephone and ensured that Hurley kept him personally informed about future tactical intentions. He visited Baidoa most weeks, spoke with Hurley's subordinates and communicated his opinions and interpretations of most aspects of 1 RAR Group's operations. Notwithstanding his penchant for vigorous discussion of all issues effecting the performance of the 1 RAR Group, Bill Mellor supported most of David Hurley's tactical decisions, and acknowledged the results the Australians were achieving in Baidoa frequently. He did disagreed with the handling of a few situations by Hurley's subordinates, but left Hurley to decide on corrective and punitive action. [198]

The only difference of opinion on tactics between Mellor and Hurley related to the techniques being used to deter road ambushes. [199] Mellor felt that deploying vehicles to patrol along roads at night might not conform to the ROE which emphasised using deadly force in self defence after weapons had been raised to fire. He felt that patrolling in blacked-out vehicles bristling with machine guns behind buses and trucks waiting for ambushes to be sprung could be interpreted as seeking the opportunity to use deadly force. Hurley continued with 'Road Runner' patrols because they were having a deterrent effect on ambushers, created a presence that was reassuring to Somali citizens and his troops were showing appropriate restrain.[200]

The issue of providing timely information once contacts with Somali gunmen had started or a serious incident was unfolding still caused some friction between the two Australian Headquarters. Caligari's view was that he did not want to send written reports until sufficient information had been received and verified to describe accurately what had gone on, or what was going on. Mellor's view was that he needed to know what was going on promptly. [201] Caligari met the expectations of HQ AFS by assigning an officer to provide updates of information or to report that there was no further information at 15 minute intervals after a contact or serious incident had been reported. He still did not provide written reports until he was satisfied that what was being written down and transmitted to Mogadishu was as accurate as possible, and had been verified and cross checked.[202]

Some of the pressure for detailed information about contacts, incidents and future intentions of the 1 RAR Group came from Australia. Mellor was being questioned closely by staff at LandHQ on behalf of General Blake and, on other occasions, to satisfy requests for information from General Gration's staff in Canberra. Richard Greville, who was often the officer directed to find out information from HQ AFS at short notice, sympathised with Mellor's need to be 'on top of things' 24 hours a day and have the latest information 'at his finger tips'.[203]

Meanwhile the new air resupply arrangements were about to receive their first test. David Creagh had requested another sortie to Mogadishu to bring the consignment of two APC engines, transmissions, track pads, track link and spare parts that had missed the mid-March C 130 flight. The meeting on 24 March may have streamlined the air resupply system by cutting out Winter's and Vale's staffs, but, without a formal allocation of air hours for OP *Solace*, there was still a requirement to 'horse trade' in air hours to make things happen. Greville still had to 'scrounge' air hours from the current Army allocation. Gary Martin had to be 'the chief scrounger' because his unit was responsible for the co-ordination of Army air hours and liaison with the Air Force. Furthermore, he was the only officer with a complete overview of the management of the Army's allocation of air hours.

Once again Martin asked Jim Wallace if spare air hours had come up from cancellations or changes to Special Forces training. He also contacted Greg Hartig from 86 Wing seeking air crew training hours for a resupply sortie to Mogadishu. However, this breach of understandings reached on 24 March did not go unnoticed. Air Commodore Bruce Searle, the Chief of Staff at AirHQ, sent a personally-signed signal to LandHQ on 26 March, with information copies to Martin's and Hartig's headquarters, stating that no air hours from the Air Force's training allocation to 86 Wing would be provided for this resupply sortie.[204] Faced with having to cancel the sortie, Greville and Martin asked Wallace if he could increase the number of hours he was planning to provide. He did the best he could without restricting Special Forces training. Martin also put the allocation of air hours for Army exercises for the remainder of the financial year well 'into the red' to make up the remainder of the air hours required so that Jamieson's staff could authorise the sortie.[205]

The combined efforts of Greville, Martin and Wallace resulted in a C 130 flight being scheduled to depart for Mogadishu from Richmond on 1 April. This flight proved to be timely. On 29 March David Creagh formally advised Barton and Stevenson, that the last spare APC engine had been used in Baidoa. Fortunately the consignment of APC engines and spare parts arrived in Mogadishu on the C 130 two days later.

While the resupply of APC spare parts by air occurred in the nick of time, the consignment of these stores sent by sea to the AFS via Mombassa and Nairobi was well overdue. Woolnough and Creagh had been told that containers from *Sabrian V* would be off-loaded at Mombassa on 24 March.[206] After some inquiries they found out that the *Sabrian V* had docked at Durban in South Africa on 24

March to pick up more cargo. After being informed of the situation, Michael King arranged for the containers with the Australian stores to be transferred to the *Angela V*, a freighter also docked in Durban. The *Angela V* was due to sail and dock in Mombassa on 11 April. Several days later, the Captain of *Angela V* advised that his ship would not be docking in Mombassa until 19 April. David Creagh assessed that the stores would probably not reach Baidoa until the end of April if the *Angela V* docked on time. He estimated that it would take 10 days for the containers to clear Kenyan customs and for the Canadians to forward the stores on their aircraft when space became available. Resupply of the AFS by commercial sea freight was proving to be quite a challenge.

David Creagh sent a signal to Andrew Barton on 7 April stating that the delay in arrival of the containers sent by sea would create another critical shortage of track link and track pads in Baidoa later in the month.[207] He requested urgent air resupply. This time Greville, Martin and Wallace did not need to find air hours from the Army's allocation. Fortunately for them the Federal Government had authorised and funded a Parliamentary delegation to fly to Somalia from Richmond in a C 130 aircraft on 14 April to visit the AFS. Greville managed to have two pallets of track link and pads loaded on this aircraft that was also scheduled to move an Operational Study Team led by Paul Retter, a group of journalists and a rotation of three key 1 RAR Group personnel to Mogadishu.[208]

Meanwhile Stevenson and his team began hastening action on unsatisfied service demands for the AFS. They had spent the first two weeks in April tracking down each service demand and had begun sending strongly-worded signals to Fleet Managers at HQ Logistic Command, Moorebank and Bandiana, requesting urgent action to satisfy them. On 13 April he had told the Bandiana Logistic Group (BLG):

> Some of the ... demands date back to Jan/Feb 93 As most demands are Priority 1, this is unacceptable. This unit [MLG] emphasises the importance of BLG providing regular updates of progress with service demands and responding to hasteners. Send update on all demands.

Three days later he wrote, 'This unit issues a further direction to BLG to send status report ASAP.' His signals continued to be ignored. [209]

In general, Stevenson found both military and civilian Fleet Managers lacked what he described as 'a sense of urgency'. [210] He also found that though he had been given authority from Colonel Thomas to act on his behalf to get the service demands moving and to order the purchase of specific items, many logistic staff either fobbed him off or ignored him. He felt that his authority had not been well communicated to older, experienced staff who found themselves being asked to take urgent action by a newly-graduated subaltern they did not know. His most vivid memory from this period was a Captain at HQ Logistic Command mocking his efforts to direct her to purchase a particular spare part urgently. She put her hand over the mouth piece of the telephone and invited her staff who were nearby to laugh with her about Stevenson invoking Thomas's authority to direct her to make the purchase. Fortunately, a subsequent telephone call from Bruce Cook to this sceptical officer produced an apology and some action. [211]

Another frustration for Stevenson was trying to stimulate creative solutions to satisfying service demands after it had been discovered that items were not available from a depot's stocks or from local commercial suppliers. He had no inhibitions about 'cannibalising' vehicles and equipment for spare parts that were needed urgently in Somalia, or directing that items should be reallocated on loan from units undergoing peacetime training in Australia to the 1 RAR Group conducting operations in Somalia. He found that some staff at depots and at logistic headquarters assessed that these solutions were administratively too difficult and that units were very reluctant to release items on loan to their compatriots in Somalia for the remaining four weeks of OP *Solace*. [212]

Stevenson's other major frustration was being given incorrect information by depot staff and Fleet Managers that items were not available or were too hard to get, only to find after further inquiries and site visits that the items were available after all. His most memorable example of this misinformation was when he was told by one member of staff at Bandiana Logistic Group that spare APC engines were not in the warehouses at Bandiana and would have to be ordered from the United States. After appealing to another officer at Bandiana, who Stevenson knew personally, to double check this information, he was told that 118 spare engines were on the shelves at Bandiana in various states of disrepair. Several would need only minor refurbishment before they could be boxed up and sent to the Moorebank Freight Terminal for forwarding to Somalia. [213]

This was also a frustrating period for Thomas. His depot was undergoing a major reorganisation under the Defence Department's Commercial Support Program. Industrial action had been taken by unions as a part of the negotiations for 'civilianising' the Moorebank Freight Depot into a commercial National Stores Freight Distribution Centre. Some of this action had been directed at the handling of freight destined for Somalia. [214] Bruce Cook gave him weekly reports on the status of resupply to Somalia and he had made many telephone calls to depots and HQ Logistic Command to remind those managing Fleet Managers that more effort was required. [215] Despite his efforts during this particularly busy time, many service demands remained unsatisfied.

Stevenson's 'paper war' with the supply depots and Fleet Managers at Moorebank, Bandiana and in Melbourne and Thomas's personal follow up action did not produce significant results. On 29 April Chris Stevenson, reported to LandHQ that over 190 service demands had not been satisfied. [216] Fortunately, it did not matter, the 1 RAR Group was winding down operations and preparing for redeployment back to Australia. In the end, just as the short duration of field exercises in Australia leaves the Army resupply system untested year after year, the slowing of the tempo of operations from the beginning of March and the improvisations of a few 'can do' personnel saved the resupply system from causing any significant operational embarrassment in Somalia.

While Stevenson had been trying to get service demands satisfied and transported as quickly as possible to Moorebank, Greville had been trying to get stores flown to Mogadishu. He was informed on 20 April that 14 tonnes of stores had accumulated at Warehouse 90 ready for forwarding to Somalia. Among the stores were high pressure steam cleaning equipment urgently needed to clean vehicles and stores to Australian quarantine standards as well as overdue Priority 1 spare parts. Greville requested that Woolnough review the unsatisfied service demands and put them into new priorities now that OP *Solace* was drawing to a close. Woolnough complied on 23 April.

Greville, Martin and Wallace combined again to find the air hours for one last resupply flight to Mogadishu before the AFS returned to Australia in mid-May. Greville was able to achieve some leverage when he pointed out that if there was no final resupply flight, several pallets of clothing and school books collected by community organisations and Army personnel in Townsville would not be distributed to the GOAL Orphanage in Baidoa. He warned that this might cause some embarrassment for the ADF in the media. The sixth and final C 130 left Richmond bound for Mogadishu on 29 April.[217]

In the previous week a visit by Major General Peter Arnison, Commander of 1st Division, timed to coincide with ANZAC Day on 25 April, prompted some explaining of the resupply difficulties being encountered by the AFS, Graeme Woolnough wrote a brief for him on 26 April. He told Arnison that 70 percent of service demands had been satisfied after intense efforts to track items through the resupply system manually. Twenty eight percent had been satisfied within the timeframe requested, but only 5.4 percent of Priority One urgent requests had been satisfied within the 7 days specified. Close to 1000 demands for supplies had been made by the 1 RAR Group during the three months of the operation. Woolnough concluded that the operational consequences of the delays in satisfying demands and the continuing problem of unsatisfied demands were minimal.[218]

Woolnough assessed correctly that the 1 RAR Group had not formally complained that operations had been significantly hampered by inefficiencies in the resupply system. Hurley and his subordinate commanders 'made do'. Nor had there been any personnel killed or injured in accidents that might have been attributed to operating vehicles and equipment beyond their servicing schedule due to a lack of spare parts or after improvised repairs. However, McKaskill, did not agree with Woolnough that the consequences were minimal. He wrote later that;

> Poor despatching of [track link] from Australia caused it to arrive in an 'ad hoc' fashion ... Consequently vehicles and safety were pushed to the limit, and grounding of some vehicles resulted.... late arrival of critical items (track link) caused unsatisfactory adjustment to tasking.[219]

Later Woolnough defended the report he made to Major General Arnison in April 1993 that the operational consequences of the unresponsiveness of the resupply system during OP *Solace* were minimal:

> ... ask yourself the questions, 'Was any operation cancelled, deferred or postponed? Did any soldier not get fed? Did any soldier not have water, ammunition and food?—the basics. The answer is, 'No'. If these criteria are to be used, then the system worked. Spare parts have always been the most important things after the basics. In Vietnam this was the situation and I don't think the system of resupply of spare parts worked all that well then.
>
> From my point of view, I was extremely disappointed with the Australian end of the supply chain. ... [However] I was grateful to the Ops staff at LHQ, particularly Dick Greville, because he did something about the situation.[220]

The final challenge for the higher headquarters supporting the conduct of OP *Solace* was to bring the AFS home. Planning had begun with a meeting convened by Graeme Woolnough on *Tobruk* on 18 February. The mission was to redeploy personnel safely and efficiently with all personal equipment and stores cleaned to Australian quarantine standards.[221] Vehicles, equipment and stores needed to be in a serviceable condition to ensure that 1 RAR and its attached sub-units could resume their required levels of operational readiness soon after arrival in Townsville and at other base locations.

In Mogadishu Woolnough turned to the Australian movements staff at HQ UNOSOM to assist with planning the redeployment. By 28 February Captain Paul Angelatos, the 2IC of the Movement Control Unit, reported to LandHQ that Australian UNOSOM staff had worked directly with the 1RAR Group in Baidoa and had developed a loading plan and a call forward program for review by HQ AFS.[222] The assistance provided by the Movement Control Unit meant that Mellor's staff were not required to do the primary planning for the air redeployment. They performed the valuable role of monitoring, commenting on and approving redeployment plans.

The primary staff work for the redeployment of the 1 RAR Group in Baidoa fell to the newly-arrived 2IC of 1 RAR, Major Mark Fairleigh, assisted by Steve McDonald. In addition, Captain Andrew Jakab, a Transport Corps officer from the BSG, was appointed the 1 RAR Group Movement Officer to co-ordinate road movement and Captain Andy Somerville, was appointed the 1 RAR Group Emplaning Officer to co-ordinate air movement. Mick Fulham and Tony Powell from HQ AFS worked with Major Paul Le Large from the Ship's Army Detachment on *Tobruk* to ensure that the loading of *Tobruk* and *Jervis Bay* went smoothly, unlike the situation when the ships were loaded in Townsville in December.[223]

In Australia Major Lance Collins, a planning staff officer on Paul Retter's staff at LandHQ wrote a formal military appreciation for getting the AFS back to Australia.[224] He submitted it to Paul Retter on 19 March with draft operations and administrative orders for the redeployment, as well as an Intelligence assessment of the threat to the AFS during the withdrawal form Baidoa and Mogadishu. This was going to be meticulously planned and coordinated phase of OP *Solace*—unlike the hasty plans that had to be made at short notice over the Christmas-New Year period to get the AFS to Somalia.

Collins favoured a staged withdrawal using Air Force C 130 aircraft to shuttle troops and personal equipment to Mogadishu from Baidoa and then chartered QANTAS 747 aircraft to fly troops and their personal equipment from Mogadishu to Townsville via Perth. All vehicles and heavy equipment would be transported back to Australia on *Tobruk* and *Jervis Bay*. A copy of his appreciation was circulated to HQ Australian Defence Force, Maritime Headquarters and AirHQ for comment.

Staff at Maritime Headquarters concurred with the use of *Tobruk* and *Jervis Bay* but staff at AirHQ questioned the use of QANTAS aircraft instead of Air Force 707 jet aircraft. Despite the challenges that Richard Greville and Gary Martin had faced finding air hours to fly resupply missions to Mogadishu, senior Air Force officers made it clear to Martin that they expected Air Force aircraft to be used to redeploy the AFS back to Australia. They did not want the Air Force to be left out of such a high profile operation.[225]

Ultimately, the staff at AirHQ had to approve strategic air movement plans in support of ADF operations. However, they knew that they would face substantial opposition from General Blake and his staff at LandHQ if they were unable to convince them that using Air Force aircraft was a operationally-sound and economical. On the evidence available, key Air Force officers became committed to convincing General Blake's staff to use Air Force aircraft for the redeployment, not QANTAS aircraft.[226]

There were certain realities to be faced before Phil McNamara and his staff would advise General Blake to support the exclusive use of Air Force aircraft. Collins had pointed out in his appreciation that a QANTAS 747 could carry 370 troops with their weapons and personal equipment while an Air Force 707 could only carry 137 personnel with the same equipment. Phil McNamara and Paul Retter liked the idea of using two 747 sorties over three days, rather than the proposed seven 707 aircraft sorties over six days to redeploy the main body of the AFS back to Australia. This seemed to be an efficient and more secure option that created a 'clean break' from

Mogadishu in half the time it would take to achieve with a Air Force 707 aircraft shuttle. They also recognised that they did not want to consume air hours that would impact adversely on the air hours available to support Land Command unit exercises in Australia.

Phil McNamara issued planning documents for the redeployment on 22 March, acknowledging that the AirHQ would have the last say on the means of air redeployment but with the caveat that, 'security and transit areas requirements in Mogadishu may necessitate use of charter 747 in preference to 707 aircraft.' McNamara requested comments to be returned to him by 31 March so he could finalise operations and administrative instructions by 7 April.

McNamara's preference for using QANTAS aircraft did not please senior Air Force officers. They were very keen to tackle the redeployment as a unique opportunity to test the operational skills of the Air Lift Group and 33 Squadron that operated the 707 aircraft. On 7 April Air Commodore Stan Clark, Commander of the Air Lift Group, in conjunction with Air Commodore Bruce Searle's operations staff, developed two options for consideration by McNamara's staff. The first option was a combined operation using two chartered QANTAS 747 aircraft with two Air Force 707 aircraft.

He knew this would be preferred by McNamara's staff at LandHQ because less air hours would be used.[227] The second option, a full military operation using Air Force 707 aircraft supported by C 130 aircraft was Clark's preference. Clark was authorised to offer sufficient 707 and C 130 air hours from Air Force's allocation to bring the AFS home. He assessed that this would make the second option very attractive to LandHQ staff.

McNamara issued a second draft of an operations instruction on 8 April based on Air Force advice that they planned to ferry troops from Mogadishu to the Seychelles by C 130 aircraft and then transfer them to 707 aircraft that would fly them onto Perth where there would be a crew change before flying them on to Townsville. He briefed Major General Blake that he had asked

Lance Corporal Dean Commons, B Squadron, posing with a steam cleaner used to clean APCs to Australian Quarantine Service standards, for return to Austalia.
Image: 1 RAR Group Photographer.

Air Force to verify in writing that, 'the additional burden on C 130 air hours ... will not have a later adverse effect on C 130 support to Land Command.'[228]

Air Commodore Searle called a summit meeting of staff from HQ ADF, LandHQ and AirHQ on 16 April at Glenbrook to discuss the Air Force plan for the redeployment. The meeting began with Colonel Gordon Hurford from HQ ADF advising those attending that the UN had requested that another Australian infantry battalion group should be deployed to Somalia as part of UNOSOM II but the Australian Government was likely to maintain its position that the 1 RAR Group would not be replaced. If the Australian Government complied with the UN request, then there would be a delay in returning the 1 RAR Group to Australia in May for up to a week while another battalion group was prepared for deployment.[229]

It became clear during discussions that the inter-Service politics of air hour allocations and the official costs of air support would be suspended for this last phase of OP *Solace*. Phil McNamara assessed that the Air Force officers present were prepared to make whatever concessions were necessary to persuade Hurford and the Chief of Staff of LandHQ, Brigadier Ian Macinnis, to support the full Air Force option for the redeployment.

Air Commodore Searle told those in attendance that Air Commander-Australia, Air Vice Marshall Geoff Beck, had directed that the redeployment be conducted using Air Force aircraft and that he had advised the ADF Chief of Operations Major General Jim Connolly, that the Air Force would only seek reimbursement for costs associated with travel allowances and fuel. Colonel Hurford confirmed that this was a 'one off' costing arrangement for OP *Solace*. The Air Force had offered a bargain basement price that was too good to refuse on the grounds of cost.[230]

After a final agreement that put Mellor in command of the Air Force personnel controlling the aircraft to be used for the redeployment, the deal was done; the Air Force would be responsible for flying the members of the AFS home and would also supplement the back-loading of stores from Somalia to Australia in conjunction with the Navy. Brigadier Macinnis closed the meeting by complimenting all those in attendance on the

Corporal Natalie Watson, Flight Attendant, 33 Squadron RAAF, handing out cans of Australian beer to members of 1 RAR Group returning to Australia on RAAF B707 aircraft. Pictured in the foreground are Privates Jason Foley and Alistair Campbell.

thorough preparations they had made before the meeting and the spirit of co-operation that had produced an excellent joint-Service plan.[231]

On 29 April Air Vice Marshall Geoff Beck signed a signal drafted by John Ward outlining the Air Force's plan for the redeployment of the Main Body of 891 personnel from Baidoa and Mogadishu back to Australia. The personnel at Baidoa would be flown to Mogadishu on C 130 aircraft based at Mombassa and then by 707 aircraft from Mogadishu to Perth, and on to Townsville.

On 30 April a Specialist Support Team arrived in Mogadishu to assist with the redeployment of the AFS in two weeks time. Captain Andrew Hall, a movements operations officer from HQ Logistic Command, led a team of representatives from the Townsville-based 8th Movement Unit, Sergeants Dominic Coiro and Andy Wormington, and a representative from 68 Ground Liaison Section, Major Doug Hasson. Hall worked closely with Mark Fairleigh and Andy Somerville to finalise loading schedules and lists of personnel travelled by air and sea. Hasson linked up with Air Force representatives from the Mobile Air Transport Unit to co-ordinate the movement of aircraft.[232]

The redeployment of the AFS from Somalia continued the Australian military tradition established at Gallipoli in 1915 for planning and executing efficient withdrawals. Personnel and vehicles began moving to Mogadishu by road progressively from 1 May. During the next two weeks an extensive cleaning operation was conducted by 1 RAR Group personnel using high-powered steam cleaning equipment flown from Australia and purchased in Nairobi.

The cleaning of vehicles and equipment to meet Australian quarantine standards was the last major challenge for the 1 RAR Group before the redeployment. The effort put in by drivers from 1 RAR and the BSG as well as the APC crews from B Squadron was herculean. Each Landrover, Unimog truck and engineer vehicle took several days to 'pull down' and clean. Materials, such as wooden trays, upholstery and rubber, were taken out and discarded. Every vehicle was then cleaned with high pressure hoses at least three times.

The pressure on the cavalrymen from B Squadron during this period was significant:

> The APC crews drove through the night, had 1-2 hours sleep then brushed and changed track all day in 40C degree heat. Every track pad had to be changed to meet the quarantine requirement—all rubber seals and padding was stripped. In effect the vehicles were gutted. It was difficult, if not impossible, to have the staff from HQ AFS make decisions on 'destruction.' ... The APC crews worked 20 hour days.'[233]

Bill Mellor summarised this period later,

> It was a time-consuming and frustrating period, with considerable duplication of effort, and little flexibility exhibited [by Australian Quarantine Inspection Service representatives]. A significant quantity of equipment was destroyed in the process.[234]

Adding to the problem was an effort by Somali authorities to transit 50 000 goats for export through the Australian 'clean' area of the port. This became a test of patience and negotiation skills. If the livestock had been allowed through where the Australian vehicles and equipment were waiting for loading onto *Tobruk* and *Jervis Bay*, all of the vehicles and equipment would have had to be cleaned again.

In the US Army the Military Police, under the supervision of US Quarantine Service officials, were responsible for inspecting US military vehicles and equipment re-entering the US after OP *Restore Hope*. When Captain Ian Westworth, the new Quartermaster of 1 RAR, described the cleaning and inspection procedures that were being insisted on by Australian quarantine officials to American Military Police personnel in Mogadishu, they thought they were hilarious, 'a good Aussie joke.'[235]

By 16 May *Tobruk* was loaded and three days later the fully-loaded *Jervis Bay* joined *Tobruk* off shore. During final cleaning and loading activities, a shuttle of two 707 aircraft flew personnel and their personal equipment back to Townsville. The first 707 left on 9 May and the last aircraft took off from Mogadishu 11 days later on 20 May. The Australians were fortunate that General Aideed saw no advantage in interfering with the Australian withdrawal. He may well have thought that the less foreign troops in southern Somalia the better. Graeme Woolnough sent a situation report to Phil McNamara on 19 May, the day before he left on the last aircraft out of Mogadishu, informing him that redeployment had gone smoothly under the guidance of his headquarters and that, 'The fat lady has mounted the podium and is preparing to sing.'[236] The take-off of the last 707 on 20 May was the cue for *Tobruk* and *Jervis Bay* to sail. Lieutenant Colonel Trevor Jones, commander of the Australian Service Contingent with UNOSOM II signalled to LandHQ, 'The fat lady has sung'.

LESSONS

OP *Solace* was divided into two operational worlds. The lower level tactical world worked well and exceeded expectations. The higher level world didn't work as well as intended and particular aspects did not meet expectations. OP *Solace* was the most important Australian operation since the end of the Vietnam War because it tested the ADF's ability to operate in both these worlds. It is important to know why success was achieved in one and the other did not meet expectations. Future ADF operations may be more dangerous and complex than protecting the distribution of humanitarian aid against an uncoordinated threat from lightly-armed, snap-shooting criminals and bandits.

The first world encompassed the operations in the Baidoa Humanitarian Relief Sector and the second encompassed the activities of the Australian headquarters in Mogadishu, and the command and support provided from Australia. By objective standards of soldiering and the opinions of NGO organisations, local Somali community organisations, and senior foreign officers who commanded the Australians, the 1 RAR Group performed exceptionally well. Somalia was an unhealthy, dangerous place with a culture and climate and that sapped the patience and energy of Western-trained combat forces. With only a few exceptions, Australian soldiers individually and collectively achieved the right balance of aggression and compassion for humanitarian operations. Under David Hurley's quiet and thoughtful leadership, initial expectations of fighting as a combat force against the warlords gave way to the disappointed recognition that, after asserting a presence through rigorous patrolling as well as countryside search and clear operations, the 1 RAR Group would be operating as a heavily-armed police force, deterring hostile political and criminal groups from interfering with UN and NGO humanitarian activities.

Ultimately the test of the effectiveness of a military operation is at its cutting edge. The first operational world of OP *Solace* tested how well the Army's individual and collective training systems had built on personal attributes to impart the military skills, knowledge and attitudes required for members of the 1 RAR Group to adapt and perform well in an unfamiliar operational environment. The Australian Army's investment in meticulous individual training, the development of cohesive, close-knit teams and career training for NCOs paid off. The 1 RAR Group was able to deploy small groups of soldiers into isolated situations under the command of junior NCOs confident that discipline would be tight, orders would be followed and common sense and initiative would apply. The cutting-edge of OP *Solace* was a success because disciplined, professional soldiers were eye-ball to eye-ball with the Somali populace, not ill-disciplined, violent thugs or timid under-trained conscripts that Somalis had seen in uniform on and off since the rest of the world began intervening in their national affairs in the 1970s.

The first operational world was also a test of unit level command and control. During the initial weeks of the operation the reins were pulled in hard on company commanders in the field. This was prompted by David Hurley's concern that at all levels within the 1 RAR Group commanders were keyed up for combat and their subordinates were looking for action. There is plenty of evidence to support Hurley's assessment. Some company commanders and platoon commanders were annoyed by the requirement to provide detailed information and, at times, justification for what there were doing or planned to do. They were also critical of the time it took their headquarters to come back to them with decisions and direction, especially in areas such as weapons confiscation, clarification of the ROE in certain circumstances and the conduct of searches. This close supervision and initial uncertainty from HQ 1 RAR was understandable, given the staff's unfamiliarity with the nature of operations and Hurley concerns about the expectations of some of his commanders. With the exception of two company commanders who was dissatisfied with the performance of HQ 1 RAR throughout the operation, the four other company commanders and battalion headquarters staff settled down and generally met each other's expectations. However, 'Zero-Alpha', [237] was often blamed for limiting the scope of operations when commanders saw opportunities to do some good by eliminating criminal elements preying on ordinary Somalis. At one level this was a conflict between a cautious headquarters and energetic, action-seeking subordinate commanders. However, the conflict was much deeper. Members of the 1 RAR Group arrived with unrealistic expectations of operational service in Somalia and a strong desire to 'do some good' while they were there. They were stymied by the dual-realities of UNITAF's limited humanitarian mission and the enduring injustices of Somali society. Hurley and his headquarters staff unfairly wore much of the blame for the intense frustration felt by members of the 1 RAR Group from this deeper conflict.

Once the tempo of operations settled down and commanders became familiar with the realities of the Baidoa Sector, Hurley and his staff trusted most companies and platoons to get on with the job. Platoon commanders and junior NCOs were the beneficiaries. Platoons garrisoned isolated villages and conducted local search and clear operations. Sections were sent out on patrols, food distributions, vehicle check points and accompanied medical teams into the field. These experiences built confidence and leadership skills in most commanders, with a small minority discovering that they were not ready to meet these challenges and responsibilities. This confidence and experience will be one of OP *Solace*'s enduring contributions to the Army.

*Colonel Bill Mellor, Commander
AFS, briefing Lieutenant Colonel
David Hurley, CO 1 RAR Group,
on his arrival in Mogadishu on
8 January 1993*
Image:Gary Ramage

The first operational world was also a test of the endurance of vehicles, weapons and equipment under operational conditions. With the exception of some communications equipment and generators, the Army's vehicles, weapons and other equipment met the demands of operations in a hot, dry and dusty environment with its occasional downpours. There were, however, some dissatisfaction some of the personal equipment and clothing on issue from the Army. Many members of the 1 RAR Group wondered why the Army designs, tests and then contracts the manufacture of personal load-carrying equipment and clothing items, such as boots, when comparable commercial products are available. Another enduring contribution from OP *Solace* was the insight the operation gave into the capabilities and limitations of ADF materiel.

The second operational world of OP *Solace* was a test of the ADF's ability to command, control and support off-shore land operations. Despite the post-World War II experience of reinforcing and deploying 3 RAR to Korea from Japan in 1950, another 1 RAR Group to Vietnam in 1965 and a 1 RAR Company Group to the international waters off Fiji in 1987, the ADF relearned many lessons again and put in place command, control and administrative arrangements that did not work as well as intended, exposing systemic logistic and strategic air movement weaknesses that were not rectified during the conduct of the operation. Why was this so?

The national command and control arrangements for OP *Solace* began on the wrong foot and, to varying degrees, remained on the wrong foot for the duration of the operation. The scene was set by the decision to divorce the 1 RAR Group from its co-located parent headquarters, HQ 3rd Brigade, in Townsville and to raise an ad hoc HQ AFS from HQ 1st Division, in Brisbane. For some senior Army staff officers, OP *Solace* was an opportunity to validate the untried concept of a Deployable Joint Force HQ (DJFHQ) commanding forces deployed overseas.[238] In reality HQ AFS was not a DJFHQ. HQ AFS coordinated the reception and redeployment of the 1 RAR Group in Mogadishu but it did not manoeuvre ground forces, nor command or task Navy or Air Force assets.[239] Aside from not being truly 'joint', HQ AFS did not have the staff or the command status for the first six weeks of operations to work like a DJFHQ. Indeed, Graeme Woolnough defined HQ AFS as not having the staff or resources to act as a tactical Brigade HQ or a third line logistic HQ. In some ways HQ AFS was better defined, as Woolnough did, by what it wasn't, not by what it was. The reason he took that approach was because, in his opinion, the 1 RAR Group had unrealistic expectations of what HQ AFS could do once problems with resupply and administration arose.

After the first six weeks, HQ AFS had the authority of a tactical Brigade HQ and a third line logistic HQ and tried to behave like both. In reality, HQ AFS performed an inspectional role, handled communications back and forward to Australia in behalf of the 1 RAR Group, double-handled pay and other administrative information, satisfied the information and reporting requirements of higher level headquarters and tried very hard to make a flawed third and fourth line resupply system work.

With hindsight, *ad hoc* liaison and reconnaissance arrangements for OP *Solace* and different interpretations of Mellor's authority as a national commander, put pressure on the command and control relationship between the 1 RAR Group and HQ AFS. The National Liaison Team was an *ad hoc* group whose mission was complicated by simultaneously being an ADF liaison team, tasked to negotiate and affect liaison with HQ UNITAF, as well as a *de facto* reconnaissance group for the 1 RAR Group. [240] This complication was evident in the dual roles of the Team leader, Colonel Bill Mellor. He primary mission was to negotiate a suitable Humanitarian Relief Sector (HRS) for the 1 RAR Group and set up arrangements for the Group's reception, resupply and subsequent command and control. In the absence of any 1 RAR Group staff, he and his staff conducted a reconnaissance of the Baidoa Sector. It is arguable whether he, his operations officer, Geoff Petersen, and his PR officer, Nigel Catchlove, [241] were the best group to do this on behalf of the 1 RAR Group.

Prior to leaving Australia Mellor was also appointed Commander of the Australian Force Somalia (CAFS) with implicit responsibilities for the success of 1 RAR Group operations in Somalia. From the beginning, his role and that of his headquarters was not well defined and had to evolve. Originally, General Gration had envisaged Mellor and his staff following a UN model and working as staff at HQ UNITAF. However, the UN model for constituting a multi-national headquarters did not apply. In reality HQ UNITAF was HQ 1st US Marine Expeditionary Force and had no vacancies for non-American staff. Unlike the UN, however, the Americans were open to having the national headquarters of donor nations co-located with their force headquarters and were happy to work with senior national liaison officers. From the American perspective, Mellor and his staff were a national liaison group monitoring the use of Australian troops by American commanders, and attending to Australian political sensitivities and media coverage of the Australian contribution to OP *Restore Hope*. This perspective was shared by the officers of the 1 RAR Group.

A conventional reconnaissance group, comprising David Hurley and key operations and administrative staff as well as representatives from other Corps sub-units allocated under command of 1 RAR, was not sent in anticipation of or immediately after General Blake's Christmas Eve approval of the American allocation of the Baidoa Sector to the 1 RAR Group. At short notice Majors Steve McDonald, Greg Hurcum and Dick Stanhope were sent to Somalia less than a week before the Advance Party deployed.

McDonald arrived in Mogadishu on New Years Day as the 1 RAR Group's reconnaissance, liaison and reception officer. Hurcum and Stanhope arrived four days later to link up with the Marines and get a feel for the Baidoa area so they could brief Hurley on his arrival. This trio deployed without their own vehicles to move about in Somalia, weapons for their own protection, or communications equipment to report back to the 1 RAR Group independently. Having received very limited information from McDonald, Hurcum and Stanhope, the 1 RAR Group Advance Party departed four days later. It was too late to use their information anyway. Pre-deployment training was well under way, and *Tobruk* and *Jervis Bay* were loaded and at sea.

These hurried, *ad hoc* arrangements for reconnaissance and liaison, as well as the absence of a shared definition of Mellor's role as a national commander, contributed to some misunderstandings between Mellor and Mc Donald, Hurcum and Stanhope. Mellor saw himself as the commander of the 1 RAR Group, releasing the Group under the operational control of the Americans after they arrived in Baidoa, but always retaining national command. The 1 RAR Group majors saw him as a senior liaison officer whose national command responsibilities were to monitor the prudent use of Australian forces. Mellor saw the majors as his subordinates who would follow his directions for the reception and subsequent deployment of the Group to Baidoa. They saw themselves as carrying out a tactical reconnaissance and linking up with the Marines in Baidoa on behalf of their commander, David Hurley. He saw them as failing to respect his authority. They perceived that Mellor was exceeding his authority.

Had a conventional reconnaissance group been sent from the 1 RAR Group with the NLT or soon after, to be on hand when the Australian HRS was approved, David Hurley and key operations and administrative staff would have accompanied Mellor to Mogadishu or met him there soon after.

This time together may have permitted Mellor and Hurley to become better acquainted and diffuse any misunderstandings by subordinates about

Mellor's role as national commander. Just as importantly, Hurley would have had first-hand knowledge of the living conditions at the Baidoa Airfield, the operational environment in the Baidoa Sector and the threat from the warlords. Consequently, stores lists could have been refined before *Tobruk* was loaded and the content of pre-deployment training would have been more relevant to subsequent operations.

With hindsight, the initial reconnaissance and liaison activities of OP *Solace* might have worked better if a reconnaissance team and a liaison team had been sent in a C 130 transport aircraft with the personnel, vehicles, weapons, communications equipment and stores to satisfy the dual requirements of liaison and reconnaissance. [242] This approach would have obviated the need to rely on the good will of HQ UNITAF for accommodation, transport, protection and other administrative support. The use of separate teams for reconnaissance and liaison may have avoided confusion about the operational and tactical responsibilities of Mellor's headquarters. Mellor and his staff could have concentrated on their operational level liaison duties as well as the requirements to co-ordinate the resupply of the 1 RAR Group and report back to LandHQ. Concurrently, a reconnaissance group led by Hurley and supported by sub unit representatives, including an Engineer officer, could have deployed to Baidoa from Mogadishu on Boxing Day.

After Hurley and his advance party arrived on 8 January Mellor made it clear to Hurley in formal directives that he would be under the operational control of the Americans, but would always remain under Mellor's national command. However, this clarification did not improve relations between the two Australian headquarters. The authority to command had been cleared up but the amount of information required to exercise national command remained unclear.

It is difficult to determine whether modern communications and sensitivity to the political consequences of media scrutiny created an appetite for frequent helpings of detailed information among senior officers and their staffs at the different levels of command during OP *Solace* or whether the appetite was already there, created by a need to closely supervise subordinates to ensure that nothing went wrong that would reflect poorly on higher levels of command. Whatever the origins, the command and control story of OP *Solace* after operations began in January was characterised by friction caused by differing perceptions about how much information was required by each level of command to do its job, and differing interpretations of the motives for demanding information.

Each level of command seemed to resent the amount of information and reporting demanded by the next level up. While this may be regarded by veterans of the previous campaigns and operations as the normal friction between subordinate and higher headquarters, the experience on OP *Solace* suggests that there were three factors contributing to increased friction— the desire to assert authority, unfamiliarity with operational reporting and a lack of trust. It was clear from the time he was appointed as CAFS that Bill Mellor wished to be kept informed and to influence the activities of the 1 RAR Group. He sensed a reluctance within the 1 RAR Group to accept that a national commander had the authority to monitor and influence 1 RAR Group operations. Hence, he deduced that he and his staff should ensure that they received sufficient information to exercise command.

Aside from differing interpretations of what national command constituted, from the 1 RAR Group's point of view, the staff at HQ AFS were unfamiliar with operational reporting procedures and their demands for information were excessive. It is a matter of professional judgement on what constitutes excessive demands for information, but the fact remains that staff at HQ AFS , with the exception of the two Captains seconded from HQ 3 Brigade were unfamiliar with current ODF reporting procedures. Finally, it should be said that both Australian headquarters were finding their way in the early weeks. HQ AFS had never exercised national command over an Australian combat unit under foreign command and HQ 1 RAR had never been under the operational control of an American divisional headquarters while remaining under national command.

Given the events leading up to their arrival in Somalia, a lack of trust between HQ AFS and the 1 RAR Group was understandable. On top of never having worked with each other, staff from HQ AFS and officers from the 1 RAR Group were not in a position to establish working relationships when the AFS concentrated in Townsville in January. Mellor and Woolnough had already deployed to Mogadishu to affect liaison with HQ UNITAF before HQ AFS deployed to Townsville. Subsequently, their leaderless subordinates arrived with no role to play during the deployment.

Another factor contributing to this lack of trust was Mellor's assessment of the temperament and expectations of the 1 RAR Group. Like Hurley, he had assessed that the expectations of many commanders for combat was at odds with the current situation in southern Somalia, the ROE and the mission of UNITAF. While Hurley had the unambiguous and undisputed authority to closely supervise his subordinates, Mellor's authority to demand information and justification as the national commander was not well understood or appreciated by the vast majority of the 1 RAR Group. He and his headquarter staff had got off on the wrong foot with the 1 RAR Group through circumstances beyond their control during the deployment phase of the operation. Unfortunately, when Mellor, like Hurley and his staff at HQ 1 RAR, pulled a tight rein on the 1 RAR Group by demanding detailed information and justification, getting off on the wrong foot became a permanent adversarial relationship.

Adversarial relations, misunderstandings about roles and demands for detailed information and justification had not characterised the exercise of national command over Australian forces under foreign command in the past. During the spectacular advances and withdrawals of the Korean War from September 1950 until April 1951, 3 RAR fought as part of a British infantry brigade. The Australian national commander, Lieutenant General Sir Horace Robertson and his staff were in Tokyo, monitoring the fortunes of 3 RAR through reports provided by the British and UN chains of command. Robertson was a mentor for the two COs of 3 RAR during this time,

Lieutenant Colonels Charles Green and Ian Ferguson[243] and a stout defender of Australian interests at the highest levels of UN strategic and operational command in face of competitive British interests and an American penchant for using 3 RAR as an advance guard to assault enemy positions or as a last ditch rear guard to protect withdrawing UN units.[244]

Similar relationships applied in South Vietnam in 1965/66 between the national commander, Brigadier David Jackson, and the two COs of the 1 RAR Group who commanded while the unit was under the operational control of the US 173rd Airborne Brigade (Separate). Jackson and his staff at the Australian Embassy in Saigon monitored the fortunes of the 1 RAR Group by listening in on the Brigade command net, receiving periodic reports from Lieutenant Colonels Lou Brumfield and Alex Preece and visiting Australian units and sub-units regularly. [245] During the reconnaissance and liaison phase preceding the deployment of the 1 RAR Group in 1965, Jackson acted as a senior liaison officer to the Americans on behalf of the inbound units. He backed up 1 RAR's Operations Officer, Major John Essex-Clark, who led the reconnaissance group, when he had differences of opinion with the US Brigade commander about the site of the 1 RAR Group base camp at Bien Hoa Airbase and whether the Australians needed two weeks of training and acclimatisation before being deployed on search and destroy airmobile operations. Jackson used his authority as the national commander to insist that the Australians be allocated a large enough area on the perimeter to ensure their security and were not sent out on operations within 48 hours of their arrival as had been planned.[246]

The demands for detailed information on the tactical operations of the 1 RAR Group were not confined to, nor did they all originate from HQ AFS. Mellor and his staff were put under pressure frequently to provide detailed information at short notice to LandHQ and HQ ADF. The issue was not the amount of information required, because there was daily routine reporting, but the short notice and urgency. Typically, the notice and urgency was driven by political imperatives related to reports in the media or the need to provide information quickly to the Minister or senior officers if there was the likelihood of media coverage of incidents or issues related to the AFS. When these requests for information came from Australia it was Bill Mellor and his staff who were in the hot seat to obtain and send information back as soon as possible. Consequently, HQ AFS needed to have detailed information about the current situation and future intentions of the 1 RAR Group on hand to meet these requirements.

There appear to be three major lessons to be learned from the command and control arrangements that were put in place for OP *Solace*. The first is to be clear about the functions of national command and to specify the role of a national headquarters when Australian units are under foreign command. This should result in national commanders, their staff and tactical commanders and their staff working together from a common understanding of each other's roles and responsibilities. This will also clarify the Australian command and control arrangements for the force headquarters having operational control over Australian units, sub units and contingents.

The second lesson is to avoid cobbling together *ad hoc* headquarters that have not commanded or trained with the units assigned to them. Unfamiliarity may not only lead to misunderstandings, but may also result in mistakes being made early in an operation when media scrutiny is intense and first impressions are being made on major allies and other donor nations to multi-national forces.[248] It would appear to be better to augment a tactical headquarters to coordinate specialist functions, such as media coverage, legal matters and visits as well as report back to Australia. This would leave the national commander and a small group of liaison officers with good communications at the force headquarters level to relieve the tactical commander of liaison duties beyond his next level of operational control and to ensure that Australian troops are employed prudently. If the creation of a national command liaison group is not authorised by the UN, then the national commander should be a senior staff officer on the force headquarters. Not ideal, but it has worked in the past.

The third lesson is about the passage of information. Since the deployment of 3 RAR to Korea in 1950 the appetite for information by higher headquarters from forces deployed overseas has increased steadily as communications technology has improved and the consequences of misemploying Australian units under foreign command has increased because of closer media scrutiny. In Korea, American and the British commanders were trusted to employ 3 RAR prudently. With hindsight, this trust was misplaced and casualties were heavy. [248] The national commander, Lieutenant General Robertson, did not have independent communications to maintain contact with the COs of 3 RAR, one of whom wrote later that he had 'the loneliest command any man could have.'[249] At the time there was no Australian journalists or military Public Relations officers accompanying 3 RAR to file stories back to Australia about the fortunes of the unit under foreign command. Those who visited 3 RAR periodically filed stories extolling the fine fighting traditions that the Australians were maintaining. Aside from possibly exercising their own form of prudence in criticising senior British and American commanders, journalists may not have heard any complaints from the Australians who were all volunteers, most were World War II veterans and the vast majority were in Korea for 'the sheer hell of fighting.'[250]

The situation in Vietnam in 1965/66 was different because communications technology enabled the national commander, Brigadier Jackson, to monitor the 1 RAR Group's operations under American command 24 hours a day. Every time Australians took casualties or there appeared to be an issue worth discussing with the American Brigade commander, Jackson, flew by helicopter to the Brigade headquarters from Saigon in a few minutes.[251] Media scrutiny of Australian operations under American command had also changed since 3 RAR was in Korea. Australia's participation the Vietnam War was more controversial back in Australia and Australian journalists covering the war were more prepared to file stories critical of higher Australian headquarters and the Australian Government.

During the 1 RAR Group's period under American command there were several political controversies in Australian generated by media reports.[252] Anecdotal evidence suggests that the careers of several Australian officers were effected adversely by these controversies. This consequence would not have gone unnoticed by the remainder of the Army's officer corps.

By 1993 when the main body of the 1 RAR Group deployed to Somalia, communications technology enabled instantaneous communications between every level of command. The media had the additional capability of reporting 'live' from Somalia into the living rooms of any Australian who had a television set turned. In addition, the ADF facilitated a comprehensive coverage of 1 RAR Group operations and encouraged continuous reporting by assigning Public Relations officers to HQ AFS and HQ 1 RAR Group. Every effort was made to keep every level of command informed before information was released to the Australian public. However, the media were often aware of incidents as they occurred. Consequently, there was a competition between the chain of command and the media to see who would inform the Australian public first. Often the headquarters caught in the middle of this competition was HQ AFS .

'Live' broadcasting and daily reporting to meet evening news broadcast deadlines will be a feature of future offshore operations by Australian forces. For political reasons, the present policy of open access by media representatives to Australian operations will probably continue for the foreseeable future. Consequently, it would appear to be better to authorise tactical headquarters in the field to report directly to the operational, and possibly even the strategic level of command, when serious incidents have occurred or serious situations are unfolding. There appears to be little to be gained by forcing intermediate headquarters, like HQ AFS , to demand instant information from subordinate headquarters, like HQ 1 RAR Group, and pass it up the line.

Resupply of the 1 RAR Group was one of the major tasks of the second operational world of OP *Solace*. The story shows that from the day the Government announced the deployment of 1 RAR Group on 15 December 1992 the second operational world had significant difficulties stocking up the 1 RAR Group, loading *Jervis Bay* and *Tobruk* and subsequently meeting deadlines for the delivery of service demands. While acknowledging that this was the first time the second operational world had been required to deploy and sustain a battalion group overseas since 1965, the resupply system for OP *Solace* was unresponsive and exposed a number of systemic weaknesses that were unable to be fixed during the operation. Why was this so ?

The first problem was that strategic logistic intent became disconnected from operational level planning. At HQ ADF logistic planning got off to a good start, but a combination of cost cutting and independent 'worst case' planning at LandHQ resulted in a weakness in 3rd line and 4th line resupply arrangements and logistic planning falling behind the operational timetable. The second problem was a lack of detail in administrative instructions contributing to a disconnect between the strategic movement system and the resupply system. The third problem was the combined effect of systemic weaknesses in how service demands were handled and how items of supply were packed for movement. OP *Solace* was a 'wake up call' for the ADF logistic and strategic movement system. Evidence from more recent overseas deployments and major field exercises suggests that some problems are still unsolved. [253] Consequently, it is very important that the reasons for the lack of response by the resupply and movement system during OP *Solace* are explained.

The separation of logistic and operations functions in military as well as at corporate headquarters, makes sense. The inputs, processes and outputs are different and involve personnel with different qualifications and experience. However, it is the blending of logistic and operational processes and outputs that should result in one well-understood operational outcome, namely, forces arrive with the wherewithal to achieve their mission and are supported thereafter to operate successfully. The logistic and operations functions appeared to blend successfully at the strategic level by not at the operational level for OP *Solace*. At HQ ADF operations and logistic staff were accommodated in offices a few doors apart from each other and attended meetings together. Even though several officers were away on leave, logisticians like Colonel Bill Traynor [later Brigadier], Lieutenant Colonel Bill Nagy and Wing Commander Barry Newham met regularly and included Lieutenant Colonel Ray Martin who was responsible for operational inputs and processes.

Logisticians at HQ ADF designed a sound resupply and movement system for OP *Solace* early in December 1992 Traynor, Nagy and Newham discussed supply options based on a system that would transport stocks via commercial air freight from Australia to Nairobi and an Air Force C 130 detachment operating from Nairobi to the nearest suitable airfield in the 1 RAR Group's area of operations. They also proposed that a small 3rd line logistic detachment be based in Nairobi to monitor and co-ordinate the maintenance of the 1 RAR Group. Though not mentioned specifically, this would have included local purchase and mail. They envisaged HMAS *Success* being located off-shore providing additional logistic, communications and movement capability as well as operating Sea King helicopters in support of 1 RAR Group operations. Their proposal also mentioned a second sailing of *Jervis Bay* if all of the stocks required by the 1 RAR Group were not able to be deployed by *Success* and *Jervis Bay* during the deployment phase. [254]

Based on these assumptions about how stocks would move, Traynor and his staff worked on negotiating a cross-servicing agreement with the Americans for OP *Solace* that would result in many classes of supply, such as water, rations, ammunition and fuel, being provided by UNITAF to simplify the sustainment of the 1 RAR Group. This was a cost effective option and left the ADF logistics system having to provide only items of supply unique to ADF vehicles, weapons and equipment. However, while Traynor and his staff continued to plan , following a concept of having the Americans provide basic necessities in Somalia and the ADF only having to provide unique items through strategic air freight and in-theatre service air, logisticians at Maritime Headquarters and LandHQ planned to resupply basic necessities across the beach near Mogadishu and, from the evidence available, did not appear to have specified how other classes of supply would move from Australia to the 1 RAR Group, except that urgently needed items would be flown from Richmond to Mogadishu by service air.

It is not clear why strategic logistic intent disconnected from operational level planning. It is also not clear why the air movement system proposed by HQ ADF and the second sailing of *Jervis Bay* did not eventuate. The Canadians used a very similar air movement system and also had a supply ship off-shore operating Sea King helicopters in support of land operations. Their system appeared to work well and was based on sound logistic principles and self-reliance. In the end, an Air Force C 130 detachment and a 3rd line co-ordination office were not established in Nairobi to support the AFS. Consequently, the Australians had to depend on the good will of the Canadians, who were not assigned formally in support of Australian operations, to transport stores, mail and local purchase items to Mogadishu when they had room on their aircraft.

The absence of a 3rd Line co-ordination office meant that Woolnough and his staff had to try and coordinate local purchase, mail and the receipt and movement of stores sent by air and ship from Nairobi and Mombassa to Mogadishu. Everyone did their best, but the results were uneven, and resources and personnel were stretched. If the reason the original system was not set up was to contain costs, then the ADF appears to have decided not to reallocate resources from training in Australia to provide better support to OP *Solace*. Peacetime training continued on schedule and the ADF's major off-shore operation since the Vietnam War received a lower priority.

Partly as a result of time being wasted planning to resupply the AFS over the beach at Mogadishu, logistic planning for OP *Solace* fell behind the operational timetable in December and did not catch up. While higher level logistic staff questioned Banister's list of logistic requirements for OP *Solace* and implied that staff in Townsville were using OP *Solace* as an opportunity to order non-essential items and 'restock the shelves in 3 Bde',[255] a week went by before the 1 RAR Group received the highest priority for the issue of stocks. The HQ ADF, after a request from the Navy, issued the FAD1 priority to *Tobruk* and *Jervis Bay* on 16 December.[256] The 1 RAR Group was issued FAD1 after a request from the Army on 22 December. There appeared to be an assumption by some senior logisticians that there were sufficient stocks in Townsville to get the 1 RAR Group away with 30 days of operating stocks. This was not so. Lieutenant Colonel Tony Ayerbe and his logistic operations staff at HQ Logistic Command hurriedly organised C 130 aircraft from Richmond and semi-trailers from Brisbane to move stocks to Townsville.

Despite the efforts of Ayerbe and his staff, logistic responses during the deployment phase of OP *Solace* fell behind the timetable for the deployment. Stores arrived too late to be loaded and all of the major administrative instructions for OP *Solace* were issued after the AFS deployed to Somalia. Subsequently, there was no second sailing of *Jervis Bay* to Somalia to transport stocks and vehicles left on the wharf at Townsville, or to respond to advice early in February that significant quantities of track pads and track link would be required to keep the 1 RAR Group's APC fleet in the field.

When it came time for the logistic system to respond to the needs of the 1 RAR Group after operations began in the Baidoa Sector, the intentions of administrative instructions appeared to have disconnected from reality what logistic and movement organisations were able to do in support of OP *Solace*. On 15 December the CDF issued a warning order specifying that the Maritime Commander and Air Commander were to be prepared to provide strategic sea and air resupply respectively to the Land Commander during the conduct phase of OP *Solace*.[257] In simple terms, the CDF's ordered that when the Land Commander needed strategic sea and air resupply for the 1 RAR Group he would get it. Subsequent meetings between members of staff from Maritime, LandHQ and AirHQ verified this intent and promises were made. On 21 December Ayerbe's staff identified the major Army supply depots that were to support the operation. Aside from medical supplies and ammunition, the Moorebank Logistic Group was assigned to provide 'all other classes of supply. ... Exact form resupply will take is to be determined and details will be advised when confirmed.'

It was the subsequent lack of detail in Administrative Instructions that contributed to resupply problems. They showed 'who was responsible for what,' the intended timeframes for resupply, and included flow diagrams for how items of supply would move. What these instructions did not show was who had the funds, resources and authority to make things happen, and exactly how resupply and movement of stores from Australia to Somalia was to be co-ordinated. With hindsight, the instructions should have also explained in detail how the system would overcome difficulties and push items of supply to the 1 RAR Group to meet deadlines. The philosophy appeared to be based on the notion that, if customers did not get items on time, it was up to them to remind suppliers that they had failed to comply with agreed resupply arrangements. There did not appear to be mechanisms for suppliers to recognise that they had not met the timeframes of resupply and to take immediate action. It was left to the 1 RAR Group, HQ AFS and LandHQ to 'pull' items of supply out of the system because there were no provisions made for suppliers to 'push' them through.

Specifically, administrative instructions should have specified what needed to be done if items could not be provided by the Moorebank Logistic Group, if items could not be found on-the-shelf at other depots or if items could not dispatched to Somalia in time for them to be received by the 1 RAR Group within the requested timeframe. There was no checklist of actions to be taken when the system failed to meet its obligations. In absence of a checklist at the lower levels of the resupply system, the buck passed upwards from customers and supply depots to HQ Logistic Command.

Administrative instructions were silent about the CDF's intentions for the strategic movement of stores, including HQ ADF's allocation of $ 2.4 million for air freight. This left those responsible for strategic movement of stores without the resources to do the job. With hindsight, one of the major weak links in the resupply and movement chain was HQ Movement Control (HQ MC) at HQ Logistic Command. Major Michael King who, with the exception of various Air Force officers made available 'on call', constituted the entire staff of HQ MC. King was 'shadow posted' [258] in this role, and was not allocated funds, air hours or clearly understood authority to do the job. In reality, his authority was specified in doctrinal publications not in the administrative instructions for OP *Solace*. The concept of a shadowed-posted Army Major from the Royal Electrical and Mechanical Engineer Corps at HQ Logistic Command selecting the mode of strategic movement, ie air or sea, and then directing Maritime and AirHQ to move stores to Africa without an allocation of resources to reimburse either HQ, had never been validated. [259]

The other weak link in the chain was Administration Branch at LandHQ, described by Woolnough as the 'deaf ears' of OP *Solace*. In reality, after Colonel Brian Vale and his staff issued the Administration Instruction for OP *Solace* on 19 January, they were not in a position to monitor the satisfaction or movement of items of resupply to Somalia on behalf of the Land Commander. All they could do was pass on the messages from HQ AFS to HQ Logistic Command. Despite having the responsibility to coordinate the logistic aspects of OP *Solace*, Vale and his staff had to depend on HQ Logistic Command, who controlled the supply depots, to move items to the Moorebank Logistic Terminal on time and then on HQ MC at HQ Logistic Command to coordinate their movement to Somalia with Maritime Headquarters and AirHQ. In effect, Administration Branch was a post office for messages to and from HQ AFS on logistic matters, not a place where action could be taken and or problems solved.

The disappearance of service demands from Somalia into the Army's logistic system and the reappearance of containers of stores, marked only with their size and weight, in Mogadishu at short notice about every two weeks thereafter was the major logistic frustration of OP *Solace*. [260] The Army had no automated means for tracking service demands from Somalia as they were handled by fleet managers in the supply depots. This 'visibility' problem was compounded when stores were packed into containers without packing notes showing what was in them. A further complication was ad hoc arrangements that had to be made between staff at LandHQ and AirHQ for the movement of stores to Somalia.

The problems of visibility, freight handling and onward movement of stores were known by every level of command in the second operational world of OP *Solace* by the mid-February, four weeks after the 1 RAR Group began operating in Baidoa and two weeks after the 1 RAR Group and HQ AFS had made it clear to LandHQ that there were severe shortages of track pads, track link as well as vehicle and equipment spare parts. For the remaining 12 weeks of OP *Solace*, LandHQ, HQ Logistic Command and AirHQ were unable to fix these problems and HQ ADF did not appear to want to 'bang heads together.'

On the evidence available, aside from the officers like Majors Peter Tweedie, Benny Woodward and Lieutenant Chris Stevenson who were assigned directly, and therefore were specifically accountable to fix the problems during these 12 weeks, some fleet managers at HQ Logistic Command and in the two depots at Moorebank and Bandiana did not give service demands from Somalia sufficient priority. This situation persisted even after authorisation was given for the immediate purchase of items if they were not on the shelves of depots, and Stevenson and his staff were following up each service demand on a day-to-day basis. [261] The evidence suggests that the buck for solving the resupply problems of OP *Solace* stayed at HQ Logistic Command for a few weeks and then went 'down' again, not 'up' to HQ ADF. This preference for the downward delegation of the problem coincided with a preference to leave the customer 'pulling' rather than encouraging suppliers to do some 'pushing'.

In the end, Logistic Command had the responsibility and the resources to resupply the AFS but not the will to fix problems when they occurred. Administration Branch at LandHQ had the authority of the Land Commander to fix the problems, but no resources. HQ ADF had the ultimate responsibility to the Minister for OP *Solace* but left HQ Logistic Command, LandHQ and AirHQ to sort it out resupply and air movement problems among themselves. The results of negotiations were uneven. At the cutting edge of the resupply problem, fleet managers continued to fob off Chris Stevenson right up until the end of OP *Solace,* seemingly unafraid of any consequences. Administration Branch, LandHQ dutifully passed on the concerns of HQ AFS to HQ Logistic Command and left Operations Branch to sort out the movement of stores to Somalia with AirHQ.

There were several worrying aspects about the interface between the LandHQ and AirHQ for strategic air resupply. The lesson learned by LandHQ was that the AirHQ operates on a 'user-pays' principle unless there is a shared understanding of an operational emergency or specific benefits can be accrued by the Air Force. On the evidence available, AirHQ had assessed correctly that the Army had not deployed the AFS to Somalia with its full compliment of stocks and were trying to overcome these deficiencies, as well as unexpected high usage rates of vehicle spare parts, through the use of unforecast air sorties. This situation did not appear to constitute a shared understanding of an operational emergency, so senior Air Force officers seemed to have assessed that the Army would have to overcome resupply problems by cancelling air support tasks planned for Army training exercises and exchange them for operational air support tasks to OP *Solace*. When it came time to bring the AFS home, the same senior Air Force officers appeared to have assessed that there was specific public relations benefits from making air services available to the Army virtually free of charge. The other lesson learned for the air resupply story of OP *Solace* is to ensure that when funds are allocated from the strategic level to the operational level to make things happen, the amount and the intent should be made known by those staff at the cutting edge who have to make an operation work.

The final and to some degree most important lesson for the story of resupply and air movement for OP *Solace* is that there are serious consequences when the supply of spare parts breaks down. Since the end of World War II the Army's obligation to supply of spare parts to deployed forces has increased significantly. In Korea in 1950/51 the Army did not have supply any spare parts to 3 RAR because the Australians were dependent on British and American logistic systems to provide and maintain vehicles and major items of equipment. Weapons were relatively simple to maintain and there was no requirement to operate strategic communications equipment or to have computer connectivity with the local area networks off higher allied headquarters or logistic agencies.

Based on Graeme Woolnough's comments, the resupply of spare parts during the Vietnam War had some difficulties. Though still dependent on the American logistic system for most classes of supply, the differences between Australian and American vehicle fleets and equipment did require some spare parts to be flown in from Australia. There appears to be little excuse, after the 1 RAR Group had completed its 12 months under American command and given the Army several months to plan ahead for the deployment of the 1st Task Force, that there should have been difficulties in the resupply of spare parts. Unlike HQ AFS , HQ 1 Task Force, HQ Australian Force—Vietnam and the Australian Logistic HQ in Vung Tau had the staff and unfettered communications back to Australia to get it right.

The requirement for the resupply of spare parts to Australian forces overseas has increased substantially in the 1980s because the Army's fleets of vehicles, weapons and equipment are different from major allies, such as the Americans. There is also a requirement to maintain more sophisticated equipment, such as satellite communications equipment and computers, in the field. Consequently, OP *Solace* was an excellent opportunity to test the Army's ability to support its vehicles and equipment fleets in the field with spare parts, and both the Army's and Air Force's ability to coordinate the timely arrival of spare parts when they were needed. On both counts, there was significant room for improvement.

The lessons for the Army would appear to be that units on high levels of readiness for off-shore deployment should not be left short of spare parts. Once operations begin, fleet managers of spare parts should be given additional resources and specific guidance. Boxes with spare parts should be clearly marked with their priority and contents. Finally, co-ordination of the onward movement of spare parts should be managed intensively.[262] Once the system is meeting customer demands than the management of the supply of spare parts may not need to be as intense.

There has been an emphasis in recent years to reallocate resources to improve the combat readiness of the ADF. Combat readiness can be measured in part by the quality of training and technology. The first world of OP *Solace* demonstrated that the Army has the individual and collective training systems as well as the technology to perform quite well on dangerous operations short of war. The performance of the second operational world suggests that the ADF should examine the lessons from OP *Solace* more closely and ensure that command and control arrangements, logistic responsiveness and the strategic air movement system will match the improvements sought in combat readiness.

7 KEEPING THE HOME FIRES BURNING

The people of Australia, have a saying we know, 'Lest we forget, those
fallen to the foe.'
Now how about remembering the one left at home, to cope with the
anguish of being alone.
No medal she's given, not a memorial in life.
But she'll walk bloody proud,
'cause she's a soldier's wife.'

Lance Corporal Darrin Miller
Excerpt from *A Soldier's Wife*
18 March 1993 [1]

The previous four months had been a great strain for many of the
spouses, partners, relatives and friends of members of the AFS. Among the
crowds waiting for aircraft to land in Townsville were wives and partners who
had given birth during OP *Solace*. One was Vicki Hill. She and her husband,
Captain John Hill, Battery Captain of 1 RAR's direct support artillery battery,
had been holidaying on Norfolk Island when the announcement came that
the 1 RAR Group would be deployed to Somalia.[2] They had been married
for just over a year and Vicki was pregnant with their first child. Initially John

had not expected to be deployed. When he called his Battery Commander, Dick Stanhope, he was told that he was to be Second-In-Command of the Civil-Military Operations Team.

For Vicki Hill the news that her husband was going to Somalia had made her angry. She did not think it was either logical or fair that John should be selected to go when there were plenty of other Captains in his Regiment who did not have wives who were pregnant with their first child. She expected John to see this logic and request to be replaced. Her anger deepened when she realised that he was ecstatic about being selected to go. Despite his acknowledgment of her situation, his sympathy and his assurances of how much he loved her, he was not going to ask to be taken off the list of those selected for service in Somalia. Vicki

Vicki Hill with her daughter Rebecca, born while her husband Captain John Hill. 107 Field Battery, Civil Military Operations Team, was serving in Somalia.
Image: Courtesy of the *Townsville Bulletin*. Jo Patterson

realised for the first time the full implications of being 'an Army wife'.[3]

Vicki Hill had not been alone with her mixed feelings about her husband's decision. Among the wives at Townsville Airport were Shirley Vine, wife of Warrant Officer Richard Vine,[4] Company Sergeant Major (CSM) of Administration Company, and Julie-Anne Hudson, wife of Warrant Officer Ken 'Rock' Hudson,[5] CSM of Delta Company. Like Vicki, Shirley, who had been married for a just over a year, was pregnant with her first child. She was devastated by the knowledge that she would be having her baby while Richard was away. She could not understand why the Army would deploy her husband and how he could be so delighted to be going. [6] Julie-Anne was having her third child and had been married to Ken for six years. She knew immediately that he would deploy and that was the way things would be.[7]

Also among the crowds waiting for the returning aircraft in Townsville was Janice McLeod, wife of Major Rod McLeod, 1 RAR's Quarter Master and Barbara O'Brien, wife of Warrant Officer Harry O'Brien,[8] 1 RAR's Regimental Quarter Master Sergeant. This was the third time they had waited at an airport for their husbands' return from overseas service. In 1965 Janice, then 19 years old, was newly-married to 20 year-old Rod when he deployed with 1 RAR to South Vietnam as a member of the Mortar Platoon. At the same time, Barbara, also 19 years old, was Harry's girlfriend when he deployed as a 19 year old riflemen with 1 RAR. Barbara and Harry married soon after his return from Vietnam in 1966 and spent their first years together in Malaysia with the newly-raised 8 RAR.

The bugle blew a second time for Rod and Harry. In 1967 Rod served with 1 RAR on a second 12 month tour of duty in Vietnam. Two years later Harry went with 8 RAR to Vietnam, leaving Barbara in Brisbane with a 6 month old baby daughter. Both Rod and Harry returned unscathed and continued with their careers in the Infantry Corps.

Before news of the deployment to Somalia broke in December 1992, both Janice and Barbara had thought that their anxious days of waiting for their husbands to return safely from overseas deployments were over. Rod had accepted a redundancy package and was due to leave the Army in a few weeks time. He and Janice had planned an 18 month caravan trip around Australia after their son, Douglas, was married in March. Harry was on his last posting in the Army before retirement. Rod, a fit 49 year old, successfully applied to have his redundancy postponed and Harry was very proud that his Army career would culminate with an overseas deployment with 1 RAR, just as it had begun 30 years before.[9]

Private Douglas McLeod (L) with his father, Major Rod McLeod, photographed in the pre-dawn light of Baidoa Airfield 1993 Rod was on his third tour of active service with 1 RAR. Douglas was on his first tour as a signaller in Support Company.
Image: ADF Electronic Media Unit.

Janice's anxiety was not just based on Rod's tempting fate a third time, but also on the imminent departure of her only son, Douglas,[10] a 20 year old signaller in Support Company. Douglas decided to bring the date of his marriage forward to 29 December so he and 20 year old Therese Hills could have some time together as husband and wife before he left for Somalia. The marriage celebrant was Major Dennis Hills,[11] Chaplain of the Army Aviation Centre at Oakey in Queensland—the bride's father.[12]

Vicki Hill, Shirley Vine and Julie-Anne Hudson all gave birth successfully. They waited among the crowds at Townsville Airport, ready to show their new-born infants to their fathers for the first time. Janice McLeod and Barbara O'Brien were there, waiting for their husbands to return from active service for the last time. Janice, standing with her daughter-in-law, knew that Therese had survived her first experience of having a young husband overseas on combat operations. She was proud of her and her son, but was also mindful of the future challenges their marriage would face as Douglas continued his career in the Army.

As the excitement grew among waiting crowds as the first 707 jet touched down, some may have reflected on the experiences of those who waited in Australia for the return of the AFS and those in the community who supported the AFS in Somalia.

Soon after the Federal Government's announcement of the deployment of the 1 RAR Group to Somalia on 15 December, Brigadier Peter Abigail, the new Brigade Commander in Townsville, and Mayor Tony Mooney, the Mayor of Townsville, had differences of opinion about what arrangements should be made to support the spouses, partners and dependents of those about to deploy to Somalia. Mooney and his staff viewed the families who would be left behind as citizens of the Townsville-Thuringowa municipalities who would need additional community support in the coming four months. From their perspective, Abigail and his staff viewed the families as part of the Townsville military community who would be supported adequately by the Army's formal and informal family support network.[13]

Aside from promising additional support from community services staff, Mooney also offered to organise a civic farewell for the 1 RAR Group to focus the community of Townsville-Thuringowa on the idea that their local units were about to embark on an important international humanitarian operation. He, and his public relations manager, Elliot Hannay, envisaged a farewell march through the city followed by a community picnic where the members of the 1 RAR Group and their families, friends and relatives would be the guests of the Council.

The Army let Mooney know in a firm but friendly way that there was no time for farewell parades and family picnics. The local administrative commander, Colonel Neil Weekes informed him that staff from the Army Community Services unit, Rear Details staff from units deploying to Somalia and a close-knit network of Army spouses would be sufficient to provide additional support to the families of those who were to deploy to Somalia.[14]

Mooney was disappointed with the Army's official position because his community services staff were advising him that there were many Army families, especially young soldiers' spouses and partners, who did not cope well with periods of separation when their husbands and partners were absent on field exercises or training courses. Some of these spouses and partners did not seek support from the Army Community Services for fear of drawing the Army's attention to their personal lives and possibly incurring criticism from their husbands or partners for not coping.[15]

The first line of support for families of members of the AFS would come from Major Helen Doyle, the newly-arrived Officer Commanding the Army Community Services section in Townsville, Ms Julie Linwood, a social worker, and Warrant Officer Neil Walsh.[16] In addition, David Hurley directed that a 1 RAR Welfare Cell be established with a free '008' telephone number to provide a 24 hour point of contact for the families of those deployed with the 1 RAR Group. Captain Gary Dick [17] ran the 1 RAR Welfare Cell and worked closely with Major Doyle and her staff.

There was also support for those members of the AFS who were not from Townsville. For example, Major David Tyler, the PR Officer, and Major Richard Clark, the AFS Financial Adviser, left their spouses and young children in Canberra while other members of HQ AFS left their families in Brisbane. Aside from local Army Community Services staff, these families could contact Major Alan Smeaton's [18] Welfare Co-ordination Cell located at Randwick Barracks in Sydney. General Blake had directed that this Cell be 'established to oversee and coordinate support to all families who had members overseas.'[19]

Smeaton's staff backed up families whose problems could not be resolved at the unit level or by local Army Community Services staff. His Cell also provided 24 hour emergency assistance to families, a free Recorded Telephone Information Service and monthly newsletters to keep families up to date with developments on overseas operations. Prompted by the increasing numbers of personnel being deployed overseas on UN operations in the early 1990s, the Australian Army Psychology Corps had produced a booklet, 'Unaccompanied Overseas Service: Preparation, Separation and Homecoming—A Guide for Members, their Families and Friends' in 1992 Smeaton's staff, Warrant Officer Doug Caple,[20] Sergeant Mark Sullivan[21] and Corporal Dan McCoy[22] produced a Family Information Package for OP Solace that included this booklet.

Hurley and his staff had kept families informed during the preparations before departure of the 1 RAR Group. On 22 and 23 December Captain Glen

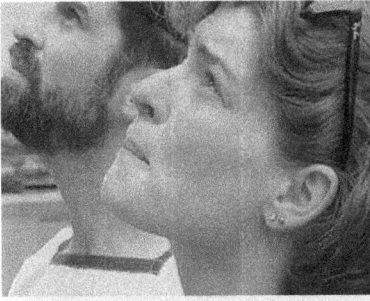

The face of a woman expressing the sadness and anxiety of those who said farewell to members of the AFS sailing on HMAS Jervis Bay on Christmas Eve 1993
Image: ADF Electronic Media Unit

Babington, Hurley's Adjutant, organised information nights at the Lavarack Barracks Area Theatre. David Hurley addressed those who attended and gave them as much information as he could about the situation in Somalia and the arrangements for family support. He emphasised that rumours could be very damaging for the morale of the troops in Somalia as well as their families in Townsville. He urged families to listen to the information they received from official Army sources and their loved ones in Somalia, and to be wary of media reports and other non-military sources of information.[23]

A group of officer and Warrant Officer's wives, Kaye Chamberlain, Louise Daiman, Robyn Dale, Debbie Fraser, Linda Hurley, Debbie Johnson, Trish Judzewitsch and Noelene Klima, re-formed the 1 RAR Ladies Group into a Spouses of Somalia (SOS) Support Group. Soon they were joined by Diane Harnwell, Lea McKaskill and Haylie Stanhope representing the families of the battalion group's supporting units. The SOS Group met before Christmas and agreed that they would not have a committee structure but would organise a function at the 1 RAR Sergeants' Mess on one Tuesday and one Saturday each month. They also decided to ask the 1 RAR Headquarters staff to assist them to distribute newsletters to the Next of Kin of those serving in Somalia. Noeline Klima agreed to be the spokesperson for the group.[24]

The SOS Support Group got to work quickly. The first 'Spouses of Somalia' newsletter, distributed in the last week of December, was a welcome reminder for the Next of Kin of members of the AFS that they were not alone with their concerns about loved ones who were about to depart.[25] The SOS Group, with the assistance of members of the 1 RAR Battalion Headquarters staff, collated and mailed out over 900 newsletters. The aim of the newsletter was, 'to inform you of up coming social events. These will include day time functions, night functions (so as the working ladies will be able to join us) and weekend functions to cater for our children.'[26] The first 'SOS' function was to be a lunchtime BBQ at the 1 RAR Sergeants' Mess on Saturday 23 January and the second a 'Games Morning' at the same venue on Tuesday 16 February.

Local media coverage of the deployment and encouraging stories of the importance of the Australian contribution to Somalia helped families and friends cope with the imminent departure of their loved ones.[27] The two local radio stations, 4TO and 4RR, and the local newspapers, the *Townsville Bulletin*, the *Townsville Independent*, the *Twin Cities Advertiser* and the *Townsville and Thuringowa News* gave priority to news from Somalia and stories of the preparations being made at Lavarack Barracks. The two North Queensland regional television stations, WIN TV and Sunshine TV, did the same. They were all positive stories with strong patriotic sentiments. Stories on the departures of *Jervis Bay* and *Tobruk* before and after Christmas made the front pages and received extensive coverage on television and radio. The *Townsville Independent* advised the SOS Support Group that the paper would print messages of 18 words or less each week from loved ones of the members of the AFS free of charge and forward copies of the newspapers containing the messages to Somalia. The other newspapers in Townsville followed suit.[28]

There was anxiety and high emotion among family members and friends at the docks when *Tobruk* and *Jervis Bay* departed and at Townsville Airport when the QANTAS 747 jets flew out later in January. The impact on children was significant. They sensed that their fathers were not just leaving for a two or three-week field exercise or a training course. The publicity, the crowds farewelling the contingents and talk at their schools of the dangers of Somalia heightened their anxiety about the safety of their fathers. In the days that followed the departure of their fathers many children reacted emotionally and misbehaved. This added to the stress of mothers who were trying to control their own emotions and set a new daily routine for their families without their husbands and partners.[29]

I sit and cry silently
nearly every day.
Why was he one of the ones
who had to go far away ?

I never got to hold him'
or tell him I was full of pride
I sat alone within my room,
trying to hold it all inside.

I didn't know he'd gone.
I found out on Christmas Day.
But then it was too late,
He was sailing away.

I never got to wave goodbye,
or share a parting embrace.
I didn't even get the chance
to talk to him face to face.

I never got to hold him close,
and reassure him, 'It was all right.'
I never got to tell him,
he'd be held in my heart day and night.

I guess I really envy
all who were there that day.
at least you had the chance to say
goodbye, and watch them sail away.

Lyndall de Vries
Manly, Sydney

Families of members of the AFS had two strong emotional needs after the ships and aircraft departed. The first was to know that their loved ones had arrived in Somalia safely and were coping with the conditions, and the second was the need to know that their letters were getting through. [30]

In many cases letters were written within hours of departure of the AFS and posted. It was almost as if news that 'all was well' and that letters were getting through would be the cue for them to get on with their lives in Australia. There were many telephone calls to the 1 RAR Welfare Cell in January from spouses and partners needing reassurance about the safety of their loved ones in the AFS.[31]

With the need to know that the AFS had arrived safely and that their loved ones were coping, the families of members of the AFS were very sensitive to media reports. A newspaper headline 'Soldier killed', referring to the accidental death of a soldier on a field exercise at Wide Bay in Queensland resulted in a flurry of telephone calls from distraught spouses, partners, family members and friends to the 1 RAR Welfare Cell.[32] Newspaper and television reports that Australian troops had taken cover from small arms fire at Mogadishu Airport on 16 January precipitated a similar flurry of telephone calls. Fortunately, local newspaper reports in Townsville of shots being fired were quickly followed by reassuring information that all the Australians had arrived safely.[33] For some families the reports of shots being fired and comments attributed to the Minister for Defence Science Personnel, Gordon Bilney, that Australians could be involved in fighting Somali warlords heightened their anxiety about the safety of their loved ones significantly. [34]

Aside from writing letters to members of the AFS, many families took the opportunity to send in messages for publication in local newspapers. In Townsville the *Townsville Bulletin* and the *Townsville Independent* published messages for loved ones in Somalia on 16 and 22 January respectively called, 'Somalia Messages' and 'Hello Somalia'.[35] The intention was for over 60 copies of each of these newspapers to be sent to Somalia via the resupply system. On local Army advice, the editors of these newspapers sent parcels of newspapers daily to the Moorebank Freight Terminal where they accumulated with containers of stores and pallets of track link and pads awaiting aircraft to fly them to Somalia. Aside from the intended benefit of having those serving in Somalia read these messages, spouses, partners and family members enjoyed reading messages from other families, realising that they were not alone in trying to encourage and communicate with their loved ones in the AFS.

The second SOS newsletter, distributed in the last days of January, also reminded families in Townsville and all around Australia that they were not alone. In Townsville over 250 spouses, partners, relatives and children attended a BBQ at the 1 RAR Sergeants' Mess on 23 January and a 'Family Fun Day' was planned for Saturday 6 February 'to entertain Mums, big kids, little kids and babies.' A letter from Brigadier Adrian D'Hage, the Director of Defence Public Relations, was attached to the newsletter, advising 'family and friends' that they could record messages on a Toll Free '008' number from anywhere in Australia that would be broadcast to the AFS from the ADF's new radio station. He wrote: 'This is the first time that ADF Radio has been able to broadcast to our servicemen and women overseas. We believe it's a service that will be enthusiastically welcomed by those in Somalia, keen to hear news from home and especially the voices of their loved ones.'[36]

While loved ones in Australia reached out to the AFS through letters, newspaper messages and the ADF radio station, members of the AFS reached back with letters. However, their all-important news on how they were coping in Somalia and their families' needs to know that letters and messages were getting through was delayed. Aside from some letters written by members of the AFS during their long flights to Mogadishu and posted back in Australia by QANTAS air crew after they returned, the vast majority of families in Townsville did not receive mail for almost three weeks after the AFS left Townsville. By the end of January and the first week of February, the 1 RAR Welfare Cell was receiving numerous calls every day asking about when mail would arrive.[37]

At the same time Army Public Relations officers and staff from some of the sub units that had been put under command of 1 RAR began calling on behalf of increasingly anxious spouses, partners and family members.

In the world of waiting for letters, many spouses, partners, and relatives, as well as members of the AFS reversed the old adage that 'no news was good news'. For them the absence of letters meant that 'no news was bad news'. For some the absence of letters was interpreted as meaning that their loved ones had not bothered to write.[38]

Despite the fundamental importance of mail as a factor in maintaining the morale of troops serving overseas and keeping up the spirits of their loved ones in Australia, planning for the postal system to support OP *Solace* was not accorded a particularly high priority, nor was an independent military mail system seriously contemplated. Planning began on 4 January when Lieutenant Colonel Graeme Woolnough in Mogadishu sent a facsimile to Major Craig Mills at Land Headquarters in Sydney asking him, 'Is a detailed postal plan to be released? If so, by whom and when?' Mills replied two days later that mail to Somalia would use the international postage system via the International Mail Exchange in Nairobi, Kenya, Woolnough's staff would be responsible for picking up mail bags from there. Mail to Australia was to have sufficient postage to take it from Sydney to its destination in Australia. Like incoming mail from Australia, it was to be sent through the international postage system via the International Mail Centre in Nairobi.

Warrant Officer Class One John Collins,[39] a senior mail clerk, and an assistant, Private David Bretherton,[40] were added to the Mellor's headquarter staff early in January after staff at Land Headquarters realised the volume of mail and its management in Somalia and Kenya would require postal specialists. No priority was given to sending Collins or Bretherton to Somalia set up the mail system. They flew out with the main body of the AFS in mid-January after hurried preparations.

As soon as Collins was settled in with the BSG at Baidoa Airfield, he flew to Nairobi with bags of outgoing mail from members of the AFS. He intended to pick up bags of mail from Australia addressed to the AFS in Nairobi. On arrival he found that the Kenyan International Mail Exchange was very inefficient by Western standards. Collins realised that getting the first consignment of mail from members of the AFS through the Mail Exchange to Australia was important to the families waiting for news. He used his own money to bribe Kenyan postal officials to free up some of the bureaucratic snarls. He put the bags of mail he had brought from Baidoa into the Kenyan mail exchange and hoped for the best.[41] Initially it took 18 days for the bulk of mail from members of the AFS to reach Townsville and other locations in Australia.[42]

Compounding the anxieties caused by the absence of mail from Somalia were the misfortunes of life. In the first three weeks after the deployment of the AFS, the 1 RAR Welfare Cell had been notified that one wife had miscarried, several wives had begun to receive obscene telephone calls, two family homes were burgled and a car repairer had intimidated one wife into paying him more money than he was due. Gary Dick made arrangements victims of the more serious mishaps to speak with their husbands and partners by INMARSAT telephone.

I'm sick of all this waiting.
The mailman slowly comes.
Doesn't he know there's women waiting
to hear from husbands, brothers and sons ?

Finally, a letter.
My eyes fill with tears,
scared the words written inside
will make real my worst fears.

But 'No', he's okay,
no bullets pass his way.
But still things are rough,
he just goes from day to day.

He says he loves me, he'll come home,
my letters give him hope.
My love will keep him going.
They fill his heart with the will to cope.

The days there slowly drip by,
my letters arrive everyday.
My words of missed love and cuddles
makes his pain flow away.

I close the letter and think of him
and what he means to me.
I don't care if he's damaged at all,
as long as he returns safely.

I don't care about the medals,
or being a rich Army wife.
Lord, just bring my soldier home
so we can get on with living 'our' life.

Rebecca Warde
Cockatoo, Victoria

In February the first letters from loved ones in Somalia brought requests for toothpaste, razors, cigarettes, magazines, books, batteries for portable cassette players, confectionary, chewing gum and personal toiletry items. Members of the AFS pointed out that there was no canteen for them to purchase these items. Spouses, partners, mothers and other relatives and friends responded by putting together parcels of 'goodies' to make life in Baidoa more comfortable. The costs of posting these parcels to Baidoa via Nairobi was high but many families were glad that they could do something tangible to assist their loved ones in Somalia. Unfortunately, some had not anticipated the time it would take for the mail to get through to Baidoa and the degree of rough handling parcels would endure on the way. Cakes went stale, other food items 'went off', parcels were torn open, and bottles and cans were broken and ruptured, saturating the rest of the parcel.[43]

Soon after the 1 RAR Group had deployed to Baidoa in January, someone in the Townsville media suggested that school children should write to members of the AFS to give them encouragement. What was dubbed in the media as the *'Dear Digger'* campaign snowballed into a spontaneous national gesture of support for the troops in Baidoa. There were hundreds of letters in the first mail bags delivered to Baidoa in mid February. *'Dear Digger'* letters arrived steadily from then on. Accompanying them were scores of letters from all over Australia, especially from Queensland, New South Wales, Queensland and the Northern Territory, wishing the AFS the best, and affirming how proud the Australian community were of their soldiers in Somalia.

The impact on members of the 1 RAR Group of *'Dear Digger'* letters and other letters and parcels from Australia was significant. The patrolling diggers found Somalia to be a hot, exhausting, brutal place. They had to deal with the constant fear of being shot by Somali gunmen, especially at night. They were engaged in a psychological test of strength with sullen and frequently aggressive Somali males, most carrying knives and some concealing firearms. For some it was difficult not to succumb to this environment and join in the cycle of violence by responding physically to the implicit and explicit malevolence of members of the Somali population. *'Dear Digger'* letters, and the letters and parcels from loved ones provided mental and emotional relief from the pressures of patrolling, guarding and escorting. These messages from home reminded them of caring human values and the promise of reunion with people whom they loved and who loved them. They also reminded them that they were fathers, brothers, uncles and sons, not just soldiers doing their job in a violent place.[44]

———◆———

Here's to our boys in a faraway land,
who are giving the starving a helping hand.
Good on you all.

Here's to our boys faithful and true,
in a faraway land that's no longer new.
Good on you all.

There's too many names to mention each one,
but they are someone's husbands, fathers and one is my son.
Good on you all, for you answered that heartfelt call,

Without moan, without groan, you all left your home and country behind
to help the people who were in a bind.
Good on you all.

Regardless of age, colour or creeds,
you will do your best to see to their needs.
Good on you all, fathers, husbands, and all mother' sons.

Olive Lewis
Adelaide, South Australia

———◆———

The third SOS newsletter was distributed in the middle of February. Once again, Kaye Chamberlain, the wife of 1 RAR's Regimental Sergeant Major, Greg Chamberlain, wrote the bulk of its contents and included two poems of her own about how she felt when Greg had departed for Somalia four weeks before. By this time over 12 businesses in Townsville had donated prizes and funds to the SOS Support Group. Prizes were raffled off or won at games conducted at the SOS functions at the 1 RAR Sergeants' Mess and the donated funds were used to buy soft drinks for the children attending.

———◆———

Dear Digger,

What is your name? My name is Ben. I have spikey hair. I hope you
don't get injured. What is it like in Somalia ? I hope you win the war.
I go to Hume Public School. I hope you write back

Your Friend,
Ben Jolley

———◆———

The next SOS BBQ was held on Saturday 20 February. Like the previous two BBQs, there was a trend for families of officers and senior NCOs from 1 RAR to attend but very few families from other sub units in the 1 RAR Group in Townsville or the young families of the soldiers of 1 RAR attended. Kaye Chamberlain made a special point in the SOS newsletter of inviting other families from the 1 RAR Group to attend the next BBQ and also encouraging soldiers' spouses, partners and girlfriends to attend.

Despite the best intentions of the founders of the SOS Support Group, three factors worked against the spouses, partners and girlfriends of junior NCOs and soldiers' attending SOS functions. Unlike the spouses and partners of officers, Warrant Officers and Sergeants who regularly attended social functions at the Officers' and Sergeants' Messes at Lavarack Barracks, the spouses, partners and girlfriends of junior NCOs and soldiers' did not socialise at the Army Barracks frequently. The John Kirby VC Club for junior NCOs and soldiers was more a 'watering hole' for single soldiers, not a family entertainment venue. Consequently, they were not used to travelling with their young children into the Barracks for social functions.

Members of the Army units and sub units in the Townsville area and their families were socially very insular. Aside from one or two combined functions in some unit Officers' and Sergeants' Messes, or at the 3rd Brigade Officers' and Sergeants' Messes, it was rare for the families of junior NCOs and soldiers from different units to socialise together in large groups. Units and sub units from different Corps in Townsville were quite tribal. Having husbands and partners away in Somalia was not sufficient incentive to bring families together socially.

Most spouses, partners and girlfriends of junior NCOs and soldiers found the prospect of attending a social function at the 1 RAR Sergeants' Mess with the spouses and partners of officers, Warrant Officers and Sergeants daunting. Many felt socially uncomfortable making conversation with who they perceived to be older and more socially sophisticated women. They preferred to socialise nearer to their homes with other young spouses, partners and girlfriends. Within these small informal groups they did not have to concern themselves with what they wore, or how they or their children behaved. Very few had ever been to a Sergeant's Mess which was regarded by most as an exclusive club, not intended for the families of junior NCOs and soldiers.[45]

Finally, some spouses and partners of junior NCOs and soldiers perceived the efforts of the spouses and partners of officers and senior NCOs to run social functions on their behalf to be patronising, especially when the 1 RAR Sergeants' Mess was nominated as the ' SOS Headquarters'.[46]

Advanced notice of functions for the 1 RAR Ladies Group highlighted in the SOS Newsletters which they, by definition, were not invited, did not help. As far as they were concerned the tradition of spouses of officers, Warrant Officers and senior NCOs supporting spouses and partners of junior NCOs and soldiers was irrelevant to them and implied social superiority. They had coped with the absence of their husbands and partners on field exercises and training courses without socialising at the Sergeants' Mess. Many felt that they could do the same for OP *Solace*. [47]

Announcements of the birth of children to spouses and partners of members of the AFS and congratulations on birthdays in the SOS newsletters unintentionally reinforced perceptions among some spouses and partners of junior NCOs and soldiers that the SOS Support Group was not relevant to them. Christian names were used when referring to officers, Warrant Officers, senior NCOs and their spouses. Junior NCOs and soldiers were identified by their rank and initials. The vast majority of birthdays acknowledged in SOS newsletters were those of officers, Warrant Officers, senior NCOs, their spouses and children.[48]

Major Martin Fleming, Chaplain of the AFS, Baidoa Airfield.
Image: ADF Electronic Media Unit

Dear Digger,

My name is Darryle Howe. I still go to school and I am in Year Six. I appreciate what you are doing in Somalia. I would like to know what the weather is like over there. Is the death rate getting any less ? Have you made any friends ? If so, tell me who they are. How do you sleep ? Do you sleep in a group under rags or a tent ? Please write back. I would appreciate if you could answer some of my questions.

Yours Sincerely,
Darryl Howe

On 22 February Major Martin Fleming,[49] the 1 RAR Group Chaplain, wrote to the Directors of State and Catholic Education in Townsville, Ms Gail Mackay and Sister Mary McDonald asking for their support and the involvement of their students in supporting the work of the Irish NGO, CONCERN. In addition to contracting with the World Food Program to co-ordinate the distribution of food, CONCERN had also re-established four Somali schools of about 1 000 students each in Baidoa and were planning to re-establish two more schools in their original buildings in the near future. Fortunately there were still sufficient Somali teachers and administrative staff available to run the schools, but there were insufficient resources such as exercise books, chalk, pens and pencils.

Fleming asked Gail MacKay and Sister Mary if they could encourage the students to ' paint posters for the Somali children's bare classroom walls. These could include Australian children's scenes and perhaps educational teaching aids like the Alphabet of the Numbers (or whatever).'[50] He also asked for donations of exercise books and writing materials. Irene Jacovos, the Communications Officer for the Northern Region of the State Education Department, wrote back to Fleming on 5 March, 'pleased to report that many of our schools have responded with books, pens, pencils and art work. The children at Vincent State School, many of whom have fathers serving with OP *Solace*, are currently making alphabet charts and other 'special' resources for the [Somali] schools.'[51] Sister Mary wrote back a week later reporting that, ' Bishop Benjamin has been very pleased to give his approval for the participation of our 24 primary and 10 Secondary Schools ... in this worthy cause.'[52]

Fleming's letters resulted in thousands of members of the north Queensland community making donations to assist schools and orphanages in Baidoa. Prompted by Fleming's letter, Irene Jacovos, 'asked Mary Vernon to place an item in her regular column in the *Townsville Bulletin* appealing for resources from the community.'[53] She also made contact with Rotary Clubs who began campaigns to raise funds to buy exercise books and writing materials for the school children of Baidoa.

The '*Dear Digger*' letter writing campaign and the call for the North Queensland community to donate funds and teaching resources for the school children of Baidoa were a great source of comfort to the waiting families of members of the AFS. It showed the community had not forgotten their loved ones and appreciated the sacrifices of their families.[54] The campaigns also kept the media interested in reporting news from Somalia. Stories from Michael McKinnon, a journalist from the *Townsville Bulletin* who had stayed in Baidoa after other journalists had left at the end of January, were eagerly read by the families of those serving in Somalia.[55]

Dear Soldier,

My name is Rebecca. I'm six year old. My Pop was a soldier too. I'm from North Mandurah School. What is it like in Somalia. Here's a song. This little pig went to market. This little pig went home.

From
Rebecca

By late February letters of complaint from spouses, partners and parents of those serving in Somalia to the Defence Minister, Senator Robert Ray, about the inefficient mail system resulted in some action being taken. The Minister's office formally asked HQ ADF in Canberra why the postal system from Somalia appeared to be malfunctioning and no canteen facilities were being provided. It fell to Brigadier Ian Macinnis, the Chief of Staff, at Land Headquarters to assure Senator Ray in a brief that everything was being done to overcome the initial delays in the provision of these services.

By the time Macinnis sent his brief to the Minister, via HQ ADF in Canberra, mail was arriving in Australia from Nairobi in less than three weeks. However, mail was arriving sporadically. Many families of members of the AFS lived in the same streets and kept in close contact in the suburbs of Townsville. It was disheartening when some could express their joy to friends when they received mail while others, who had not received mail, imagined the worst. Often families would receive mail and keep it a secret for fear of upsetting other families who had not received mail at the same time. The random nature of the return mail from Somalia caused much heartache among those desperately waiting for progress reports on how their loved ones were faring.[56]

Once the decision was made to channel all incoming and outgoing mail through the Kenyan Mail Exchange in Nairobi for OP *Solace* there was little that could be done to significantly improve the efficiency of the postal system. John Collins and David Bretherton worked very long hours to ensure that bags of mail for the AFS went through the Exchange as quickly as possible and they personally escorted all mail bags on available Canadian, UN and New Zealand aircraft to and from Nairobi, Mogadishu and Baidoa. No bags of mail were lost but Collins and Bretherton were unable to guarantee that every mail bag would move through the Kenyan system smoothly.[57]

In addition to disappointment with the postal system during February, there was confusion over when the additional pay and allowances of members of the AFS would arrive in bank accounts. Wives and partners knew that substantial allowances were about to be paid for service in Somalia and that no tax was going to be deducted from pay packets. In many cases these special conditions for service in Somalia doubled the 'take-home pay' of members of the AFS. Expectations were raised in the second and third SOS Newsletters that these allowances and tax-free payments would be paid on 11 February. On that day some allowances and increased payments were made into the bank accounts of some members of the AFS, but not for others. This led to a sharp increase in requests for access to Army pay slips, called Salary Variation Authorities, by spouses and partners.[58]

On 11 February bank staff and others made several calls to the 1 RAR Welfare Cell stating that some spouses, partners and girlfriends of members of the AFS were attempting to, or had cleared out, bank accounts after the increased payments had been credited by the Army pay system. Prompted by this information, a decision, based on legal advice, was made that copies of the pay slips would not be given, as they had been in the past, to spouses and partners. This led to further turmoil and consternation. The uncertainty about pay rates and the inconsistency in their payment on 11 February led to an understandable curiosity among spouses and partners about what amount would be paid and when they would be paid. Despite all of the members of the AFS receiving their allowances and new rates of pay on 25 February, the problems of access to information about pay by spouses and partners continued.

The lack of direct access by Pay staff at the 1 RAR Group to ADF authorities in Australia created further complications with pay and allowances. All pay administration from Baidoa had to directed through HQ AFS, the authorised direct link back to Australia. From the 1 RAR Group's perspective, this was inefficient double-handling that resulted in several incidences of soldiers not being paid their correct allowances, and some frustration and delay in sorting out pay matters after those serving in Baidoa received letters from Australia containing concerns from spouses and partners about pay, allotments to bank accounts or the payment of allowances.[59]

By the time the sixth SOS newsletter was distributed in early March, scores of heartfelt letters from families from all over Australia had been received thanking the SOS Support Group for keeping them in touch with others who were enduring the absence of their loved ones in Somalia. There was little news of Somalia in the newspapers and on the television to sustain these families. Aside from letters from Somalia, Armed Forces Radio and taped information on the '008' information telephone number, the SOS Newsletters provided information and the comfort of knowing that the SOS was working hard to keep up the spirits of families of members of the AFS.

One of the initiatives introduced at the 1 RAR Welfare Cell was very well received by the Next of Kin of those serving in Somalia. Duty Officers manning the Welfare Cell telephone had to make at least 20 telephone calls to Next of Kin during their shift. These calls of reassurance and enquiries about how 'things were going' were welcome surprises for many Next of Kin, many of whom wrote letters of thanks to the Welfare Cell and to the Spouses of Somalia subsequently.[60]

Dear Digger,

Hi ! My name is Duane Fernandez. I am writing this letter from St John's Primary School, Campbelltown. I have just begun Year 6, and we have been talking about Somalia. Somalia came up under a religion topic about the needy, and you came under the topic of people who help them. Our teacher has recently read a newspaper article requesting us to write a letter to "A Digger". I was obliged. By the way, if you have time, could you please make an effort to write back?

Anyway. I have been meaning to ask you a few questions. Have you made any friends yet? Are they nice? Is what we (Australia) have been sending enough for everyone there? I suppose not. We will be sending as much as possible to you just to make your difficult task a little easier. Anyway. Now I will tell you what's happening in Australia. There was recently a "Great" Debate between Paul Keating and John Hewson. The election is on now. If I were 18, I would vote for Paul Keating. The footy has just started but not the Winfield Cup yet. I support Manly. Do you?

We will be thinking about you every single day, especially in school. We are praying for you.

Yours Thankfully,
Duane Fernandez

———————◆———————

By the end of March OP *Solace* was taking its toll on relationships in Australia. David Hurley had written to his wife, Linda, on 13 March expressing concern that there had been an increase in the number of 'Dear John' letters being received in Baidoa. These letters, announcing that a spouse, partner or close friend had ended a relationship as well as letters from spouses, partners and loved ones imploring soldiers to return home because they could not cope with them being away, added to the pressure on those serving in Baidoa.

In Support Company, a 21 year old digger assaulted another over a remark about the pressure he was receiving from his wife to return home. He was disarmed and isolated from his section for a day so he could cool down. Two days later he put his rifle barrel under his chin and threatened to kill himself. He was talked out of taking this action, disarmed and sent to Mogadishu for psychological assessment, returning to Australia soon after.[61]

During March the morale of many members of the 1 RAR Group was put to the test. A combination of the 'mid-tour blues', 'Dear John' or other 'bad news' letters, and an increase in telephone contact with Australia contributed to this situation. Commanders had to spend more time monitoring and maintaining the morale of their subordinates. They also had to 'dig deep' themselves to overcome the effects of the 'mid-tour blues' and any difficulties they were having with their own relationships:

> ... a young officer has just received a letter from his fiance saying their engagement was off, and she has a new boyfriend. They were to be married three days after he got back to Australia, and he was terribly distraught ... It was a terrible thing for her to do to him when he's on the other side of the world. She could have waited until he got home.[62]

By mid March the 1 RAR Group had been on operations for eight weeks and had nine weeks to go. Aside from some members of 1 RAR who had served for three months in Butterworth on the north-western coast of the Malay Peninsula or members of the Group who had served on UN missions for six to nine months unaccompanied, this was the longest time they had been in the field and away from home. It was also the longest absence from home for loved ones that many spouses and partners had had to endure. For many girlfriends and boyfriends the eight week absence either 'made the heart grow fonder' or ended relationships. There were rumours of spouses, partners and fiances having affairs and 'going out on the town.'[63]

A growing number of obscene and 'crank' calls to spouses and partners had a significant unsettling effect on the morale of husbands and partners who were unable to protect their families and loved ones from this type of harassment. A Platoon Commander and a Sergeant had to cope with knowing that their spouses were regularly receiving obscene telephone calls from a person or persons who knew their spouses' Christian names. David Tyler wrote at the time:[64]

> There are also some loonies in Townsville causing problems by ringing spouses and saying they have had affairs with their husbands, and someone in Townsville has produced T-shirts with offensive slogans relating what's been happening to spouses while their husbands have been in Somalia. There are some sick people in the world.[65]

By this time a leave program for 450 members of the 1 RAR Group to Mombassa had been arranged by Graeme Woolnough. He had negotiated with New Zealand and Canadian squadron commanders for groups of Australians to be flown to and from Baidoa and Mombassa. While Hurley was not happy that all of his troops could not go on leave, he decided to give the soldiers and Platoon Commanders in the rifle companies and Support Company, as well as the 1 RAR Group drivers the priority for this program. He assessed that they were doing most of the demanding and dangerous work and should be given the opportunity for a two-day break in hotels on the beach front in Mombassa.

The reaction of the spouses, partners and families of members of the 1 RAR Group back in Australia to this leave program was mixed. Many were glad that their loved ones were getting a break but some had reservations about sending them to a place where a high percentage of the African prostitutes were HIV positive. One wife commented that it was 'like sending starving dogs to a butcher's shop.'[66]

The risk of HIV infection was well understood by the medical staff in the 1 RAR Group who ensured that all of those going on leave were well informed about the risks they would take if they engaged the services of the local African prostitutes. Condoms were issued to all going on leave. In addition, the soldiers were sent in their platoon and section groups to ensure that they were accompanied by officers and NCOs.

Probably the most prominent characteristic of the behaviour of Australians on leave in Mombassa was drunkenness. Notwithstanding a possible penchant for exaggerating the amount of alcohol consumed, many sections approached the opportunity to drink with exceptional enthusiasm. As a consequence, several groups and individuals were reported to have behaved immaturely on a number of occasions by tourists and expatriate NGO staff who were also taking time off in Mombassa. Many officers and NCOs commented that this binge drinking may have been unavoidable, given the build up of stress from soldiering in Baidoa. Others reflected that the short duration of leave and the many weeks of not being able to have a drink, or completely relax in recreational facilities at the Baidoa Airfield, added to the zest for drinking displayed by members of the 1 RAR Group on leave in Mombassa. [67]

Up until March tight restrictions had applied to telephone contact with loved ones in Australia. The 1 RAR Group's INMARSAT telephone had broken down in January, leaving the Australians in Baidoa without direct communications back to Australia. If subordinates received bad news for home in a letter or from welfare support agencies, commanders assessed the merit of allowing the subordinate to call home using David Tyler's INMARSAT telephone provided to him for communications to Australia on PR and media matters. Often Martin Fleming, or the person's commander, would accompany them to Tyler's tent. On other occasions Tyler was there alone to share a cup of tea or coffee and listen after the call had been made. Many of those who sat with Tyler cried as they told him of their pain: Tyler wrote to his wife at the time:

> Because the battalion's satellite phone has broken, I am getting all the welfare cases down here. There have been an average of 30 soldiers a week needing to call home for one reason or another. Some have money problems and more serious worries about relationships.[68] ...

> One was a Fijian, whose mother was dying in a Fiji hospital, and it took me some time to get a connection for him. ... He too was very distressed, and had said earlier that he didn't want to be taken off the OP. ... She told him that she was just waiting to hear his voice before dying.[69]

The setting up of booths with AUTOVON telephones[70] for American personnel in Baidoa in March prompted a request by John Caligari to the US unit responsible for the telephone facilities to allow members of the 1 RAR Group to be given access to the American system on a rostered basis between midnight and 6 am. Around the same time he also managed to convince an American communications officer to set up an American AUTOVON telephone link in his office. Like the AUTOVON telephones used by the members of the 1 RAR Group, this link enabled him to call into the Australian STD system via the Fort Wachuca switchboard in Texas.

Tina Fletcher, partner of Private Graham Fletcher, 1RAR, pregnant with their third child who was due to be born two weeks before Private Fletcher returned from Somalia.Pictured on 5 February 1993 with their daughters Natasha, 6 and Sandra 2.
Image: *Townsville Bulletin*, Michael Chambers

Caligari used this telephone as an emergency facility for any member of the 1 RAR Group who needed to call someone in Australia urgently for personal reasons. He and the Adjutant, Glen Babington, were also able to speak directly with Majors Mark Fairleigh and Warwick Austin [71] at 1 RAR's Rear Details headquarters in Townsville to sort out administrative issues and also obtain first hand information on welfare issues effecting the families of members of 1 RAR. This direct contact with Australia was kept secret from staff at HQ AFS who insisted that all contact with Australia be made through them. Caligari persisted with the deception because the direct telephone link sorted out many routine administrative problems efficiently and effectively. He directed important administrative and welfare issues requiring higher headquarter support through HQ AFS .

By 30 March over 500 calls had been made back to Australia using the AUTOVON telephones. Coinciding with the access to these facilities were persistent rumours of an extension to the duration of OP *Solace*. Tyler wrote:

> There are apparently stories at home about the UN asking Australia to extend our stay by three months, and that is under consideration. Many of the spouses are pretty upset about the prospect, but even if there's no intention of extending, the Government has to say it's

considering the request, it can't just say, 'No', outright. Tomorrow I will try to get some answers from Canberra about the likelihood [of an extension]. If it were to happen, there would be a lot of very unhappy diggers in Somalia, and I would be one of them. The prospect of another three months here would be bad news for us all.[72]

For many spouses and partners in Australia telephone calls from Baidoa were emotional 'two-edged swords'.[73] Most did not expect direct telephone contact. Many were shocked when the first call came. Sometimes the shock was magnified when the calls came through from Baidoa in the early hours of the morning. The sound of the telephone ringing at these times produced a stab of extreme anxiety about whether it was a notification of death or injury. Due to time differences between Somalia and Australia, these calls also came through in the afternoon. Some were so clear that the spouse, partner or parent thought it was a local call and found it hard to believe that they were speaking to a loved one in Baidoa.

Initially, the sound of their loved one's voice was exhilarating and provoked a flood of emotion and excitement. After settling down, there was pressure to say 'all the right things' in the short time available. Typically conversations were rushes of positive and negative news as well as affirmations of love and longing. It was difficult to know how much the loved one in Baidoa needed to know about how their family was coping without them. The dilemma was to balance positive and negative news. Letters were easier because they could be thought out and carefully worded; a telephone conversation was immediate, emotionally-charged and spontaneous.

After putting down the receiver, many spouses and partners struggled with a leaden feeling of depression about whether all that could have been said was said, whether their loved one had been supported by the call or now felt worse about being away from their family. Most thought of things they could have said, should have said, or regretted saying. In many cases these regrets caused emotional turmoil for many hours after the call.[74] Those calling from Baidoa also suffered the same emotional turmoil because they did not want anything they said to increase pressure on their families. However, it was difficult to hide the 'mid-tour blues' over the telephone. I was also very difficult not to take the opportunity to talk about these feelings with a loved one.[75]

There were different attitudes among spouses, partners and close friends about what to tell their loved ones in Baidoa in either letters or during telephone calls. [76] Many felt that they should only pass on good news and not write or talk about the crises they had, or their feelings of anguish about the safety of their loved one in Somalia. Just as many others advocated telling the truth about the ups and downs of their lives in Australia. They felt that their spouse, partner or friend would detect any false reporting. Most agreed that it was particularly difficult to hide negative news over the telephone when the tone of voice or the repetition of cliches about everything being all right would give a loved one the feeling that they were not being told the truth about their family's circumstances.

Initially, Hurley assessed that those who were rendered emotionally unsuitable for duty because of the impact of a family or relationship situation should be returned to Australia. In all cases, with the exceptions of crisis situations requiring immediate family support, such as the death of a child, a person's emotional state was assessed by his or her commander. Typically, Martin Fleming would counsel the person and make his own assessment. Before there was a formal request for a member of the 1 RAR Group to be returned to Australia, a professional assessment was made by an American psychologist in Mogadishu. In the end, Hurley was faced with the option of retaining emotionally disturbed personnel in Baidoa until the end of the operation or sending them home.

The issue of returning members of the AFS to Australia on the grounds that their marriages or significant relationships were breaking up, or that their spouses or partners had left to live elsewhere, caused concern both in Somalia and in Australia. It began to take on significance after three members of the 1 RAR Group were returned to Australia in the last week of March. General Blake decided to intervene with a letter to Colonel Bill Mellor entitled 'Some Personnel Issues':

> I am concerned that 1 RAR appear to have sent home a couple of NCOs because their marriages were breaking up. My concern in twofold. Firstly, if a marriage is breaking up after a relatively short separation it was not in good shape in the first instance. Second, the ODF must be prepared for a lengthy deployment at any time and the break up of a marriage, while regrettable, is not grounds for RTA [Return to Australia]. ... If we take a soft line now it will establish the pattern for future operations and I am concerned about our need to make it clear to all ranks what obligations are entailed in operational service.[77]

Blake's letter made it clear that he expected Mellor to take a closer interest in all subsequent requests from the 1 RAR Group for a member to be returned to Australia for compassionate reasons. This placed further pressure on Hurley and his staff to ensure that all recommendations for a person to be returned to Australia were thoroughly explained. They had to explore all the options for retaining individuals in Baidoa while their domestic circumstances were resolved or stabilised by staff from Army Community Services in Australia.

Blake's clarification of his policy for returning personnel from Somalia to Australia came at an interesting time. On 31 March before departing for the US, the Australian Foreign Minister, Senator Gareth Evans pre-empted speculation about whether he would agree to a request by the UN to extend the tour of the 1 RAR Group. 'There may well be requests of that kind coming from a number of sources, but our answer is very clear and firm. It was a short term commitment of that number of troops. We will make it clear as we have in the past that the battalion will be returning on the scheduled date.' Despite these assurances, many spouses and partners feared that Evans might be put under sufficient international pressure to extend the tour of the 1 RAR Group. [78]

The death of Lance Corporal Shannon McAliney on 2 April was a shock for members of the AFS in Somalia as well as their families in Australia. The 1 RAR Welfare Cell was flooded with telephone calls from distraught relatives and friends, many offering sympathy and support for McAliney's family, especially for his mother who had bravely announced that she not only felt very sad about the death of her son but also for the young forward scout and his family who were suffering because of this tragic accident. David Tyler comforted those in Townsville with an article published in his regular 'Letters from Somalia' column in The *Townsville Bulletin*:

> There is little that is glorious about death. This week its shadow was cast not only over the Australian camp at Baidoa, but across continents, and no glib phrases can erase the heartache over the loss of Lance Corporal Shannon McAliney.
>
> Even for those of us in the battalion Group who didn't know him, the tragedy is personal one, as it is for the hundreds of families who have loved ones serving here. As deeply saddened as we are, we must remember why we are in Somalia, a country that has endured years of unimagined human misery. To dwell on this tragic setback will benefit nobody, and the memorial service will enable many of his comrades to deal with their own feelings, and to re-focus on the professional completion of the mission.
>
> ... Lance Corporal McAliney died while serving his country, and helping the innocent victims of a cruel civil war. He will not be forgotten.[79]

Early April was an emotionally hard time for Delta Company as they dealt with the shock of Shannon McAliney's death Their morale took another hit on 4 April when one of the soldiers in the Company received news in a letter from his wife in Townsville that she had been raped. Blumer mused in his diary:

> She is apparently OK now -the poor girl. I am shocked and stunned at what gives someone the right to violate someone so intimately. Poor _____ is beside himself. He phoned twice and will 'soldier on' as there is little to be done other than just be there. It's really hurting the company now. It's a difficult time for Delta.[80]

A week later Mick Moon had to pass on news to one of his NCOs that it had been reported that his wife had been raped. Like Blumer, he also faced the challenge of counselling a member of his company about his wife's situation and assessing whether he should remain in Baidoa, or be returned to Australia. He recommended that the NCO be sent home. Moon knew that if his wife was raped while he was on operations he would apply to return to Australia to provide support.[81]

However by 9 April it became clear that getting the NCO home would not be easy. Staff at HQ AFS in Mogadishu asked for more evidence to support a recommendation to return him to Australia. The Army Community Services unit in Townsville had assessed that the husband should be returned home but there was no medical or police report available to substantiate their recommendation.[82]

At Battalion Headquarters, John Caligari knew that if he did not send the NCO to Mogadishu while the issue of his return to Australia was being sorted out he would miss the return flight of a RAAF resupply aircraft. He decided to send the NCO to Mogadishu without authorisation and leave Bill Mellor with no other choice but to interview him personally to ascertain whether he should be sent back to Townsville. After interviewing the NCO, Mellor decided to send him home immediately. However, the initial enquires from Mellor's staff and the delay in confirming whether he would be sent home to comfort his wife caused some ill-feeling between the two Australian headquarters.[83]

The news in the seventh SOS Newsletter distributed early in April was encouraging. The end of OP *Solace* was in sight and activities in Baidoa were described as routine. There was also information in the newsletter advising that the staff of the Army Community Services, the Salvation Army (Red Shield), the Townsville detachment of 11 Psychology Unit and Army Chaplains were had planned a series of 'reunion talks'. These talks were being put on, 'to raise awareness of issues that may affect family relationships on the members' return to Australia', focussing on 'hints to make homecoming more harmonious for all involved ... and ways to identify signs and symptoms of stress as well as sources of assistance within the civilian and military community.'[84]

The focus for the SOS Support Group in April was on a $ 25-a-head Mothers Day Anzac Day lunch organised by John Salter, a previous Commanding Officer of 1 RAR and his wife Christine. Staff from the Sheraton Hotel in Townsville, led by Gerald Therron and his wife Yvonne, worked hard to ensure the day went well. John and Christine canvassed their network of friends and associates in the business community to donate gifts and raffle prizes to the 56 'Spouses of Somalia' who attended. Steve Price, a DJ from 4RR, who had been particularly supportive of the families of those serving in Somalia was the Master of Ceremonies. Louise Cook, a local vocalist, did not leave 'a dry eye in the place' after she sang American vocalist Whitney Houston's No 1 Hit 'I Will Always Love You' from the film 'The Body Guard'.[85]

Two days after the ANZAC Day lunch, an appeal to raise $25 000 for the Goal Orphanage in Baidoa, was launched jointly by Deputy Mayor of Townsville, Alderman Anne Bunnell and Brigadier Peter Abigail. The idea for the appeal had come from Jackie Braddock, whose brother, Warrant Officer Michael Robinson,[86] had been writing to his wife in Townsville telling her about the orphanage and asking for items to be sent to him that he could donate them. Ms Braddock contacted the Townsville Council, proposing a community effort to send donations of money, clothing and blankets.[87]

The Somalia Orphanage Appeal gathered momentum in the first week of May. Students for local primary schools began collecting children's clothing. The 'hat' was passed around Lavarack Barracks raising $3 500 for the orphans. Employees of four local clothing manufacturing firms, Fourex Clothing, Aranka, Cueldee and Ezifit, persuaded their managers to donate 'seconds' to the orphanage. 'Spokesman Dennis Cutting said the companies wanted to get involved because it was

*Cranbrook State Primary
School vice-captain Andrew
Kaus (L) and captain Carly
Hislop, with clothing and
footwear for orphans in Baidoa,
donated by fellow students in
Townsville*
Image: *Townsville Bulletin*,
Jason South

basically a Townsville effort.'[88]

In the final weeks before the main body of the AFS flew back from
Somalia, families continued to count off the days. There were a different
feelings abroad as the return dates of the AFS approached. Many spouses
and partners attended briefings at the Lavarack Barracks Theatre about how
to assist their loved ones to readjust to their return. These briefings did raise
some concerns about the impact service in Somalia would have on their
loved ones, but they were also tangible evidence that the Army was bringing
the AFS home on schedule.

All of these concerns were put aside for the joyous welcoming of the
AFS as they flew into Townsville Airport on the Air Force 707 jets. The first
jet with about 120 personnel aboard, including David Hurley and his advance
party arrived in Townsville on 10 May. The next group of 600 personnel
were shuttled in over the next eight days. The final group of 120 personnel,
including a Rear Party of just under 100 personnel who had been employed
cleaning vehicles and stores at Mogadishu Port arrived on 20 May. They had
just under 48 hours to rest and prepare for a medal parade and a 'Welcome
Home' march through the streets of Townsville.

After he touched down, Bill Mellor assured the waiting media that
the Australians had performed very well in Somalia. The *Townsville Bulletin*
reported that, 'he dismissed any suggestion that the withdrawal of the

Lieutenant Colonel David Hurley with his family at Townsville Airport on his return home from Somalia. His wife, Linda was a member of the Spouses of Somalia group that distrubuted newsletters and organised functions for the families of Townsville-based members of the AFS. Hurley is holding his daughter Amelia and in front are his son Marcus and daughter Caitlin wearing his hat.
Image: David Hurley

Australians could mean a return to anarchy in Somalia, 'No, it won't. I'm very confident that the UN has got the right structures now. They are getting the forces in and I believe the job started by *Restore Hope* will continue.'[89]

David Hurley told the waiting media, 'It [OP *Solace*] has shown the Australian Army as a competent force that can hold its own internationally. From that perspective, for the Army alone, it has been a good job.'[90] Rod McLeod echoed Hurley's sentiments when asked by reporters to give a personal perspective on the nature of Australian soldiers in Somalia compared to when he joined the Army 30 years before in the early 1960s, 'The operation showed me that Australian soldiers are as well trained, flexible, and as good humoured today as they were in 1963.'[91]

Despite the difference of opinion over the merit of a farewell parade for the 1 RAR Group in January, there was unanimous support for a co-operative project in Townsville to welcome the AFS home. Peter Abigail appointed Major Ian Hughes,[92] Gary Banister's replacement as DQ of 3rd Brigade, to co-ordinate the Army's involvement and Mooney appointed Elliot Hannay to co-ordinate civic support. Hughes and Hannay formed a working group comprising the President of the Townsville Returned and Services League, John Doyle, a State Police representative and member's of

District Support Group administrative staff and the Council's logistic staff and Public Address contractor.

Major service club support came from the Townsville Women's Apex Club. Hannay also arranged for the Council to employ members of Rotary to cook sausages and distribute free soft drink and ice cream to the families and relatives of members of the AFS. John Doyle arranged for Vietnam War veterans to distribute free beer to the soldiers and their families, and for veterans from World War II and later conflicts to line the Saluting Dias area as an acknowledgment of their support for the returning troops from Somalia. The Vietnam War veterans were very conscious of ensuring that previous generations who had served overseas would be there in person to affirm the service of another generation of Australian troops in Somalia. [93]

Radio Stations 4TO and 4RR as well as the local newspapers responded enthusiastically to Elliot Hannay's requests for promotion of the 'Welcome Home Parade' for the AFS. He pointed out that all proceeds from the sale of food, soft drinks, ice cream and beer to the general public after members of the AFS and their families had enjoyed a free lunch and family entertainment provided by bush balladeer, Peter Dymond Ramplin, would be directed to the Mayoress Benefit Fund for needy groups, mainly disadvantaged or terminally-

Private Terry Abdul-Rahman being welcomed home from Somalia by his parents Albert and Nellie on 15 May 1993
Image: *Townsville Bulletin*, Jason South

ill children, and War Veterans' Homes. He asked all the local media 'to do their bit' to ensure that the 'Welcome Home Parade' was a success.[94]

The weather was perfect for the parade on Saturday 22 May 1993 In the morning several hundred families and friends of members of the AFS attended a parade where members of the AFS were decorated with their Australian Active Service Medals. After some differences of opinion with Army ceremonial staff in Canberra, those on parade were dressed in their field dress with slouch hat, side turned down. They wore their Corps badges in the front of their slouch hats for the first time.[95] Soldiers carried their Steyr rifles and officers wore pistols. McKaskill's cavalrymen wore emu plumes attached to their slouch hats, signifying their unit's proud lineage back to the World War I Regiments of the Australian Light Horse.

At 1 pm the newly-decorated members of the AFS stepped off down the main street of Townsville where over 10 000 citizens waited to call out and clap a fervent 'Welcome Home'. Bill Mellor led the parade behind a simple sign which read 'Australian Force-Somalia.' He was followed by his headquarters staff. Behind Mellor's headquarters came David Hurley and his headquarters staff, including CMOT, followed by Support Company, Administration Company, and then the four rifle companies in order. Supporting sub unit's came behind 1 RAR led by the BSG, followed by B Squadron and 17 Field Engineer Troop. Parade halted and turned to face the dais where a group of visiting military and political guests sat.[96] Dressed in his robes and the chain of office, Tony Mooney expressed the sentiments of the community of Townsville-Thurwingowa:

> This parade has been more than a symbolic return home for the Australian contingent. It has also provided this community, along with national and State officials, the opportunity to say 'thanks' for a job well done. Because the contingent comprised such a large number who were based in Townsville, many with their families, it is also a fitting occasion for your fellow citizens ... your civilian friends and neighbours, to welcome you back to this very special part of Australia.

Members of the AFS then handed in their weapons and joined their families and well-wishers for a BBQ lunch, drinks and entertainment.[97] This was a significant event in the history of Townsville and its military community.

The 'Welcome Home' parade on 22 May did not mark the end of the redeployment phase of OP *Solace*. *Tobruk* and *Jervis Bay* were still at sea heading for Townsville after stopping over at Diego Garcia and Singapore. The ships docked in Townsville on 16 June 1993, bringing most of the stores and vehicles of the AFS home and just over 40 Army personnel who had volunteered for a longer sea voyage to Australia rather than flying home.

Soon after the ships of the AFS docked, the crews were on parade being awarded medals. Afterwards Mayor Tony Mooney presided over a civic reception for the crews and the last Army personnel to return from Somalia. Straight after the reception, the Ships Army Detachments on *Tobruk* and *Jervis Bay*, assisted by members of the crew and soldiers from 1 RAR and other units based at Lavarack Barracks, unloaded all of the vehicles and stores onto the wharves. The next day *Tobruk* and *Jervis Bay* sailed together for Sydney where most of the families of crew members lived. The reception in Townsville was greatly appreciated, but the crews of both ships were looking forward to seeing their loved ones in Sydney and having a beer in familiar surroundings.

The ships entered Sydney Harbour for their home port welcome at 10 am on 21 June. The fleet base at Garden Island was festooned with balloons, signs welcoming the crews home and flowers. By 2 pm single members of both crews were in Frisco's Hotel, Woolloomooloo, toasting the end of a successful deployment. *Tobruk* had been away from home port for 177 days.

LESSONS

The experiences of the families and friends of members of the AFS during OP *Solace* had been mixed. Many marriages and relationships stood the test of separation and matured, changed forever: others collapsed—

gone forever. Several spouses and partners gave birth: others miscarried or terminated foetuses under the stress of separation. Many spouses, partners and friends bonded together, sharing the vigil while loved ones were away: others maintained the vigil alone or with the support of immediate family members. Some sexual assaults were reported along with a number of obscene telephone calls. Many spouses and partners had appreciated the unique opportunity the extra allowances paid for service in Somalia gave their families for a more secure financial future. A few waited until sufficient money had accumulated in joint bank accounts before cleaning them out, packing up their belongings and leaving.

A major lesson learned from OP *Solace* was that communication between loved ones and members of the community in Australia, and those serving in Somalia was a very important factor heightening or lowering the morale of members of the AFS. Many diaries kept in Baidoa testified to the importance of mail. Those waiting in Australia and those in Baidoa had a very strong emotional need to know that their letters were getting through. Failure to set up an efficient mail system at the beginning of OP *Solace* resulted in unnecessary anxiety and frustration during the first three weeks after deployment. Nothing much appeared to have changed since another 1 RAR Group had deployed at short notice to South Vietnam in 1965. On that occasion it took over three months for the mail system to operate effectively after several journalists and cartoonists publicised the problem in the Australian media. Some observers believed that the lack of an efficient mail system to and from South Vietnam in 1965 contributed to many marriage and relationship break downs. [98] There was evidence that this oversight did the same in 1993. [99]

Community communications with the AFS through 'Dear Digger' letters, messages in newspapers and on talk-back radio, as well as other letters and gifts helped maintain morale and reminded soldiers of home. Community organisations, such as the Returned and Services League, church groups, Rotary and the Lions Club, combined with hundreds of school children to send messages of encouragement and support to the AFS. Michael McKinnon and David Tyler made significant contributions from Baidoa through their columns in the *Townsville Bulletin*.

The Army's network of community service agencies, including the Welfare Co-ordination Cell set up for deployed forces and the 1 RAR Welfare Cell, gave the families of members of the AFS 24 hour service and did much to quell anxiety and to scotch rumours. These agencies also supported families with news, advice and information. The far north Queensland media were significant disseminators of news and promotors of community campaigns in support of the AFS. Newspapers also provided the means for two-way communications between those waiting in Australia and those on operations in Somalia. The public nature of these communications helped to maintain a profile for the AFS and reminded those waiting in Australia that they and their loved ones were appreciated by the Australian community and there were many others in the community maintaining a vigil for members of the AFS.

In March, family communications to and from Somalia took on a new dimension when AUTOVON telephone facilities were arranged in Baidoa. In hindsight, these facilities were timely. Many members of the 1 RAR Group had the mid-tour blues and the ability to communicate directly with their families by telephone boosted morale. Though difficult to prove, direct telephone communications may have pre-empted serious problems with morale in Baidoa and family welfare back in Australia. On the down side direct communications between Baidoa and Australia may have also facilitated immediate communications about family problems and frustrations being endured in Somalia that might have been down-played or not mentioned in letters.

The Spouses of Somalia Support Group was a well-intended gesture by the wives of officers, Warrant Officers and senior NCOs. The seven newsletters that were published and distributed connected those who were maintaining the vigil for members of the AFS. The SOS social functions also served to bring many spouses and partners together. The SOS Group provided opportunities and many families benefited from the efforts of a few volunteers. However, the bulk of the spouses and partners of junior NCOs and soldiers formed informal support groups or went to live with their parents, passing up the opportunities presented by the SOS Group.

For the families of members of the AFS, OP *Solace* did not end after their loved ones arrived home in May 1993. It took several weeks for sleep patterns to settle down. It was common for veterans to sit 'bolt upright' in bed re-living some scene in Somalia or not knowing where they were. A comforting word or a sharply delivered reminder soon brought the situation under control. Some spouses and partners reported that there husbands would leave their beds for no reason in the middle of the night or early morning.

Some Op Solace veterans changed forever and battled with mental illness, either immediately or soon after returning to Australia, or years later when further operational service or other stressful events triggered persistent mental illness that may have had its origins in Somalia.

There were rumours that, after experiencing overseas service, many veterans of the AFS would leave the Army. This exodus did not occur. Much of the stress and frustration of serving in Somalia subsided during leave. Many families decided to take the holidays away from Townsville. Time spent with loved ones or drinking with friends was the preferred way to relax and adjust from the dangers of Baidoa back to the routine of life in Australia

POSTSCRIPT

The French battalion that took over from the Australians in the Baidoa Sector focussed on maintaining their security at the Baidoa Airfield and escorting convoys, ceasing the occupation of NGO compounds and foot patrols in Baidoa. Though the French maintained a 24 hour radio watch linking all NGOs on one radio net, they 'often failed to answer'. Rather than continue the system of gun registration for all NGO personnel, the French 'ordered all except four guns per agency to be collected'. The NGOs objected and complained to UNOSOM headquarters. 'The seizure was overruled. The French handed over to a UNOSOM Indian battalion several weeks later—the Blue Helmets were back.'[1]

From the perspective of the distribution of humanitarian aid, OP *Restore Hope* had achieved its mission. UNITAF units met little resistance and virtually stopped the overt looting and extortion of humanitarian aid. Covertly, the redirection of humanitarian relief supplies into the black market economy of southern Somalia continued unabated. Ambassador David Shinn, Director of East African Affairs, US Department of State in an interview with Robert Patman, a Australian academic, on 24 January 1994, estimated that 250 000 lives had been saved during the period of OP *Restore Hope*. While creating the secure environment that saved these lives, UNITAF Army and Navy engineering units, supported by the civilian contractors, Brown and Root, repaired more than 1 800 km of roads, restored two airfields and established 14 major water wells and numerous minor wells.

Within a few weeks of the departure of UNITAF combat units in May 1993, the situation facing UNOSOM II grew more dangerous. The US had agreed to leave a 1 200-man Quick Reaction Force from the 10th Mountain Division and another 3 500 troops, including Special Force Ranger units, in southern Somalia to back up the 18 000 UNOSOM II troops. In the first week of June 1993 fighting broke out in Mogadishu after four Pakistani soldiers were taken hostage and had to be rescued by Italian troops.

On 5 June several truckloads of Pakistani soldiers were deployed to inspect stockpiles of General Aideed's weapons in southern Mogadishu. Aideed struck decisively when Somali machine gunners opened fire on the trucks killing 24 Pakistani soldiers and wounding 59 others. There were several subsequent interventions by Italian troops mounted in armoured personnel carriers and supported by tanks to disperse crowds that surrounded the ambush site. The UN Security Council passed Resolution 813 several days later calling for the arrest of General Aideed and those responsible for the killing of the Pakistanis.[2]

In an editorial on 7 June 1993, *The Sydney Morning Herald* commented:

> Somalia was to be something of a testing ground for humanitarian intervention in the post-Cold War world—an 'unprecedented enterprise aimed at nothing less than the restoration of an entire country', as Dr Boutros-Ghali put it. But instead of that, Somalia is shaping up to be a killing field for UN peacekeepers.

On 17 June US Cobra attack helicopters and AC-130 Hercules gunships, capable of firing 105 mm artillery shells with great accuracy, conducted air assaults against General Aideed's weapons storehouses, his radio station and a hospital where he and 150 armed supporters were reported to be occupying. While fire support was provided from the air, French, Italian and Pakistani troops assaulted what turned out to be abandoned hospital buildings. Several days later, after suffering five killed and several other wounded, the UN forces withdrew from Aideed's cantonment without apprehending him or his cohorts.

In the following days doubts were expressed in the media about what HQ UNOSOM would do if Aideed was captured. It would be difficult to convene a Somali court. Proposals for an international tribunal were criticised because a trial would make Aideed a hero in Somalia for once again standing up to the intervention of foreigners in Somali affairs. Pakistan's ambassador to the UN, Mr Jamsheed Marker, ruefully commented, ' I truly do not know what to do with Aideed. Maybe we should exile him to Nigeria so he would share the same room with Siad Barre [the Somali dictator ousted by General Aideed in 1991].'[3]

By early July 1993, five weeks after the last UNITAF units had left, 35 UN troops had been killed and over 130 injured in clashes with Somali gunmen and during violent demonstrations against the presence of foreign troops in Somalia. In a two hour confrontation on 2 July, over 800 Italian soldiers had been surrounded by a crowd of Somalis during a large-scale weapons search. In an ensuing fightfight with Somali gunmen, three Italians were killed and over 30 wounded or injured. Two weeks later on 14 July Italy's Defence Minister, Mr Fabbio Fabbri, commented, 'We must reconsider the overall objective of the Somalia mission. We suggest a cooling-off period to reduce tension, the suspension of combat operations and a renewed effort to renew dialogue.'[4]

Several weeks later the Commanding General of the 2 600-strong Italian contingent serving with UNOSOM, General Bruno Loi, who had never really co-operated with the US-led Quick Reaction Force, resigned and left Mogadishu. He was highly critical of Admiral Jonathan Howe, the UN Special Envoy to Somalia, for his policy of keeping pressure up on General Aideed and his forces with threats of arrest and trials as well as the conduct of weapon searches. There was also discontent among other participating nations in UNOSOM II about US forces using Cobra helicopters and AC-130 aircraft to fire into suburban areas and heavy-handed raids by US Ranger units. In the latter months of 1993 armed clashes continued sporadically amidst efforts by UN diplomats to bring the clan leaders together to form a viable government. In August, three months after UNITAF had left Somalia, 26 NGOs sent a joint letter to Dr Boutros-Ghali stating that the UNOSOM tactics were hindering the distribution of humanitarian aid.

The end of the US involvement in Somalia was humiliating. On 3 October, 1993, 120 Delta Force Commandos and Army Rangers were dropped into the heart of Mogadishu. Their mission was to capture Mohammed Farrah Aideed. The mission failed and two Blackhawk helicopters were shot down and 18 American servicemen were killed in combat with Aideed's gunmen. Several American bodies were then dragged through the streets, and humiliated by crowds of Somalis. These images of the remains young American servicemen being mocked and mutilated were broadcast over the national television networks in the US.

President Clinton and his advisors assessed that the American people had had enough. He announced that the US would be out of Somalia in six months and his administration was reviewing the criteria for supporting UN military operations in the future.

In November 1993 the UN Security Council suspended any arrest actions for Aideed and anyone involved in the killings of the 24 Pakistani soldiers in June. Aideed was on the comeback trail. Two weeks later the UN flew Aideed to Addis Ababa for informal talks with representatives from Ali Mahdi's faction. These talks resulted in a deadlock. Subsequently, fighting between Aideed's and Ali Mahdi's militia escalated significantly. Concurrently, there was an outbreak of bandit activity in southern Somalia. UNOSOM troops in the Humanitarian Relief Sectors vacated by UNITAF units six months before in May withdrew into fortified camps and watched as southern Somalia returned to the Dark Ages again. Several UN agencies and NGOs stopped humanitarian operations and withdrew expatriate personnel in face of the return of bandits, extortionists and marauding militia units.

On Friday 6 May 1994 the US Government announced a new policy on the participation of its armed forces in UN operations. The prerequisites for participation included, American command of American troops, the need to advance American interests, the availability of personnel and funds, the need for clear objectives and an exit strategy, and whether US participation was a prerequisite for the operation's viability. The US Ambassador to the UN, Mrs Madeleine Albright, said the policy was designed to ensure that, 'we refrain from asking the UN to undertake missions it is not equipped to do and to help the UN succeed in missions we would like it to do. ... The UN has not demonstrated the ability to respond effectively when the risk of combat is high and the level of cooperation is low.'[5]

On 26 August 1994 the UN Security Council decided to reduce the numbers of UN troops in Somalia by 1 500 and foreshadowed a complete withdrawal of all international troops supporting UNOSOM II in March 1995 if no substantial progress was made towards a political settlement. These announcements were made four days after General Aideed's militiamen had killed seven Indian peacekeepers and wounded nine others in an ambush on the road between Baledogle and Baidoa. Anti-aircraft guns, mortars and heavy machine guns mounted on 'technicals' had been used in the ambush. A Swedish nurse, Ms Lena Thelander, who had just been released after being held as a hostage for two weeks by Aideed's forces in southern Somalia commented, 'I regret the situation appears deeply hopeless. There is a grave distrust of each of the political groups, a distrust of the clans, an uncertainty over whether the long wounds of war can be healed.'[6]

Underscoring this pessimistic assessment and the deterioration of the situation in the Baidoa Sector , three Indian military doctors were killed during an attack on their field hospital at the Baidoa Airfield in the first week of September. Presumably the hospital had been set up on the concrete floor of the aircraft hanger where the BSG had been located. By the end of September 1994 the US Government had withdrawn all of its diplomatic and military staff from Mogadishu, taking with them a significant amount of the military hardware that had been loaned by UNITAF to UNOSOM.

On Sunday 23 January 1995, the Royal Canadian Regimental Airborne Combat Team was disbanded after 27 years of service. The Canadian Government had overruled advice from its Defence Force to maintain the Regiment in order to protect Canada's reputation as a responsible participant in peacekeeping operations. The Prime Minister, Mr Chretien, ordered the disbandment as a consequence of incidents that occurred during the Regiment's operations in Somalia in March 1993 and subsequent revelations about the Regiment's culture of violent racism and humiliating induction training methods.[7]

A multi-national operation commanded by US Marine, Lieutenant General Anthony Zinni, who had been the senior operations officer for OP *Restore Hope* in 1993, was mounted in February 1995 to withdraw UNOSOM II troops and UN civilian staff from Mogadishu. The operation, called OP *United Shield*, was also mounted to retrieve 30 M-1 tanks, 80 armoured personnel carriers, 13 military helicopters and a vast quantity of vehicles, weapons, equipment and stores that the US had loaned to the UN for use in Somalia. OP *United Shield* was conducted amidst fierce fighting between rival factions within General Aideed's Somali National Alliance, jostling to capture the Mogadishu Airport and adjacent port facilities.

On 28 February over 2 500 US and Italian Marines were offshore in 20 warships and troopships sent by the US, Britain, France, Italy, Pakistan and Malaysia to cover the final withdrawal of UNOSOM troops and materiel. Heavy machine gun and anti-aircraft gun fire from Aideed's and Mahdi's fighting factions 'raked across the airport, hitting a jet chartered to take the remaining 2 400 Pakistani and Bangladeshi troops out of Mogadishu.

Fighting spread into the town along the access road to the airport in the late afternoon, trapping the dozen journalists on the floor of the Sahafi Hotel for two hours while rival clans attacked one another with heavy weapons, shattering many of the buildings windows.'[8] The warlords were unable to control rival factions so General Zinni ordered Marines to secure the airfield and port.

At midnight on 28 February 1995 (local time) the hovercrafts of the US Marines hit the beaches of Mogadishu again. A rearguard of 500 Marines moved from their beachhead and secured the airfield and port. On 3 March the last 1 500 Pakistani soldiers filed onto a chartered troopship. As soon as the ship pulled away from the wharf, the last Marines moved back to their beachhead under the cover of three Cobra attack helicopters, an AC 130 Spectre gunship and several armoured personnel carriers, poised to fire over the heads of any Somalis who looked as if they were moving in too close to the beachhead. The day before Marine snipers had killed two Somalis who had ignored warnings broadcast from helicopters, 'to leave the airport alongside the Marines departure point or 'risk being shot dead.'[9] The last Marines were gone by sundown on 3 March 1995.

In Mogadishu in March 1995 as the last UNOSOM troops left, '... welding shops were doing a brisk trade turning four-wheel drive vehicles into battle wagons mounted with machine guns.' A local garage owner, Mr Hassan Abdi Ahmed commented, 'The UN is leaving and everyone says it has failed here. The people all want an end to the anarchy and war, but the faction leaders do not. Therefore it is not so much that the UN's operations failed, but that the Somalis failed.'[10]

The UN co-ordinator of humanitarian assistance to Somalia, Mr Erling Dessau, confirmed that nine UN agencies and 38 NGOs would continue to operate in the north-west and north-east of the country. He hoped that one day humanitarian operations would resume in the south. However, for the time being, 'after years of warnings and squandering chances, the Somalis have been left to sort out their own problems.'[11]

The blood price since the departure of UNITAF units in May 1993 had been high. The community of nations who had intervened in the Somali tragedy under the UN flag had lost 132 peacekeepers, including Americans not under command of the UN, killed, and over 400 hundred personnel wounded or injured. The UN experience in Somalia proved to be a significant deterrent to international support for military intervention into Central Africa in November 1996, and may prove to be so for some time to come.

Honours and awards for Operation Solace

Distinguished Service Cross

Colonel W.J.A. Mellor AM ADC
For distinguished command and leadership as Commander,
Australian Force-Somalia

Lieutenant Colonel D.J. Hurley
For distinguished command and leadership as
Commanding Officer, 1st Battalion,
The Royal Australian Regiment Group in Somalia

Legion of Merit (US)

Colonel W.J.A. Mellor AM
For exceptionally meritorious conduct in the performance
of outstanding services

Distinguished Service Medal

Corporal T.A. Aitken
For distinguished leadership as a section commander
in C Company, 1st Battalion, The Royal Australian Regiment Group
in Baidoa, Somalia, during Operation *Solace*

COMMENDATION FOR DISTINGUISHED SERVICE

Major D.G. McKaskill
For distinguished performance of duties as
Officer Commanding, B Squadron, 3rd/4th Cavalry Regiment
in Baidoa, Somalia, during Operation *Solace*

Major M.J. Moon
For distinguished leadership as
Officer Commanding, C Company, 1st Battalion,
The Royal Australian Regiment Group
in Baidoa, Somalia, during Operation *Solace*

Major R.H. Stanhope
For distinguished performance of duties as Officer Commanding the Civil-
Military Operations Team
in Baidoa, Somalia during Operation *Solace*

Captain S.J. Dodds
For distinguished performance of duties as Officer Commanding Mortar
Platoon, 1st Battalion,
The Royal Australian Regiment Group
in Baidoa, Somalia during Operation *Solace*

Corporal P.J. Martin

For distinguished leadership as a section commander in D Company, 1st
Battalion, The Royal Australian Regiment Group
in Baidoa, Somalia during Operation *Solace*

Private C.J. Day
For distinguished performance of duties as a patrol signaller
in Baidoa, Somalia, during Operation *Solace*

Chief of the Defence Force Commendation

Major J.G. Caligari

Lieutenant C.J. McDonald

Sergeant P.H. Von Kurtz

Sergeant P. Watson

Chief of the General Staff Commendation

Lieutenant Colonel G.T. Woolnough

Major M.J. Kelly

Lieutenant W.R. Bowyer

Warrant Officer Class One W.F. Bowser

Warrant Officer Class One J.D. Collins

Warrant Officer Class Two M.E. Robinson

Sergeant D.B. Callaghan

Sergeant G.W. Wilkes

1 RAR Group Nominal Roll

990 names have been identified as serving in Somalia during Operation *Solace* between Jan-May 1993 This includes members of 1 RAR, attachments from 2/4 RAR, 103 Signals Squadron, 107 Field Battery, 17 Field Troop (including 3 dogs), B Squadron 3/4 Cavalry Regiment, BSG, HQ AFS , Defence Public Relations and Visitors. Not included are members deployed with the UNOSOM 1 or UNOSOM 2 missions. The visitors group includes the Operations Study Team, the psychological debriefing team, legal officers and others sent over for short term specific tasks.

This nominal roll was compiled over a period of 18 months from various sources including Army Records, personal diaries and various data bases. It has been thoroughly cross-checked and while some names may have been omitted or are in error the number should be minimal as all known sources have been consulted. One area of possible discrepancy is the identification of those attachments from 2/4 RAR. For those from 2/4 RAR not acknowledged or those from 1 RAR wrongly attributed to 2/4 RAR the error is regretted.

The placement of personnel into sub-units rather than an alphabetical listing of those who served was felt to be more personal and should be more useful in the years to come. Thanks to all of those who reviewed the rolls and helped fill in the gaps. Special thanks goes to Ms Debbie Robinson at the Directorate of Infantry who collated and set up the data base.

Ranks shown are those worn on departure from Somalia. Those personnel who served in more than one sub-unit are generally listed where they served the most time with a note detailing the other sub-unit. In most cases the members employment in Somalia is also indicated. Any errors are regretted.

Unit

1 RAR	1st Battalion, The Royal Australian Regiment
HQ AFS	Headquarters Australian Forces Somalia
103 SIG SQN	103 Signals Squadron
107 BTY - CMOT	107 Field Battery, 4 Field Regiment, Royal Australian Artillery, Civil and Military Operation Team
17 FD TP 3 CER	17 Field Troop 3 Combat Engineer Regiment
BSQN 3/4 CAV REGT	B Squadron 3/4 Cavalry Regiment
BSG	Battalion Support Group

Sub-unit

ADMIN	Administration
BHQ	Battalion Headquarters
CHQ	Company Headquarters
COY	Company
DET	Detachment
DFSW	Direct Fire Support Weapons
ENGR	Engineers
FRG	Forward Repair Group
HQ	Headquarters
INT	Intelligence
OST	Operational Study Team
PL	Platoon
QM	Quartermaster
RECON	Reconnaissance
RPs	Regimental Police
SECT	Section
SHQ	Squadron Headquarters
SPT	Support
SST	Specialist Support Team
SUP	Supply
TP	Troop
TPT	Transport

Rank

MAJGEN	Major General
RADM	Rear Admiral
BRIG	Brigadier
COL	Colonel
LTCOL	Lieutenant Colonel
CMDR	Commander
MAJ	Major

MAJ(T)	Defence Civilian made Temporary Major for duration of deployment
CAPT	Captain
LEUT	Lieutenant (Navy)
SREP	Salvation Army Representative
LT	Lieutenant
WO1	Warrant Officer Class 1
WO2	Warrant Officer Class 2
SSGT	Staff Sergeant
SGT	Sergeant
CPL	Corporal
BDR	Bombardier
LCPL	Lance Corporal
LBDR	Lance Bombardier
PTE	Private
TPR	Trooper
GNR	Gunner
SPR	Sapper
SIG	Signaller
CFN	Craftsman

Employment

1 SIG REG	1st Signals Regiment (attached for redeployment of AFS back to Australia)
8 MU	8th Movement Unit (attached for redeployment of AFS back to Australia)
AADJT	Assistant Adjutant
ADJT	Adjutant
AMMO TECH	Ammunition Technician
ARMY HIST SECT	Army Historical Section
ASM	Artificer Sergeant Major
ATO	Ammunition Technical Officer
BC	Battery Commander
BN 2IC	Battalion Second in Command
CAFS	Command Australian Forces Somalia
CBT HIST	Combat Historian
CCLK	Chief Clerk
CLK	Clerk
CO	Commanding Officer
COFS	Chief of Staff
COL OPS LHQ	Colonel Operations LHQ
COMD 3 BDE	Commander 3rd Brigade

COMD 1 DIV	Commander 1st Division
COMMS	Communications
COY 2IC	Company Second in Command
CP CPL	Command Post Corporal
CPL ARMT	Corporal Armourment
CQMS	Company Quartermaster Sergeant
CSM	Company Sergeant Major
DVR	Driver
EDE	Engineering Design Establishment
EOD	Explosive Ordnance Disposal
FIN	Finance
FIT ARMT	Fitter Armourment
FIT & TUR	Fitter and Turner
GLO	Ground Liaison Officer (attached for redeployment of AFS back to Australia)
HQ ADF	Headquarters Australian Defence Force
HQ LOG COMD	Headquarters Logistic Command (attached for redeployment of AFS back to Australia)
IO	Intelligence Officer
LEGAL	Legal Officers (attached to conduct Court Martial before return to Australia)
LESSONS	Operational Study Team (attached to research and produce an operational analysis of Operation SOLACE)
LHQ	Land Headquarters
LO	Liaison Officer
LPO	Local Purchasing Officer
MED	Medical
MEDIC	Medical Assistant
MFC	Mobile Fire Controller
MLO	Mortar Line Officer
MO	Medical Officer
MOVT/R&F	Movements/Records and Files
MP	Military Police
OC	Officer Commanding
OPS	Operations
OPS SGT	Operations Sergeant
OPSO	Operations Officer
OR CPL	Orderley Room Corporal
PERS SGT	Personnel Sergeant
PERS/LOG	Personnel/Logistics
PL SGT	Platoon Sergeant
PL COMD	Platoon Commander
PR	Public Relations

PSYCH	Psychological Debriefing Team
PTL MASTER	Patrol Master
PUR OFFR	Purchasing Officer
QM	Quartermaster
RAD ELEC	Radio Electronics
RAP	Regimental Aid Post
REC CLK	Records Clerk
RECOV MECH	Recovery Mechanic
RFN	Rifleman
RMO	Regimental Medical Officer
RQMS	Regimental Quartermaster Sergeant
RSM	Regimental Sergeant Major
RSO	Regimental Signals Officer
SALLY MAN	Salvation Army Red Shield Services Representative
SECT 2IC	Section Second in Command
SECT COMD	Section Commander
SIG	Signaller
SO2	Staff Officer Grade 2
SO3 AIR	Staff Officer Grade 3 Air
SO3	Staff Officer Grade 3
SQMS	Squadron Quartermaster Sergeant
STMN	Storeman
STWD	Steward
TECH	Technical
TECH ELEC SYS	Technical Electrical Systems
TOCO	Transport Officer
TP LDR	Troop Leader
TP SGT	Troop Sergeant
TP COMD	Troop Commander
TPT SPVR	Transport Supervisor
UPR	Unit Pay Representative
VEH MECH	Vehicle Mechanic

Other

(BAND)	From Band Platoon
(2/4)	From 2/4 RAR
RTA	Returned to Australia
OJT	On Job Training

1 RAR

Rank		Name	Position

BHQ

LTCOL	D.J.	Hurley	CO
MAJ	J.G.	Caligari	OPSO
MAJ	G.A.	Crosland	BN 2IC
MAJ	M.C.	Fairleigh	BN 2IC
MAJ	M.J.	Fleming	Chaplin
CAPT	G.P.	Babington	ADJT
WO1	G.C.	Chamberlain	RSM
WO1	J.R.	Sales	RSM
WO2	M.K.	Judzewitsch	Chief Clerk
SGT	M.D.	Johnson	Ops SGT
SGT	J.A.	Lougheed	Pay SGT
SGT	A.T.	Mccarthy	Pers SGT
CPL	T.A.	Britton	REC CLK
CPL	G.J.	FitzAllen	OR CPL
CPL	A.	McKellar	UPR
LCPL	D.J.	Miller	Ops Clk OA
PTE	M.J.	Dale	UPR
PTE	W.J.	Grey	CO Batman
PTE	M.J.	Searle	UPR
PTE	R.J.	Voss	Movt/R&F

INT SECT

CAPT	J.R.	Burns	IO
SGT	D.A.	McKay	INT SGT
CPL	J.L.	Griffiths	INT
LCPL	A.D.	Craig	INT
LCPL	S.A.	Taylor	INT
PTE	J.C.	Attrill	INT
PTE	M.S.	Handley	DVR
PTE	P.J.	Macmillan	INT

RPs

SGT	P.R.	Dunton	RP SGT
CPL	D.C.	Buddle	RP CPL
LCPL	L.M.	Meulengraaf	RP
PTE	B.R.	Boreham	RP
PTE	S.G.	Dingley	RP
PTE	K.A.	Noon	RP
PTE	M.J.	Scanlon	RP

A Company

CHQ

MAJ	D.J.	Fraser	OC
CAPT	M.H.	Cooper	COY 2IC
CAPT	P.J.	Short	COY 2IC
WO2	K.R.	Symmons	CSM
SGT	P.W.	Galvin	CQMS
CPL	P.J.	Brown	COY CLK
PTE	T.J.	Buswell	STMN
PTE	D.B.	Hoffman	RFN

1PL

LT	R.J.	Adams	PL COMD
LT	G.M.	Gray	PL COMD/AADJT
SGT	D.A.	Wilson	PL SGT
CPL	A.M.	Dowling	SECT COMD
CPL	L.R.	Eason	SECT COMD
CPL	P.D.	Slattery	SECT COMD
LCPL	M.D.	Fisher	SECT 2IC
LCPL	T.A.	Hocking	SECT 2IC
LCPL	C.V.	Stevens	SECT 2IC
PTE	S.I.	Baldwin	RFN
PTE	P.A.	Breen	RFN
PTE	R.L.	Conway	RFN
PTE	T.J.	Cook	RFN
PTE	M.A.	Eversden	RFN
PTE	R.J.	Fletcher	RFN
PTE	S.M.	Folie	RFN
PTE	B.R.	Hargraves	RFN
PTE	W.J.	Hennessey	RFN
PTE	S.J.	Hey	RFN (2/4)
PTE	P.	Krasevskis	RFN
PTE	J.C.	Locke	RFN
PTE	T.A.	Lockyer	RFN
PTE	T.J.	Paton	RFN
PTE	S.A.	Squires	RFN
PTE	S.J.	Starks	RFN
PTE	C.E.	Stevens	RFN
PTE	P.J.	Thompson	RFN
PTE	T.F.	Vaiciulevicius	RFN (BAND)
PTE	S.A.	Walker	RFN

2PL

LT	R.J.	Worswick	PL COMD

Rank	*Name*		*Position*
SGT	G.S.	Burns	PL SGT
CPL	T.R.	Conner	SECT COMD
CPL	D.W.	Jobson	SECT COMD
CPL	L.J.	Moessinger	RFN (BAND)
CPL	S.	Moore	SECT COMD
LCPL	D.L.	Braban	SECT 2IC
LCPL	S.G.	Lavin	SECT 2IC
LCPL	P.J.	Madden	RFN (BAND)
LCPL	T.S.	Martin	RFN (BAND)
LCPL	P.G.	Smith	SECT 2IC
PTE	T.W.	Batge	RFN
PTE	G.M.	Brown	RFN
PTE	L.M.	Carlill	RFN
PTE	A.S.	Coll	RFN
PTE	S.T.	Donald	RFN
PTE	J.G.	Fooks	RFN
PTE	D.T.	Gollop	RFN (2/4)
PTE	A.R.	Harrison	RFN
PTE	D.E.	Hickey	RFN
PTE	M.J.	Howarth	RFN
PTE	S.J.	Hughes	RFN
PTE	J.J.	Lidgerwood	RFN
PTE	O.	Namata	RFN
PTE	P.R.	Robins	RFN
PTE	K.S.	Smith	RFN
PTE	D.G.	Stainer	RFN (BAND)
PTE	I.W.	Turner	RFN
PTE	K.G.	Went	RFN

3PL

Rank	*Name*		*Position*
LT	G.W.	Butler	PL COMD
SGT	P.H.	Von Kurtz	PL SGT
CPL	C.A.	Cunningham	SECT COMD
CPL	S.E.	Hosking	SECT COMD
CPL	C.J.	Townson	SECT COMD
LCPL	S.M.	Duggan	SECT 2IC
LCPL	P.D.	Johnson	SECT 2IC
LCPL	P.J.	Simeon	SECT 2IC
PTE	A.L.	Boyd	RFN
PTE	J.W.	Clelland	RFN
PTE	A.L.	Davies	RFN
PTE	B.A.	Douglas	RFN
PTE	Q.R.	Gainer	RFN
PTE	D.J.	Huckerby	RFN
PTE	T.W.	Jeffree	RFN
PTE	A.J.	Milne	RFN
PTE	S.A.	Moore	RFN
PTE	K.C.	Murrells	RFN (2/4)
PTE	M.J.	Norris	RFN
PTE	S.J.	Orme	RFN
PTE	M.J.	Presbury	RFN
PTE	R.B.	Ralphs	RFN
PTE	D.R.	Randall	RFN (2/4)
PTE	S.A.	Robinson	RFN
PTE	L.C.	Smith	RFN
PTE	J.L.	Smith	RFN
PTE	M.C.	Townson	RFN
PTE	D.J.	Whereat	RFN
PTE	H.F.	Wicker	RFN

B Company

CHQ

Rank	*Name*		*Position*
MAJ	J.D.	Simpson	OC
CAPT	T.P.	Dewhurst	COY 2IC
WO2	N.A.	Clarke	CSM
SSGT	D.J.	Mclean	CQMS
CPL	I.R.	Maud	COY CLK
PTE	D.N.	Baxter	RFN
PTE	J.H.	Kelly	RFN
PTE	B.C.	Little	RFN

4 PL

Rank	*Name*		*Position*
LT	V.C.	Creagh	PL COMD
SGT	W.A.	Bruce	PL SGT
CPL	K.S.	Alloway	SECT COMD
CPL	S.J.	Francis	SECT COMD
CPL	W.B.	Prosser	SECT COMD
LCPL	G.M.	Corbett	SECT 2IC
LCPL	R.W.	Gasson	SECT 2IC
LCPL	D.A.	Webb	SECT 2IC
PTE	S.A.	Bradshaw	RFN (2/4)
PTE	N.A.	Brook	RFN
PTE	D.A.	Buckton	RFN (2/4)
PTE	R.E.	Clements	RFN
PTE	B.T.	Connolly	RFN
PTE	J.K.	Costello	RFN
PTE	R.	Djokic	RFN
PTE	S.L.	Dotter	RFN
PTE	A.S.	Eeles	RFN
PTE	N.J.	Elms	RFN (2/4)
PTE	M.B.	Greene	RFN
PTE	S.C.	Griese	RFN (2/4)
PTE	D.J.	Hawkins	RFN (2/4)
PTE	D.J.	Jarrett	RFN
PTE	G.H.	Kesby	RFN
PTE	A.O.	Mace	RFN

Rank		Name	Position	Rank		Name	Position
PTE	B.D.	Miller	RFN	PTE	B.K.	Carroll	RFN
PTE	D.T.	O'Shea	RFN	PTE	T.J.	Clarke	RFN
PTE	R.T.	Outtrim	RFN	PTE	J.L.	Cox	RFN (2/4)
PTE	M.E.	Rowlingson	RFN	PTE	R.T.	Grant	RFN
PTE	B.A.	Wason	RFN	PTE	L.A.	Harris	RFN
				PTE	P.M.	Hedge	RFN

5 PL

				PTE	R.J.	Hemmings	RFN
				PTE	D.A.	Hindle	RFN
LT	J.A.	McTavish	PL COMD	PTE	R.A.	Holmes	RFN
SGT	P.A.	Anderson	PL SGT	PTE	A.S.	Keating	RFN
CPL	S.S.	Armstrong	SECT COMD	PTE	D.A.	McGrath	RFN
CPL	G.C.E.	Jones	SECT COMD	PTE	K.J.	Perry	RFN (2/4)
CPL	A.R.	Snell	SECT COMD	PTE	S.W.	Pollock	RFN (2/4)
LCPL	B.J.	Conroy	SECT 2IC	PTE	C.S.	Rhodes	RFN (2/4)
LCPL	S.K.	Harding	SECT 2IC	PTE	D.R.	Richards	RFN
LCPL	R.A.	Tufekcic	SECT 2IC	PTE	C.A.	Richardson	RFN (2/4)
PTE	D.C.	Blair	RFN	PTE	M.J.	Rosier	RFN (2/4)
PTE	A.C.	Cooper	PL SIG	PTE	S.M.	Staton	RFN
PTE	S.M.	Harris	RFN	PTE	G.M.	Whitehead	RFN
PTE	T.W.	Ingleby	RFN				

C COMPANY

PTE	B.A.	Kipping	RFN (2/4)				
PTE	S.A.	Kreusler	RFN				

CHQ

PTE	V.D.	Law	RFN				
PTE	M.P.	Lewis	RFN				
PTE	B.G.	Lumby	RFN (2/4)	MAJ	M.J.	Moon	COY COMD
PTE	T.A.	McGill	RFN (2/4)	CAPT	M.D.	Hankinson	COY 2IC
PTE	A.J.	McGuinness	RFN (2/4)	WO2	P.J.	Angus	CSM
PTE	J.I.	Minton	RFN	SSGT	L.J.	Jackson	CQMS
PTE	M.P.	Nicholson	RFN	CPL	D.M.	Heaslip	COY CLK
PTE	A.R.	Parker	RFN (2/4)	PTE	B.R.	Dickinson	STMN (2/4)
PTE	M.J.	Quinn	RFN	PTE	B.R.	Pieschel	RFN
PTE	M.J.	Rowe	RFN	PTE	B.	Piper	RFN
PTE	T.A.	Simmonds	RFN	PTE	P.S.	Smith	STMN
PTE	G.J.	Stone	RFN				

7 PL

PTE	P.R.	Tofield	RFN				
PTE	S.L.	Wicks	RFN	LT	A.C.	Swinsburg	PL COMD
PTE	M.K.	Williamson	RFN	SGT	J.	Johnson	PL SGT
				CPL	G.J.	Baker	SECT COMD

6 PL

				CPL	M.F.	Braeckmans	SECT COMD
				CPL	R.W.	Chapman	SECT COMD
LT	A.P.	Pritchard	PL COMD	LCPL	J.B.	Abbot	SECT 2IC
SGT	M.L.	Bobbin	PL SGT	LCPL	M.W.	Edwards	SECT 2IC
CPL	G.E.	Holston	SECT COMD	LCPL	A.J.	Usher	SECT 2IC
CPL	S.A.	Kleingeld	SECT COMD	PTE	M.J.	Allen	RFN
CPL	K.M.	Lewis	SECT COMD	PTE	W.R.	Blake	RFN
LCPL	J.S.	Ranson	SECT 2IC	PTE	G.M.	Brown	RFN (2/4)
LCPL	J.P.	Reeves	SECT 2IC	PTE	G.A.	Fletcher	RFN
LCPL	R.P.	Robinson	SECT 2IC	PTE	S.I.	Gale	RFN
PTE	J.	Agg	RFN	PTE	A.K.	Glebow	RFN
PTE	A.C.	Bilston	RFN (2/4)				

Rank		Name	Position
PTE	D.C.	Hogan	RFN
PTE	G.W.	Hopgood	RFN
PTE	P.E.	Laws	RFN (2/4)
PTE	J.C.	McDonald	RFN
PTE	S.D.	Milligan	RFN (2/4)
PTE	S.A.	Norman	RFN
PTE	S.L.	O'Meara	RFN (2/4)
PTE	J.L.	Slattery	RFN (2/4)
PTE	M.H.	Sloman	RFN
PTE	M.S.	Smith	RFN
PTE	P.J.	Turnbull	RFN
PTE	A.N.	Williams	PL SIG
PTE	D.T.	Williams	RFN
PTE	G.S.	Woodham	RFN
PTE	S.A.	Woolley	RFN

8PL

Rank		Name	Position
LT	I.A.	MacGregor	PL COMD
SGT	P.N.	Gibbs	PL SGT
CPL	T.A.	Aitken	SECT COMD
CPL	M.R.	McCulloch	SECT COMD
CPL	W.A	Perkins	SECT COMD
LCPL	J.R.	Conway	SECT 2IC
LCPL	R.R.	Free	SECT 2IC
LCPL	G.	Lively	SECT 2IC
PTE	P.T.	Abdul-Rahman	RFN
PTE	D.A.	Ball	RFN (2/4)
PTE	M.C.	Beresford	RFN
PTE	J.	Blakeman	RFN
PTE	A.	Campbell	RFN (2/4)
PTE	C.J.	Day	RFN
PTE	S.J.	Eaton	RFN
PTE	K.W.	Farmer	RFN
PTE	J.K.	Flatley	RFN
PTE	J.A.	Foley	RFN
PTE	M.R.	Graichen	RFN
PTE	J.R.	Guevorts	RFN
PTE	R.A.	Heard	RFN
PTE	S.D.	Hewitt	RFN
PTE	S.J.	Jackson	RFN
PTE	S.C.	Lowry	RFN
PTE	M.J.	Meehan	RFN
PTE	S.M.	Nalder	RFN
PTE	J.L.	Rogash	RFN
PTE	A.J.	Shinner	RFN

9PL

Rank		Name	Position
LT	T.A.	Everett	PL COMD

Rank		Name	Position
LT	G.E.	Steer	PL COMD
			(from RECON PL)
SGT	M.T.	Morrissey	PL SGT
CPL	D.A.	Caple	SECT COMD
CPL	A.G.	Hodges	SECT COMD
CPL	S.P.	Robinson	SECT COMD
LCPL	A.T.	Combs	SECT 2IC
LCPL	M.G.	Salom	SECT 2IC
LCPL	A.	Tessieri	SECT 2IC
PTE	S.P.	Baker	RFN
PTE	S.A.	Besgrove	RFN
PTE	A.J.	Bishop	RFN
PTE	S.T.	Cooper	RFN
PTE	J.D.	Donald	RFN (2/4)
PTE	M.B.	Dyer	RFN
PTE	A.C.	Everson	RFN (2/4)
PTE	P.J.	Ingram	RFN
PTE	P.R.	Lundberg	RFN
PTE	R.J.	McCormick	RFN (2/4)
PTE	C.A.	Miller	RFN
PTE	D.P.	Moore	RFN (2/4)
PTE	D.J.	O'Neill	RFN
PTE	M.W.	Parsons	RFN (2/4)
PTE	J.C.	Payne	RFN
PTE	J.M.	Roberts	RFN
PTE	D.	Ryan	RFN
PTE	D.C.	Searle	RFN
PTE	A.D.	Wilson	RFN

D Company

CHQ

Rank		Name	Position
MAJ	A.D.	Blumer	OC
CAPT	M.	Lewis	COY 2IC
WO2	K.G.	Hudson	CSM
SSGT	G.	Lofthouse	CQMS
CPL	K.M.	Brown	COY CLK
PTE	M.J.	Cook	CLK/RFN (2/4)
PTE	C.R.	Elliott	RFN (2/4)
PTE	M.R.	Nancarrow	RFN (2/4)
			RTA Med Reasons
PTE	T.A.	Stephenson	RFN

10 PL

Rank		Name	Position
LT	J.M.	Haddock	PL COMD
SGT	N.M.	Gathercole	PL SGT
CPL	W.F.	Davis	SECT COMD
CPL	G.G.	Evans	SECT COMD

Rank		Name	Position
CPL	P.W.	Nunan	SECT COMD
LCPL	S.E.	Allen	SECT 2IC
LCPL	R.R.	Baker	SECT 2IC
LCPL	D.A.	Coleman	SECT 2IC
PTE	J.A.	Adams	RFN (2/4)
PTE	J.W.	Blunden	RFN
PTE	D.J.	Burnett	RFN
PTE	B.P.	Dalle-Nogare	RFN
PTE	S.J	Farrell	RFN
PTE	S.J.	Goodwin	RFN
PTE	D.J.	Harrison	RFN
PTE	A.W.	Hickson	RFN
PTE	H.A.	Hobbs	RFN
PTE	S.J.	Hollis	RFN (2/4)
PTE	R.	Hughes	RFN
PTE	S.D.	Innes	RFN (2/4)
PTE	M.D.	Newton	RFN
PTE	J.A.	Nuske	RFN
PTE	M.E.	O'Neill	RFN
PTE	S.J.	Sneddon	RFN
PTE	P.A.	Stephens	RFN (2/4)
PTE	D.J.	Thompson	RFN
PTE	W.A.	Wanigasekera	RFN
PTE	P.W.	Zernicke	RFN

11 PL

Rank		Name	Position
LT	D.G.	Tulley	PL COMD
SGT	J.E.	Garbellini	PL SGT
CPL	W.G.	Chawner	SECT COMD
CPL	P.J.	Martin	SECT COMD
CPL	E.J.	Mimi	SECT COMD
LCPL	S.J.	Chinner	SECT 2IC
LCPL	S.	McAliney	SECT 2IC
LCPL	B.D.	Murtagh	SECT 2IC
LCPL	A.K.	Overton	SECT 2IC
PTE	P.A.	A'Bell	RFN
PTE	R.T.	Dann	RFN
PTE	P.	Denby	RFN (2/4)
PTE	S.F.	Fanning	RFN
PTE	A.M.	Franklin	RFN
PTE	J.I.	Gerhardt	RFN
PTE	A.J	Kowarik-Berger	RFN
PTE	D.A.	Lamkin	RFN
PTE	M.A.	Lumby	RFN
PTE	C.B.	McGarrity	RFN
PTE	A.J.	McNamee	RFN (2/4)
PTE	S.J.	O'Loughlin	RFN
PTE	S.M.	Perry	RFN
PTE	J.L.	Putland	RFN

Rank		Name	Position
PTE	T.R.	Rial	RFN
PTE	C.S.	Russell	RFN
PTE	J.G.	Scamp	RFN
PTE	A.S.	Stagg	RFN
PTE	S.A.	Vassallo	RFN
PTE	J.L.	Watene	RFN (2/4)

12 PL

Rank		Name	Position
LT	J.M.	Van Der Klooster	PL COMD
SGT	R.J.	Elmy	PL SGT
CPL	A.B.	Parnaby	SECT COMD
CPL	A.J.	Pavlic	SECT COMD
CPL	G.T.	Wehmeier	SECT COMD
LCPL	D.A.	Solonec	SECT 2IC
LCPL	J.J.	Verschelden	SECT 2IC
LCPL	J.	Wells	SECT 2IC (from DFSW PL)
PTE	R.A.	Booth	RFN
PTE	S.G.	Bracknell	RFN
PTE	L.T.	Carson	RFN
PTE	P.A.	Dabrowski	RFN
PTE	J.R.	Hollingworth	RFN (2/4)
PTE	G.J.	James	RFN
PTE	K.J.	Kennedy	RFN
PTE	C.D.	Keyte	RFN
PTE	K.	Kilpatrick	RFN
PTE	M.J.	Lange	RFN
PTE	W.J.	Lovis	RFN
PTE	T.T.	Navusolo	RFN
PTE	J.C.	Readdy	RFN (2/4)
PTE	W.O.	Redlich	RFN (2/4)
PTE	T.A.	Shields	RFN
PTE	B.P.	Szabo	RFN
PTE	J.D.	Walliker	RFN
PTE	J.E.	Watson	RFN
PTE	J.L.	Williams	RFN
PTE	C.J.	Willis	RFN (2/4)
PTE	A.J.	Wilson	RFN (2/4)

Support Coy

CHQ

Rank		Name	Position
MAJ	G.P.	Hurcum	COY COMD/ PTL MASTER
CAPT	D.J.	Campbell	COY 2IC/ WATCHKEEPER
CAPT	A.M.	Somerville	COY 2IC/ WATCH/SO3 AIR

Rank		Name	Position
WO2	R.J.	Klima	CSM
SSGT	T.A.	Damen	CQMS
CPL	K.E.	Snell	OPS CLK OA
PTE	J.P.	McLouchlin	SPT COY
			STMN

DFSW PL

Rank		Name	Position
LT	T.D.	Webb	PL COMD
SGT	M.A.	Thorp	PL SGT
CPL	C.W.	Hughes	SECT COMD
CPL	D.L.	Wright	SECT COMD
LCPL	P.J.	Daniels	SECT 2IC
LCPL	A.L.	Rule	SECT 2IC
PTE	G.A.	Batty	DFSW
PTE	S.G.	Doma	DFSW
PTE	S.R.	Hudson	DFSW
PTE	B.G.	Hughes	DFSW
PTE	S.P.	McKay	DFSW
PTE	B.P.	Paganoni	DFSW
PTE	G.S.	Seymour	DFSW
PTE	J.P.	Tauri	DFSW
PTE	A.J.	Wallace	DFSW (2/4)
PTE	P.	Weston	DFSW

Mortar PL

Rank		Name	Position
CAPT	S.J.	Dodds	PL COMD
LT	P.J.	Connolly	MLO
SGT	S.W.	Boye	MFC C COY
SGT	S.P.	Brown	MFC B COY
SGT	A.R.	Hollamby	MFC D COY
SGT	G.A.	Rodgers	PLSGT
CPL	I.A.	D'Arcy	CP CPL
CPL	B.D.	Field	MFC A COY
CPL	A.B.	Fisher	SECT COMD
CPL	I.D.	Johnston	SECT COMD
CPL	D.G.	Mitchell	SECT COMD
CPL	J.B.	Moore	SECT COMD
LCPL	T.J.	Bell	SECT 2IC
LCPL	R.J.	Chandler	SECT 2IC
LCPL	D.J.	Lowrey	SECT 2IC
LCPL	B.C.	Ragh	SECT 2IC
PTE	S.A.	Ansell	MOR
PTE	D.A.	Berrill	MOR
PTE	J.R.	Boddington	MOR
PTE	C.G.	Brock	MOR
PTE	M.J.	Brydon	MOR
PTE	R.J.	Burford	MOR
PTE	J.G.	Burton	MOR

Rank		Name	Position
PTE	A.D.	Christensen	MOR
PTE	J.A.	Collins	MOR
PTE	C.T.	Dugal	MOR
PTE	G.W.	Hunter	MOR
PTE	P.R.	Kingston	MOR
PTE	S.D.	Lenz	MOR
PTE	T.A.	Lines	MOR
PTE	T.C.	Logan	MOR
PTE	X.	Lopez	MOR
PTE	S.A.	McCaig	MOR
PTE	A.J.	Morrison	MOR
PTE	P.F.	Newson	MOR
PTE	J.A.	Pain	MOR (2/4)
PTE	D.A.	Pearson	MOR
PTE	M.W.	Reid	MOR
PTE	P.J.	Russell	MOR
PTE	A.M.	Smith	PL STMN
PTE	J.A.	Smith	MOR
PTE	B.L.	Stanley	MOR
PTE	S.E.	Vuetibau	MOR
PTE	D.N.	Woodrow	MOR

Assault Pioneer PL

Rank		Name	Position
LT	J.A.	Simeoni	PL COMD
SGT	S.A.	Colman	PL SGT
CPL	B.A.	Blackhurst	SECT COMD
CPL	C.	McSwan	SECT COMD
CPL	J.R.	Pickett	SECT COMD
LCPL	S.R.	Jenson	SECT 2IC
LCPL	D.A.	Petrie	SECT 2IC
LCPL	B.E.	Shearer	SECT 2IC
PTE	M.J.	Abbott	PNR
PTE	B.S.	Ayrton	PNR
PTE	B.P.	Barrett	PNR
PTE	K.B.	Carlin	PNR (2/4)
PTE	M.D.	Cooke	PNR (2/4)
PTE	J.R.	Davis	PNR
PTE	S.J.	Heath	PL SIG
PTE	J.B.	Hilbourne	PNR
PTE	A.T.	Jackson	PNR
PTE	M.	Jayet	PNR
PTE	A.F.	Malaspina	PNR STMN
PTE	S.M.	Malone	PNR
PTE	K.G.	McKay	PNR
PTE	D.B.	Moon	PNR
PTE	M.C.	Parker	PNR
PTE	E.Q.	Seruitukana	PNR
PTE	A.C.	Smith	PNR
PTE	R.J.	Somerton	PNR (2/4)

Rank		Name	Position
PTE	C.A.	Yates	PNR

Recon PL

Rank		Name	Position
LT	T.A.	Everett	PL COMD (from 9PL)
LT	G.E.	Steer	PL COMD
SGT	W.H.	Thomson	PL SGT
CPL	G.S.	Blackman	SECT COMD
CPL	M.	Retallick	SECT COMD
CPL	D.P.	Rule	SECT COMD
CPL	C.J.	Wainwright	SECT COMD
LCPL	C.M.	Andersen	RECON
LCPL	J.L.	Badman	RECON
LCPL	S.R.	Flynn	RECON
LCPL	D.C.	Wilson	RECON
PTE	J.G.	Brendecke	RECON
PTE	S.L.	Brown	RECON
PTE	B.F.	Diddams	RECON
PTE	D.P.	Goddard	RECON
PTE	J.C.	Herrington	RECON
PTE	A.R.	Plater	RECON
PTE	C.A.	Reeve	RECON
PTE	J.A.	Roderick	RECON
PTE	M.D.	Van Droffelaar	RECON

Signals PL

Rank		Name	Position
CAPT	J.L.	Bryant	RSO
SGT	G.L.	Pearsall	SIG SGT
SGT	A.D.	White	SIG SGT
CPL	D.A.	Ahmelman	SIG (B COY DET COMD)
CPL	A.	Brown	SIG
CPL	N.J.	Haddock	SIG
CPL	M.C.	Logue	SIG
CPL	B.W.	Newberry	SIG (B COY DET COMD)
CPL	S.J.	Osbaldstone	SIG
CPL	C.J.	Peck	SIG STMN
CPL	A.G.	Skinner	SIG (CCOY DET COMD)
LCPL	S.A.	Alexander	SIG (DCOY/ A ECH DET COMD)
LCPL	J.A.	Hammett	SIG
LCPL	J.G.	Mackenzie	CO SIG
LCPL	A.S.	Robertson	SIG (A ECH DET)
LCPL	J.S.	Wilson	SIG (A ECH DET)
PTE	S.G.	Armstrong	SIG

Rank		Name	Position
PTE	A.G.	Barnes	SIG (A ECH DET)
PTE	M.A.	Bryant	SIG
PTE	A.K.	Burns	SIG
PTE	G.T.	Green	SIG
PTE	A.P.	Harding	SIG
PTE	D.R.	Kemp	SIG
PTE	A.J.	Luttrell	SIG
PTE	I.N.	MacDonald	SIG
PTE	D.N.	McLeod	SIG (B COY DET)
PTE	B.G.	Melbourne	SIG
PTE	S.A.	Smits	SIG (B COY DET)
PTE	T.J.	White	SIG

Sniper Sect

Rank		Name	Position
SGT	L.W.	Nayda	SNIPER SGT
CPL	D.K.	Mortimer	SNIPER CPL
LCPL	D.S.	Loftus	SNIPER
PTE	C.C.	Ahmelman	SNIPER
PTE	J.P.	Cridland	SNIPER
PTE	R.D.	Maitland	SNIPER
PTE	S.	Seath	SNIPER

Admin Company

CHQ

Rank		Name	Position
MAJ	S.J.	McDonald	OC
SREP	A.E.	Daly	SALLY MAN
CAPT	S.J.	Grace	2IC
WO2	R.J.	Vine	CSM
SSGT	J.J.	Daniels	CQMS
CPL	W.	Haynes	OPS CLK OA
PTE	P.	Moss	ADMIN COY STMN
PTE	A.J.	Pattullo	REC\PERS

Catering PL

Rank		Name	Position
WO2	A.E.	Stow	CATERER
SGT	T.S.	Herrmann	COOK
SGT	F.J.	McGrath	COOK
SGT	G.	Thompson	STWD
CPL	D.S.	Creece	COOK
CPL	G.R.	Garnett	COOK
CPL	M.W.	Rochford	COOK
LCPL	T.J.	Crowe	COOK
PTE	C.E.	Cartwright	RATION STMN
PTE	A.R.	Davis	COOK
PTE	I.R	Mackie	COOK
PTE	S.P.	McLauchlan	COOK

Rank	Name	Position
PTE	E.J. Mulligan	COOK
PTE	J.S. Roscow	COOK
PTE	S.J. Trezise	COOK
PTE	S.A. Wadwell	COOK

Medical PL

Rank	Name	Position
CAPT	D.W. Keating	RMO
SSGT	M.D. Liebeck	RAP SSGT
SGT	S.M. Marles	RAP DESK SGT
CPL	G.E. Bowyer	HYGIENIST A ECH
CPL	L.B. Johnson	MEDIC (C COY)
CPL	K.R. Pearson	MEDIC (A COY)
CPL	P.D. Vigar	MEDIC (B COY)
LCPL	R.A. Osborne	MEDIC (D COY)
PTE	M.G. Spearpoint	MEDIC (SPT COY) (3 BASB)

Main Q

Rank	Name	Position
MAJ	R.D. McLeod	QM
CAPT	I.G. Westworth	QM
WO1	H.J. O'Brien	RQMS WAR
SSGT	K.G. Caldow	TECH SGT
SSGT	G.J. Stanworth	AUTO Q MAIN Q LEDGER SGT
SGT	D.B. Sim	AUTO Q LEDGER SGT
CPL	A.C. McNamara	STMN
CPL	R.B. Piggott	STMN
LCPL	N.J. Crothers	STMN
LCPL	D.R. Day	STMN
LCPL	A.R. Kilby	STMN
LCPL	T.J. McInnes	STMN
PTE	W.L. Campbell	STMN
PTE	P.M. Harrison	STMN

Technical SPT PL

Rank	Name	Position
WO2	P.G. Jarratt	ASM
WO2	D.C. Lancaster	ASM
CPL	H.T. Berry	VEH MECH
CPL	A.W. Clarke	CPL ARMT
LCPL	J.D. Grindrod	VEH MECH
CFN	A.W. Bennett	FIT ARMT
CFN	A. Bruhn	TECH ELEC SYS
CFN	S.E. Burden	FIT ARMT
CFN	N.R. Davis	STMN
CFN	D.R. Dick	FIT & TUR

Rank	Name	Position
CFN	A.J. Goss	VEH MECH
CFN	D.M. Hall	VEH MECH
CFN	P.D. Salter	TECH ELEC SYS (3 BASB)
CFN	S.B. Scott	VEH MECH
PTE	G.N. Kirkpatrick	OJT AS VEH MECH

Transport PL

Rank	Name	Position
LT	C.J. McDonald	TOCO
SGT	W.M. Ball	TPT SPVR
LCPL	A.S. Burke	TPT SECT COMD
LCPL	P.J. Masters	DVR (3 BASB)
LCPL	R.P. Peacock	TPT SECT COMD
PTE	M.C. Bolton	DVR
PTE	P. Busson	DVR
PTE	R. Eisenhuth	DVR
PTE	J.R. Faint	STMN
PTE	G.N. Geary	DVR
PTE	P.M. Harvey	DVR
PTE	M.M. Lally	DVR
PTE	S. Littlechild	DVR
PTE	P.K. Renshaw	DVR
PTE	A.P. Smith	DVR
PTE	M.D. Smith	DVR
PTE	P.J. Tulk	DVR
PTE	D.S. White	DVR
PTE	S.E. Wood	DVR

103 SIG SQN

Rank	Name	Position
LT	D.P. Copeland	TP COMD
SGT	R.J. Kent	TP SGT
CPL	B.R. Fry	SIG
CPL	D.A. Fuller	SIG
CPL	W.J. Gest	SIG
CPL	K.P. Jennings	SIG
CPL	D.S. McKenzie	SIG
CPL	R.D. Moyle	SIG
CPL	D.P. Waller	SIG
CPL	M.D. Woods	SIG
LCPL	A.S. Connolly	SIG
LCPL	G.B. Fill	SIG
LCPL	A.M. Goldner	SIG
LCPL	C.P. Latham	SIG
LCPL	G.E. Middis	SIG
LCPL	N.A. Mullins	SIG
SIG	P.A. Donlon	SIG
SIG	M.B. Kesby	SIG

Rank		Name	Position
SIG	L.R.	Martin	SIG
SIG	P.L.	Nettleton	SIG
SIG	D.T.	Wilson	SIG

107 BTY–CMOT

Rank		Name	Position
MAJ	R.H.	Stanhope	BC
CAPT	S.A.	Bagnall	LO MOG
CAPT	M.	Carrodus	
CAPT	J.C.	Hill	
CAPT	D.W.	Reid	
SGT	D.B.	Callaghan	
BDR	E.R.	Connor	
BDR	D.S.	Free	
BDR	P.B.	Manoel	
LBDR	S.G.	Gittoes	
LBDR	J.J.	Lafferty	
LBDR	D.J.	Matthews	
LBDR	C.D.	Peet	
LBDR	C.P.	Riggs	
LBDR	T.P.	Whitwam	
GNR	A.S.	Butler	
GNR	W.L.	Byrne	
GNR	J.M.	Frankcombe	
GNR	P.D.	Henry	
GNR	P.T.	Malone	
GNR	M.B.	Peters	
GNR	S.W.	Swan	
GNR	M.A.	Voormuelen	

17 FD TP 3 CER

Rank		Name	Position
LT	W.R.	Bowyer	TP COMD
SSGT	K.D.	Darcy	TP SGT
SSGT	W.B.	Schoer	TP SGT (EOD)
CPL	G.C.	Condon	
CPL	G.J.	Cottier	
CPL	A.L.	Koschel	
CPL	R.J.	McDougall	
CPL	D.	Robinson	
LCPL	D.C.	Davis	
LCPL	D.J.	Fullgrabe	
LCPL	J.N.	Trass	
LCPL	J.N.	Watson	
SPR	S.J.	Anderson	
SPR	K.	Bawden	
SPR	C.S.	Campbell	
SPR	S.L.	Doherty	
SPR	M.F.	Easton	
SPR	J.A.	Faint	

Rank		Name	Position
SPR	M.J.	Gardiner	
SPR	C.B.	Gisborne	
SPR	S.	Harvey	
SPR	B.A.	Johnson	
SPR	D.L.	Johnston	
SPR	P.D.	Kember	
SPR	D.J.	Kuilboer	
SPR	G.D.	Lloyd	
SPR	J.S.	Maitland	
SPR	J.G.	Mayes	
SPR	J.A.	McDonald	
SPR	P.D.	Postma	
SPR	S.M.	Power	
SPR	J.C.	Rolleston	
SPR	R.J.	Rye	
SPR	A.R.	Smith	
SPR	S.A.	Taylor	
SPR	Dog	Duke	DOG
SPR	Dog	Mick	DOG
SPR	Dog	Tia	DOG

B SQN 3/4 CAV REGT

SHQ

Rank		Name	Position
MAJ	D.G.	McKaskill	OC
CAPT	C.A.	Websdane	2IC/OPSO
LT	M.J.	Toohey	LO
WO2	G.T.	Hooper	SSM
SGT	I.G.	Jones	CCLK
SGT	P.J.	McCanna	TP SGT
CPL	C.G.	Clarkson	CREW COMD
CPL	M.	Freischmidt	CREW COMD
LCPL	T.A.	Dillon	DVR
LCPL	R.J.	Lee	PAY CLERK
LCPL	F.G.	Radman	COOK
TPR	M.G.	Smith	DVR
TPR	C.D.	Yourell	DVR

1 TP

Rank		Name	Position
LT	M.P.	Hanna	TP LDR
SGT	W.J.	McDonald	TP SGT
CPL	D.A.	Barlow	SECT COMD
CPL	C.R.	Cook	TP CPL
CPL	J.R.	Hepburn	SECT COMD
CPL	H.S.	Smelt	SECT COMD
LCPL	D.L.	Beck	CREW COMD
LCPL	C.E.	Berghofer	CREW COMD
LCPL	A.R.	Bradbrook	CREW COMD

Rank		Name	Position	Rank		Name	Position
LCPL	S.A.	Campbell	CREW COMD	TPR	K.A.	Welch	DVR
LCPL	S.R.	Gibbs	DVR	TPR	C.G.	Wilkinson	DVR
LCPL	M.I.	Joliffe	SECT 2IC				
LCPL	A.O.	Millers	CREW COMD	**Admin TP**			
LCPL	T.E.	Orme	CREW COMD				
LCPL	S.M.	Raynor	SECT 2IC	WO2	S.G.	Bingham	SQMS
LCPL	D.A.	Williams	SECT 2IC	CPL	D.P.	Brakovskis	MEDIC
TPR	S.M.	Abdoo	DVR	LCPL	D.P.	Neale	STMN
TPR	C.S.	Andersen	DVR	TPR	A.K.	Michalak	STMN
TPR	J.P.	Arthur	DVR	TPR	D.A.	Mooney	DVR
TPR	J.P.	Brown	DVR				
TPR	J.P.	Burke	DVR	**SPT SECT**			
TPR	G.J.	Fisher	DVR				
TPR	W.D.	Herbert	DVR	CPL	L.J.	Entink	SECT COMD
TPR	D.B.	O'Brien	DVR	LCPL	P.	Quentin	CREW COMD
TPR	G.P.	Peters	DVR	TPR	C.A.	King	DVR
TPR	S.C.	Prince	DVR				
TPR	J.S.	Schubert	DVR	**Technical SPT TP**			
TPR	S.	Spencer	DVR				
TPR	D.J.	Webb	DVR	WO2	M.E.	Robinson	ASM
				SGT	K.C.	McAteer	WELDER
2 TP				SGT	G.W.	Seale	RPS
				CPL	P.A.	Bartlett	VEH MECH
LT	S.A.	Voss	TP LDR	CPL	B.J.	Niemann	RECOV MECH
SGT	S.C.	Wakley	TP SGT	CPL	S.J.	Reynolds	VEH MECH
CPL	D.R.	Commons	SECT COMD	LCPL	S.R.	O'Neil	VEH MECH
CPL	D.J.	Ferriday	SECT COMD	CFN	D.M.	Christensen	RAD ELEC
CPL	G.A.	Frankel	TP CPL	CFN	M.J.	Gurney	VEH MECH
CPL	R.	Moore	SECT COMD	CFN	C.A.	Lindsay	VEH MECH
LCPL	M.R.	Allen	CREW COMD	CFN	S.J.	Rose	RECOV MECH
LCPL	W.P.	Cooper	CREW COMD	TPR	D.T.	Page	DVR
LCPL	A.R.	Golding	SECT 2IC				
LCPL	W.J.	Hall	CREW COMD	**BSG**			
LCPL	M.A.	Holmes	CREW COMD				
LCPL	S.J.	Hope	CREW COMD	**HQ**			
LCPL	P.J.	Larsen	SECT 2IC				
LCPL	K.B.	Nelliman	SECT 2IC	MAJ	M.L.	Harnwell	OC
LCPL	P.M.	Soppitt	CREW COMD	CAPT	A.J.	Anetts	ADJT
TPR	L.S.	Beveridge	DVR	CAPT	A.N.	Jakab	OPS CELL
TPR	W.N.	Chetcuti	DVR	CAPT	D.F.	Nicholl	OPS CELL
TPR	C.L.	Donnellan	DVR	WO1	J.D.	Collins	POSTAL
TPR	A.J.	Edwards	DVR	WO2	C.M.	Young	CSM
TPR	G.A.	Gough	DVR	SSGT	R.W.	Blair	QMS
TPR	P.W.	James	DVR	SGT	K.P.	Glithro	CLK TECH
TPR	A.A.	Johnston	DVR	SGT	R.P.	Moignard	CLK PROD
TPR	D.J.	Minton	DVR	CPL	C.T.	Brennock	
TPR	P.A.	Reeves	DVR	CPL	D.M.	Card	
TPR	C.G.	Richmond	DVR	CPL	D.J.	Muskee	
TPR	J.G.	Robinson	DVR	LCPL	S.A.	Thurlow	
TPR	T.	Siliato	DVR	PTE	D.R.	Bretherton	POSTAL

Rank		Name	Position
PTE	D.J.	McLeod	

Dental Sect

Rank		Name	Position
CAPT	S.P.	McCallum	DENTAL OFFR
CPL	T.S.	Thompson	

FRG

Rank		Name	Position
LT	J.M.	Lawson	OC
WO2	P.A.	King	ASM
SGT	R.J.	Conn	VEH MECH
SGT	D.G.	Wheelhouse	FITTER ELEC
CPL	P.W.	Angus	
CPL	P.B.	Chalker	
CPL	J.M.	Nicholls	
CPL	W.B.	Nutchey	
CPL	G.A.	Seccull	
CPL	J.B.	Stow	
LCPL	S.F.	Cooke	
CFN	D.N.	Ford	
CFN	D.L.	Goninan	
CFN	M.A.	Lawson	
CFN	D.	Maggenti	
CFN	S.C.	Mitchell	
CFN	B.A.	Savidge	
CFN	I.T.	Stewart	
CFN	D.J.	Telford	

Medical Sect

Rank		Name	Position
MAJ	D.J.	Duncan	MO
CAPT	P.	Jongeling	HEALTH OFFR
LT	D.R.	Werda	NURSE OFFR
SGT	A.	Bannister	ASST HEALTH
SGT	J.	Cook	MEDIC
SGT	B.	Dick	MEDIC
CPL	M.L.	Drysdale	
CPL	K.	Felmingham	
CPL	G.J.	Wilsen	
PTE	M.D.	Franks	
PTE	R.J.	Goldsmith	
PTE	C.J.	Hagedoorn	
PTE	S.E.	Hall	
PTE	K.P.	Low	
PTE	L.J.	Low	
PTE	C.	MacLean	
PTE	J.R.	McCarthy	
PTE	D.J.	Smolt	
PTE	S.L.	Steen	

Rank		Name	Position
PTE	D.A.	Walker	
PTE	A.J.	Woodward	

SUP PL

Rank		Name	Position
LT	D.W.	Stanley	OC
WO2	H.	Hampton	CLK TECH
SGT	H.M.	Lintern	STMN TECH ORD
SGT	R.I.	Morgan	AMMO TECH
SGT	D.A.	Wilson	CLK TECH
CPL	G.N.	Cribbs	
CPL	P.G.	Herbert	
CPL	N.B.	Snape	
CPL	J.J.	Thompson	
CPL	V.G.	Turner	
LCPL	B.K.	Cox	
PTE	L.J.	Appleton	
PTE	D.S.	Green	
PTE	B.A.	Johnstone	
PTE	S.D.	Splatt	
PTE	D.G.	Thackery	
PTE	A.J.	Wickham	

Transport TP

Rank		Name	Position
SGT	G.W.	Wilkes	SPVR TPT
CPL	D.W.	Hilton	
CPL	D.S.	Russell	
LCPL	B.J.	Anderson	
LCPL	M.J.	Holden	
LCPL	J.P.	Hughes	
LCPL	P.J.	Nelson	
PTE	P.D.	Allen	
PTE	J.	Farraway	
PTE	M.A.	Kay	DVR
PTE	B.M.	O'Donnell	
PTE	R.J.	Reeves	
PTE	P.K.	Reilly	
PTE	D.K.	Smart	
PTE	P.N.	Smith	
PTE	A.J.	Steene	
PTE	A.C.	Taylor	
PTE	G.F.	Walsh	
PTE	B.L.	Wilton	

HQ AFS

Rank		Name	Position
COL	W.J.	Mellor	CAFS
LTCOL	G.T.	Woolnough	COFS
MAJ(T)	R.J.	Clarke	SO2 FIN

Rank		Name	Position
MAJ	D.A.	Creagh	SO2 PERS/LOG
MAJ	M.J.	Kelly	SO2 LEGAL
MAJ	A.O.	Pagett	SO2 COMMS
MAJ	G.T.	Peterson	SO2 INT
MAJ	C.P.	Powrie	SO2 MED/PERS
MAJ	B.A.	Scott	SO2 OPS
CAPT	N.H.	Catchlove	PR OFFR/ SO3 OPS
CAPT	M.R.	Fulham	SO3 PERS/LOG
CAPT	M.A.	Hiscox	SO3 FIN
CAPT	S.C.	MacNish	SO3 PERS
CAPT	D.M.	McKeon	PUR OFFR
CAPT	M.T.	O'Neill	SO3 OPS
CAPT	I.A.	Young	SO3 OPS
LEUT	A.R.	Powell	NAVY LO
WO2	A.P.	Allison	PAY
WO2	A.W.	Harrison	SIG
WO2	K.W.	Yeo	CCLK
SGT	R.B.	Phillips	LPO
SGT	P.J.	Watson	MP
CPL	J.M.	Baranowski	MP
CPL	D.J.	Cripps	MP
CPL	L.	Leppens	MP
CPL	D.T.	Rayner	MP
CPL	G.J.	Simon	OPS CLK
CPL	T.D.	Warke	MP

INT

Rank		Name
WO1	W.F.	Bowser
SSGT	B.G.	Thomson
SGT	A.	Barrie
SGT	M.K.	Bramley
SGT	G.C.	Bull
SGT	W.J.	Douglas
SGT	M.A.	Thompson
CPL	D.A.	Martyn
CPL	D.G.	Moore

Defence Public Relations

Rank		Name	Position
CMDR	T.N.	Bloomfield	Escort Offr
LTCOL	A.S.	Reynolds	Escort Offr
MAJ	D.J.	Tyler	PRO
WO1	B.R.	Buckley	WO TECH, ENG and Stills
WO2	J.	Burns	CSM
WO2	T.R.	Dex	ENG/STILLS

Rank		Name	Position
SSGT	B.R.	Johnson	Q
CPL	S.D.	Mackenzie	DVR
CPL	G.D.	Ramage	PR TEAM
CPL	D.	Rickets	SIG/COMMS
SPR	R.A.	Guyatt	DVR
MS	L.	Keen	Det Comd
MS	A.	Forman	Escort Offr
MR	R.	Lester	Comms/Tech

Operational Study Team

Rank		Name	Position
LTCOL	P.B.	Retter	
LTCOL	R.J.	Breen	CBT HIST
MAJ	M.L.	Clifford	LESSONS
MAJ	A.J.	Craig	LESSONS
MAJ	J.H.	Kirkham	LESSONS
LT	J.M.	Baker	LESSONS
MR	P.	Vandenburg	EDE

Specialist Support Team

Rank		Name	Position
COL	L.J.	Young	LEGAl
LTCOL	J.W.	Kelley	PSYCH
MAJ	R.H.	Abbott	LEGAL
MAJ	M.A.	Griffin	LEGAL
MAJ	D.N.	Hasson	GLO
MAJ	K.	Parker	ATO
MAJ	M.P.	Walters	PSYCH
CAPT	A.J.	Hall	HQ LOG COMD
CAPT	M.	Levey	PSYCH
CAPT	C.P.	Manktelow	PSYCH
WO1	T.	Murray	HQ LOG COMD
WO2	I.A.	Kuring	Army Hist Sect
SSGT	G.	Britton	1 SIG REGT
SGT	D.A.	Coiro	8 MU
SGT	A.G.	Wormington	8 MU
CPL	M.T.	Hembury	PSYCH
MR	G.	Gittoes	Artist/Photographer

Visitors

Rank		Name	Position
RADM	A.R.O	Rowlands	HQ ADF
MAJGEN	P.M.	Arnison	COMD 1 DIV
BRIG	P.J.	Abigail	COMD 3 BDE
COL	P.J.	McNamara	COL OPS LHQ
LTCOL	G.J.	Cartledge	LHQ
WO1	P.W.	Rosemond	RSM 1 DIV
WO2	T.R.	Marwick	LHQ

SHIP'S COMPANY HMAS *TOBRUK*

Commanding Officer	**CMDR**	**K.B. Taylor**	**0108557**
Executive Officer	**LCDR**	**B.J. Cowden**	**0122224**

Rank	Name	Number	Rank	Name	Number
LCDR	Bramwell	0126819	WO2	MacDonald	1733480
LCDR	Gazzard	0134459	POATA	Allen	S125588
LCDR	Pay	0105968	POETC	Bailey	S112481
LCDR	Siegmann	0124011	POETS	Beer	S129356
MAJ	Le Large	225742	POPT	Chapman	R120553
CHAP	Williams	0142057	POMTP	Chouffot	S123711
LEUT	Dick	0132817	POETP	Clegg	S117980
LEUT	Faulkner	0125547	POWTR	Collicut	R130486
LEUT	Fowler	L122778	PORS	Davis	R120763
LEUT	Garrard	0125114	POMTP	Dearing	S123784
LEUT	Gray	0144142	PORP	Dodsworth	R118251
LEUT	Hardy	0135403	POSTD	Dunbar	R117498
LEUT	Healy	0147281	POMTP	Graham	S133055
LEUT	Mill	L150557	POMTP	Gray	S123517
LEUT	O'Malley	0130370	POCK	Harnett	R130669
LEUT	Page	0141474	POFC	Hewitt	R115741
LEUT	Roberts	0137209	POMED	Hope	S127423
LEUT	Ryan	0115278	POQMG	Howard	R121969
LEUT	Waddell	0126716	POSY	Hoctor	R119611
LEUT	Woods	0124128	POQMG	Jones	R131220
SBLT	Boyle	L144593	POMTP	Lewis	S114942
SBLT	Edmistone	0143577	POSN	McLean	R117790
SBLT	Giffard	0132335	POMTP	Pearson	S114987
SBLT	Grieg	0139146	POATWL	Ricetti	S121411
SBLT	Langford	L149259	POA	Smith	R133791
SBLT	List	0132457	POWTR	Springhall	R136013
SBLT	Waye	0142209	POA	Tahn	R127663
SBLT	Wiltshire	0142444	POETP	Tovey	S115554
CPOMTP	Cole	S120209	POCD	Wilson	R120396
CPOMTH	Franke	S116954	SGT	Joyce	227970
CPOQMG	Hill	R124925	SGT	McMillan	112241
CPONPC	Kaleta	R106160	LSMET	Anderson	R135688
CPOETP	Ledlie	S117827	LSETP	Beattie	S120991
CPOATA	Rendle	S147793	LSQMG	Bryant	R136790
CPOSY	Perryman	R127781	LSSN	Buckingham	W138125

Rank	Name	Position	Rank	Name	Position
LSRP	Connors	R141379	ABCK	Goldstein	R148330
LSSE	Corney	R136159	ABATC	Gorden	S144331
LSSTD	Druett	S135984	ABCK	Gunn	R140610
LSRO	Dunn	W137784	ABGD	Harding	R148609
LSA	Dutton	S137583	ABCK	Harvey	R144575
LSSIG	Elder	R141459	ABETS	Hobman	S143990
LSMTH	Fuller	S134058	ABATA	Holloway	S141302
LSCK	Hammond	R121833	ABRP	Huckstadt	R148435
LSMTP	Hawley	S136758	ABSIG	Hunt	R147721
LSETS	Henderson	S142546	ABWTR	Hutchinson	W146167
LSMED	Hetherington	W136164	ABSN	Johnson	S147371
LSATWL	Jones	S132255	ABSIG	Jones	R147909
LSA	Livingstone	R134483	ABETS	Kalman	S139626
LSMTP	Love	S115809	ABDEN	Krueger	W143515
LSETW	McHugh	S13563?	ABQMG	Lee	R136168
LSATC	Miller	S132440	ABCD	Leis	R141987
LSSN	Phillips	W137556	ABFF	Lynch	R133773
LSMTP	Phillips	S129517	ABMTP	Macinnis	S144861
LSMTP	Phillips	S110294	ABQMG	Manser	R143845
LSRP	Quinn	R125074	ABSIG	Maskell	R142640
LSETC	Smith	S130331	ABMTP	Matchet	S147980
LSRP	Thomson	R137437	ABMTP	McCudden	S148539
LSETC	Treacy	S134837	ABMTP	McIntosh	S147245
LSMTH	Wheeler	S132279	ABMED	McIntyre	W136481
LSMTH	Worland	S122761	ABMTP	Meek	S147321
LSNPC	Wright	R136478	ABCK	Munoz	W145336
LSQMG	Youell	R128733	ABSIG	Nicholas	R140932
LSSTD	Zionzee	R118861	ABETC	Oakley	S140913
CPL	Burke	327853	ABSTD	Obree	R134456
CPL	Duncan	186432	ABSTD	Perkins	W146473
CPL	Harrison	234288	ABMTH	Robb	S144864
CPL	Le Boydre	556277	ABQMG	Rollinson	R139725
ABCK	Amphlett	R148672	ABSE	Ross	R145297
ABETP	Asplin	W146105	ABSN	Rouse	R147004
ABATWL	Balmforth	S141775	ABETP	Scott	S130075
ABQMG	Barnes	R147309	ABSN	Scott	W146328
ABMTP	Berry	S146110	ABQMG	Simpson	R144533
ABMTP	Bloomfield	S144832	ABMTH	Sinclair	S148392
ABQMG	Bourne	R141370	ABSTD	Townsend	R143017
ABCK	Brady	W146703	ABMTH	Vickery	S143993
ABETS	Chandler	S149708	ABRO	Wells	R139921
ABMTP	Clark	S142736	ABATA	Wheeler	S139278
ABATA	Crowe	S140447	ABQMG	Woller	R146529
ABFF	Daniels	W149199	ABRP	Wykes	R141626
ABMTP	Emptage	S143571	ABETC	Zoneff	S148130
ABSTD	Evens	W143411	LCPL	Bell	237150
ABQMG	Farrow	R150111	LCPL	Pata	328350
ABRP	Fischer	R149698	PTE	Barratt	557314
ABSTD	Gibbs	W147465	PTE	Buck	240540
ABMTH	Godfrey	S144852	PTE	Climas	7397

Rank	Name	Position	Rank	Name	Position
PTE	Henrichs	186745	SMNETC	Kricker	S150217
PTE	Smith	2308812	SMNGD	Leach	R149760
PTE	O'Neill	522252	SMNATW	Ledger	S150723
SMNATA	Becker	S150634	SMNETC	Morrison	S149800
SMNMTP	Blundell	S147968	SMNMTP	Rogers	S148103
SMNRO	Burke	W150387	SMANETP	Sherry	S149163
SMNATA	Ciraolo	S150714	SMNGD	Watterson	A151012
SMNGD	Heycock	R151076	SMNGD	Wright	A151010
SMNATA	Holmes	W150719			

GLOSSARY

Action. This is the second state of weapon readiness. The three states are 'Load', 'Action' and 'Instant'. 'Load' is when a magazine with rounds is on the weapon without a round in the chamber. When a weapon is at 'Action', a magazine containing ammunition is on the weapon, and the weapon is cocked to place a round in the firing chamber. The weapon's safety catch remains on. 'Instant' is when a magazine with rounds is on the weapon, a round is in the chamber and the safety catch is off. Excepting the forward scout, all weapons within Australian patrols in Somalia were on 'Action'.

Administration Order (ADMINORD). An Administration Order is a document specifying the administrative arrangements for an operation.

AK/AK 47. Russian made 7.62 mm calibre automatic assault rifle.

AICF. Action Internationale Contre La Faim [International Action Against Hunger] International aid agency based in France that had been working in Somalia since late 1991. The AICF had a regional base at Buurhakaba in the Baidoa Humanitarian Relief Sector.

Aeromedical Evacuation [AME]. Transport of patients to medical facilities by air transport. In Somalia, AME was usually by helicopter configured for the purpose.

Airmobile. A type of operation involving the insertion of troops by air, usually supported by armed aircraft providing fire support. Airmobile units are those which have assigned aircraft for transport, fire support, aero-medical evacuation and resupply.

Airmobile Assault. This is an assault on an objective by troops deployed in helicopters. The assault is initiated by attack helicopters suppressing any enemy in the area of the objective, then helicopters insert the ground force to carry on with the assault. Attack helicopters provide fire support to the ground force.

AN/PVS 4A. A second generation image intensifying night sight introduced to the ADF in the mid-80s. The sight takes in all the available light (from stars, moon and reflected light) and magnifies it many thousands of times. On a dark night, it allows for observation and engagement of targets out to about 200 metres. The AN/PVS4A are due to be replaced in 1998 with a new fleet of image intensifiers which have greatly enhanced resolution and amplification.

AN/PVS 5. A second generation image intensification night vision goggles (NVG), providing night vision out to 50–100 metres depending on conditions. The AN/PVS 5 are due for replacement in 1998.

Anti-tank Rocket. The anti-tank rocket referred to in the book was a 66 mm Short Range Anti-Armour Weapon. A tubular device which fires a rocket from the shoulder, used once and then discarded. Also used against an entrenched enemy, and against enemy forces in bunkers and buildings.

Area of Operations (AO). Areas marked on maps to define the geographic limits of designated units' operations.

ARFOR. Acronym for Army Force, the Army component of the Unified Task Force. HQ ARFOR was, HQ US 10th Mountain Division.

Armoured Personnel Carrier, M113 (APC). A 1960s/70s, tracked, lightly-armoured, amphibious vehicle, armed with .30 and sometimes .50 calibre light machine guns. APC variants include fitters vehicles (mechanical repair vehicles), ambulances and command vehicles. The basic configuration has room to carry one infantry section of nine men with two crew.

Auxiliary Security Force (ASF). A Somali police force paid, trained and equipped by UNOSOM and initially established and protected by UNITAF forces during Operation *Restore Hope*. The 1 RAR Group built detention centres and refurbished other facilities for the ASF whose constables accompanied Australian troops on security operations. Members of the ASF in the Baidoa Humanitarian Relief Sector were unarmed but had the power to arrest criminal suspects and place them in detention.

0A–(zero alpha). Radio call sign for Battalion Headquarters. Often also used in conversation as an abbreviated way of identifying Battalion Headquarters.

Battalion (Bn). 1 RAR in 1993 was an infantry battalion with an establishment of 713 personnel, consisting four rifle companies, a support company and an administration company.

Battalion Support Group (BSG). The BSG in 1993 was group of logistic elements structured to s upport an infantry battalion on operations with an establishment of 92 personnel.

'Bric' Patrol. This is a four-man patrol trained for operations in urban or built up areas. The Bric Patrol was developed by British forces in Northern Ireland.

Brigade (Bde). The 3rd Brigade in 1993 was a combined arms force consisting of two Infantry battalions, a Cavalry Regiment, a Combat Engineer Regiment, an Artillery Regiment, a Signals Squadron, an Aviation Squadron, a Field Supply Company and a Transport Squadron.

'Bug Out'. This is a slang term originating in the Korean War which means to leave an area in a hurry, possibly without having offered strong armed resistance. It is generally used to describe a hasty withdrawal from an area during or immediately after contact with another force.

C130. Hercules transport aircraft.

CARE International/CARE Australia. Internationally and nationally based relief organisation. This is a non-government, non-profit aid organisation with 11 different member countries around the world. They provide immediate food relief, self-help farming and construction programs.

Catholic Relief Service. This is an American-based relief organisation sponsored by the Catholic Church, providing world-wide humanitarian relief.

CENCOM. US Armed Forces Central Command, based in Tampa Bay, Florida.

CH 53. US Marine Sea Stallion medium lift troop/cargo helicopter

Civilian and Military Operations Team (CMOT). In Somalia, 1 RAR Group's CMOT comprised the Battery Commander and Forward Observer parties from 107 Field Battery, Royal Australian Artillery. CMOT were responsible for liaison and coordination with military and civilian agencies and individuals on behalf of the 1 RAR Group.

Claymore. Electrically detonated, directional blast, close protection weapon, convex rectangular shape, approximately 200 mm x 100 mm, and stands just above ground level on spike legs. When detonated, 700 grams of explosive charge fires 700 ball bearings in a forward arc. The mine is detonated by a hand-held static electricity firing device which is connected to the weapon by cable, known colloquially as a 'clakker', which is connected to the mine by cable. Often used in ambushes and defensive positions, it can be arranged in 'banks' with other Claymores to ensure total coverage of any expected enemy approach.

COBRA, AH1. This is a US Armed Forces attack helicopter, crewed by a pilot and a gunner, and armed with anti-tank rockets, free flight rockets and 30 mm cannon.

Designed for use on the conventional battlefield as an anti-tank helicopter and to provide fire support for ground troops.

Combat Body Armour. See Flak Jacket.

Combat Loaded. Combat loading means that vehicles, weapons, ammunition, equipment and stores required to begin operations on arrival are loaded onto ships and aircraft last so that they can be unloaded first.

Company (Coy). The basic rifle company comprises approximately 120 men based around three rifle platoons and a HQ. A company is commanded by a Major. There are four rifle companies, one support company and one administration company in a Battalion.

Company Sergeant Major (CSM). The CSM is a Warrant Officer Class 2, and is the senior soldier in a company. He is responsible for discipline, training and some administrative matters.

CONCERN. International relief organisation based in the Republic of Ireland. This is a non-denominational voluntary organisation which is devoted to all forms of relief assistance in the Third World, especially in the relief of famine conditions.

CONEX. Air and water-tight metal shipping container used by the ADF, coming in different sizes compatible with the current fleet of ADF heavy vehicles.

'Contact'. This term is used by Australian forces to describe an engagement and exchange of fire between own and hostile forces.

Convoy Master. On Operation *Solace*, the convoy master was responsible for coordinating convoy protection tasks, including the issuing of orders and reconnaissance of the route and destination.

Counter Intelligence (CI). Information which was gained on hostile groups and individuals in the Australian HRS was gathered by Intelligence Corps personnel from Australia who were established as two-man Counter Intelligence teams accompanied by a Somali interpreter. This information was collected from informants, Somali police, representatives from Somali organisations and local Somalis.

Crack/thump. This term is used to describe the effect created when a live round passes overhead. The 'crack' refers to the sound caused as the round breaks the sound barrier and the 'thump' refers to the sound of the actual weapon firing. The 'crack' always precedes the 'thump'.

Data Processor (PACE 10). A Data Entry Device (DED) that allows messages of up to approximately 700 characters to be typed in and stored. The message can then be sent by radio or telephone by placing the handset to the PACE. Another PACE is required at the other end to receive and decode the message. The 1 RAR Gp deployed to Somalia with six PACE 10s but after two months only two were working.

Dead Ground. This term is used to describe a depression in the ground, small or large, which provides both cover from fire and concealment.

Deployed Forces Support Unit (DFSU). This is a combined Army, Navy and Air Force unit responsible for the pre-deployment training of personnel who are deploying overseas on operational service, support for Next of Kin of members of deployed forces and information services on the progress of overseas operations. Before its tri-service restructuring, the unit had been called Reinforcement Holding Branch (RHB) and Reinforcement Holding Company (RHC)

Defence Intelligence Organisation (DIO). An ADF organisation responsible for providing strategic intelligence support to the ADF on world-wide security concerns. The organisation is manned by tri-service and civilian personnel.

'**Digger**'. This term originated during the First World War as was used to describe Australian soldiers. It is still used as a slang term for Australian soldiers.

Director—Joint Operations (DJOPS). Appointment at HQ ADF in charge of joint operations involving Navy, Army and Air Force forces.

Disposal of Malfunctioned Explosive Ordnance (DMEO). This task includes the safe disposal of unexploded ordnance, including aerial bombs, artillery and mortar shells, grenades, mines, rockets and anti-tank ammunition. In Somalia, only qualified personnel, particularly members of the Engineer, Ordnance and Infantry Corps, were permitted to undertake DMEO. The task usually involved destroying the ordnance with an explosive charge.

Duty Officer. An officer whose role was to remain on duty or be available in the absence of the other HQ staff, usually overnight.

Extended Line. This is an assault formation of a straight line of troops which is designed to provide maximum firepower directly to the front along a wide frontage. '**Firm**'. Shortened from 'firm on the ground' this term is used indicate that a force which has moved location is now in their new location and has adopted a defensive perimeter, ready for further operations.

First Light. This is used to describe the pre-dawn half-light.

First Line. Unit level holdings of vehicles, weapons, ammunition, equipment and other stores. Typically, units carry enough 'first line' stores to operate for a prescribed number of days. First line resupply happens within units. First line holdings are topped up by units designated to hold a quantity of items of supply designated as 'second line'. During conventional operations, second line supplies are held some distance away from frontline units. In Somalia, second line supplies for the 1 RAR group were held by the BSG at the Baidoa Airfield. HQ Australian Force Somalia (AFS) coordinated third line resupply of the 1 RAR Group through local purchase, supply from US logistic units in Somalia and from fourth line logistic units in Australia. HQ Logistic Command in Melbourne was responsible for fourth line resupply of the AFS from Australia.

Fitted For Radio (FFR). A term describing a four-wheel drive Land Rover vehicle configured to carry and operate radio equipment.

Flak Jacket. A vest of combat body armour constructed of protective composite materials and light pliable sheets of metal. The jacket covers the vital areas of chest, abdomen, and back. It will stop light-low velocity shrapnel and other fragmentation, but is not bullet proof. In Somalia, all Australian personnel wore flak jackets outside the Baidoa Airfield.

Force Activity Designator 1 (FAD 1). The FAD is a notation placed on all logistic requests for an operational deployment which indicate that the request from that unit is to have the highest priority over any other request for the same item. A suffix of 1,2, or 3 is added to the FAD indicating the priority over other FADs, with 1 being the highest.

Forward Recovery Team (FRT). The FRT was a small team of mechanics, armourers and technicians from the Technical Support Platoon of Administration Company, usually comprising two personnel travelling in specially-equipped vehicles.

Global Positioning System (GPS). GPS is a satellite-based world-wide navigation system. The GPS receiver is a hand-held device approximately the same size as a hand-held radio. It must receive at least three satellite signals to function correctly using triangulation to compute its exact position on the earth's surface. Depending on the number of satellite signals it is receiving, the system can be accurate to within a few metres.

GOAL. International relief organisation based in the Republic of Ireland providing famine relief, medical and educational services including the establishment of orphanages. The GOAL orphanage in Baidoa was the focus of some of the 1 RAR Group's civic aid activities.

Grenade Launcher, M79. This is a 40 mm grenade launcher, firing a high explosive contact detonation grenade to an effective range of 150 metres. The M79 dated from the Vietnam War and is on interim issue until a replacement launcher can be found.

Gun Group. In Somalia, there were three groups in an Australian infantry section—the scout group, gun group and rifle group, controlled by the section commander. The gun group consisted of two machine gunners carrying 5.56 mm Minimi Light Machine Guns.

Hercules, C130. Four engine turbo-prop aircraft, with the capability for short take off and landing on both sealed and unsealed runways, used by the Royal Australian Air Force to resupply the AFS.

Humanitarian Relief Sector (HRS). An Areas of Operations in southern Somalia assigned to battalion-sized units from donor nations to UNITAF and UNOSOM.

Helicopter Squadron 817 (HS 817). A Royal Australian Navy Helicopter Squadron equipped with Sea King helicopters with an anti-submarine warfare role. During Operation *Solace*, one helicopter from this squadron was deployed on HMAS *Tobruk*. It was tasked by Maritime Headquarters in Sydney for ship-to-ship and ship-to-shore sorties.

Hummer/Hum Vee. Standard US Armed Forces four wheel drive vehicle. It is lightly armoured in areas around the driver and passenger, armed with a machine gun or grenade launcher on a pintle mount.

International Committee of the Red Cross (ICRC). This organisation operates out of Geneva, Switzerland and acts as a neutral intermediary to protect and assist victims of war, guided by the Geneva conventions.

International Medical Corps (IMC). International aid agency based in the US providing emergency medical aid.

Intelligence Officer (IO). The IO worked with a team at HQ 1 RAR Group to gather information on hostile elements in Somalia. The IO is responsible for providing threat assessment to the Commanding Officer to inform the planning and conduct of future operations.

Jervis Bay. In 1993, HMAS *Jervis Bay* was the Royal Australian Navy's training ship. Prior to its commissioning into the Navy, it was a roll-on, roll-off passenger ship capable of carrying vehicles and cargo.

Joint Activities Planning Group (JAPG). A group of key ADF and Departmental personnel convened to plan joint service activities.

Joint Administrative Planning Group (JAPG). A group of key ADF and Departmental personnel convened to plan joint service administration of operations.

Joint Operations Planning Group (JOPG). Similar to the JAPG, the JOPG is responsible for planning joint operations.

KANGAROO '92 (K92). The KANGAROO series of joint service exercises are held in the north of Australia periodically. They are planned to exercise the ADF in conjunction with the armed forces of regional allies.

Killing Ground. A term to describes an area of terrain to the front of a defensive position or an ambush site, where approaching enemy forces can be fired on effectively. Commanders will site weapons to fire into the killing ground to cause maximum casualties.

Land Commander, Australia (LC AUST). A Major General, who commands all forces within Land Command. Land Command includes the Army's combat forces. In 1993, the other two functional commands commanded by Major Generals were Logistic Command and Training Command.

Land Headquarters (LHQ). Land Command's headquarters located in Victoria Barracks, Sydney.

Landing Craft Mechanised—Series Eight (LCM 8). This is a flat-bottomed, ramp-loading craft designed to embark vehicles, personnel and stores. The LCM 8 has a 54 tonne capacity, enabling it to carry one main battle tank or two armoured personnel carriers or 100 fully-equipped troops.

Landing Ship Heavy (LSH). HMAS *Tobruk* was the Royal Australian Navy's only LSH in 1993 It was purpose built to carry up to 550 troops, vehicles and stores. It is also capable of carrying a helicopter. With ramps at both bow and stern, *Tobruk* can beach itself to off-load or use terminal facilities.

Lighter, Amphibian Resupply Cargo (LARC V). A five tonne capacity, boat-shaped, wheeled and propeller-driven, amphibious resupply vehicle, designed to carry stores from ship to shore. It has a maximum water speed of seven knots and maximum land speed of 48 km/h.

'Load'. This is the first state of weapon readiness. The three states are 'Load', 'Action' and 'Instant'. 'Load' indicates that weapons have a magazine containing ammunition attached, but they have not been cocked and there is no round in the firing chamber.

Logistic Support Force (LSF). A combined Regular Army and Army Reserve Land Command unit is responsible for providing logistic support to Land Command units on operations and exercises in Australia.

'Marry-Up'. This term describes the process of units meeting at a pre-arranged location, often involving confirming radio communications and issuing orders.

Medecins Sans Frontieres (MSF). International aid organisation based in France, comprising several European countries providing emergency medical services.

Media Support Unit (MSU). This unit consisted of Defence Public Relations personnel whose role was to accompany and administer journalists deployed with the AFS to Somalia. Later, this unit was involved in producing photographic and video record of the AFS in Somalia.

Minimi. A 5.56 mm calibre, belt or magazine-fed light machine gun designed in Belgium. Two were carried by Australian patrols in Somalia.

Mobile Army Surgical Hospital (MASH). A mobile surgical facility.

Movement Control Unit (MCU). In Somalia, an Australian MCU served with UNOSOM I and II, coordinating movements of personnel and materiel in and out of Somalia.

National Serviceman. This was the term given to those who were conscripted into the Army during the 1960's to bolster numbers for the Australian commitment in Vietnam.

Night Vision Goggles. See ANPVS 5

Non-Government Organisation (NGO). Humanitarian aid organisations, such as CARE Australia, ICRC, CONCERN and GOAL, providing relief services in Somalia were collectively known as NGOs.

Observation Post. An observation post is usually manned by at least two personnel, positioned to observe a designated area. Personnel report back to a headquarters, typically by radio or telephone, on activity in the designated area.

Orders for Opening Fire (OFOF). Instructions given to military personnel describing the circumstances under which they are authorised to fire their weapons.

Operational Deployment Force (ODF). The ODF is an air mobile Brigade based in Townsville 3rd Brigade, comprised of 1,2, and 3 Battalions of the Royal Australian Regiment and a number of combat and support units. The ODF is maintained at a high level of readiness to deploy at short notice.

Operations Officer (OPSO). In Somalia, the OPSO was responsible for the planning, coordination and execution of 1 RAR Group operations.

Patrol Master. In the 1 RAR Group, the Patrol Master was responsible for planning and coordinating the unit's patrol program.

Platoon (Pl). Each rifle platoon in Somalia comprised three sections of 7-9 infantrymen and a headquarters made up of the Platoon Commander, Platoon Sergeant and a signaller. There were three platoons in each rifle company. There were other specialist platoons of varying sizes and compositions making up Support and Administration Companies in the 1 RAR Group.

Patrol Ambush Light (PAL). Australian-designed lightweight device comprising a battery box and powerful light on an arm. The light is connected to a power supply and switch by a lightweight wire. Problems with the connecting wire has diminished user confidence in the PAL.

Point Of Entry (POE). The designated location for the entry of personnel and material into an Area of Operations, typically serviced by an airfield and/or port. Mogadishu was the POE for Operation *Solace* but items of supply were delivered to Nairobi and Mombassa during the operation.

Quick Reaction Force (QRF). This is a force which remains on alert in the base area in the event that a unit conducting operations requires immediate support. The 1 RAR Group constituted an platoon-sized QRF mounted in trucks, and later in APCs, at the Baidoa Airfield.

Regimental Sergeant Major (RSM). The RSM is a Warrant Officer Class 1 responsible for advising the CO on discipline and morale, the planning and conduct of ceremonial activities and the work of the Regimental Police.

RPD. Russian made 7.62 mm calibre light machine gun

Rocket Propelled Grenade (RPG). The RPG is a grenade, fired from a recoilless launcher positioned on the firer's shoulder and designed to explode on impact with armoured vehicles, bunkers and buildings. The RPG has similar capabilities to the 66 mm anti-tank rocket carried by the Australians.

Rules of Engagement (ROE). Instructions issued to military personnel describing the circumstances under which they are able to use force, including the Orders for Opening Fire (OFOF). Typically, personnel will carry a card at all times that specifies the ROE and OFOF. In an operation where the force is not involved in open warfare, ROE must be observed. The ROE for operations in Somalia were based on UNOSOM guidelines and the Geneva Conventions on Human Rights. The ROE constitutes a legal direction to personnel. Those who do not adhere to the ROE can be charged and made subject to legal proceedings.

Sapper. Nickname for members of the Corps of Royal Australian Engineers (RAE). Sapper is also the Engineer Corps equivalent to the rank of Private.

Scouts. Soldiers comprising the Scout group of an infantry section. The scouts move in front of a section, covering each other, during patrols and are responsible for observing and reporting back, often using hand signals.

Sea King Helicopter, SK 50. This is an anti-submarine warfare helicopter, used also to carry personnel and cargo, and manned by two pilots, two observers and two crewmen. It is armed with a light machine gun mounted in the doorway, used for self defence.

SEA BEES. Nickname for US navy construction battalions.

Ship's Army Detachment (SAD). This unit comprises Royal Australian Army Corps of Transport (RACT) personnel who are attached to Navy ships carrying Army personnel and cargo. Their role is to provide technical advice and assistance on the loading, storage and off loading of personnel and materiel.

Sight Picture. The sight picture is the view of a target as seen through a rifle sight. It is also used by soldiers as a slang term to indicate that a target is in view and they are ready to fire.

Signaller (Sig). In an infantry battalion context, a signaller or Sig is a soldier trained in the operation of radio equipment who is attached to platoon, company and battalion headquarters. In a broader context, it is any soldier or NCO trained to operate a radio. Signaller is also the Royal Australian Signals Corps equivalent of the rank of private.

SKK. Russian/Chinese made 7.62 mm calibre semi-automatic rifle.

Sniper Pair. A sniper is a specialist rifleman trained to stalk, observe and hit targets accurately at long ranges, employing self discipline, stealthy movement and sophisticated surveillance and camouflage techniques. Snipers usually operate in pairs, covering each other's movement and backing up after targets have been engaged.

Somali Democratic Movement (SDM). Dominant Western-style Somali political party in the Baidoa Region, representing the sub-clans of the Rahaweyne Clan. The SDM collaborated with other Somali political groups within the United Somalia Congress (USC), founded in the 1991 to overthrow of the dictator Siad Barre. The SDM advocated regional autonomy with the power to levy taxes and to establish and administer essential services.

Somali Liberation Army (SLA). Para-military group, sponsored by General Mohammed Farah Aideed, set up in the Baidoa area to counter opposing factions of the SDM.

Somali National Alliance (SNA). This group was an alliance of political factions sponsored by General Aideed's faction of the USC, incorporating members of several other Somali factions.

Somali National Front (SNF). This group was sponsored and controlled by General Hersi Morgan, a supporter and relative of the deposed dictator Siad Barre based in the port of Kismaayo.

Somali National Movement (SNM). This group conducted widespread operations and originated among the northern clans in Somalia.

This group was formed in the early 1980s and was committed to the overthrow of the dictator Siad Barre. They created an alliance with the SSDF against Barre.

Spook. A slang term which refers to members of intelligence teams and in particular members of the Australian Army Intelligence Corps.

Standard Operating Procedures (SOPs). Documents describing the inputs and processes used by units and formations to perform their role.

Steyr. A 5.56 mm calibre automatic assault rifle with optical sight, designed in Austria and manufactured in Australia for the ADF. The Steyr was carried by most Australian personnel in Somalia.

6 x 6 (pronounced Six-by-six). A six-wheeled variant of the basic four wheel drive Land Rover vehicle.

Target Indication. This term refers to a verbal procedure used to indicate to others in the vicinity where something is located on the ground.

'Technicals'. Slang term used to describe wheeled vehicles, usually utilities, that were fitted with pintle-mounted machine guns and anti-armour weapons.

Technical Support Vehicle. A 6 x 6 Land Rover configured for use as a repair and maintenance vehicle by forward recovery teams.

'Termites'. Slang term for Royal Australian Corps of Transport (RACT) personnel serving with a Terminal Regiment. Terminal regiments load and off-load materiel at air, road transport, rail and port facilities. These units also operate amphibious vehicles, such as LCM 8 and LARC-V, for ship-to-shore and ship-to-ship transport operations.

Thermal Imager (TI). A thermal imager is a surveillance and observation device that operates by detecting the differences between the heat being given off by a target and its surroundings. Any object or living organism which is emitting heat can be detected by the TI at night as well as under limited visibility conditions, such as in fog or smoke.

Tobruk. See Landing Ship Heavy (LSH).

Trip Flares. A magnesium illumination flare, initiated by a trip wire or by pulling a wire. Often used in defensive positions and ambushes to illuminate anyone approaching or trying to infiltrate.

Unauthorised Discharge (UD). The accidental firing of a weapon, with either live or blank ammunition. Treated as a serious breach of safety regulations and a military offence, normally resulting in disciplinary action being taken against the person responsible.

Unified Task Force (UNITAF). The forces of the donor nations assigned under US command for Operation *Restore Hope*. HQ UNITAF comprised HQ 1st Marine Expeditionary Force. The 1 RAR Group was a UNITAF unit under command of ARFOR [Army Force Component] whose HQ was HQ US 10th Mountain Division.

Unimog. The Mercedes Benz Unimog is the standard four tonne medium utility truck used in the ADF.

United Nations Operations in Somalia (UNOSOM). This was the United Nations mission in Somalia. Member nations assigned military units to UNOSOM I in September 1992 but this phase of UNOSOM was suspended while Operation *Restore Hope* was conducted to create a secure environment for the distribution of humanitarian aid and enable UNOSOM II to take over and resume peace keeping operations. Eventually, all UNOSOM forces were withdrawn from Somalia in March 1995.

United Somali Congress (USC). A group of clans that combined to overthrow the dictator Siad Barre.

Vehicle Check Point (VCP). A VCP was a check point restricting the flow of vehicles so they could be inspected, manned and protected by troops.

Wait Out. A radio message directing other call signs on a radio net to wait for further information.

Warning Order (WNGO). A Warning Order is the formalised version of being warned out. A Warning Order indicates approximately when an operation will begin, what it will be and any other special preparation instructions.

World Food Program (WFP). United Nations organisation whose role was to deliver food to other organisations, such as NGOs, who were contracted to distribute food and other humanitarian aid to towns and villages.

World Vision. International aid organisation based in the US.

BIBLIOGRAPHY

DEPARTMENTAL FILES

AIR HEADQUARTERS

AHQ 'Overseas resupply procedures—Air Mode' 24 Mar 93

AHQ 'Captain's report—ATO 094 / Mar' Scott, I.K. WGCDR CO 37 Sqn 25 Mar 93

AHQ 'Australian Forces Somalia—Resupply'. Ward J. WGCDR COPSO 26 Mar 93

AHQ 'Withdrawal from Somalia—Air Plan' Clark, S. AIRCDRE CDRALG 7 Apr 93

AHQ 'OP *Solace*—Accommodation of Army Personnel en route Somalia' Martin, G.K. LTCOL CO 1 GL GP 13 Apr 93

AHQ 'A brief on RAAF strategic tpt options for OP *Solace*—Phase 4' Clark, S. 15 Apr 93

AHQ 'OP *Solace* resupply tasks—passenger bookings' 18 Apr 93

AHQ 'OP *Solace* resupply tasks—passenger bookings' 22 Apr 93

AHQ 'JOPG discussion notes' -16 April 1993' Treloar, R.B. GPCAPT SOPSO 22 Apr 93

AHQ 'AHQ OPORD 9/93—OP *Solace* Phase 4' Ward J. 29 Apr 93

AHQ 'Operation Order 6/93—Operation *Solace* Phase Four Redeployment' 3 May 93

AHQ 'OP *Solace* C130 Deployment /Shuttle' 6 May 93

AHQ 'OP *Solace*—RAAF Support, resupply / log support system' undated brief

AHQ 'OP *Solace* Phase Four—Post Operation Report' Lumsden G.I. 4 Jun 93 AIRCDR

AHQ 'OP *Solace* phase four-redeployment Post Operation Report' I.K. Scott WGCDR 3 Jun 93

HEADQUARTERS AUSTRALIAN DEFENCE FORCE

HQ ADF 'Chronology of lead up to OP *Solace*.' Notes provided to author

HQ ADF Activity Brief. 'Activities outside the Australian area, 25 Nov—1 Dec 92'

HQ ADF Activity Brief. 'Activities outside the Australian area, 2 Dec—8 Dec 92'

HQ ADF Activity Brief. 'Activities outside the Australian area, 9—15 Dec 92'

HQ ADF Evans, G., Minister for Foreign Affairs and Trade, and Ray, R, Minister for Defence.'Cabinet Agenda items for 15 Dec 1992— Letter to the Hon. P.J. Keating, Prime Minister.' 14 Dec 92

HQ ADF 'Statement by the Prime Minister, the Hon. P.J. Keating MP on commitment of Australian Forces to Somalia' 15 Dec 92

HQ ADF 'Provision of a Battalion Group (BN GP) to a Multinational Coalition Force (Operation *Restore Hope*)' 23 Dec 92

HQ ADF 'Operation *Solace*—Legal considerations and additional ROE Guidance' Fisher L.B. AVM ACOPS 23 Dec 92

HQ ADF 'Geography Brief for Operation *Solace* Contigent to Somalia' 24 Dec 92

HQ ADF 'AS LO USCENTCOM Report—Period 26-28 Dec 1992' Murray, J.S.COL Australian Liason Officer 28 Dec 92

HQ ADF 'Relinquishment of Cat One air hrs' 26 Mar 93

HQ ADF 'OP *Solace* resupply—hours accounting' 29 Mar 93

HQ ADF. 'Commendation of Major Dan Ryan.' Gration, P.C. GEN, CDF, 13 Apr 93

HQ ADF 'Redeployment OP *Solace*' 29 Apr 93

HQ ADF 'OP *Solace*: Planning for Phase Four' 30 Apr 93

ARMY OFFICE

Army Office (DSC-A)., 'OP *Solace* Somalia—Conditions of service' Heyde, R. LTCOL 12 Jan 93

Army Office DSC-A PERS DIV 'OP *Solace*—Conditions of Service' Kirby, B. 12 Jan 93

Army Office 'Financial Arrangements' 29 Jan 93

HEADQUARTERS AUSTRALIAN FORCE-SOMALIA

DEC 92

HQ AFS 'Logistical support of Australian Coalition Forces during Operation *Restore Hope*.' Woolnough, G.T. LTCOL CofS 24 Dec 92

HQ AFS 'Notes 26 Dec 92, Future OPS / Transition.' Mellor W.J.A. COL CAFS 26 Dec 92

HQ AFS '3 BDE information requirements.' Peterson, G.M. MAJ SO2 OPS/INT 27 Dec 92

HQ AFS 'Report on visit by CAFS to Baidoa 27—28 Dec 1992 Mellor, W.J.A. 28 Dec 92

HQ AFS 'OP *Solace*—Admin report No 6' Woolnough, G.T. 29 Dec 92

HQ AFS 'Report on briefing on transition arrangements' Mellor, W.J.A. 29 Dec 92

HQ AFS 'OP *Solace*: identifiable emblem for AS troops' Mellor, W.J.A. 31 Dec 92

JAN 93

HQ AFS 'Report on meeting with BG Zinni: 2 Jan 1993' Mellor, W.J.A. 2 Jan 93

HQ AFS 'CAFS responsibilities for PI/PR planning and implementation' Catchlove, N.H. CAPT PRO 4 Jan 93

HQ AFS 'Consignment address- OP *Solace*' Woolnough G.T. 8 Jan 93

HQ AFS 'AFS Routine Orders. Issue number 001.' Mellor, W.J.A. 10 Jan 93

HQ AFS 'Directive by Commander Australian Forces Somalia to LTCOL D.J. Hurley' Commanding Officer 1st Battalion Royal Australian Regiment Operation *Solace*' Mellor, W.J.A. 10 Jan 93

HQ AFS 'Conditions of Service' Mellor, W.J.A. 12 Jan 93

HQ AFS 'Allocation of funds to HQ AFS for Somalia and other issues' Clarke R. MAJ Financial Adviser 13 Jan 93

HQ AFS 'Directive by Commander Australia Forces Somalia to LTCOL D.J. Hurley Commanding Officer 1st Battalion Royal Australian Regiment Group Operation *Solace*' Mellor, W.J.A. 16 Jan 93

HQ AFS 'Report on meeting with US Special Envoy, Mr Robert Oakley, 18 Jan 1993' Mellor, W.J.A. 18 Jan 93

HQ AFS 'Report on meeting with COMD UNOSOM BRIG Shaheen, 19 Jan 1993' Mellor, W.J.A. 19 Jan 93

HQ AFS 'Request for information.' Peterson, G. M. 20 Jan 93

HQ AFS 'Reply to requests for information 002/93' Paget, A.O. MAJ. 23 Jan 93

HQ AFS 'Reports and returns' Woolnough G. T. 29 Jan 93

Feb 93

HQ AFS 'HRS COMD briefings to CINCCENT—4 Feb 93' Mellor, W.J.A. 4 Feb 93

HQ AFS 'CAFS meeting with as High COM Nairobi-9 Feb 1993' Mellor, W.J.A. 11 Feb 93

HQ AFS 'Service demands' Fulham M.R. CAPT SO3 LOG 13 Feb 93

HQ AFS 'Brief for CAFS-Local Purchase' McKeon, D. M. CAPT LPO 13 Feb 93

HQ AFS 'Report on Shooting Incident—Baidoa' 17 Feb 93

HQ AFS 'Sustainment—OP *Solace*' Woolnough G.T. 18 Feb 93

HQ AFS 'Flight information—ASY 097' Fulham, M.R. 20 Feb 93

HQ AFS 'Force Stockstate as at 22 Feb 1993' 22 Feb 93

HQ AFS 'Notes from briefing with AS staff re Transition Arrangements' Anon Undated

HQ AFS 'Australians Army's 92nd birthday' Mellor, W.J.A. 26 Feb 93

HQ AFS 'Resupply and maint of AFS' Creagh D. S. MAJ SO2 LOG 27 Feb 93

HQ AFS 'Log SPT to OP *Solace*' Creagh, D.S. 28 Feb 93

HQ AFS 'NZDF offer to prov Andover acft in spt of AFS.' Woolnough, G.T. 28 Feb 93

HQ AFS 'Points for BSG' Creagh, D.S.Undated

HQ AFS 'Brief Points for 21C 1RAR Class 6' Anon Undated

HQ AFS 'Lessons Learned, Anon Feb 93

Mar 93

HQ AFS 'Satisfaction of demands raised by 1 RAR BN GP' Fulham M.R. 1 Mar 93

HQ AFS 'SVC demand 28/ 93- Notebook computer and printer' Creagh, D.S 2 Mar 93

HQ AFS 'George Gittoes—Artist' Mellor, W.J.A. 5 Mar 93

HQ AFS 'George Gittoes—credentials to 1RAR' Bagnall, S. CAPT LO
    5 Mar 93

HQ AFS 'Letter to CMOT 1RAR requesting assistance.' Bagnall, S. CAPT
    5 Mar 93

HQ AFS 'RNZDF Aircraft support to—AFS'. Woolnough, G.T.
    6 Mar 1993

HQ AFS 'Peacekeeping questionnaire' Catchlove N.H. 9 Mar 93

HQ AFS 'Outstanding service demands-OP *Solace*' Mellor W.J.A.
    11 Mar 93

HQ AFS 'Resupply of AFS- Civil air via Nairobi' Creagh D. S.
    19 Mar 93

HQ AFS 'Request for issue of cayenne pepper spray'. Woolnough, G.T.
    22 Mar 93

HQ AFS Joint Task Force Support Command. 'Cayenne pepper spray'
    23 March 1993

HQ AFS 'Service demands as at 23 March 93' Fulham, M.R.
    23 Mar 93

HQ AFS 'List of unsatisfied service demands' Woolnough, G.T. Undated

HQ AFS 'Threat Assessment—BRIG Abigail's Visit 21-25 Mar 93' Peterson
    G.M. Undated

APR 93

HQ AFS 'OP *Solace* Planning—Phase 4' Woolnough, G.T. 1 Apr 93

HQ AFS 'The award of the Australian active service medal to 1 RAR
    Battalion Group—Somalia' Mellor, W.J.A. 6 Apr 93

HQ AFS 'Press release—Australian active service medal' 14 Apr 93

HQ AFS 'Information requirements.' Peterson, G.M. 15 Apr 93

HQ AFS 'Information requirements.' Peterson, G.M. 20 Apr 93

HQ AFS 'An account of the prosecution of Gutaale by the Somali Bay
  Regional and Appeal courts and his subsequent execution by the
  Baidoa ASF ' [Sent to Schofield, J. of ABC Nairobi] Mellor, W.J.A.
  Apr 93

HQ AFS 'Operation *Solace*' Mellor, W.J.A. 23 Apr 93

HQ AFS OPORD 1/93 23 Scott B.A.R., MAJ OPSO, Apr 93

HQ AFS ADMINORD 02/93 24 Apr 93

HQ AFS 'Overseas movement advice—cargo' Fulham, M.R. Apr 93

HQ AFS 'Overseas movement advice—pax' . Fulham, M.R Apr 93

HQ AFS 'Brief for COMD 1 DIV, OP *Solace*- Supply matters' Woolnough,
  G. T. 26 Apr 93

HQ AFS 'Passenger manifest 02 May C130' 27 Apr 93

HQ AFS File 33A—Welfare 19 Feb—27 Apr 93

MAY 93

HQ AFS Organisational charts-Canadian Task Forces Somalia. 12 Dec 92,
  1 May 93

HQ AFS 'Post operation report—Operation *Solace*' Watson, P.J. SGT
  RACMP 2 May 93

HQ AFS 'HQ AFS Guard Commitment to Embassy Front Gate.'
  Solmazturk, H. MAJ CO, Turkish Unit Somalia 3 May 93

HQ AFS 'Provision of additional quarantine officer for OP *Solace*' Creagh
  D.S. 3 May 93

HQ AFS 'Message from COMD AFS' Mellor, W.J.A. 4 May 93

HQ AFS 'Compensation claims.' Anon 10 May 93

HQ AFS 'Deployment to Cambodia' Mellor, W.J.A. 11 May 93

HQ AFS 'Message from COMD AFS' Mellor, W.J.A. 15 May 93

HQ AFS 'Operation *Solace*—overseas movement advice' 16 May 93

HQ AFS 'Operation *Solace*—overseas movement advice' 19 May 93

HQ AFS File 18 'Operation *Solace*, Logistics-Postal aspects' Notes made by Author

Keen, L. Director Defence Electronic Media Unit. 'Media support to Operation *Solace*' 12 Feb 93

LAND HEADQUARTERS OCT-NOV 92

LHQ RHB 'Brief for LCAUST—21 Oct 92' Chidgey, C. MAJ 20 Oct 92

LHQ 'Record of conversation—LTCOL Nagy and LTCOL P.B. Retter re RECON GP Report' Retter, P.B. LTCOL SO1 PLANS 21 Oct 92

LHQ 'Record of conversation between Mr. Peter Hooton and LTCOL P.B. Retter re ASC UNOSOM Deployment of Advance Party and Financial Adviser' Retter, P.B. 21 Oct 92

LHQ 'Letter to LTCOL P.B. Retter' Nagy, W.E.,. LTCOL 24 Oct 92

LHQ 'Record of conversation between LTCOL Nagy and LTCOL P.B. Retter re ASC UNOSOM RECON and Adv Party SITREP' Retter, P.B. 26 Oct 92

LHQ 'Weekly SITREP, ASC UNOSOM' Jackson, G.W. MAJ COMASC 1 Nov 92

LHQ 'Debrief LTCOL B Nagy—2 Nov 1992' Black, A. MAJ SO2 INT/ PLANS 6 Nov 92

DEC 92

LHQ 'Brief on ODF BN GP option for commitment to peace-making operations in Somalia.' Retter, P.B. 12 Dec 92

LHQ 'LHQ WNGO 2/92—OP *Solace*' Crouch, M.K. SO2 Plans (MOV) 15 Dec 92

LHQ 'LHQ OPINST 48/92—OP *Solace*' Crouch, M.K. 17 Dec 92

LHQ 'OP *Solace*—Recon Tasks.' Bucholtz, D.C. MAJ for COLADMIN
    18 Dec 92

LHQ 'Provisional Directive by Land Commander Australia to 55634
    Colonel W.J.A. Mellor Commander Australian Forces Somalia.'
    Blake, M.P. MAJGEN.LCAUST 18 Dec 92

LHQ 'LHQ OPINST 48/92—OP *Solace* AMDT 2' Retter, P.B.
    28 Dec 92

Jan 93

LHQ 'Weekly SITREP 001/93 of 030700Z Jan 93' Jackson, G.W. 3 Jan 93

LHQ Letter from MAJGEN M.P. Blake to MAJ G.W. Jackson, Blake, M.P.
    5 Jan 93

LHQ 'OP *Solace*—JAPG. List of attendees' 5 Jan 93

LHQ 'Further guidance for media coverage of Operation *Solace*' Blake, M.P.
    9 Jan 93

LHQ 'LHQ OP INSTR 1/93—OP *Solace* SPT to media SPT DET (MSD)
    andaccredited correspondents (ACCOR)' Retter, P.B. 9 Jan 93

LHQ 'Australian Force Somalia (AFS), Administrative Instruction.' 19 Jan 93

Feb 93

LHQ RHC Family Information Package. Feb 1993

LHQ 'Admin concept OP *Solace*' 1 Feb 1993

LHQ 'OP *Solace*—Embarked Forces' Undated

LHQ 'OP *Solace*-Equip and deployment stocks' Mills, C.J. LTCOL SO1
    PERS 12 Feb 93

LHQ 'Aircraft and stores movt to Somalia' Yeaman, J.W. CAPT DUTY
    OPSO 14 Feb 93

LHQ 'OP *Solace*- Class Six items' Baker, R.M. 17 Feb 93

LHQ 'Brief for Minister for Defence Senator the Honourable Robert Francis
   Ray, support to Operation *Solace*-mail and canteen stores' acinnis, I.
   BRIG CofS 22 Feb 93

MAR 93

LHQ 'OP *Solace* Orbat' 1 Mar 93

LHQ 'Draft LCAUST OP Instructions /93 OP *Solace* Phase Four' . Blake,
   M.P. Mar 93

LHQ 'An appreciation on the return to Australia of

HQ AFS and 1RAR BN GP (OP *Solace* Phase Four)' Collins, L. MAJ SO2
   (Plans) 22 Mar 93

LHQ 'OP *Solace*: Planning for Phase Four'. McNamara, P.J. 22 Mar 93

LHQ 'LHQOPS—Co-ord of air movt in spt of OP *Solace*' 29 Mar 93

LHQ 'Australian Force Somalia OP *Solace*—Phase 4 Return to Australia
   Admin Instruction' Vale, B.L. Mar 93

APR 93

LHQ 'Directive by Land Commander Australia to 55634 Colonel W.J.A.
   Mellor Commander Australian Forces Somalia.' Blake, M.P. 2 Apr
   93

LHQ LCAUST 97/93 Op *Solace*—Some Personnel Issues Blake M.P.
   5 Apr 93

LHQ 'Brief for LCAUST on OP *Solace* Phase Four.' McNamara, P.J.
   8 Apr 93

LHQ 'LHQOPS—OP *Solace* resup tasks—passenger bookings' 20 Apr 93

LHQ OPORD 28/93 OP *Solace* Phase Four 21 Apr 93

LHQ 'AFS OP *Solace*—Phase 4 Return to Australia Admin Instruction'
   Vale, B.L. 27 Apr 93

MAY 93

LHQ 'OPORD 35/93 Spec spt to OP *Solace* Phase Four.' Collins L. 4 May 93

LHQ 'Passenger manifest for OP *Solace* medal parade' 6 May 93

LHQ 'OP *Solace* Phase Four—Logistics' Collins L. 20 May 93

LHQ 'Visibility of service demands for Australian Force Somalia' Barton, A.J. LTCOL SO1 ADMIN OPS/PLANS. 26 May 93

LHQ 'Brief for LCAUST' Vale, B.L. 26 May 93

JUN-AUG 93

LHQ OST Report 'OP *Solace*-Post operation report' Jul 93

LHQ OST Report 'OP *Solace*—An analysis of comd, control and communications' Jul 93

LHQ OST Report 'OP *Solace*—An analysis of equipment performance' Jul 93

LHQ OST Report. 'OP *Solace*—An analysis of Infantry Battalion Gp Security Ops' Jul 93

LHQ OST Report 'OP *Solace*, An analysis of administration and logistics' Jul 93

LHQ OST Report 'OP *Solace*, An anaylsis of Intelligence Support and Civil/ Military Operations' Jul 93

LHQ 'OP *Solace*—Post Operation Health Services report.' Undated

LHQ 'OP *Solace*- Admin and Log analysis' Hartney, D.W. LTCOL SO1 MM 2 Jun 93

LHQ 'OP *Solace* summary of staff / service demands- (list of unsatisfied service demands)' 6 Jul 93

LHQ 'OP *Solace*-Summary of recommendations and key issues requiring action' Blake, M.P. 3 Aug 93

LHQ 'OP *Solace*-Comments on MHQ PXR' Vale, B.L. 13 Aug 93

LHQ 'OP *Solace*-Recommendations and key issues requiring action' Barton, A.J. 18 Aug 93

LHQ 'OP *Solace*; key issues' Hartney, D.W. 24 Aug 93

LHQ 'Post Operation report (POR)—1 ASC UNOSOM' McNamara P.J. 5 Dec 93

FILES

LHQ K92-00463 UNOSOM Extraction of Contingent—Planning Aspects

LHQ K92-00653 UN Considerations

LHQ K92-00656 UNOSOM

LHQ K92-00672 UNOSOM Medical Aspects

LHQ K92-00663 OP Iguana—Health

LHQ K92-00704 UNOSOM Pers Aspects

LHQ K92-00708 UNOSOM Log Aspects Parts 1-6

LHQ K92 00931 UNOSOM Postal Spt (AFPO 6)

LHQ K92-00964 UNOSOM OP Iguana med / health aspects

LHQ K92-00965 UNOSOM Health

LHQ K92-00966 UNOSOM Health/Logistics

LHQ K92-00720 UNOSOM—Communications Aspects

LHQ K92-00-704 UN OPS Somalia—Pers Aspects Parts 1-4

LHQ K 92-00997 OP *Solace*—Pers Aspects

LHQ K92 00998 OP *Solace*—Log Aspects Parts 1-9

LHQ K 92-00999 OP *Solace*—Intelligence Aspects

LHQ K92-01023-25 OP *Solace* Medical/Health

LHQ K93-00030 OP *Solace*—Service Demands

LHQ K93-00044-49 OP *Solace* Medical/Health

LHQ K93-00082 OP *Solace* Planning for Phase Four—Withdrawal

LHQ K93-00178 UNOSOM II

LHQ K93-00272 OP *Solace* Commander's Diaries

LHQ K93-00538 OP *Solace* Profile Documents

LHQ K93-00743 OP *Solace* Ministerial Audit Conditions

HEADQUARTERS LOGISTIC COMMAND

LOG COMD 'Movement of svc demand stores to AFS' 11 Feb 93

LOG COMD 'Resupply and Maintenance to AFS' 23 Feb 93

LOG COMD 'Brief for supply spt OP *Solace*' Tweedie P.M. MAJ SO2 OPS
21 May 93

LOG COMD 'Outstanding service demands'
2 Mar 93

LOG COMD 'Supply Support to Somalia' Tweedie, P.M. 3 Mar 93

LOG COMD 'Supply support to OP *Solace*-Somalia' Tweedie, P.M.
10 Mar 93

LOG COMD 'Supply support to Somalia' Tweedie, P.M.
15 Mar 93

LOG COMD 'Supply support—OP *Solace*' Tweedie, P.M. 23 Mar 93

LOG COMD 'Log Comd SPT to OP *Solace*' . Tweedie, P.M. 24 Mar 93

LOG COMD 'Brief for GOC Spt to OP *Solace*' Ayerbe, A.W. LTCOL SO1
OPS 30 Mar 93

LOG COMD DRV1-T-24. 'OP *Solace*—Logistic Command comments
on Maritime Command's Post Operation Report' McLachlan, D.J.
MAJGEN GOC 13 Aug 93

HQ LSF 'Ships Army Detachment (SAD)—*Jervis Bay*' 27 Apr 93

Packaging Development Centre. 'Operation *Solace*;- Report on the packaging
of equipment on return from Somalia' 16 Jul 93

MARITIME HEADQUARTERS

MHQ 'United Nations Security Council Resolution—Operation *Solace*' 18 Dec 92

MHQ Directive to Commander E.L. Morgan. Commanding Officer HMAS *Jervis Bay*—Operation *Solace* Walls, R.A.K. RADM, MCAUST 18 Dec 92

MHQ 'Directive to Lieutenant S.J. O'Keefe'. Scarce, K.J. CAPT, RAN 22 Dec 92

MHQ 'HMAS *Tobruk*- Cargo carried during Operation *Solace*' Undated

MHQ 'HMAS *Jervis Bay*'s role in Operation *Solace*.' Undated

MHQ 'Operation *Solace*—HMA ships *Tobruk* and *Jervis Bay* Welcome Home' Barrie, C.A. CDRE, RAN for MCAUST 28 May 1993 MHQ 'Operation *Restore Hope*' Walls, R.A.K. 6 Jul 93

MHQ 'Operation *Solace*—Post-Operation Report' Walls, R.A.K. 9 Jul 93

HEADQUARTERS 1 AND 2 MOVEMENT UNITS

HQ 1 MU 'OP *Solace*—Post activity report'. Thompson, P.D. LTCOL CO 1 MU 16 Feb 93

HQ 2 MU 'Overseas movement advice.' 11 Jan 93

HQ 2 MU 'Resupply and maintenance to AFS' 26 Feb 93

HQ 2 MU 'Tasking—Resupply and maintenance to AFS' 1 Mar 93

HQ 2 MU ' OP *Solace* support-stores available' 22 Mar 93

HEADQUARTERS 1 RAR GROUP

HQ 1 RAR Gp 'ROE for Operation *Restore Hope/Solace*' 23 Dec 92

JAN 93

HQ 1 RAR Gp OPORD 1RAR GP—Warning Order 2 Jan 93

HQ 1 RAR Gp Copy of 'Somalia Mission Report' Larbey T.N.C., SBLT
3 Jan 93

HQ 1 RAR Gp OP *Solace*—1 RAR Rear Details Component (AS) 3 Jan 93

HQ 1 RAR Gp 'ESBS Manning for BN GP OP *Solace*' 5 Jan 93

HQ 1 RAR Gp OPORD 01/93—Prelim Orders for deploy Mogadishu to
Baidoa' 7 Jan 93

HQ 1 RAR Gp 'Rules for Dealing with 'A Somali at the Front Gate' Undated

HQ 1 RAR Gp OPORD 02/93—Relief in Place of 3/9 Marine Bn USMC
15 Jan 93

HQ 1 RAR Gp OPORD Operations in HRS Baidoa from 15 Jan—
20 Feb 93

HQ 1 RAR Gp OPORD 02/93—Humanitarian Relief SEC OPS in HRS
Baidoa Jan 93

HQ 1 RAR Gp OPORD 03/93—Relief in Place of 3/9 Marine Bn USMC
15 Jan 93

HQ 1 RAR Gp Operation Log—OP Snow Storm—House bust by
B Company' 28 Jan 93

FEB 93

HQ 1RAR Gp Baidoa. 'Preparation for HRS Briefing-4 Feb 93' Babington
G.P. CAPT ADJT 2 Feb 93

HQ 1RAR Gp Operation Log 'Cookhouse burns down' 4 Feb 93

HQ 1 RAR Gp 'CMOT SITREP.' Stanhope, R.H. MAJ OC CMOT
4 Feb 93

HQ 1 RAR Gp Operation Log 'Conex Incident' 6 Feb 93

HQ 1 RAR Gp FRAGO 4 to 1RAR OPORD 02/93 7 Feb 93

HQ 1 RAR Gp 'Personality information requirements.' Burns, J.R.CAPT IO
    9 Feb 93

HQ 1 RAR Gp FRAGO 5 to 1RAR OPORD 02/93 10 Feb 93

HQ 1 RAR Gp OPORD 04/93—Orders for Cordon and Search of
    Dunuunay OP Cassowary

HQ 1 RAR Gp OPORD Warning Order 12/02 Kookaburra 3 12 Feb 93

HQ 1 RAR Gp OPORD 05/93—OP Kookaburra 3-12 Feb 93

HQ 1 RAR Gp 'Seed Distribution in Baidoa Region' Burns, J.R.
    18 Feb 93

HQ 1 RAR Gp OPORD Warning Order 18/02 18 Feb 93

HQ 1 RAR Gp Operation Log Extracts 21 Feb 93

HQ 1 RAR Gp 'Stores deployed on OP *Solace*' Harnwell, M.L. MAJ OC
    BSG 23 Feb 93

HQ 1 RAR Gp 'Information requirements.' Burns, J.R. 24 Feb 93

HQ 1 RAR Gp FRAGO 7 to 1RAR OPORD 02/93 28 Feb 93

Mar 93

HQ 1RAR Gp 'Conditions Brief OP *Solace*' Judzewitsch, M.K. WO2 Chief
    Clerk 2 Mar 93

HQ 1 RAR Gp 'CJTF COMD Policy Guidance—Civilian Detainees'
    Hurley, D.J. 3 Mar 93

HQ 1 RAR Gp 'Information requirements.' Bowser, W.F. WO1 3 Mar 93

HQ 1 RAR Gp OPORD 06/93—OP Dottral 5 Mar 93

HQ 1 RAR Gp 'Information requirements.' Burns, J.R. 8 Mar 93

HQ 1 RAR Gp 'Duties for Tasks ALOC to COY for Rotation PD 7'
    9 Mar—1 April 1993

HQ 1 RAR Gp OPORD 02/93 (OP *Solace*) 10 Mar 93

HQ 1 RAR Gp FRAGO 8 to 1RAR OPORD 02/93 10 Mar 93

HQ 1 RAR Gp Routine Orders Part One' Hurley, D. J. LT COL 11 Mar 93

HQ 1 RAR Gp Patrol Report. 'Contact Gasarta' 11 Mar 1993

HQ 1 RAR Gp Patrol Report. 'Operation Bushranger' Pritchard, S.P. LT 13 Mar 93

HQ 1 RAR Gp OPORD 07/93—Preparation for the Defence of Baidoa
    AIRFD 14 Mar 93

HQ 1RAR Gp Operation Log—Chase and Capture 15 Mar 93

HQ 1RAR Gp '1 RAR History Collection' Caligari, J.G. MAJ 16 Mar 93

HQ 1RAR Gp '1 RAR on OP *Solace* in Somalia' Caligari, J.G. MAJ
    16 Mar 93

HQ 1 RAR Gp Patrol Report. 'Bandit Chase Incident' 18 Mar 93

HQ 1 RAR Gp FRAGO 9 to 1RAR OPORD 02/93 20 Mar 93

HQ 1 RAR Gp D Company. 'Duty Officer Log' 25 Mar 93

HQ 1RAR Gp Operation Log—'MSF Compound Shooting' 26 Mar 93

HQ 1RAR Gp Operation Log—'Capture of Gutaale' 29 Mar 93

HQ 1 RAR Gp D Company Duty Officer Log 29 Mar 93

HQ 1 RAR Gp 'An estimate on the capture of Degeweyne.' Burns, J.R.
    31 Mar 93

Apr 93

HQ 1RAR Gp Operation Log Extracts 2 Apr, 19 Apr, 21 Apr, 25 Apr 93

HQ 1RAR Gp D Company. 'Signals Log' 2 Apr 93

HQ 1RAR Gp Operation Log—APC Accident in Town' 3 Apr93

HQ 1RAR Gp 'Eulogy for LCPL Mc Aliney killed during OP *Solace*.'
    3 Apr 93

HQ 1RAR Gp '32c Contact Debrief' 5 Apr 93

HQ 1RAR Gp 'An estimate on General Morgan's options in Baidoa.' Burns,
    J.R. 6 Apr 93

HQ 1RAR Gp 'Incident Report 15/93' Cooper CAPT Duty Officer
    7 Apr 93

HQ 1RAR Gp FRAGO 11 to 1RAR OPORD 02/93 11 Apr 93

HQ 1RAR Gp 'An estimate on bandit activity in the eastern sector of HRS
    Baidoa.' Burns.R. 11 Apr 93

HQ 1RAR Gp C Company. 'Contact Debrief' 13 Apr 93

HQ 1RAR Gp 'Information requirements.' Burns, J.R. 15 Apr 93

HQ 1RAR Gp FRAGO 12 to 1RAR OPORD 02/93 21 Apr 93

HQ 1RAR Gp OPORD Redeployment Planning Dates—as at 22 Apr 93

HQ 1RAR Gp OPORD Rotation on 25 April 1993 22 Apr 93

HQ 1RAR Gp OPORD Warning Order Redeployment 24 Apr 93

HQ 1RAR Gp Administration Company. 'Patrol Debrief' 30 Apr 93

HQ 1RAR Gp OPORD 28/93 OP *Solace* Phase Four 30 Apr 93

HQ 1RAR Gp ADMINORD 1/93 OP *Solace* Phase Four 30 Apr 93

HQ 1RAR Gp 'Operation *Solace*' Hurley, D.J. LTCOL 30 Apr 93

HQ 1RAR Gp 'Redeployment Planning Dates—as at 30 Apr 93'

HQ 1RAR Gp 'Rules for the Holding, Storing, Issuing and Use of Weapons
    for the Auxiliary Security Force HRS, Baidoa' Undated

POST OPERATION REPORTS

HQ 1RAR Gp 'Infantry Methods of Operation and Lessons Learnt from
    Somalia' Blumer, A.D. MAJ 28 Oct 93

HQ 1RAR Gp 'Lessons in Low Level Operations from Operation *Solace*'
    Undated

HQ 1RAR Gp 'PL COMD Lessons Learnt' Simpson, J.D. MAJ
    6 Nov 1993

HQ 1RAR Gp 'COY COMD Lessons Learnt—OP *Solace*' Hurley, D.J.
    6 Dec 93

FILES AND LOGS

HQ 1RAR Gp OP *Solace* Operation Log

HQ 1RAR Gp OP *Solace* Situation Reports

HQ 1 RAR Gp Op *Solace* Intelligence Summaries

HQ 1RAR Gp Convoy Log Register

HQ 1RAR Gp Welfare Log

HQ 1RAR Gp File 227-1-2 part 1. 'General Correspondence.'
     22 Feb—19 Apr 93

HQ 1RAR Gp File 445-10-3 part 1 Primary Information Requirements

HQ 1RAR Gp File 445-10-13 Part 1 Intelligence Estimates

HQ 1RAR Gp File 445-10-14 part 1. 'CI Reports.' 19 Jan—3 Feb 1993

HQ 1RAR Gp File 445-10-14 part 2. 'CI Reports.' 3 Jan—21 Mar 1993

HQ 1RAR Gp 'Counter Intelligence Reports—Food Distribution' Undated

HQ 1RAR Gp 'Counter Intelligence Reports—HRS Reports' Undated

HQ 1RAR Gp 'Baidoa Auxillary Security Forces (ASF) chronology
     12 Feb—3 May 1993' Burns, J.R.Undated
     [Notes provided to author]

HEADQUARTERS 1 DIVISION

HQ 1 DIV 'COMD 1 DIV Visit Report Somalia 22-29 April 1993' Arnison,
     P.M. MAJGEN COMD 7 May 93

HQ 1 DIV Enoggera. 'WNGO—RTA of OP *Solace* Personnel'
     6 May 93

HEDQUARTERS 3RD BRIGADE

HQ 3rd Brigade. 'Post-Deployment Report—Operation *Solace*' Abigail, P.J.
     BRIG. COMD 9 Apr 93

HEADQUARTERS UNIFIED TASK FORCE

HQ UNITAF. 'Background to the Somali Conflict' Undated

HQ UNITAF. 'Unified Task Force Somalia—Commander's Policy Guidance #1 (weapons collection procedures) Undated

HQ UNITAF 'Commander's Policy Guidance # 3 (Weapons Confiscation and Disposition)' 8 Jan 93

HEADQUARTERS UNITED NATIONS OPERATION—SOMALIA

HQ UNOSOM. 'United Nations Operation in Somalia—UNOSOM (Somalia)' Undated

HQ UNOSOM Barrow, Dr. J.M., Mahamed, Dr. H., Mohamed, Prof. A.SH. 'Report on Digil/Mirifle Communities.' Undated

HQ UNOSOM Mukhtar, M.H. 'Ergada Mission to Somalia (Inter-river Regions).' Undated

HQ UNOSOM. 'Departure of AFS' 20 May 93

LETTERS

Brown, G. PTE. Letter to wife Sharon 7 Feb 93

Catholic Relief Services. Letter to General J.P. Hoare, USMC Pezzullo, L.A. Executive Director 8 Mar 93

Catholic Relief Services—USCC. 'Medal of Appreciation.' Curtin, J.S. Country Representative Undated

Community Leaders of Upper Juba.Letter to UNOSOM Office, Baidoa. 18 Feb 93 Community Leaders of Upper Juba. Letter to UNOSOM Office, Baidoa, on Security Issues. 23 Feb 93

Community Leaders of Upper Juba. Letter of Appreciation and Thanks. Hassan, M.M. Chairman of the Community Leaders 13 Mar 93

Community Leaders of Upper Juba. Letter to UNITAF Office, Baidoa, on
     Appreciation and Thanks. 13 Mar 93

CONCERN Letter to COL D.J. Hurley. Kilkenny, J. Assistant Field
     Director 27 Apr 93

CONCERN Letter to the Australian Forces in Baidoa. Mulrooney, Lyn
     Anne 24 Apr 93

Hurley D.J., Letters to family dated 7 Jan, 3, 4, and 26 April, 8 May 93

Mc Kinnon, M. Letter to LTCOL D.J. Hurley 15 Feb 93

NGO Community in Baidoa. Letter to LTGEN Johnston 10 Mar 93

Regional Arts and Folklore Club Baidoa. 'The Song of Australia' 13 May 93

Relief Workers Bay Region, Somalia. Letter to Mr. B.B. Ghali
     Secretary-General U.N. Headquarters, N.Y. 8 Mar 93

Returned and Services League. Letter of Thanks to the Officer
     Commanding Australian Contingent Somalia. 19 Feb 93

Returned and Services League. Letter to the Commanding Officer
     and Servicemen of the Somalia U.N. Contingent. Smith, H.
     President Undated

Returned and Services League of Australia. Letter to COL Mellor.
     Garland, A. National President 14 Apr 93

Returned Services League of Australia., Letter to the Commanding
     Officer of the Somali Contingent Hill, F. HON.SEC 10 Feb 93

Save the Children Fund Australia. Thank-you Note Rose, W. Executive
     Director 10 Feb 93

Spouses of Somalia Newsletters 1-7

UNICEF. Letter to COL Mellor, Stirling, M. Representative 4 Apr 93

HQ UNITAF Letter to COL Mellor, Johnston, R.B. LTGEN USMC 6 Apr 93

HQ UNITAF Letter of Condolence. Johnston, R.B. Lieutenant
     General, USMC, 6 Apr 93

HQ UNOSOM. Letter to LTCOL D.J. Hurley, Howe, J.T. Special Representative of the Secretary-General 18 Apr 93

World Vision International. Letter of Appreciation to LTGEN Johnston 10 Mar 93

World Vision International. Letter of Condolence 4 Apr 93

REPORTS AND OTHER PUBLIC DOCUMENTS

Jane's Defence Weekly. 'Seeking to Replace Sea Kings.' 21 Feb 96

BOOKS, MAGAZINES, PAMPHLETS

Crowther, Geoff, (1986) Africa on a Shoe String, *Lonely Planet*

D'Hage, A.S. 'The Military and the Media' Undated United Service.

Enver, Carim, (1983) *African Guide, Encyclopedia of the Third World (1987). Third Edition Volume III.* 'Somalia'

International Committee of the Red Cross., (1983) Basic Rules of the Geneva Conventions and their Additional Protocols, Geneva: ICRC

Medicins Sans Frontieres, (1993) 'Life, Death and Aid', MSF Annual Report

SHORT STORIES AND RECOLLECTIONS

Anon. 'Pers Cell visit to Bartamaha School Baidoa.' 22 Apr 93

Anon. 'Untitled' Undated

Anon. 'Untitled' Undated

Anon. 'Untitled' Undated

Anon. 'R & R in Mombasa' Undated

Anon. 'Somalia.' Undated

Anon. 'Somali Feast.' 12 Feb 93

Boye, S. SGT. 'Statements of Events' Undated

BCampbell, W.L. PTE. 'Untitled.' Undated

Chinner, S. 'Untitled' Undated

Crosland, G. 'A Short Story of the Deployment.' Undated

Dabrowski, P. PTE. ' The Front Gate Riot' Undated

Dodds, S.J. MAJ. 'Comments on *'Restore Hope'*.' Undated

Elmy, R.J. SGT. 'Notes on contact' Undated

Fraser, D.G. MAJ. 'Alpha Company in Buurhakaba' Undated

Fraser, D.G. MAJ. 'Notes from 16 Feb 1993—13 Mar 1993'

Fraser, D.G. MAJ. 'The Difference Between Companies' Undated

Hankinson, M. CAPT. 'Statement of Events.' 7 Apr 93

Hollamby, A.R. SGT MFC D COY. 'Information on Operation *Solace*' 9
    April 1993

O'Loughlin, PTE. 'Untitled' Undated

Parnaby, A.B. CPL. ' The Blue' 16 Aug 93

Parnaby, A.B. CPL. ' The Contact' 17 Aug 93

Parnaby, A.B. CPL. 'A Night In Baidoa' 17 Aug 93

Pavlic, A. CPL. 'The Blue' Undated

Piggot, R.B. Corporal. 'Untitled.' Undated

Elmy, R.J. SGT. 'The Contact' Undated

Everett, T.A. LT. 'Air and Food Distribution to Molill' 9 Aug 93

Garbellini, SGT. 'Untitled' Undated

Gerhardt, PTE. 'Short Stories from Somalia' Undated

Goss, CFN. 'Untitled' Undated

Haddock, J.M. LT. 'Untitled' 13 Aug 93

Hall, D. CFN. 'Admin Company-Night Patrol in Baidoa.' 7 Aug 93

Hannay E., 'Somalia Some Background' 17 Aug 93

Harvey, P.M. PTE. 'Through My Window.' Undated

Herrmann, T.S. SGT. 'Operation *Solace*.' Undated

Hobbs H.A., PTE. 'Somalia' Undated

Jackson G.W., Recollections 30 Apr 96

Keyte C.D. PTE. 'The Blue' Undated

Kilby, A.R. LCPL. 'First Night Patrol.' Undated

Lange M.J. PTE. 'The Warehouse Shooting' Undated

McDonald, R. 'OP *Solace*:—Salient Points (Financial Aspects Operation
      Solace).' Undated

Mcdonald, P.W. WO2. 'Ship's Army Detachment HMAS *Tobruk*'

____., 'HMAS *Tobruk*—Key Personnel' Undated

____., 'HMAS *Tobruk*—The SAD during OP *Solace* 26 Dec 92—21 Jun 93'

Martin, G.K. LTCOL. 'Notes on Air Resupply' 10 Mar 94

Miller, D.J. LCPL. 'A Soldiers Wife.' Undated

Mimi, E.J. CPL. 'Untitled' Undated

Nickson, A.W. PTE. 'Whoops!' Undated

Pickett J., CPL C/S 62A 1 Sect Assault Pioneers

Shields, T.A. PTE. 'The Front Gate Riot' Undated

Simeoni J.A, LT Assault Pioneer Platoon Diary Extracts

____., Assault Pioneer Platoon Somalia Article- Mine Incident

____., Assault Pioneer Platoon Somalia Article—Dinsoor

Short, P.J. CAPT. 'An Essay on Military Protection of Food Convoys'
      Undated

____., 'Convoy Log Register, Explanatory Notes.' Undated

____., 'Personal Notes' 6 Aug 93

Taylor, LCPL. 'Patrol Debrief on 'Bandit Shoot out.' 7 Apr 93

Thorp, M.A. WO2. 'Notes on Somalia' May 96

Van Der Klooster, J. LT. ' A Stoning.' Undated

____., 'Further Information on 12 Platoon Contact 18 Mar 93'

____., 'The Blue' Undated

_____., 'The Contact' 17 Aug 93

_____., 'The Warehouse Shooting.' Undated

Verschelden J.J., LCPL. 'The Warehouse Shooting' Undated

Watson, J.E. PTE. 'The Blue' Undated

Westworth, I. 'Observations on pack up and return to Australia of the 1RAR
     Gp.' 27 Feb 96.

Wehmeier G.T., CPL. 'The Contact' Undated

Williams J.L., PTE. 'The Contact' Undated

Worswick, R. LT. ' Urban Patrolling–A Coy Perspective' Undated

MONOGRAPHS, ARTICLES
AND ESSAYS

James, N.F., 'A Brief History of Australian Peacekeeping.'
     Jan/Feb 1994

Peterson, G.M., 'Human Intelligence and Somalia.' Jan/Feb 1994. Defence
     Force Journal

HMAS *Tobruk.*, '*Tobruk's Solace*—Pictorial Record Undated

Hurley, D.J., 'Operation *Solace.*' Defence Force Journal, Jan/Feb 1994

Mellor W.J.A., 'Operation *Restore Hope* The Australian Experience', US Army
     War College, 23 May 95

Peterson. G.M., 'Psyops and Somalia', Defence Force Journal. Jan/Feb 1994

Morrice, A.J., 'The Mogadishu Express', Defence Force Journal. Jan/Feb 1994

Harnwell, M.L, 'Supporting Hope—Logistic support to 1 RAR BN GP
     Operation Solace.' Army Ordnance Jun 93

Samatar, Said S. (1991) 'Somalia: A Nation in Turmoil', Minority Rights
Group Aug 91

NEWSPAPERS AND NEWSLETTERS

Townsville Bulletin

The Australian

The Sydney Morning Herald

Spouses of Somalia Newsletters 1-6

HQ 1RAR Gp 'Battle Hard—1st Edition.' Undated

HQ 1RAR Gp 'Battle Hard—2nd Edition.' Feb 93

HQ 1RAR Gp 'Battle Hard—3rd Edition.' 22 Apr 93

HQ 1RAR. Gp 'Big Blue One.' Undated

INTERVIEWS

Abigail, P.J., 27 Jul 93

Ayerbe A.W., 4 Nov 93

Bagnall, S. CAPT 26 Jul 93

Blake, M.P. MAJGEN 25 Jul 93

Clarke, R. MAJ 15 Dec 93

Fleming M.J., MAJ 28 Jul 93

Fulham, M.R. CAPT and Young, I.A., CAPT 27 Jul 93

Hall W., CAPT 4 Nov 93

Hurley D.J., LTCOL 28 Jul 93

Hurley Linda, 27 Jul 93

Kelly M., MAJ and Peterson G.M., MAJ 9 Nov 93

McDonald R., 15 Dec 93

Martin R.J., LTCOL 24 Nov 93

Martin, G.K. LTCOL. 4 Mar 94.

Mellor, W.J.A. COL. 10 Nov 93

Mooney A., 28 Jul 93

Nagy, W.S.,. LTCOL. 15 Dec 93

Newham, B.C. WGCDR 15 Dec 93

O,Neill M., CAPT, 10 Nov 93

Retter, P.B. LTCOL 2 Jul 93

Ryan, D. MAJ 24 Nov 93

Schafferius, W. SGT 5 Mar 96

Simon, G.J. CPL 10 Nov 93

Stanhope, R.H., MAJ In Somalia

Stanhope R.H. and H., 26 Jul 93

Stevenson, C.R. LT.' Undated

Taylor, K.B. CMDR' 25 May 95

Tweedie P.M.MAJ and Woodward B., MAJ, 3 Nov 93

Vial P., and Terry F., 15 Jul 96

Watson, P.J. SGT., 10 Nov 93

Woolnough, G.T. LTCOL 9 Nov 93

Yeo, K.W., WO2.' 10 Nov 93

INTERVIEWS RECORDED IN BAIDOA, MAY 1993

ADMIN COMPANY

Berry, H.T. CPL.

McDonald, C. J. LT.

McLeod, R. D. MAJ.

O'Brien, H. J. WO1.

Scott, S. B. CFN.

Stow A. E. WO2.

A COMPANY

Brown, G. M. PTE.

Fraser, D. J. MAJ.

Hennessey, WJ. PTE.

Jeffree, T.W. PTE.

Moore, S. CPL.

Simmonds, T.A. PTE.

Stevens, C. V. LCPL.

Worswick, R. J.

B COMPANY

Grant, R. T, PTE.

Holmes, R. A. PTE.

McGuinness, A. J. PTE.

Pritchard, A. P. LT

Prosser, W. B. CPL.

Rhodes, C. S. PTE.

Simpson, J. D. MAJ.

Snell, A. R. CPL.

C COMPANY

Aitken, TA. CPL.

Beresford M. C. PTE.

Brown, G. M. PTE.

Fletcher, G. A. PTB.

Hodges, A. G. COL.

Moon, M. J. MAJ.

Tessieri, A. LCPL.

D COMPANY

Blunden, J. W. PTE.

Blumer A. D. MAJ.

Gathercole, N. M. SGT

Keyte, C. D. PTB.

Nunan, PW CPL.

Van Der Klooster J. M. LT

B SQN 3/4 CAVALRY REGT

Chetcuti, WN. TPR.

Commons, D. R. CPL.

Hepburn, J. R. CPL.

McDonald, W. J. SGT

McKaskill, D. G. MAJ.

Toohey M. J. LT

Williams, D. A. LCPL.

SPT COY

Blackhurst, B. A. CPL.

Daniels, P. J. LCPL.

Davis, J. R. PTE.

Dodds, S. J. CAPT

Hughes, B. G. PTE.

Hughes, C. W. CPL.

Loftus D. S. LCPL.

Mortimer, D.K. CPL.

Thorp, M. A. SGT

Webb, TD. LT

SMALL GROUP INTERVIEWS

Group 1. Fraser D.J., MAJ, Butler G., LT Creagh V. LT Campbell CAPT
Lewis M CAPT Webb T LT Hughes CPL (SPT) 27 Jul 93

Group 2. Dobson Megan, Blunden Virginia, 28 Jul 93

Group 3. Caligari Narelle, Hurcum Jenny, Simpson Sue, McDonald Simone,
27 Jul 93 Schoer Gabriel, Robinson Emma, Daniels Catherine,
Tessieri Lynda,

Mortimer Georgia, 28 Jul 93

Group 3 Formilan Liza, Keyte Simone, Brown Sharon, Fitzallen Patricia,
19 Aug 93

Group 4. Fletcher Tina, Holmes Jane,Thornely Lisa, Stevens Leanne, Renshaw
Lisa, 19 Aug 93

Group 5. Vine Shirley, Kima Noelene, Hudson Julieanne, O'Brien Barbara,
OB'rien Lianie, 28 Jul 93

Group 6. Brahban LCPL, Burke LCPL, Dewhurst CAPT, Everett T.A., LT,
Hudson WO2 17 Aug 93 re Food Distribution

PERSONAL DIARIES

Blumer, A.D. MAJ

Burns, J.R. CAPT

Breen, R.J. LTCOL

Caligari, J.G. MAJ

Connolly, P.J. LT

Dodd, S. MAJ

Everett, T.A. LT

Fraser, D.J., MAJ

Gittoes G., Artist

Hill, J.C. CAPT

Hurcum, G.P. MAJ

Hurley D.J., (Letters)

Lewis, M. CAPT

Mcdonald, P.W., WO2

Mc Donald, S. MAJ

Mc Leod, R.D. MAJ

Moon, M. MAJ

Pritchard, A. LT

Ramage, G. CPL

Simpson, J. MAJ

Snell, K.E. CPL

Somerville, A. CAPT

Stanhope, R.H., MAJ

Tessieri, A. PTE

Toohey, M., LT

Tyler, D. MAJ

Tulley, D. LT

Wood, S. PTE

Woolnough, G.T. LTCOL

Tabular Compilation of Diaries of Blumer A.D., MacDonald S., Hurcum G., Stanhope R.H., Caligari J.G.,

SCHOOLS/YOUTH ORGANISATIONS—
DEAR DIGGER LETTERS

NEW SOUTH WALES

Birchgrove Primary School, Balmain

C.J. York School, Penrith

Coolongolook School, Coolongolook

Granville Public School, Granville

Hume Public School, Lavington, Albury

Hunters Hill High School, Hunters Hill

Ingleburn North Primary School, Ingleburn

John Paul II Catholic High School, Marayong

Lithgow High School, Lithgow

Mortdale Public School, Mortdale

Narraweena Primary School, Narraweena

Port Macquarie High School, Port Macquarie

Redfield College, Dural

St Andrews Catholic Primary School, Malabar

St John's Catholic Primary School, Campbelltown

St Joseph's Primary School, Macquarie Fields

Sydney Grammar, Edgecliff Preparatory School, Edgecliff

The Entrance High School, Bateau Bay

Thornleigh West Public School, Thornleigh

Tuncurry Public School, Tuncurry

VICTORIA

Cambridge Primary School

Lilydale Secondary College, Lilydale

Maroondah Secondary College, Croydon

Parade College, Bundoora

QUEENSLAND

Cairns North State School, Cairns
Holy Spirit Catholic Primary School, Cranbrook
Kawangan State School, Scarness
Kirwin State High School, Kirwin, Thuringowa
Mackay North High School, North Mackay
Ryan Catholic Community School, Thuringowa
Central Sacred Heart School, Booval
St Andrews Catholic Primary School,
St Joseph's Catholic Primary School, Mundingburra
St Mary's College, Charters Towers
Second Gordonvale Brownie Pack, Gordonvale

WESTERN AUSTRALIA

Allandale Primary School, Geraldton
Bullsbrook District High School, Bullsbrook
Falcon Primary School, Falcon
Pemberton District High School, Pemberton
St Mary's School, Muredin
South Mandura Primary School, Mandura
South Padbury Primary School, Padbury

NORTHERN TERRITORY

Holy Spirit School, Casuarina, Darwin

AUSTRALIAN CAPITAL TERRITORY

Majura Primary School, Watson

Notes

Chapter 1

1. FLTCOL W.S. Nagy and CAPT P.L. Angelatos. Nagy was from the Headquarters, Australian Defence Force (HQ ADF) in Canberra and Angelatos was the Second-In-Command (2IC) of 30 ADF personnel who were scheduled to arrive in Mogadishu in a few weeks time.

2. UNOSOM had an authorised strength of just over 4 600 personnel made up of a Force HQ, five infantry battalions, a composite logistics battalion, 200 military engineers and a civil affairs branch. The Australian MCU was to comprise 7 Navy, 15 Army and 8 Air Force personnel. Force HQ planned to deploy detachments of Australians to Mogadishu and four other regional cities to coordinate the arrival of UNOSOM infantry battalions and supporting units.

3. There were many abandoned mansions in Mogadishu belonging to Somali businessmen. Some still operated in Mogadishu, but many had evacuated their families to Italy. Typically, the mansions were surrounded by high walls with a sturdy steel gate. The UN agencies and NGOs leased them and used them for their headquarters, accommodation, parking and storage facilities.

4. LHQ K90-00304 'Debrief LTCOL W.S. Nagy—02 NOV 92' dated 6 Nov 92

5. Ibid.

6. Nagy W.S., Record of conversation with LTCOL P.B. Retter dated 21 Oct 92 LHQ K 92-00653 (5) Folio 19 Nagy W.S., Interview on 24 Nov 93

7. *Khaat* was originally intended to assist Somali men in their study of the Koran. The drug acted as a stimulant which heightened a user's senses and awareness. It was now more widely used as a recreational drug.

8. FLHQ K90-00304 'Debrief LTCOL W.S. Nagy—02 Nov 92', 6 Nov 92

9. Carim E., *Africa Guide 1983*, Essex World of Information, 1983; p.351

10. Samatar S.S., *Somalia: A Nation in Turmoil*, The Minority Rights Group; London 1991, p.6

11. Ibid, p.19

12. The Somali Armed Forces had been rearmed by the US after Barre was dumped by the Soviet Union. The US had built naval and other military facilities for the US Rapid Deployment Force at Berbera.

13. Kurian G.T., *Encyclopedia of the Third World*, Third Edition Vol III Philippines to Zimbabwe, Facts on File Inc, New York; 1987, p.1800.

14. Manthrope S., 1991 *The Montreal Gazette*, p.21, quoted in Samatar S., op. cit., p.22

15. G. W. Jackson, Situation Report, dated 22 Nov 1992

16. Peter Hanson's visit was pursuant to Paragraph 14 of UN Resolution 767 adopted by the Security Council on 27 July 1992 which called for an evaluation of the situation in Somalia and recommendations for UN action. He submitted a report, 'Report to the Security Council, *The Situation in Somalia*. on 24 August 1992

17. LHQ K92-00653 'Somalia UN Participation Considerations' Parts 1, 2 and 3.

18. Major G.W. Jackson. Retter P.B., Record of Conversation 'Recon Gp Report, LTCOL Nagy' dated 21 Nov 92

19. Jackson G.W., Situation Report dated 6 Nov 92

20. Jackson G.W., 'Situation Report' dated 22 Nov 92

21. Vial P., and Terry F., Interview with author of 15 Jul 96. Patrick Vial was the Manager of *Medicins Sans Frontieres* in Mogadishu for 18 months in 1991/92 and Fiona Terry was Deputy Manager of CARE Australia in Baidoa in 1992

22. Wagner T., 'Clansmen strangle Somali city', *The Weekend Australian* 21-22 Nov 92, p.7

23. Fraser M., 'Somalia awaits stronger hand', Letter to the Editor, *The Australian*, 24 Nov 92

24. Wagner T., 'Clansmen strangle Somalia city', *The Weekend Australian*, 21-22 Nov 92

25. COL B.V. Osborn. Martin R.J., Interview with author on 7 Jul 93

26. Retter P.B., Interview with author on 5 Jul 93, Martin R.J., Interview with author on 7 Jul 93

27. FLTCOL R.J. Martin. LTCOL P.B. Retter.

28. Retter P.B., Interview with author on 5 Jul 93

29. MAJ D. Ryan.

30. GEN P.C. Gration, AC.

31. Ryan D., Interview with author of 14 Dec 93 Ryan received a Chief of the Defence Force's Commendation on 13 Apr 93 for his work at HQ ADF in general and during the preparatory phase of Operation *Solace* in particular.

32. CDR D.J. White. MAJ M.K. Crouch.

33. Despite this formal blanket of secrecy, Commander Kevin Taylor [CMDR K.B. Taylor] the newly-appointed Captain of HMAS *Tobruk* was told confidentially around 16 Nov 92 that plans were being made to deploy a force to Somalia and it was likely that his ship would be involved. This was a wise security breach by someone because Taylor was then able to accelerate the schedule of major repairs his ship was undergoing at the time. Taylor K.B., Interview with author on 25 May 95.

34. The author has not been able to ascertain why this level of secrecy was imposed. Earlier in 1992 a far more politically sensitive contingency plan for the overseas deployment of units of the ODF had been developed. This planning had involved all of LHQ's subordinate HQ, including 1 RAR and 2/4 RAR. McDonald S.J. Annotations on first draft dated 11 Dec 95.

35. Retter P.B., Interview with author on 5 Jul 93

36. Ibid.

37. LHQ K92-00653 Somalia UN Participation Considerations. LHQ K93-00085 Post Operational Report Operation *Solace*, 2 Jul 93, p.4

38. Chapter 7 of the UN Charter permits the use of military force to enforce peace. It had been used to authorise US-led military intervention into Korea in 1950 and into the Persian Gulf area in 1990.

39. Trofimov Y., 'UN to control Somalia mission' *The Australian* 4 Dec 92, p.7

40. *The Times.*, 'Civilians want troops to arrive yesterday' quoted in *The Australian* 4 Dec 93, p.7

41. *The Sydney Morning Herald.*, 'Across the Styx to Somalia', Editorial 3 Dec 92

42. MAJGEN M.P. Blake AO MC. Blake M.P., Interview with author, 5 Jul 93

43. One option was to deploy a battalion group with Armoured Personnel Carriers (APCs). The other was to deploy the same group without APCs. The major factors influencing which option would be accepted was cost, not tactical capability.

44. Retter P.B Interview with author of 5 Jul 93

45. Boswell B., 'Marines face threat of sniper fire', *The Weekend Australian* 5-6 Dec 92

46. Hill A., 'Shooting forces relief staff out.' Reuter quoted in *The Australian* 7 Dec 92

47. Martin R.J., Interview with author on 7 Jul 93

48. Blake M.P., Interview with author on 5 Jul 93

49. Hill A., 'Shooting forces relief staff out', *The Sydney Morning Herald*, 7 Dec 92, p.11. McCarthy Phillip, 'US warned: If you liked Beirut, you'll love Mogadishu', *The Sydney Morning Herald*, 7 Dec 92, p.11.

50. Barnes K., 'Troops must stay to reform gangsters: UN' *The Australian*, 8 Dec 92, p.7

51. Blake M.P., Interview with author of 5 Jul 93 Martin R.J., Interview with author of 7 Jul 93

52. Ibid.

53. Barnes K., 'Guns fall silent as Marines storm Mogadishu', *The Australian*, 10 Dec 92, p.1 and p.9

Chapter 2

1. MAJ G.A. Crosland. COL P.A. Sibree

2. LTCOL D.J. Hurley

3. Crosland G.A. Undated letter to author describing the events leading up to and after the deployment of 1 RAR on OP *Solace*.

4. Cameron S., 'Australians asked to join operation 'The Australian 10 Dec 92, p.1

5. Crosland, op cit

6. COL N.H. Weekes AM MC ADC, LTCOL D.J. Murray, LTCOL M.D. Saw, MAJ G.R. Banister AM

7. On 10 December, the day the Marine landings on the beaches of Mogadishu were reported in the Australian press, General Blake approved an outline organisation of 910 personnel for the Australian Force-Somalia (AFS). It consisted of: a national headquarters of 10 personnel from HQ 1st Division in Brisbane and other units, an infantry battalion of 650 personnel (1 RAR), a Battalion Support Group of 100 personnel from 3rd Brigade Administrative Support Battalion in Townsville, two troops, a HQ, Admin and Tech elements and two Mortar APCs (36 APCs and 90 personnel) from B Squadron, 3rd/4th Cavalry Regiment in Townsville, a Civil-Military Operations Team of 22 personnel from 107 Field Battery, 4th Field Regiment Royal Australian Artillery, a field engineer troop of 35 personnel from 17 Field Troop, 3 Combat Engineer Regiment in Townsville, and a communications element of 15 personnel from 103 Signals Squadron. All these units and sub-units had been 'down-sized' to fit a limit of 910 personnel.

8. Crosland asked Hurley whether his acceptance of the redundancy package could be postponed until after the battalion's deployment. Hurley replied that he would support a postponement and advised Crosland to ask the Land Commander in person when he was in Sydney on Friday if he wanted to get a decision that would stick. Crosland asked Blake and was told that, if Hurley wanted him, he could go. Hurley decided to retain Crosland for OP *Solace*.

9. COL W.J.A. Mellor AM ADC, MAJGEN P.M. Arnison AO. LTCOL G.T. Woolnough

10. This planning group comprised mostly administrative and planning

personnel. The move of the 1 RAR Group to Somalia was seen as an administrative move of personnel, vehicles and equipment with sufficient stocks for initial operations. There was little discussion of operational capabilities or resupply. There were no representatives from HQ ADF. LEUT A.R. Powell. WO2 P.W. Macdonald

11.   The American Defence Secretary, Mr Dick Cheney, appeared on television that night to warn the American public that , 'penetration into the interior could be dangerous because of the warring gangs of Somalis are fighting bloody battles in several towns, including the vital southern port of Kismaayo. Boswell B., 'Cheney warns troops of warring Somalis' *The Australian,* 11 Dec 92, p.5

12.   HMAS *Tobruk,* known as a Landing Ship Heavy, was the first purpose-built major amphibious ship in the RAN. *Tobruk* was commissioned in 1981 and designed to carry troops (350-550), vehicles and stores, and put them ashore without the aid of port facilities. In port, *Tobruk* can use a 70 tonne capacity derrick and cranes as well as over the bow and stern ramps onto a 'roll-on roll-off' terminal. Otherwise, *Tobruk* can beach itself and lower its bow ramp to a beach causeway or onto pontoons, landing craft or amphibians. *Tobruk,* powered by two diesel engines and capable of a speed of 17 knots, can carry two 60 tonne Army landing craft as deck cargo or two self-propelled pontoons alongside. The ship is also able to carry and operate a helicopter from a stern pad and is equipped with a small hospital and sophisticated communications facilities. HMAS *Jervis Bay* was the RAN training ship. Formerly MV AUSTRALIAN TRADER, a roll-on-roll-off passenger vehicle vessel. Sold to the RAN in 1977, *Jervis Bay* was capable of a speed of 17 knots. 111 personnel were in the ship's company with accommodation for 43 trainees.

13.   Land HQ K93-00085 Post Operational Report—OP *Solace* dated 2 Jul 93, p.6. Retter P.B., Interview with author on 2 Jul 93

14.   Ibid

15.   Ibid

16.   Blake M.P., Interview with author on 5 Jul 93. Capping the force at 1 000 personnel and the cost at $ 20 million was a 'guesstimate' by one of the planning staff at HQ ADF in Canberra who was under pressure to provide 'a cost to numbers' estimate for 17 weeks. He advised General Gration that this

was the best estimate he could give. At this point there had been no detailed analysis of the costs of a 17 week operation involving a battalion group of 1 000 personnel supported by APCs. HQ ADF had no automated means for costing operational contingencies.

17. Ibid

18. Another version was that Sembler asked Evans whether Australia would support UNITAF and a subsequent expansion of UNOSOM II with military forces and/or provide funds to help the US finance the military involvement of Islamic and African nations. Martin R.J., Interview with author on 7 Jul 93

19. With the Director Joint Operations, Captain Russ [R.E.] Swinnerton, RAN, away on leave, Martin chaired the Immediate Planning Group made up of JOPS 1, Commander Brian Nye, RAN, JOPS 3 Wing Commander Chris Spence, RAAF, Director Joint Logistic Operations, Lieutenant Colonel Bill Nagy, and Martin's assistant, Major Dan Ryan.

20. McDonald R., Interview with author on 15 Dec 93. COL C.E. Stephens, Director of Army Activities and Public Affairs [acting DGLWP]. This JAPG had the job of estimating the costs of OP *Solace* based on Retter's planning proposals. Also in attendance was MAJ M.J. Prain, Acting SO1 Army Activities, (normally Staff Officer Logistics). Prain and McDonald were from the sections of Army office responsible for advising on financial management of Army operations and activities, and developing the Army's financial estimates.

21. MacDonald R., op cit. MacDonald also provided two and a half pages of typed notes entitled 'OP *Solace*—Salient Points'. McDonald later received a Chief of the General Staff's Commendation for his work on the financial estimates of OP *Solace*.

22. Ibid

23. MAJ G.L. Birch, CAPT M.R. Fulham, CAPT I.A. Young, CAPT N.D. Roundtree

24. Fulham M.R., and Young I.A., Interview with author on 20 Aug 93

25. Hurley D.J. Interview with author on 27 Jul 93. Phase One would be the recall of essential operations and administrative staff to begin detailed planning. Phase Two the personnel required to provide the essential services, such as

medical, catering, transport and vehicle maintenance would be recalled to prepare for the final phase which would be the recall of the remainder of the 1 RAR Group by 31 Dec 92

26.   Ibid

27.   Barnes K., 'Rival Somali warlords agree to ceasefire, *The Weekend Australian* 12-13 Dec 92

28.   McDonald R., op cit. The circumstances surrounding the provision of estimates for OP *Solace* were interesting. Navy and Air Force representatives were not available for Stephen's initial JAPG meeting on 9 Dec 92 Within days of the Cabinet submission being approved and costs capped at $ 20 million, McDonald found out from Army Activities staff at Army Office, Joint Logistic Operations staff at HQ ADF and from Brian Vale at Land HQ that the Minister had committed the Defence Department to hosting up to 80 media representatives in Somalia, a special allowance of $ 68 per day was to be paid to those serving there, the estimates from Vale's staff were for 60 days of peace enforcement operations based on World War II consumption rates, rather than on 120 days which was the duration of OP *Solace*, a new estimate of the costs for American rations was four times higher than the price given to him earlier, and that no one had thought to estimate the additional travel costs of recalling just over 900 personnel from leave and concentrating them in Townsville. With hindsight, estimating the cost under $ 20 million without this additional information probably helped convince the Cabinet to approve the battalion Group proposal, but it was a significant, although unintended, under-estimation.

29.   AVM L.B. Fisher

30.   Blake M.P., *Somalia Book I*, Notes made on 14 Dec 92

31.   Retter P.B., Interview with author on 2 Jul 93

32.   Wright A., 900 Diggers to join Somalia peace force, *The Sydney Morning Herald,* 15 Dec 92 p.1

33.   Stewart C., Troops face high risk role *The Sydney Morning Herald,* 16 Dec 92 p.1

34.   MAJ D.J.Fraser

35.   Fraser D.J., Interview with author on 2 May 93

36. Interviews and conversations with members of the 1 RAR Group in Somalia Apr-May 93 and in Townsville Jul 93

37. Ibid

38. Closed camp is a period of time before deployment when all personnel would be required to stay in barracks for final pre-embarkation administration. Interviews with groups of wives by the author on 27/28 Jul 93

39. Interviews and conversations with wives, partners, relatives and loved ones by the author in May and July

40. Somerville A.M. Diary entry of 16 Dec 92. Many members of 1 RAR who were due to leave on posting were told to stay in Townsville so they could be officially recalled. Others were allowed to leave.

41. McDonald S.J., Annotations on first draft dated 11 Dec 95

42. Phelan A., Aid workers angry at US troops' delay, *The Sydney Morning Herald*, 14 Dec 92

43. Neales S., Trapped Aust aid workers face a tense wait for relief, *The Sydney Morning Herald*, 15 Dec 92 p.1

44. Ibid

45. Boswell B., 'Warlord agrees to turn on Mogadishu's water', *The Australian*, 15 Dec 92 p.7

46. Blake M.P., *Somalia Book I*, Notes dated Wed 16 Dec 92

47. After their discussions, Gration and Blake directed that a Liaison Officer be sent to the United States Central Command in, Tampa Bay, Florida, to find out what American intentions for the 1 RAR Group were and how the 1 RAR Group could fit into the United Task Force deployment plan. The officer selected was Lieutenant Colonel Rod Margetts,[LTCOL R.J. Margetts] who was the Australian Liaison Officer with the United States Marine Corps, based in Quantico, Virginia. He arrived within 48 hours, equipped with portable satellite communications equipment which enabled him to talk securely to Australia. The Americans were impressed with Margett's prompt deployment as well as his autonomous and reliable communications. Despite earlier acceptance by other countries to send forces to Somalia, Australia was the first country to have a Liaison Officer working at Central Command. Margetts was replaced by Colonel 'Jock'

Murray [COL J.S. Murray] just before Christmas as the Australian Liaison Officer for OP *Restore Hope*.

48. Headquarters 1st Division had exercised its deployable Joint Force Headquarters (JFHQ) staff at the ADF Warfare Centre, Williamtown, during the period 8-9 Dec 92. This exercise was intended to test the JFHQ element of the headquarters in planning and deployment procedures in response to a short notice contingency. Most of Mellor's staff from 1st Division were to come from those who were on this exercise. ACOPS Weekly Brief (as at 0830L 7 Dec 92) DJOPS Aspects. Yeo K.W. Interview with author on 9 Nov 93

49. Mellor W.J.A., Interview with author on 10 Nov 93

50. Blake M.P., *Somalia Book I*, Notes dated Fri 18 Dec 92

51. Mellor W.J.A., op cit

52. BRIG O.D. Jackson OBE (RL)

53. There had been a number of precedents for appointing national commanders for overseas deployments. Colonel Kevin [F.K.] Cole had been appointed to lead a contingent of 152 Australian Army personnel who were part of the Commonwealth Monitoring Force maintaining law and order in Zimbabwe Dec 79-Mar 80. In 1989-90 Colonels Dick [R.D.] Warren and John [J.A.] Crocker were appointed as national commanders of Australian contingents of 309 Australian Army personnel, mostly engineers, serving as part of the UN Transition Assistance Group in Namibia.

54. Neales S., Signs from the sky in Baidoa, *The Sydney Morning Herald*, 16 Dec 92 p.6

55. Neales S., 'Bring out your dead', is cry on streets of Baidoa *The Sydney Morning Herald*, 18 Dec 92 p.3

56. After the announcement on 15 December planning activity at LHQ in Sydney became frenetic. LHQ began receiving bids from headquarters and Corps Directorates in the Army suggesting additional capabilities and personnel for deployment to Somalia. Given Cabinet direction specifying a force size 'of about 900 'and the need to impose some discipline in face of this stampede to send additional personnel to Somalia, Blake imposed an upper limit of 930 on the AFS.

57. HQ 1st Division was not involved in mounting the AFS but ended up providing a National Commander, Colonel Bill Mellor, and raising a small ad hoc National HQ of nine personnel to support him. Mellor's HQ comprised personnel from HQ 1st Division, Army Office in Canberra and HQ 3rd Brigade. A Navy officer, Leutenant Andy Powell, joined Mellor's HQ just prior to Mellor's deployment to Somalia on 21 December.

58. BRIG M.J. Keating AM, MAJ M. Kelly. Brigadier Keating had been posted overseas and took no part in OP *Solace*.

59. BRIG P.J. Abigail AM

60. Abigail P.J., Interview with author on 27 Jul 93

61. The previous Colonel Operations, Colonel Peter Sibree, had been posted and the new Colonel Operations, Colonel Phil McNamara [COL P.J., McNamara] was due to arrive early in the New Year from War College in the US, two weeks before the planned deployment of the main body of the AFS. Blake M.P., Interview with author on 5 Jul 93

62. The Director of Joint Operations, Captain Russ [R.A.] Swinnerton, RAN, was on leave.

63. MAJ J.G. Caligari. MAJ R.D. McLeod AM. Caligari J.G., Diary Entry dated 19 Dec 93

64. 1 RAR Battalion Group Post OP Report—OP *Solace* Phase 1 Pre-deployment Phase.

65. CMDR E.L. Morgan, CO HMAS *Jervis Bay*. CMDR K.B. Taylor, CO HMAS *Tobruk*.

66. Taylor K.B., Interview with author of 25 May 95. Taylor blocked about 50 percent of postings.

67. MHQ 'OP *Solace*—Post OP Report' dated 9 Jul 93

68. Macdonald P.W., Essay written for author entitled, 'The SAD HMAS *Tobruk*' during OP '*Solace*' 26 Dec 92—21 Jun 93

69. Perryman D.J., Essay written for the author entitled, 'OP *Solace* 1992-1993' in April-May 1995. CPOSY (Chief Petty Officer Signals Yeoman) D.J. Perryman, Signals Yeoman, HMAS *Tobruk*.

70. Taylor K.B., op cit

71. Ibid

72. The Court Martial was related to charges brought against the Captain and several officers from HMAS *Swan* about the alleged sexual harassment of a female medical officer aboard the ship.

73. LCDR B.J. Cowden

74. Banister G.R., 'Logistic Requirements—OP *Restore Hope* dated 16 Dec 92

75. Banister G.R., Annotations on second draft

76. Mills C.J., Brief for Chief of the General Staff's Advisory Committee held on 12 Feb 93 LHQ 92 00998 OP *Solace* Log Aspects Part 1 Folio 81. 1 RAR Group Post OP Report-OP *Solace*. 1 RAR had 1950 outstanding or pending service demands, some of which dated back to June 1992 McLeod resubmitted all of them as Priority One demands so they could be satisfied before *Jervis Bay* and *Tobruk* docked. Just two months before, in October 1992, an Operational Readiness Report had been submitted stating that there were no shortfalls in stock holdings in 3rd Brigade.

77. MAJ C.J. Mills, Acting SO1 Administration Operations and Plans, MAJ R.M Baker, SO2 Administration Plans, and MAJ D.C. Bucholtz, Logistic Operations.

78. LHQ 92 00998 OP *Solace* Log Aspects Part 1 Folio 6 and 8. Banister's quantities were crossed out and new numbers inserted. The following were typical reductions: GPS Hand held from 35 to 5, NVG from 50 to 'various', Thermal Imagers from 3 to nil, squad radios from 55 to nil, PACE 10 (data transfer) equipment from 10 to 2, Flood light sets from 5 to 2, Notebook Computers from 5 to 1 (Pay), photocopier from 1 to none.

79. The 'Trimble' Global Positioning System (GPS) was a particularly useful navigation aid. The Course Acquisition (CA) coded systems gave a Grid Reference accurate to 150 metres anywhere in the world and worked off a series of 21 satellites. The Mortar Fire Controllers in 1 RAR had been trained on GPS. A few members of the Reconnaissance and Signals Platoon purchased their own commercially-available GPS. Others would probably have done so if too few had been provided by the Army logistic system.

80. The Army only had limited numbers of Thermal Imagers and Night Vision Goggles. These items were on issue to specialist units for training purposes

so that there was a pool of personnel able to use the equipment. The options for equipping the 1 RAR Group as Banister had requested were to buy more equipment or loan the equipment from specialist units to the 1 RAR Group for 17 weeks.

81. MAJGEN G.D. Carter AM. LHQ K92 00998 OP *Solace* Log Aspects Part 1. Several folios relate to the debate about Banister's list of logistic requirements and the author presumes that there was a lot more said but not recorded on file in the days before Christmas. Unfortunately, none of the key logistic officers were interviewed about these matters but the folios on file verify the issues that were discussed and the attitudes of certain officers. The provision of 50 of the latest Hand held 'Trimble' NavStar GPS navigation receivers was more straightforward. However, it still took two weeks of administrative action and justification before Wing Commander Keith McPherson from the Space and Joint Systems Project Office in Canberra advised the Army that they were available for issue. WGCDR K.W. McPherson. LHQ K 92 00998 OP *Solace* Log Aspects Part 1 Folio 17.

82. LHQ 92 00998 OP *Solace* Log Aspects Part 1 Folio 34. Signal from Banister to Bucholtz at LHQ asking him to follow up the request for FAD 1 he had sent on 16 Dec 92

83. MHQ Minute RAN Log Spt to OP *Solace* dated 22 Dec 92. According to this reference, *Tobruk's* turn around for Mogadishu-Mombassa would be five days for rapid resupply and seven days for routine resupply.

84. LHQ K 92 000 998 OP *Solace* Log Aspects Part 2 Folio 56. MHQ Minute RAN Log Spt to OP *Solace* dated 22 Dec 92

85. There was confusion over whether two LCM 8 (Land Craft Medium) or older LARC V amphibious craft would be loaded. Eventually, Taylor was told that an American Marine and Watercraft Detachment would have the ship-to-shore role for OP *Restore Hope*.

86. *New York Times* and *Agency France-Presse* reports quoted in *The Sydney Morning Herald*, 19 Dec 92 p.6

87. MAJ C.P.T. Powrie LCDR A. Naughton. The composition of the Team had unusual inclusions and exclusions. Normally, Woolnough would have remained in Australia to raise and equip HQ AFS and establish procedures for its role in Somalia. This task was overtaken by the priority for establishing

resupply arrangements with HQ UNITAF. It was vital for administrative planners in Australia to know what support would be available from UNITAF in Somalia so they could set up a resupply system and deploy the 1 RAR Group with the right mix of major equipment and stocks. Despite the space that was available in the Falcon 900 (up to 15 passengers), there was no one from the 1 RAR Group included in the Team. Much of the additional space was taken up with personal equipment, communications equipment, rations and water.

88.   Originally, Brigadier Adrian D'Hage had made strong representations to accompany the NLT. He felt that public relations and the management of the media was one of the most important aspects of OP *Solace*, requiring his personal presence in Mogadishu.

89.   In reality HQ UNITAF was HQ 1st US Marine Expeditionary Force with attached UN, US Army, US Air Force and US State Department personnel

90.   Belgian forces were to have a difficult time in Kismaayo and suffer a number of casualties. Like Mogadishu, it was the scene of a ferocious contest between two powerful warlords. The city environment made it particularly difficult for the Belgians to assert control over warring factions and criminal gangs.

91.   Mellor W.J.A., Interview with author on 10 Nov 93. While Mellor had not been given specific criteria for Areas of Operations that would suit the 1 RAR Group, he assessed that Baidoa was key inland centre, located on a major road, serviced by an airfield with a more benign climate that some other regions of southern Somalia.

92.   BRIG A.S. D'Hage AM, MC

93.   D'Hage A.S., (1993) *The Military and the Media, United Service, The Royal United Service Institution of NSW Vol 46*, No 4 April 1993

94.   LHQ K92 00998 OP *Solace*—Log Aspects Part 1. Folio 19. Several other folios specified that the MSU vehicles and stores were to be loaded onto *Jervis Bay* on 24 Dec 92 even if there was disruption to 1 RAR Group preparations and 1 RAR Group vehicles and stores were not able to be loaded. For example, 'The CGS [Grey] has directed that the stores are to be provided from within Land Command resources. The stores are of the highest priority and MUST be loaded on HMAS *Jervis Bay*. HQ 3 Brigade is to provide the stores and equipment from within 3 Brigade resources.' Folio 5. Signal from LHQ to HQ 3 Brigade.

95. Ibid

96. CAPT M.R. Fulham. LHQ K92 00998 OP *Solace* Log Aspects Folio 11
     dated 18 Dec 92. If Fulham followed Brigadier D'Hage's list of requirements
     the 1 RAR Group would have to off-load four trucks loaded with rations and
     repair parts, four Land rovers and trailers (two of them fitted for radio) and
     10 pallets of stocks. This would have the effect of reducing ration holdings
     from 30 to 26 days, reduce stocks of spare parts, and impair First Line
     replenishment and command and control.

97. LHQ K92 00998 OP *Solace*—Log Aspects Part 1 Folio 19 dated 22 Dec 92.
     The final stores list for the MSU was sent by facsimile by Ms Keen on 29
     Dec 92 (Folio 79). Thus the loading of both *Jervis Bay* on 23/24 Dec 92
     and *Tobruk* on 29 Dec 92 were disrupted by the requirement to load MSU
     vehicles and stores. In the end, the 1 RAR Group allocated vehicles from
     its own stocks for the MSU rather than load additional vehicles at short
     notice.

98. LEUT A. Powell, RAN, SGT D.W. McMillan, Ships Army Detachment,
     HMAS *Tobruk*. Unfortunately Powell, who was intimately aware of the
     capacity and configuration of *Jervis Bay,* was seconded at short notice to be
     the RAN Liaison Officer to HQ AFS and was told he would fly out with
     the National Liaison Team to Mogadishu on 21 December. Major Glenn
     Crosland recalled later, 'This decision [to deploy Powell to Mogadishu] took
     away a very important person from the planning process and should not have
     been allowed to occur at such a critical time.'

99. CAPT J., Atkinson. Thompson P.D., OP *Solace*—Post Activity Report dated
     16 Feb 93. Thompson, CO 1 Movement Unit, wrote a covering note for
     a report 'compiled by the Logistic Command Movement Representative,
     CAPT J. Atkinson'

100. Ibid.

101. MAJ D.C., Bucholtz.

102. Taylor K.B., Annotations on second draft on 4 Apr 96

103. Atkinson J., op cit

104. Crosland G.A., Undated essay written for the author on the deployment of
     the AFS.

105. Blake M.P., Interview with author on 5 Jul 93

106. McCarthy P., 'UN increases effort for marines to disarm Somalis', *The Sydney Morning Herald*, 23 Dec 92

107. Ibid

108. Martin R.J., Interview with author of 7 Jul 93

109. Ibid. Mellor W.J.A., Minute,' Report of Meeting with BG Zinni: 2 Jan 93' dated 2 Jan 93

110. COL B.L. Vale, Colonel Administration. LHQ 92 00998 OP *Solace*—Log Aspects Part 1 Folio 66. Minute signed by MAJ D.C. Bucholtz dated 24 Dec 92

111. HQ 3 BDE ADMINORD 08/92 dated 19 Dec 92. This document was based on a contingency plan for the deployment of an ODF battalion group overseas at short notice.

112. Mellor W.J.A., 'Report on Visit by CAFS to Baidoa 27-28 Dec 9,' dated 28 Dec 92

113. CAPT J.J.R. Burns

114. This map had been photocopied from a book obtained from the Townsville Municipal Library

115. Somerville A.M., Diary Entry dated 16 Dec 92

116. McDonald S.J., Diary entry of 27 Dec 92

117. Agence France Press and Reuters reports quoted in *The Sydney Morning Herald* 29 Dec 92

118. Blake M.P., Interview with author on 5 Jul 93

119. Taylor K.B., Interview with author on 25 May 95. Cowden B.J., Interview with author on 25 May 95. Perryman D.J., op cit. Macdonald P.W., op cit.

120. MAJ P.G. Le Large. WO2 P.W. Macdonald. SGT D.W. McMillan. SGT L.A. Joyce. CPL J.P. Bell. CPL D.J. Burke. CPL A.J. Le Boydre

121. LT J.M. Lawson

122. 1 RAR Group Post OP Report—OP *Solace*. 12 tonnes of small arms ammunition was loaded by hand into *Tobruk's* magazine. The rest was loaded

on the upper deck. Just under 40 000 combat rations were loaded by hand into unoccupied accommodation areas.

123. Maritime Headquarters 'OP *Solace*—Post OP Report dated 9 Jul 93. 'The Box Metal Stores Shipping (BMSS) containers used by the Australian Army worked well, but unfortunately not enough were available for the deployment. Had all loose equipment been stowed in such containers, the security and transport of stores in the AO [Area of Operations] would have been much simpler.'

124. Five members of Ship's company were awarded Commendations from the Maritime Commander-Australia for their performance during this incident.

125. RADM R.A.K. Walls, RAN

126. MacDonald P.W., Essay, 'The SAD HMAS *'Tobruk'*. During OP *'Solace'* 26 Dec 92-21 Jun 93' prepared for author in Jun 95

127. Somerville A.M. Diary entry of 31 Jan 92. Somerville had been 1 RAR's Embarking Force Adjutant for Exercise Kangaroo 92 and was involved in loading and unloading *Tobruk*. He was left out of the loading planning for both *Jervis Bay* and *Tobruk* in Dec 92 despite being the only officer in 1 RAR with recent experience in deploying 1 RAR personnel, vehicles and stores by sea. Somerville felt that all vehicles and stocks could have been loaded if Le Large had not been given incorrect information on the dimensions of Landrover vehicles, and stores had been better organised before the ships arrived.

128. The Australian Defence Force had signed an Acquisition and Cross Servicing Agreement with the US Armed Forces on 30 Aug 1990. This was the authority for raising a Standard Order/Receipt Procedures for Mutual Logistics specific to combined operations. Brigadier Bill Traynor [BRIG W. Traynor], Director of Logistic Operations and Plans at HQ ADF flew to CENTCOM at Tampa Bay over the Christmas period and negotiated to have US supply agencies in Somalia respond to Australian requests for resupply. However, it was taking some time for the wording to be finalised before signature. The ADF were not used to working with CENTCOM. Typically, the point of contact with the US Armed Forces was usually the HQ of the Commander in Chief in the Pacific (CINCPAC) in Hawaii.

129. Blake M.P., Interview with author on 5 Jul 93 and Hurley D.J., Interview with author on 27 Jul 93

130. MAJ S.J. MacDonald

131. McDonald S.J., Diary entry dated 2 Jan 93

132. McDonald S.J., Diary entries dated 2/3 Jan and 9/10 Jan 93

133. MAJ G.P. Hurcum. MAJ R.H. Stanhope.

134. Hurcum G.P., Diary entry dated 31 Dec 93. This was the second time Stanhope and Hurcum had farewelled their wives in the early hours at an airport. Both had served for six months in Iran as part of the UN Iran-Iraq Military Observer Group.

135. There were 12-15 soldiers, who had not returned from leave at that time, arrived later. Crosland G.A., op cit

136. Flak jacket called Combat Body Armour was a vest of flexible light armour plating encased in a thickly-woven plastic 'skin'. Wearing a flak jacket in a hot climate was like wearing a tightly-fitted, thick plastic raincoat around the torso.

137. MAJ C. Chidgey, OC, Reinforcement Holding Branch (RHB), Logistic Support Force. Major General Murray Blake, after discussions and papers dating back to May 1992, had ordered the RHB to be raised on 9 Sep 92 in response to the increasing number of ADF personnel being deployed on UN operations overseas. The name of the RHB changed to Reinforcement Holding Company and then to Deployed Forces Support Unit under the command of Chidgey's successor, Major Alan Schmidt. RHB 611-1-10 Minute /92, 'Brief for LCAUST Visit—21 Oct 92' dated 20 Oct 92

138. MAJ B.M. Oswald CSC

139. Hollamby A.R., 'Information on OP *Solace*'. Notes written for MAJ A.D. Blumer, Officer Commanding D Company to assist with his Post OP Report. There were a number of diary entries as well as comments during interviews emphasising the irrelevance of the DIO briefings and their unnecessary duration.

140. Holmes P., 'Somali Protesters denounce UN chief', *The Australian*, 4 Jan 93 p.6

141. Bone J., UN chief defies the fears of his critics, *The Sunday Times* quoted in *The Australian*, 4 Jan 93 p.6

142. Holmes P., op cit

143. Ibid

144. COL J.S. Murray

145. LHQ AS LO Report US CENTCOM Period 26-28 Dec 92 Briefing by CINCCENTCOM of 28 Dec 92

146. Holmes P. op cit

147. CENTCOM Signal to LHQ dated 5 Jan 93

148. MAJ M.L. Harnwell

149. HQ Movement Control comprised a Major from Ayerbe's staff.

150. LTCOL A.W. Ayerbe, SO1 Logistics Operations, Logistic Command. LHQ 92 0098 OP *Solace* Log Aspects Part 2 Folio 42. Minutes of JAPG of 5 Jan 93 dated 7 Jan 93 Appendix 3 to Annex B to LHQ K92-00997 dated 19 Jan 93. The Air Force officers present, Wing Commanders Coleman, Jamieson (AirHQ) and Hartig (86 Wing, ALG) stated that 'Air Force aircraft, preferably 707 jet aircraft [payload of 45 000 pounds], could be made available at relatively short notice for the movement of stores.' Minutes of JAPG of 5 Jan 93 dated 7 Jan 93. The flight time of two days by the 707s from Sydney to Mogadishu and their payload compared favourably to the flight time of three days for C 130 transport aircraft with a payload of 20 000 pounds

151. Minutes of JAPG of 5 Jan 93 dated 7 Jan 93

152. LHQ 92 0098 OP *Solace* Log Aspects Part 2 Folio 41. Copy of Cross Servicing Agreement signed by BRIG Traynor on 7 Jan 93

153. McDonald S.J., Diary entry dated 5 Jan 93

154. Ibid

155. Stanhope R.H., CMOT Sitrep dated 4 Feb 93

156. McDonald S.J, Diary entry dated 7 Jan 93

157. Hurcum G.P, Diary entry dated 10 Jan 93

158. McDonald S., Diary entry dated 7 Jan 93

159. Reuters, AFP., 'Marines die storming warlord's armoury, *The Australian* ,8 Jan 93

160. Ibid

161. Movement Control Office Liverpool SIC WAQ OPS 100 Overseas Movement Advice of 11 Jan 93. Escort Officers: LTCOL A. Reynolds, CMDR T. Bloomfield. Media Representatives: Mr M. Maher, Mr B. Wilson, Mr R. Quinn, Mr D. McKinnon, Mr F. Kemp, Mr M. Toal, Mr C. Stellart, Mr D. Bone, Mr J. Frennick, Mr D. Wild, Mr M. Waly, Mr P. Mohane, Mr G. Bevilaqua, Mr A Gowing, Mr C. Adams, Mr G. Carroll, Mr. T. Decesare, Mr T. Robson, Mr M. Blenkin, Mr M. Landy, Mr B. Rueless, Ms J. Brough, Miss A. Foreman

162. Porter J., Somali [sic] contingent can shoot in self defence, *The Weekend Australian,* 9-10 Jan 93

163. Hurley D.M, Interview with author on 27 Jul 93

164. Hurley had received a report from Colonel W.J. Mellor entitled, 'Report on Visit by CAFS to Baidoa 27-28 Dec 92' which gave a concise and accurate description of the situation in Baidoa. Later Major S.J. McDonald sent a diagram by facsimile of unit dispositions at the Airfield and gave a verbal briefing to Hurley. However, without video footage or photographs, it would have been difficult to visualise the physical environment at Baidoa accurately.

165. *The Times.*, 'Somali factions agree to ceasefire', quoted in *The Australian,* 12 Jan 93, p.10

166. McDonald S.J, Diary entry of 14 Jan 96

167. MAJ J.D. Simpson. MAJ M.J. Moon

168. McDonald S.J., Annotations of second draft. McDonald had been unable to book US Army trucks before the arrival of the first contingent of Australian troops. Vehicles were made available some hours later after they had finished other tasks.

169. Annotations to First Draft, WO2 P.J. Angus. Angus witnessed this incident. This reporting resulted in a number of frantic telephone calls to the Welfare Cell at 1 RAR in Townsville from wives, partners, parents and friends. Entries in the Telephone Log Book of the 1 RAR Welfare Cell.

170. McDonald S.J., Annotations on second draft of 30 Mar 96

171. COL P.J. McNamara. Discussion with author on the second draft and his visit to Mogadishu and Baidoa 15-18 Jan 93 on 3 May 96. No one from Mellor's staff, Jackson's staff or McDonald had checked the transit lines

within 24 hours of the arrival of the first flight. The transit lines had been visited and arrangements had been made several days before. McDonald S.J., Diary Entry of 16 Jan 93

172. MAJ A.D. Blumer. MAJ D.G. McKaskill. MAJ M.L. Harnwell. LT W.R. Bowyer

173. Mckaskill D.G., Annotations on second draft dated 1 Apr 96

174. Macdonald P.W., op cit, p.13

175. Ibid, p.15

Chapter 3

1.   Hurley D.J., Interview with J.D. Simpson on 20 Jun 95

2.   Operational Study Team Report, Operation *Solace*—An Analysis of Intelligence Support and Civil/Military Operations submitted in June 1993

3.   'Busting' was a term used to describe a house or building search. Typically, a group of soldiers would surround the premises to be searched. After smashing the door open with a sledge hammer, several soldiers would run into the premises yelling out loudly and pushing occupants to the floor to cause maximum shock and surprise. The purpose of 'busting' was to ensure that those conducting the search retained the initiative.

4.   HQ AFS Intsum 12/93 to 13 Jan 93 dated 13 Jan 93

5.   Fraternal nickname given to the 3/9 Bn by fellow Marines.

6.   CI Report—Team 1 dated 18 Jan 93

7.   Stanhope R.H., Diary entry of 16 Jan 93

8.   John Caligari, Operations Officer, Greg Hurcum, Patrol Master, Dick Stanhope, Civil and Military Operations and Jim Burns, Intelligence Officer.

9.   Fraser D.J., Diary Entry dated 16 Jan 93

10.   LT R.J. Worswick.

11.   Worswick R.J., Undated essay, 'Urban Patrolling—A Company Perspective'

12.   CPL T.R. Connor.

13.   Worswick R.J., op cit. Note. Journalists accompanied the second patrol from A Company that day. Worswick wrote, 'Of course in true media fashion they

simply adjusted the facts to suit themselves by reporting in the Australian press that they had in fact escorted the first Australian patrol into Baidoa.'

14.   Worswick is talking about pulling back on the cocking handle of the Steyr rifle and allowing a round from the magazine to be picked up and pushed into the beginning of the rifle barrel, ready to be fired when the trigger is pulled.

15.   Worswick R.J., op. cit. Hurcum G.P., Annotations on First Draft. Hurcum who was the Patrol Master at the time annotated that the early patrols were 3 to 5 kilometres in length and would only exceed 3 hours in duration by exception. They were certainly conducted under very hot conditions, exhausting the soldiers who were acclimatising.

16.   Stanhope R.H., Diary Entry dated 16 Jan 93. Keen L., Videotaped interview with IMC Hospital theatre sister of Feb 93

17.   Hurley D.J., Interview with author on 27 Jul 93

18.   Worswick, op. cit.

19.   CPL D.W. Jobson

20.   1 RAR Operational Log Book Entry dated 17 Jan 93

21.   Probably not this long. See Footnote 15

22.   Worswick R.J., op. cit. During a research trip to Somalia in April-May 1993 the author spoke formally and informally with scores of officers, Warrant Officers, NCOs and soldiers. Though the ANZAC legend may have meant different things to different people, it was a source of inspiration for many of those who served in Somalia. Mixed in with the notion of an ANZAC legend was a fierce pride in themselves as members of the Australian Army. They felt that they had an obligation to the reputation of their Army to be the best UNITAF unit in Somalia

23.   CI Report Team 1 dated 19 Jan 93

24.   Worswick R.J., Annotations on first draft Dec 95

25.   Interviews with author in Somalia 29 Apr—12 May 93

26.   CPL W.B. Prosser. Simpson J.D, Diary Entry 18 Jan 9

27.   Simpson J.D, Diary Entry 20 Jan 93

28.   After some reluctance by logistic staff in Australia, seven thermal imagers were

deployed. Some were used at the airfield and others on night road patrols.

29. Pritchard A.P., Diary Entry dated 19 Jan 93

30. Pritchard A.P., Diary Entry dated 23 Jan 93

31. CAPT S.J. Dodds commanded the Mortar Platoon and LT P.J. Connolly was his Mortar Line Officer and Second In Command. Dodds and Connolly shared the work of commanding the Mortar Platoon on specific operations. The battalion had deployed with eight 81 mm mortars and several thousand rounds of mortar ammunition. The Platoon did set up and maintain a mortar base plate position at the airfield

32. 'sniper pair' 'schedule'. Two Australian snipers—long range marksmen—had occupied a concealed position to maintain surveillance of an area of interest. They had not sent in a routine radio report that they were OK. 'OA' is the call sign on a radio net for Battalion Headquarters.

33. CPL D.G Mitchell

34. CPL I.D. Johnson

35. PTE G.W. Hunter

36. Jargon for, 'I have a target and I am about to fire 'A sight picture is a term used to describe what firers see through the sights of their rifle when they aim it at a target, ie the 'picture' in the sight.

37. Connolly P.J., Diary Entry dated 21 Jan 93

38. 'BHQ' Battalion Headquarters

39. 'CARE' CARE Australia compound

40. Peter Kieseker was the manager of CARE Australia operations in Baidoa at the time.

41. Hurley had authorised pairs of Australian snipers to infiltrate into town and set up secret surveillance posts. One of these pairs had observed two suspicious groups of armed Somalis moving in a military manner on the outskirts of town through thermal imaging equipment.

42. OPSO Operations Officer. Caligari called off a pursuit of the Somali squads because at night with little information on their exact location, such a pursuit would have been extremely dangerous

43. Lieutenant T.A. Everett

44. IO—Intelligence Officer, Captain J.J.R. Burns

45. Connolly P.J., Diary Entry dated 22 Jan 93

46. Pritchard A.P., Diary Entry dated 2 Feb 95

47. PTE K.G. Went

48. PTE D.T. Gollop

49. Worswick., op cit

50. Other sources do not agree that most Somalis who carried weapons had not received military training. Years of conscription under the dictator Barre before 1990 and the subsequent formation of clan armies to fight the civil war suggests the opposite was probably true. WO2 P.J. Angus, CSM C Company, annotated on the First Draft that he had never met a Somali male who had not undergone some form of military training.

51. Petersen G.T, op cit. p.14.

52. Review of HQ AFS Intelligence Summaries from Operation *Solace* by author.

53. SSGT B.G. Thomson, SGT W.J. Douglas.

54. Thomson B.G., Annotations to first draft.

55. CI Report Team 1 dated 18 Jan 93

56. Ibid, p.10

57. Hurley D.J., Interview with author on 27 July 1993

58. Thomson B.G. Annotations on first draft

59. Over the operation WO1 W.F. Bowser, SGT A. Barrie, SGT M.K. Bramley, SGT G.C. Bull, CPL D.A. Martyn and CPL D.G. Moore arrived from Australia to increase the CI capabilities of the 1 RAR Group

60. CI Reports dated 21 Jan, 30 Jan 93 and 28 Feb 93

61. CI Report dated 17 Mar 93

62. Thomson B.G., Annotations on second draft dated 13 Mar 96

63. Gutaale, pronounced Goo Ta Lee. Thomson B.G., Annotations on first draft

64. Ibid

65. Warsame, pronounced Wa Sar Me

66. CI Report—Team 1 dated 21 Jan 93

67. CI Report—Team 1 dated 20 Jan 93

68. ADFA Australian Defence Force Academy.

69. 'Crack/thump' is Australian Army slang for the sound a passing bullet makes. The 'crack' is the sound the bullet makes as it breaks the sound barrier. The 'thump' is the distant report of the weapon that fired it following soon after.

70. Connolly P.J., Diary Entry dated 22 Jan 93

71. Somerville A.M., Diary Entry dated 3 Feb 93. Originally Hurley had complied with the American practice of wearing helmets. He stopped this after a few days. For a brief period he also permitted the wearing of flak jackets to be optional. However, he soon returned to making them compulsory attire after he received reports that the bandits were more likely to shoot at Australians now that they were not wearing their body armour.

72. CPL T.A. Aitken, C Company

73. Moon M.J., Diary Entry dated 29 Jan 93

74. CPL M.F.M. Braeckmans, C Company

75. Moon M.J., Diary Entry dated 30 Jan 93

76. The firing of a weapon accidentally or without an appropriate reason is deemed to be an unauthorised discharge. The abbreviation 'UD' is pronounced phonetically in conversation. Australian military personnel who have a 'UD' are charged immediately with an offence. Depending on the degree of negligence proven in a military court, the offender is fined and may spend 14 days in detention.

77. Pritchard A.P., Diary Entry dated 29 Jan 93. Annotations on the first draft by LCPL M.E. Rowlinson states that the rounds did not impact close to this soldier's body but impacted safely into a door where the shot appeared to have come from.

78. Moon M.J., Diary Entry dated 30 Jan 93. Most soldiers in the rifle companies worked from dawn until dusk. During the night they would have their sleep broken by having to stay awake with another soldier as a sentry for two hours. Those who went out on night patrols found it difficult to sleep the next day in the heat and with the noise of the camp.

79. Hurcum G.P. Diary entry of 23 Jan 93

80. There is a disagreement between Dodds and McKaskill on whether there were one or two APCs used in this activity. Dodds thought there was one but McKaskill disagrees. He believes that one APC would never deploy into a night location without the support of another APC.

81. Dodds S.J., Diary Entry of 23 Jan 93

82. Dodds S.J., Diary entry dated 23 Jan 93

83. Harnwell M.L., Diary Entry dated 1 Feb 93

84. CO—Commanding Officer.

85. Captain J.J.R. Burns, Intelligence Officer.

86. Dodds S.J. Diary Entry dated 1 Feb 93

87. Thomson W.H., Written description of 'Ambush Activity 31 Jan—1 Feb 93' and annotations on first draft

88. Ibid

89. Night vision goggles.' The goggles give a weird greenish view of the scene through two mini TVs over each eye. The world is made artificial like an advanced virtual reality computer game. The device is heavy—its straps and angles cut into the face—and as the face begins to perspire the goggles work their way into uncomfortable positions and seem to want to slip off—adding to the sense of disorientation. ... After wearing them for a short time I began to get a headache.' Hart D., (1995) *The Realism of Peace*—George Gittoes, Museum and Art Gallery of the Northern Territory: Darwin p. 68

90. Thomson W.H., op cit

91. Ibid

92. Moon M.J., Diary Entry dated 14 Feb 93. Comments made in various interviews about Hurley's warnings at this time.

93. Hurley D.J., Letter to wife dated 13 Feb 93

94. Blumer A.D., Diary Entry dated 3 Feb 93

95. CI Reports dated 30 Jan, 2 Feb and 3 Feb 93

96. CI Report—Team 1 dated 5 Feb 93

97. CI Report—Team 1 dated 30 Jan 93. The MSF organisation in Baidoa was made up of Dutch personnel who were not as experienced as the major MSF organisation in Somalia comprised of experienced French managers who had been in Somalia for some time. Unfortunately, this inexperience had resulted in the hiring of non-Baidoan criminals as guards. Vial. P., Interview with author on 15 Jul 96. Vial was the French manager of MSF in southern Somalia who had been in Mogadishu for over a year.

98. CI Report—Team 1 dated 3 Feb 93

99. Slouch hats and equipment belonging to members of C Company were on sale in town within a few hours.

100. Stow A.E., 1 RAR Group Caterer, Entry in George Gittoes' Journal dated 27 Mar 93

101. An ISO (International Shipping Organisation) container is a large metal freight container able to be loaded and unloaded from the back of semi-trailers. These containers were attractive items in Somalia because they could provide secure storage for valuable items.

102. CPL P.A.G. Bartlett, CFN C.A. Lindsay. A Fitter's vehicle is a mobile workshop used to maintain armoured and other military vehicles. It is equipped with a crane that had the capacity to load and unload heavy loads, such as APC engines. In this case, the crane was to be used in conjunction with a forklift to load the empty ISO container onto the back of the truck.

103. Worswick R.J., Annotations on first draft Dec 95. Worswick was in command of the QRF.

104. CPL W.B. Prosser

105. SGT G.S. Burns

106. CPL S. Moore

107. Caligari J.G., Diary Entry dated 6 Feb 93

108. Worswick R.J., Annotations on first draft Dec 95

109. CI Report dated 8 Feb 93

110. Blumer A.D., Annotations on first draft dated 9 Dec 95

111. Blumer A.D., Letter to author dated 20 May 96

112. Hurley D.J., Personal notes dated 16 Feb 93

113. Hurley D.J., Interview with author on 27 Jul 93 and other conversations.

114. Pritchard A.P., Diary Entries 11, 13 and 14 Feb 93

115. Pritchard A.P., Diary Entry dated 14 Feb 93

116. Kieseker P., Relationships between Non-Government Organisations and Multinational Force in the Field in Smith H., (ed) (1993) *Peacekeeping—Challenges for the Future*, Australian Defence Studies Centre. Kieseker pointed out that 'dogs in Somali culture are the lowest of the low. According to tradition ... they must wash seven times if they touch a dog'.

117. LCPL A.R. Golding. TPR L.S. Beveridge.

118. 1 RAR Operational Log Entries over the period 1-15 Feb 93 The incident involving the grenade being rolled under the APC occurred on the afternoon of 14 Feb 93

119. CI Report—Team 1 dated 11 Feb 93

120. PTE M.H. Sloman.

121. 1 RAR Operational Log 14 Feb 93

122. The author had numerous discussions with junior NCOs and soldiers about the use of appropriate force in face of provocation such as being spat on or having rocks thrown at them. Many put their support for prompt retaliation in the context of keeping the psychological upper hand, like a police force. They felt they had to be respected in Baidoa to do their job. They knew that if they seriously injured an unarmed Somali, even one who was just carrying their personal knife, while 'adjusting Somali attitudes', they would be charged and punished. Many junior NCOs and soldiers commented that quick, violent and brief retaliation raised morale, 'let off steam' and contributed to teamwork and individual self confidence.

123. Fraser D.J., Annotations on first draft dated 8 Dec 95

124. Blumer A.D., Letter to author dated 20 May 96

125. Hurley D.J., Interview with author of 27 Jul 93

126. LT A.C. Swinsburg. CPL G.J. Baker. CPL R.W. Chapman. 1 RAR Operational Log of 16 Feb 93 and Dodds S.J., Diary Entry dated 16 Feb 93

127. Swinsburg A.C., Diary entry dated 17 Feb 93 and annotations on first draft

128. Ibid

129. CPL A.G. Hodges

130. An Australian section in Somalia was made up of two groups. A Command/ Scout Group, comprising the Section Commander and a signaller, moved behind a Scout Group comprising two senior riflemen who had above average observation skills, and could react quickly and fire accurately. At night the scouts wore Night Vision Goggles. Moving behind the Command/Scout Group was a Machine Gun Group comprising the Section 2IC, two machine gunners equipped with F89 5.56 mm Minimi Light Support Weapons, each supported by one or two riflemen. The Section 2IC directed the fire of the two Machine Gun Fire Teams in the Machine Gun Group. Typically, the Command/Scout Group and MG Fire Teams only moved when they were covered by each other. The patrol commander and 2IC carried small Motorola squad radios to keep in contact.

131. SGT S.W. Boye. Boye was normally assigned as C Company's Mortar Fire Controller. He appears to have been 'moonlighting' as a scout on Hodges' patrol. PTE R.J. McCormack.

132. LCPL A.T. Combs. PTE J.C. Payne. PTE A.D. Wilson. PTE D.C. Searle. PTE P.J. Ingram.

133. PTE M.B. Dyer.

134. 1 RAR Contact Report, 'Report on Shooting Incident—Baidoa 170146C Feb 93' undated but probably written on 17 Feb 93 by the 1 RAR Intelligence Officer CAPT J.R. Burns. MAJ J.G. Caligari, the Operations Officer, was in Mogadishu.

135. Moon M.J., Diary Entry dated 17 Feb 93

136. PTE K.P. Low, PTE C.J. Hagedoorn. Harnwell M.L., Diary Entry dated 17 Feb 93 1 RAR Operational Log Entry of 17 Feb 93

137. MAJ D.J. Duncan.

138. 1 RAR Operational Log Entry of 17 Feb 93

139. Caligari J.G., Annotations on first draft Jan 96

140. Harnwell M.L., Diary Entry dated 17 Feb 93 1 RAR Operational Log Entry of 17 Feb 93

141. CI Report—Team 1 dated 19 Feb 93

142. 1 RAR Operational Log Entry of 19 Feb 93

143. CPL W.A. Perkins, PTE J.K. Flatley.

144. PTE C.J. Day. Entry in George Gittoes Journal dated 19 Mar 93. LCPL G. Lively

145. PTE J. Blakeman.

146. LCPL D.A. Williams Tpr D. Bell.

147. Day's wound was stitched up at the Swedish MASH but became infected a few days later. The wound was then unstitched and left open to heal from the inside out. He had to have the hole through his shoulder cleaned out and redressed daily for several weeks. Day C.J., Entry in George Gittoes' Journal dated 19 Mar 93

148. CI Report—Team 1 dated 19 Feb 93

149. CI Report—Team 4 dated 20 Feb 93

150. CI Report—Team 1 dated 19 Feb 93

151. LCPL A.R. Bradbrook.

152. PTE G.M. Brown.

153. Swinsburg A.C., Diary entry dated 19 Feb 93

154. Dodds S.J., Diary Entry dated 18 Feb 93

155. Moon M.J., Diary Entry dated 20 Feb 93

156. PTE M.J. Meehan.

157. LT I.A.G. MacGregor. CPL D.J. Ferriday.

158. Moon M.J., Diary Entry dated 21 Feb 93

159. LT T.D. Webb.

160. LCPL M.J. Holden.

161. WO2 C.M. Young. Harnwell M.L., Diary Entry dated 19 Feb 93

162. One section of APCs comprised four vehicles and could lift one infantry platoon

163. Blumer A.D., Diary Entry dated 22 Feb 93

164. LT M.J. Toohey CPL H.S. Smelt. CPL J.R. Hepburn.

165. Thomson W. Notes made about first draft.

166. Ibid

167. Thomson B.G., Annotations on first draft.

168. McDonald S.J., Diary Entry dated 22 Feb 93

169. McDonald S.J., Annotations on first draft dated 11 Dec 95. Sergeant Alan Barrie, an ex-military policeman from CI Team 2, had interviewed the teacher, Norkay, on McDonald's recommendation to obtain further information. From McDonald's point of view, the information provided by Norkay was clear about where the restaurant was located. However, he was not impressed with Barrie's 'very aggressive interviewing style'. He assessed that either Barrie misinterpreted Norkay's directions or Blumer did not receive them in sufficient detail later on.

170. 1 RAR Operation Log Entry of 23 Feb 93

171. Blumer A.D., Diary Entry dated 23 Feb 93

172. ICRC International Committee of the Red Cross.

173. Blumer A.D., Diary Entry dated 23 Feb 93

174. Blumer A.D., Diary Entry dated 24 Feb 93

175. Blumer A.D., Annotations on first draft dated 8 Dec 95

176. Blumer, op cit. Moon M.J., Several diary entries from this time. Caligari J.G., Several diary entries from this time.

177. Blumer, op cit.

178. Blumer A.D., Diary Entry dated 25 Feb 93

179. Ibid

180. Blumer A.D., Annotations to first draft dated 8 Dec 95.

181. Ibid

182. Moon M.J., Diary Entry dated 27 Feb 93

183. Blumer A.D., Diary Entry dated 27 Feb 93

184. Ibid.

185. Hurcum G.P., Diary Entry dated 2 Mar 93. Hurcum felt that Blumer's approach was too proactive.

186. Caligari J.G., Diary Entry dated 28 Feb 93

187. Moon M.J., Diary Entry dated 26 Feb 93. Blumer pointed out that there were incursions into D Company's area at times which gave him reason to complain. Blumer A.D., Annotations on first draft. It would seem that both company commanders found patrols from the other company in their areas.

188. CPL D.K. Mortimer.

189. A 'Bric' patrol was a four-man formation used in urban operations. The term had come into usage as a result of these types of patrols being used in Northern Ireland by British units.

190. LCPL D.S. Loftus, PTE R.D.R. Maitland, PTE S. Seath.

191. Loftus D.S., Entry in George Gittoes' Journal dated 26 Mar 93

192. Blumer A.D., Diary Entry dated 28 Feb 93

193. Mortimer D.K., Annotations on First Draft.

194. The body was wrapped in a plastic shelter half (hutchie) and carried in the trim vane of Corporal C.G. Clarkson's APC.

195. Blumer A.D., op cit.

196. Ibid

197. Moon M.J., Diary Entry dated 27 Feb 93

198. Hurcum G.P., Diary Entry dated 2 Mar 93 and Blumer A.D., Diary Entry dated 2 Mar 93

199. Caligari J.G., Diary Entry dated 2 Mar 93

200. Fraser D.J., Diary Entry dated 5 Mar 93

201. Blumer A.D., Diary Entry dated 5 Mar 93 and Hurcum G.P., Annotations to First Draft.

202. Blumer A.D., Diary Entry dated 6 Mar 93

203. WO2 R.J. Vine, CSM Administration Company. PTE A.K. Burns.

204. McDonald had also come into town to buy cartons of cigarettes for the 1 RAR canteen. McDonald S.J., Diary entry dated 6 Mar 93. The account of this contact is based on annotations on the first draft of this chapter by SSGT B.G. Thomson, a diary entry dated 6 Mar 93 from MAJ S.J. McDonald and conversations with both Thomson and McDonald in Feb 96. McDonald's action was in accordance with standing operating procedures that stated that when a group reacted to an incident, the senior person present was to take control. McDonald felt that to ignore the woman would have been both immoral and would have lost face in front of the locals. McDonald Annotations on second draft.

205. Thomson B.G., Annotations on first draft. 'ululating' is defined as 'howling' in the dictionary. This implies a fairly deep low-pitched sound. When they were afraid or issuing a warning of pending danger to others, Somali women made a high-pitched sound that was a cross between a howl and a scream.

206. McDonald S.J., Diary Entry dated 6 Mar 93

207. Thomson, op cit.

208. Ibid

209. Blumer A.D., Diary Entry dated 7 Mar 93

210. Hurley D.J., Diary entry dated 3-6 Mar 93

211. Letter signed by representatives from MSF (Holland), CARE, IMC, GOAL, ICRC, CONCERN, CRS World Vision International dated 10 Mar 93

212. CPL T.R. Conner. Fraser D.J., Diary Entry dated 12 Mar 93

213. Worswick R.J. Annotations on first draft Dec 95

214. Dodds S.J., Diary Entry dated 15 Mar 93

215. CPL L.J. Entink. LCPL P. Quentin.

216. CPL S. Moore, CPL C.G.J. Townson.

217. Fraser D.J., Annotations on first draft dated 8 Dec 95

218. Caligari J.G., Diary Entry dated 26 Mar 93

219. CPL P.J. Martin.

220. LCPL S.J. Chinner.

221. PTE A.M. Franklin. The primary source for the description of Martin's contact is the Contact Debrief prepared by CAPT J.R. Burns, the Intelligence Officer. The author read the typed final copy and the notes taken by Burns during his interviews with members of Martin's patrol. Other sources include: Caligari J.G., Diary Entries dated 26 and 27 Mar 93, Blumer A.D., Diary Entry dated 26 Mar 93 and 1 RAR Operational Log 26 Mar 93 Pages 824-826.

222. PTE A.B. Kowarik-Berger.

223. LCPL P.J. Larsen. LCPL G.A. Gough PTE R.T. Dann , PTE C.S. Russell.

224. McKaskill D.G., Annotations on second draft.

225. PTE P. Denby.

226. Blumer A.D., Annotations on first draft dated 8 Dec 95. It was standard procedure for Australian soldiers to call out 'Grenade'! as a grenade is thrown as a warning to their own troops in the vicinity.

227. Blumer A.D., Diary Entry dated 26 Mar 93

228. CI Report—Team 1 dated 27 Mar 93

229. Caligari J.G., Diary Entry dated 27 Mar 93

230. Blumer A.D., Diary Entries dated 28-29 Mar 93

231. Thomson B.G., Annotations on first draft

232. MAJ M Kelly. SGT P.J. Watson.

233. Thomson B.G., Annotations on first draft

234. CAPT S.J. Grace.

235. McDonald S.J., Diary Entry dated 1 Apr 93, Blumer A.D., Diary Entry dated 1 Apr 93

236. CAPT A.J. Annetts. Harnwell M.L., Diary Entry dated 2 Apr 93

237. LCPL S. McAliney.

238. CPL P.W. Nunan, CPL G.T. Wehmeier.

239. Nunan P.W. Letter to wife dated 3 Apr 93

240. WO2 K.G. Hudson, LCPL R.A. Osborne. CPL K.M. Brown.

241. Toohey M.J., Diary Entry dated 2 Apr 93

242. TPR M.G. Smith, Toohey's driver.

243. Toohey M.J., op cit.

244. CAPT D.W. Keating, Regimental Medical Officer, 1 RAR. >

245. Toohey, op cit.

246. CPL C. McSwan. He was 'Swany' referred to in McAliney's last words.

247. Blumer A.D., Diary Entry dated 2 Apr 93

248. The author conducted 109 interviews with members of the AFS in Baidoa during a research visit in April/May 1993 Lance Corporal McAliney's death featured in the vast majority as the single event that caused them the most sadness aside from incidents involving women and children.

249. Typed copy of LT Don Tulley's eulogy.

250. Blumer A.D., Diary Entry dated 5 Apr 93

251. TPR C.A. King.

252. 1 RAR Operational Log Entries dated 5 Apr 93

253. CPL G. Lively.

254. Incident Report and Intelligence Summary dated 5 Apr 93 and Moon M.J., Diary Entry dated 5 Apr 93

255. Moon M.J., Diary Entry dated 6 Apr 93

256. Moon M.J., Diary Entry dated 8 Apr 93 Vial P., and Terry F., Interview with author on 15 Jul 96.

257. The following narrative on the actions of LT Everett's platoon are based on the following sources: 1 RAR Operational Log Entries dated 7 Apr 93, Moon M.J., Diary Entry dated 7 Apr 93, Everett T.A., Diary Entry dated 7 Apr 93, Incident Report dated 7 Apr 93

258. SGT S.W. Boye CPl D.M. Heaslip.

259. Hankinson M.D., Annotations on second draft dated Apr 96

260. CPL D.A. Caple.

261. SGT M.T. Morrissey.

262. CPL R. Moore. LCPL P M. Soppitt.

263. CAPT M.D. Hankinson.

264. Moon M.J., Diary Entry dated 7 Apr 93 and 1 RAR Operational Log Entries dated 7 Apr 93

265. Moon M.J., Annotations on first draft

266. Moon M.J., Diary Entry dated 11 Apr 93 Thomson B.G. and Annotations on first draft

267. Blumer A.D., Diary Entry dated 12 Apr 93

268. Pritchard A.J., Diary Entry dated 22 Apr 93

269. MAJGEN P.M. Arnison AO.

270. LT S.A. Voss.

271. CPL W.B. Prosser.

272. PTE B.T. Connolly.

273. This was an amazing piece of luck for Connolly. 'Hard struck' meant that the firing pin had hit the base of the round, but the round did not fire. The malfunctioning firing pin probably saved Connolly's life.

274. PTE M.B. Greene, PTE D.J. Hawkins.

275. The 'pistol' he drew on Private Greene turned out to be a plastic imitation that he probably carried to frighten off would-be assailants. No one will ever know why he drew it and pointed it at Greene. It may have been a gesture of anger because his friend was being searched.

276. CPL P.D. Vigar.

277. 1 RAR Operational Log Entries dated 25 Apr 93, Prosser W.B. Statement dated 25 Apr 93, Craig A. Statement dated 25 Apr 93 and 1 RAR Incident Report dated 25 Apr 93

278. Sergeant A. McCarthy, Corporal T.A. Britton, and Privates M. Dale and A.J. Patttullo.

279. Britton T.A., Essay 'Pers Cell Visit to Bartamaha School Baidoa' 22 Apr 93

280. Van Der Klooster J.M., Annotations on first draft Feb 96

281. Ibid

282. LCPL J.J. Verschelden. PTE J. R. Hollingsworth. PTE M.J. Lange.

283. Lange M.J., Report entitled 'The Warehouse Shooting' Undated

284. Verschelden J.J., Report entitled 'The Warehouse Shooting' Undated

285. Harris I., Letter to Senator Robert Ray, Minister of Defence received 29 Jun 93

286. On 5 June, several truckloads of Pakistani soldiers were deployed to inspect stockpiles of weapons in southern Mogadishu as had been agreed under the terms of the ceasefire agreement in force at the time. Aideed struck decisively when Somali machine gunners opened fire on the trucks killing 24 Pakistanis and wounding 59 others. Eighteen American servicemen were killed in October in an ambush and their bodies dragged through the streets of Mogadishu by Aideed's supporters.

287. Unfortunately, the following analysis of the factors contributing to the Australian success has not been informed by Somali sources, except via CI Reports of Somali reactions and attitudes. The evidence to support conclusions is based on Australian sources and establishing correlations between Australian intentions and probable outcomes. Expressed in another way, Australian characteristics, techniques, tactics and attitudes have been analysed to assess which of them appeared to contribute to or detract from the successful outcomes of urban security operations in Baidoa.

288. On 17 Feb Canadian troops killed one Somali and wounded another when confronted and provoked by a rock throwing crowd. On the night of 4 March, one Somali was killed and another wounded approaching the Canadian compound at Belet Uen.

Chaprter 4

1.   Hurley D.J., Interview with author of 27 Jul 93

2.   Interviews by author with many members of the 1RAR Group in Somalia 29Apr—12 May 93

3.   Thomson B.G. CI Report, Intelligence Estimate—Buurhakaba dated 20 Jan 93

4.   Ibid

5.   LT M.P. Hanna

6.   SGTD. A. McKay. The rock monolith was about a quarter of the size of Uluru in Central Australia.

7. Thomson B.G., CI Report 'Operations at Buurhakaba 23-27 Jan 93' dated 28 Jan 93

8. Ibid

9. Thomson B.G. Annotations on first draft

10. Blumer A.D., Diary entries 22-23 Jan 93

11. Thomson B.G., Annotations on second draft dated May 96

12. Blumer A.D., Various diary entries from this period

13. Blumer A.D., Conversations with the author. Thomson B.G., Annotations on first draft.

14. LT R.J. Adams. Lieutenant G.W. Butler.

15. Fraser D.J, Diary entry dated 28 Jan 93 and Fraser D.J., Undated essay, 'Alpha Company in Buurhakaba.'

16. The APCs still had their shrouds on which slowed their movement. They were removed soon after.

17. Fraser D.J., Undated essay, 'Alpha Company in Buurhakaba'

18. CPL I.D. Johnston

19. CPL A.B. Fisher

20. 'Target indication' is a series of directions given by someone referring to key natural or man-made features to where an enemy group is located so that fire can be returned accurately.

21. 'Visual' meant moving the line of supporting fire away from advancing line of soldiers when they are seen.

22. 'Firm' meant no one else will move until the fire support section is on the ground in fire support positions and able to engage the enemy.

23. Connolly P.J., Diary entry dated 27 Jan 93

24. Connolly P.J., Diary entry dated 29 Jan 93

25. NVG Night Vision Goggles

26. A Patrol Ambush Light was an Australian-designed light-weight device comprising a box, an extendable arm and a light globe. It was small but capable of instantly throwing light out to 50 metres for 10 minutes.

27. Dodds held the 'clakkers' at the end of the wires running from the Claymores and also was able to operate the Patrol Ambush Light.

28. 'Killing ground' is a term used to describe the area in front of an ambush group where the enemy is expected to enter and be killed. Typically, the killing ground is focus of the fire power of the ambush group and will be criss-crossed by arcs of fire from weapons as well as being the impact area for thrown hand grenades.

29. '50s' and '30s' .5 inch and .3 inch calibre machine guns on APCs. "Minimis" were light machine guns carried in infantry sections.

30. Dodds S.J., Diary entry dated 29 Jan 93

31. Hurley D.J., Interview with author on 27 Jul 93

32. CI Report—Team 2 dated 2 Feb 93

33. Moon M.J., Diary entries dated 31-2 Feb 93

34. Ibid

35. LT I.A.G. MacGregor

36. Thomson B.G., Annotations on first draft

37. Summary of Operational Outcomes prepared for author by 1 RAR Intelligence Section

38. C I Report Team 2 dated 7 Feb 93

39. Thomson B.G., Annotations to first draft

40. Moon M.J. Diary entry dated 7 Feb 93. Moon understood the reasons for the show of force by the Australians. He felt his men demonstrated an ability to adapt to cordon and search operations and initially showed imagination and zeal. However, he also felt that the battalion expended significant resources and energy on these operations for modest returns. Moon M.J. Annotations on first draft

41. Blumer A.D., Diary entry dated 14 Feb 93

42. Dodds S.J., Diary entry dated 13/14 Feb 93

43. Hurley D.J., Interview with author on 27 Jul 93

44. Ibid

45. Moon later stated that he did not 'seek opportunities to test my men. My men had enough danger to contend with on security operations in Baidoa and in the HRS. C Company did not seek action nor did they shirk it when it came their way.' Moon M.J. Annotations on first draft. Blumer agreed with Moon and asserted that, 'Delta sought to be effective and decisive, but NOT to chase opportunity on the chance of creating conflict situations.' Blumer A.D., Letter to author dated 20 May 96

46. The author conducted over 100 interviews in Somalia in early May 1993 with a cross section of officers, WOs, NCOs and soldiers in each of the sub-units of the 1 RAR Group. He also had numerous informal conversations during that time. WOs, NCOs and soldiers had expected more opportunities to test themselves in combat by finding and fighting bandits. They were disappointed that their senior officers had not created these opportunities.

47. CPL A.B. Fisher

48. The Assault Pioneers are a specialist platoon in Support Company responsible for light engineering tasks and the laying and destruction of mines.

49. Dodds S.J., Annotations to First Draft. Dodds emphasises that he did not direct or order Somali children to find and retrieve mines. However, they did receive coffee, rations and encouragement when they found them and pointed them out to his men. Somali children already knew not to touch mines.

50. Dodds S.J., Diary entry dated 7 Mar 93

51. SGT M.A. Thorp

52. Thorp M.A., Notes made for second draft in Apr 96

53. Cupola—a circular hatch on the passenger's side of the driver's compartment permitting a soldier to stand on the seat with his head and shoulders above the roof.

54. Fraser D.J., Undated essay 'Alpha Company in Buurhakaba.'

55. Ibid. Conversations with author

56. CI Report dated 3 Mar 93

57. LT T.D. Webb

58. Only Infra Red headlights on

59. Dodds S.J., Diary entry dated 19 Mar 93

60. Moon M.J., Diary entry dated 14 Mar 93. Moon felt that the Australians had lost their operational security by this time and bandits were being forewarned about operations. Moon M.J., Annotations on first draft.

61. Bowser W.F., CI Report, 'Security Risks via Interpreters' dated 4 Mar 93

62. Blumer A.D., Entry in George Gittoes' Journal dated 27 Mar 93. Conversations by the author with other company commanders and officers confirmed that with only one or two exceptions, the Somali interpreters were crucial to the success of Australian operations.

63. LT J.M. Haddock

64. CPL D.R. Commons. TPR J.G. Robinson

65. CI Report—Team 1 dated 5 Mar 95

66. Dhegeweyne actually had ears that stuck out prominently. Many a turban was taken off Somali men in the region by the Australians during this time as they hoped to discover 'ears like wing nuts'.

67. Mayow was released by the ASF in Baidoa soon after his arrival there before he was able to be questioned further about Degeweyne's whereabouts and method of operation. CI Report Team 1 dated 21 Mar 93

68. MSF–*Medicens Sans Frontieres*, a French medical NGO. '33B and uniform elements' was an APC with a section of infantry.

69. Everett T.A., Entry in George Gittoes' Journal dated 17 Mar 93

70. Thomson B.G., Annotations on second draft dated 12 Mar 96

71. Mortimer D.K., Interview with author 12 May 93 Thomson B.G., Op cit

72. Thomson B.G., op cit

73. Interviews and discussions by author with a cross section of members of the 1 RAR Group in Somalia 29 Apr-12 May 93

74. CI Report—Team 1 dated 21 Mar 93 and Thomson B.G. Annotations on first draft

75. Dodds S.J., Annotation on second draft dated Apr 96

76. Dodds S.J., Diary entry dated 11 Mar 93

77. Thomson B.G. Annotations on first draft dated 1 Dec 95

78. Vial P., Interview with author on 15 Jul 96. Vial was the manager of the French MSF in Mogadishu.

79. Gavin Gray's [LT G.M. Gray]1 Platoon and Gavin Butler's 3 Platoon,

80. Ibid

81. Fraser D.J., Undated essay, 'Alpha Company at Buurhakaba.'

82. In the jargon of OP *Solace*, Road Runner techniques were daylight variants of Night Rider techniques.

83. The distance of two kilometres behind vehicles was a compromise between being seen or heard by ambushers and being in a position to react to an ambush that had been sprung. The ROE prevented Fraser from attacking ambushers until after they had sprung their ambush.

84. CPL S. Moore

85. Bandits 'shot' many people in the hand or leg as an act of terror.

86. Fraser D.J., op cit

87. Ibid

88. Ibid

89. Dodds S.J., Diary entry dated 15 Apr 93

90. 1 RAR Contact Debrief No 448 dated 13 Apr 93

91. LCPL D.A. Coleman

92. CPL G.G. Evans. CPL W.F. Davis.

93. 1 RAR Contact Debrief No 448

94. Blumer A.D., Diary entry dated 10 Apr 93

95. CPL J.B. Moore. Dodds S.J. Annotations on first draft Jan 96. Woodrow D.N., Record of interview with S.J. Dodds

96. PTE C.T. Dugal

97. PTE M.W. Reid

98. 6 x 6 Landrover fitted out as a mobile vehicle repair workshop

99. CPL H.T. Berry. CFN A.J. Goss. CFN S.E.J. Burdon. PTE G.N. Kirkpatrick

100. Goss A.J., Undated Incident Report

101. Ibid. 1 RAR Patrol Brief No 510 dated 30 Apr 93

Chapter 5

1. Stanhope broke up his 22-man team into four Forward Observer (FO) parties of an officer and three others. Typically FO parties would be attached to combat units to co-ordinate the use of artillery and other forms of joint offensive support, such as close air support or naval gunfire support. For OP *Solace* each FO Party became a liaison team with independent communications and transport, and were thereafter dubbed 'CMOT' [pronounced See-mot].

2. GNR P.T. Malone. GNR M.B.T. Peters

3. CAPT S.A. Bagnall

4. Stanhope R.H., BC/019 Sitrep dated 29 Jan 93

5. Stanhope R.H., CMOT Sitrep dated 4 Feb 93

6. Vial P., and Terry F., Interview with author on 15 Jul 96. Terry was Deputy Manager of CARE Australia in Baidoa until end of Dec 1992

7. MAJ D.J. Duncan

8. Dodds S.J., Diary entry dated 6 Jan 93

9. Everett T.A, Interview with author 29 Jul 93

10. Ibid

11. This CARE manager may have been Lockton Morrissey, not Martin Morris.

12. 15th Counter Intelligence Team, 7th Regimental Combat Team, Report 'Theft of Food by Gunmenat Daynuunay.' dated 3 Jan 93

13. Connolly P.J., Diary entry dated 19 Jan 93

14. Dodds S.J, Diary entry dated 19 Jan 93

15. FFR Land Rover 'Fitted For Radio'

16. Moon M.J., Diary entry dated 20 Jan 93

17.   Michael Willesee, Australian journalist and television presenter

18.   Moon M.J., Diary entry dated 21 Jan 93

19.   Moon M.J., Diary entry dated 25 Jan 93

20.   ²Moon M.J., Diary entry dated 23 Jan 93

21.   CAPT J.C., Hill

22.   CAPT P.J. Short

23.   Short P.J., Letter to author dated 6 Aug 93

24.   LZ Landing Zone

25.   Everett T.A., Essay entitled, 'Air Food Distribution to Molill' dated 9 Aug 93

26.   15th Counter Intelligence Team, 7th Regimental Combat Team, Report 'Theft of Food by Gunmenat Daynuunay.' dated 3 Jan 93

27.   OA-radio call sign for Battalion Headquarters

28.   Moon M.J., Diary entry dated 25 Jan 93

29.   Deduced by the author from interviews with members of the 1 RAR Group in Baidoa 1-10 May 93

30.   Blumer A.D., Diary entry dated 4 Feb 93

31.   Short P.J., Letter to author dated 6 Aug 93

32.   Deduced by the author from interviews conducted with members of the 1 RAR Group in Baidoa 1-10 May 93

33.   WO2 P.J. Angus

34.   Moon M.J., Diary entry dated 23 Jan 93

35.   Van Der Klooster J.M., Undated report 'The Blue' dated 16 Aug 93

36.   PTE C.D. Keyte

37.   Keyte C.D., Undated report 'The Blue' (probably 16 Aug 93)

38.   Parnaby A.B., Undated report 'The Blue' (probably 16 Aug 93)

39.   PTE J.E. Watson

40.   Watson J.E., Undated report 'The Blue' (probably 16 Aug 93)

41. LCPL A.J. Pavlic. The iron bar was a truck tyre lever.

42. PTE L.T.A. Carson

43. Pavlic A.J., Undated report 'The Blue' (probably 16 Aug 93)

44. Deduced by author from interviews with members of the 1 RAR Group in Baidoa 1-10 May 93

45. 'stroppy' slang for angry

46. Blumer A.D., Diary entry dated 10 Feb 93

47. Moon M.J., Diary entry dated 25 Jan 93

48. Fraser D.J., Diary entry dated 15-16 Feb 93

49. CI Report-Team 1,'Bandit Activity-Daynuunay' dated 2 Feb 93

50. Ibid

51. Gittoes G.N., Entry in Journal dated 17 Mar 93

52. Hurley D.J., Interview with author on 27 Jul 93

53. 'Jumale' pronounced Jewmalee. CI Report—Team 1 dated 21 Jan 93

54. CI Report—Team 1 dated 30 Jan 93

55. Ibid

56. Stanhope R.H., Diary notes of 24 Jan 93 and Vial P. and Terry F. Interview with author on 15 Jul 96.

57. Stanhope R.H., Annotations on final draft 28 Jul 96

58. Stanhope R.H., BC/019 CMOT Sitrep dated 29 Jan 93

59. Hurley D.J., Interview with J.D. Simpson of 20 Jun 96.

60. Stanhope R.H., Annotations on final draft 28 Jul 96

61. Stanhope R.H., Diary entry of 4 Feb 93

62. Stanhope R.H., Several diary entries and notes in February. Interview with author on 6 May 93 and Conversations with author during visit to Baidoa 2-9 May 93

63. Non-Government Organisations, Bai Region., Letter to Dr Boutros Ghali, Secretary General, United Nations dated 8 Mar 93

64. Hurley D.J., Interview with J.D. Simpson of 20 Jun 95

65. Hurley D.J., Interview with J.D. Simpson of 20 Jun 95

66. Stanhope R.H., Notes in working diary of 11 and 12 Feb 93

67. Abraham R., Entry in George Gittoes Journal dated 1 Apr 93

68. Van Der Klooster J.M. Annotations on first draft Feb 96

69. Van Der Klooster J.M. Annotations on first draft Feb 96

70. Ibid

71. SGT R.J. Elmy

72. PTE J.L. Williams

73. Van Der Klooster J.M., Report 'The Contact' dated 17 Aug 93

74. Williams J.L., Undated Report 'The Contact' (probably 17 Aug 93)

75. Van Der Klooster J.M., Report 'The Contact' dated 17 Aug 93

76. Ibid

77. PTE W.J. Lovis. LCPL J.J. Verschelden.

78. Williams, op cit

79. CPL G.T. Wehmeier

80. CPL P.W. Nunan

81. CPL A.B. Parnaby

82. Stanhope R.H., Diary entry of 31 Mar 93

83. LCPL T.J .Bell

84. Dodds S.J., Diary entry dated 8 Mar 93

85. Blumer A.D., Diary entry dated 14 Mar 93 'do a bolt'—leave quickly without warning.

86. Somali women could carry a 50 kilogram sack of grain by supporting it across their foreheads with a strap that went round the base of the sack.

87. Dodds S.J., Diary entry dated 6 Apr 93

88. Abraham R., Entry in George Gittoes Journal dated 1 Apr 9

89. Numerous interviews and conversations with members of the 1 RAR Group by the author during 1-10 May visit to Baidoa.

90. Harris. I., Letter to Senator Robert Ray, Minister of Defence, received 29 Jun 93

91. 15th US CI Team Report dated 3 Jan 93

92. Numerous CI reports. Eye witness accounts gained through interviews with members of the 1 RAR Group.

93. Mellor W.J.A., (1995) OP Restore Hope—The Australian Experience, Personal Experience Monograph, US Army War College, Carlisle Barracks, p. 29

Chapter 6

1. Initially, Mellor was the National Commander of the AFS. It was not until 3 March that he assumed direct operational command over the 1 RAR Group.

2. McNamara's operations staff comprised Lieutenant Colonel Richard Greville, Majors John Withers and Nick Welch, and Captains John Yeaman, Dick [R.E.] Siebert, Robert Fiske and Chris Nees. Vale's operations staff comprised Lieutenant Colonel Andrew [A.J.] Barton, and Majors R. Baker (UK), Mark [M.W.] Wall, Jim [J.H.] Kirkham, Bob [R.] Easton (NZ) and Craig [C.J.] Mills (promoted and posted to Personnel Section after Deployment Phase)

3. COL T.B. Winter

4. The depot at Randwick in Sydney was responsible for medical resupply and the depot at Woolangara in western New South Wales was responsible for ammunition resupply.

5. COL G.R. Thomas

6. COL J.K.H. Campbell

7. WGCDR I. Jamieson

8. WO2 K.W. Yeo

9. Interview WO2 K.W. Yeo with author on 10 Nov 93. Fortunately Yeo had been on Exercise Aussie Assist earlier in December which involved setting up a small Joint Force Headquarters. He was recalled from a holiday on Norfolk Island to return to Enoggera on 13 Dec 92 to help raise HQ AFS .

10. Most 1 RAR officers assessed that an ad hoc staff assembled at short notice in Brisbane would not meet the demands of overseas operational service. Several Townsville-based officers interviewed by the author felt that Mellor's headquarters staff should have come from HQ 3 Brigade personnel who were familiar with the requirements of 1 RAR and the ODF sub-units comprising the 1 RAR Group. Many officers ignored the HQ 1st Division officers after they arrived.

11. Interview with CAPT M.J. Fleming on 25 Jul 93. Several diary entries from 1 RAR officers. Interviews with 1 RAR officers. Interview CAPT M.R. Fulham and CAPT I.A. Young with author on 26 Jul 93

12. Mellor had graduated from Duntroon in 1971 and was allocated to the Australian Army Aviation Corps. Before beginning flying training he had served for one year as a platoon commander in 3 RAR. Hurley had graduated from Duntroon in 1975 and was allocated to the Royal Australian Infantry Corps. He had served as a platoon commander in 1 RAR and a company commander and operations officer in 5/7 RAR.

13. Interview LTCOL D.J. Hurley DSC with author on 27 Jul 93 and Mellor W.J.A., Comments on first draft of 8 Apr 96

14. Op *Solace* ORBAT as at 1 Jan 93. Fourteen signallers were assigned to provide 24 hour local and strategic communications. Six Military Police personnel were assigned to provide security and administrative support. A specialist communications detachment was assigned to liaise with US specialists in UNITAF. Two postal clerks and two Public Relations representatives were added to enhance the capability to manage mail and Public Relations.

15. The cap was raised to 930 to accommodate a three-man specialist communications detachment but from then on any additional personnel added to HQ AFS meant that someone in the 1 RAR group did not deploy.

16. Numerous conversations with members of the 1 RAR Group revealed deep resentment about the reduction of 1 RAR Group numbers to accommodate the increase in the numbers for the National Headquarters. Interestingly, the military personnel deployed from Canberra to form the Media Support Unit (MSU) were not included in the AFS establishment. The MSU was not a part of the AFS. Consequently, there was some ambiguity about who its commander, Ms Lisa Keen, was answerable to. In reality, she reported directly to Brigadier Adrian D'Hage in Canberra.

17. McDonald S.J., Diary entry of 2 Jan 93

18. Stanhope R.H. Diary entry dated 4 Jan 93

19. Hurley D.J., Interview with author 25 Jul 93 and Mellor W.J.A., Comments on first draft of 8 Apr 96

20. Hurley D.J., Interview with author 25 Jul 93

21. Hurley D.J., op cit. Retter P.B., Interview with author of 7 Jul 93 and Mellor W.J.A., Interview with author of 10 Nov 93

22. Interviews and conversations with author

23. McDonald S.J, Diary entry dated 7 Jan 93

24. Several diary entries made by Hurcum, Stanhope and McDonald from this period.

25. Mellor W.J.A., (1995) 'OP *Restore Hope*—The Australian Experience', Personal Experience Monograph, US Army War College, Carlisle Barracks, PA

26. 'CAFS' Commander Australian Force Somalia. Mellor W.J.A., Directive No 1/1993 'Directive by Commander Australian Forces Somalia to Lieutenant Colonel D.J. Hurley, Commanding Officer, 1st Battalion Royal Australian Regiment Group OP *Solace*' dated 10 Jan 93

27. Mellor W.J.A., Directive No 2/1993 'Directive by Commander Australian Forces Somalia to Lieutenant Colonel D.J. Hurley, Commanding Officer, 1st Battalion Royal Australian Regiment Group OP *Solace*' dated 10 Jan 93 '110001CJan93' means one minute past midnight, local Somali time (ie Time Zone 'Charlie') 11 January 1993. Time Zone 'Charlie' was 3 hours behind Greenwich Mean Time which is Time Zone 'Zulu'.

28. Mellor W.J.A., op cit

29. On the evidence available, there was no formal definition of what constituted a 'serious incident' warranting immediate notification to Mellor's headquarters.

30. 1 RAR Operational Log Book Entry dated 17 Jan 93

31. Communications to HQ 10th Mountain Division by telephone, e-mail, data transfer and facsimile from Baidoa were convenient and reliable. Communications from the 1 RAR Group to HQ AFS by RAVEN High Frequency radio over 240 km was unreliable and time consuming and

Inmarsat which was expensive, all information having to be passed by voice. Caligari annotations on first draft Mar 96. After some time the 1 RAR Group were able to communicate directly with HQ AFS by telephone, courtesy of the US communications system.

32. McDonald S.J., Diary Entry dated 19 Jan 93

33. Mellor W.J.A., Comments on first draft of 8 Apr 96

34. Mellor W.J.A, Interview with author on 19 Nov 93. He felt that this policy established his authority and exemplified the principle that it was easier to loosen the reins of command after they had been pulled tight from the beginning rather than tightening them later if they had been left loose at the beginning of a relationship.

35. Ibid.

36. Hurley D.J., Interview with J.D. Simpson of 20 Jun 95. 'In Retrospect—An Oral History on an OP Other Than War.' Command and Staff College, Queenscliff

37. Caligari J.G., Interview with author on 2 May 93. Interviews and conversations with 1 RAR Group officers verified that most had a perception that HQ AFS was kept busy by seeking information from HQ 1 RAR Group. Hurley's Headquarters (excluding Stanhope and his four Captains from CMOT) comprised 10 Majors and Captains including Major David Tyler, the PR Officer for the AFS, who was 'on the books' as a HQ AFS officer but was based in Baidoa.

38. Mellor W.J.A., op cit

39. HQ AFS LOG 5 Minute 245/93 dated 29 Jan 93

40. Kerin's visit had been planned by his Department six months before but had been postponed due to the security situation in Somalia. Now that the AFS was deployed he wished to combine his trip to visit NGO staff in Mogadishu with a visit to the 1 RAR Group in Baidoa.

41. CAFS Minute No 85 'Report on Meeting with Special Envoy, Mr Robert Oakley, 18 Jan 93' dated 18 Jan 93

42. The difference between the two ROE lay in the circumstances for the use of 'deadly force'. Peace enforcement allowed for the use of 'deadly force' in response to observing persons carrying crew-served weapons and operating

armoured vehicles. Peace keeping only allowed the use of 'deadly force' as a final act of self defence.

43. CAFS Minute No 93 'Report on Meeting with Comd UNOSOM BRIG Shaheen, 19 Jan 93' dated 19 Jan 93

44. Ibid

45. Catchlove N.H. Minute 'CAFS Responsibilities for PI/PR Planning and Implementation.' dated 4 Jan 93. Catchlove used the Joint Service Publication 41, Public Information Policy as his reference. Major David Tyler, the AFS PR officer, was due to arrive with Hurley's advance party on 10 Jan 93

46. The intention in Major General Blake's provisional directive to Mellor was that 'CO MSU is responsible to me'. In reality, the MSU was commanded and directed from Canberra by Brigadier Adrian D'Hage. He communicated directly to Lisa Keen, and she to him.

47. Interview Lisa Keen with author on 10 Aug 93 and Interview MAJ D.J. Tyler with author on 10 Aug 93

48. Lisa Keen and a video camera crew remained in Somalia for several weeks gathering footage for a documentary on Australian operations in Somalia, 'Every Mother Should be Proud.'

49. Harnwell M.L., Diary entries 15-26 Jan 93

50. Harnwell M.L., Diary entry of 18 Jan 93. As a consequence of loading *Jervis Bay* and *Tobruk* with the ammunition and stores for conventional operations, there was insufficient power generation, tentage, lighting, water pumps, tarpaulins and flooring to establish a tented, serviced camp at Baidoa.

51. Davis R.L., Message on the Internet E Mail dated 28 Mar 96

52. Harnwell M.L., Diary entry of 18 Jan 93

53. OP *Solace*—An Analysis of Equipment Performance'. Report prepared by the Operational Study Team deployed to Somalia in Apr 93

54. Ibid.

55. Ibid. Hurley did authorise a small number of RAVEN radios to be repaired in Baidoa by replacing faulty modules 'to overcome the severe shortage of HF radios.'

56. Ibid.

57. Major Gary Banister, DQ 3 Brigade, had made senior logistics staff aware of the stores and vehicles left on the wharf in Townsville on 5 Jan 93 at the Joint Administrative Planning Group meeting. He recommended a second sailing of *Jervis Bay* to transport these stores to Mogadishu. His recommendation was not supported. Banister G.R., Conversation with author on 15 Jul 96

58. COL T.B. Winter

59. A Service Demand is a request for stores to make up a unit's entitlement to an inventory of items to enable it to do its job. Logistic Command was responsible for satisfying these demands. Staff Demands were requests by units for items that were not on their published entitlements. These demands required justification and were forwarded to Army Office in Canberra for satisfaction. For example, Major Gary Banister's requests for the 1 RAR Group to be issued with extra GPS Navigation equipment, Motorola squad radios and notebook computers were sent to Army Office in Canberra, not to Logistic Command in Melbourne.

60. The Moorebank Logistic Group is now a Defence Department organisation called the Defence National Storage and Distribution Centre. HQ Log Comd Minute 1/93 'Op *Solace* Maintenance' dated 26 Jan 93. Ayerbe A.W., Interview with author on 4 Nov 93, Tweedie P., (SO2 Logistic Operations) and Woodward B., US Army Exchange Officer, (SO2 Materiel Management), interview with author on 3 Nov 93. Timeframes for satisfaction of demands varied from 7 days for Priority 1 to 28 days for Priority 4 in accordance with the Australian Services Materiel Issue and Movement Priority System (AUSMIMPS).

61. LTCOL A. W., Ayerbe

62. Thomas G.R., Interview with author on 15 Aug 96. Thomas had been a senior logistics operations officer with Log Div during the Gulf War.

63. LTCOL P. Neuhaus, Acting SO1 Operations.

64. LHQ Minute 'Assessment of Stores Requirements' dated 18 Jan 93 This aircraft left Townsville on 24 Jan 93 carrying five pallets of 77 Radio Set batteries, an APC engine, differential and spare parts, 25 portable generators (several generators had arrived in Baidoa in an irreparable condition), two Landrover engines, 6 000 sandbags and six Motorola radios. Neuhaus was

saying that from now on the resupply of the AFS was in the hands of Logistic Command's resupply system. He stated that Operations Branch at LandHQ would no longer intervene and dispatch stores by RAAF aircraft at short notice unless it was 'absolutely necessary'.

65. Although they had been reconditioned several times, most of the APCs dated back over 25 years to the Vietnam War. On field exercises in Australia APC fleet operations had been severely restricted for years to prolong the life of each vehicle and to reduce the costs of operating them.

66. McKaskill D.G., Interview with author on 5 May 93

67. Harnwell M.L., Diary Entry dated 9 Feb 93

68. Harnwell M.L., Diary Entry dated 4 Feb 93

69. Woolnough G.T., Annotations on second draft dated 24 Jul 96

70. Woolnough G.T., Interview with author on 11 May 93. Woolnough was right. There was and is a systemic problem during field exercises based on overseas deployment scenarios. In 1996 the following extract from a report supports his view about ODF units and also notes larger systemic problems. 'Units made up shortfalls in the stocks they took on the exercise by driving back and forward to supply depots in Townsville. ... The concern with CSS [Combat Services Support] provision centres on the sustainability to units over the longer term. Follow on logistic spt will be required shortly after deployment or a dedicated support ship off the coast for the ground forces will be essential for operations of the type practised in ... [this] scenario. Few problems observed by the evaluation team were systemic in nature with an ability to significantly affect ... operations."

71. At that time Woolnough's point was academic. With the exception of Peter Neuhaus's intervention in late January, none of the 1 RAR Group's Priority One demands were being satisfied in seven days.

72. Woolnough G.T., Annotations on first draft

73. Harnwell M.L., Several diary entries in Feb 93, McLeod R.D., Several diary entries in Feb 93 and McDonald S.J., Several diary entries in Feb 93

74. Harnwell M.L., Diary Entry dated 8 Feb 93

75. 3rd Brigade Administration Support Battalion, Harnwell's parent unit at Townsville. On operations supporting an independent battalion group

Harnwell would be expected to submit his service demands to 3 BASB if it was acting as 3rd Line support for the operation.

76. 2nd Field Supply Battalion was a 4th Line logistic unit in Townsville. Like 3 BASB, this unit could act as Harnwell's next line of support to satisfy his service demands.

77. The point Woolnough is making is that HQ AFS was not set up to be a higher headquarters for either the operations (HQ 1 RAR) or logistic (HQ BSG) units of the 1 RAR Group. He may as well have added that HQ AFS was not set up to be a higher headquarters for personnel functions either. MAJ Nigel Powrie who had been 'double hatted' as the staff officer responsible for medical matters as well as personnel matters commented that 'the world of personnel will be a new experience for me.'

78. Woolnough G.T. Diary Entries dated 6 an 8 Feb 93

79. Harnwell M.L., Diary Entry dated 9 Feb 93 and 1 RAR Operational Log Entry of 8 Feb 93. The million dollar price tag appears to have been an initial estimate. A 'trickle system' of resupply meant a small-scale steady flow of items to keep up with consumption, not infrequent bulk supply. McKaskill D.G., Annotations on second draft. Woolnough G.T., Annotations on second draft

80. Harnwell M.L., Diary Entry dated 9 Feb 93 and Harnwell M.L., Interview with author of 1 May 93. Aside from being ultimately responsible for APC operations, Hurley had a total of four years experience with mechanised operations as an exchange officer with the Irish Guards and as a company commander and operations officer with 5/7 RAR, a mechanised unit.

81. Woolnough had set up a local purchase office in Kenya, located in Mombassa, but it took Captain Mark McKeon some time to get used to locating and purchasing items in an African city, despite his office being located with a US local purchase supply office. He was also ordering in quantities that were too small to compete with the 'bulk' purchases of the American Local Purchase Officers.

82. Woolies' is an abbreviation for 'Woolworths', the name of a chain of retail stores that sell low priced items for the mass market.

83. Harnwell M.L., Diary Entry dated 7 Feb 93

84. Ibid

85. Mellor W.J.A., Annotations on third draft. Woolnough G.T., Annotations on second draft

86. Several diary entries from this period. This issue was mentioned by almost every officer and NCO who the author spoke with or interviewed. From the 1 RAR Group perspective it was a prime example of the lack of support they received from HQ AFS. From the HQ AFS perspective they had done their best in an imperfect resupply situation. For those in Baidoa the gap between the intentions and outcomes of their higher Australian headquarters was too great.

87. McKaskill D.G., Annotations on first draft of Chapter 6 dated Apr 96. While there are other explanations for this situation, it is clear that the reason for the delay and eventual arrival of the larger screwdrivers was not communicated well to McKaskill and his subordinates. They appeared to "dine out" on this story as an exemplar of the performance of the local purchase system.

88. Stanhope was in charge of liaison with Somali civilian authorities in Baidoa. He had ascertained from his staff working with Somali elders and political factions that the ISO container was abandoned

89. Mellor W.J.A., Comments on first draft of 8 Apr 96

90. CAPT A.W. Reid

91. Caligari J.G., Diary Entry dated 6 Feb 93

92. Mellor W.J.A., Interview with author on 10 Nov 93 and annotations on first draft of Chapter 6

93. HQ AFS NAM LOG 19 'Service Demands' dated 13 Feb 93. By this time the 1 RAR Group had sent 15 Priority 1 service demands. LHQ K93 00030 Op *Solace*—Service Demands Folios 12-21

94. Ayerbe A.W., Interview with author on 4 Nov 93, Nagy W.S., Interview with author on 15 Dec 93 and Tweedie P., Interview with author on 3 Nov 93. The Australian Army's peacetime resupply system was based on plenty of warning, overseas purchase, and resupplying and maintaining units in barracks or on short-duration field exercises where well-defined usage rates and predictable operating conditions applied.

95. An 'air hour' was the estimated cost of operating a particular type of aircraft for one hour, some types are more expensive to operate than others.

96. Despite having about half the load capacity and range of a 747 jet aircraft, an Air Force C 130 transport aircraft costs twice as much to operate per hour

97. Newham B.C., Interview with author on 15 Dec 93 and Nagy W.S., Interview with author on 16 Dec 93

98. CPL J. Rickard

99. SGT W. Schafferius.

100. Schafferius W. Interview with author on 5 Mar 96

101. CAPT J.W. Yeaman.

102. LHQ Operations Branch Minute 'Aircraft and Stores Movt to Somalia' dated 14 Feb 93 and HQ LOG COMD SIC E3L/IYJ OPS 'Movt of Svc Demand stores to AFS' dated 110610Z FEB 93. According to this signal, the initial quotes for flying this 66 000 pound load to Nairobi, which was the closest point of entry used by commercial air freight companies, was $AUD120–140 000. The cost of using 150 air hours to transport this load in two RAAF C 130 aircraft was estimated to be $AUD 800 000.

103. The operation of all Australian Air Force aircraft is governed by who has been allocated the air hours to operate aircraft. Each financial year air hours are allocated through tri-service negotiations to support Navy, Army and Air Force operations and training.

104. Martin G.K., Annotations on first draft.

105. CAPT J.W. Yeaman. LHQ Ops Branch Minute 'Aircraft and Stores MOVT to Somalia' dated 14 Feb 93. There had been a flurry of signals and telephone calls beginning on the weekend of 8/9 Feb 93 Yeaman prepared a brief on 14 Jan 93 for Administration Branch to clarify what he assessed to be the situation was with the movement of stores to Somalia.

106. Yeaman took a personal interest in this problem because he had just spent two years at 1 RAR as the Assistant Quartermaster.

107. Schaffarius W. Interview with author on 5 Mar 96

108. The author could find no evidence on file that anyone at LandHQ or HQ Logistic Command was aware of these funds being set aside for OP *Solace*. Subsequent events suggest that staff were not aware for these funds at this time.

109. Mellor W.J.A., Annotations on third draft of Chapter 6

110. Fortunately for HQ AFS staff Richard Clarke was authorised to use public funds to purchase a refrigerator, a television and a video cassette player.

111. MAJ R.J. Clarke. Clarke was a Canberra-based public servant commissioned as a Second Lieutenant and promoted to Temporary Major on the same day, and then posted to HQ AFS for OP *Solace*.

112. The regulations governing the use of public funds at the time did not permit 'investment' in commercial transactions such as starting up a unit canteen

113. Harnwell M.L., Diary entries of 15 Feb 93 and 25/26 Feb 93

114. Ibid

115. Woolnough G.T., Diary Entry dated 14 Feb 93

116. Mc Kinnon M., 'Red tape puts troops lives in jeopardy', *Townsville Bulletin*, 15 Feb 93

117. LHQ ADMIN/OP NAM 1389/93 'Op *Solace*—Class Six Items' dated 17 Feb 93

118. Macinnis I. Brief for Minister of Defence—Support to OP *Solace*—Mail and Canteen Stores dated 22 Feb 93

119. Harnwell M.L., Diary Entry dated 15 Feb 93

120. Ibid

121. Annex A to HMAS *Tobruk* Minute 5/7/15 dated 4 Jul 93. This annex summarised *Tobruk*'s activities during OP *Solace*. McDonald P.W., 'The Ships Army Detachment (SAD) HMAS *Tobruk* During OP *Solace* 26 Dec 92-21 Jan 93'. This document was a chronology of *Tobruk*'s activities with descriptions of cargo carried.

122. Mellor W.J.A., Annotation on first draft of Chapter 6 dated 8 Apr 96

123. Culleton J.J., Interview with author on 19 Jan 96. Culleton was an Australian exchange officer working as a member of the Directing Staff at the Canadian Command and Staff College, Kingston. He and the author reviewed the Canadian organisation and command and control arrangements for OP *Deliverance*. The Canadian Government had made a bigger investment in supporting OP *Restore Hope* than the Australian Government. Labbe had

medium range aircraft for flying troops and stores in Somalia, as well as to and from Nairobi. He also had helicopters to support his troops with Aero-Medical Evacuation, urgent resupply and command and liaison.

124. Taylor K.B., Interview with author on 25 May 95. Caligari J.G., Interview with author on 3 May 93. The three Canadian Sea King helicopters on HMCS *Preserver* not only moved 226 800 kg of materiel ashore in support of Canadian troops in Belet Weyne, but also acted as gunships and reconnaissance aircraft.' *Jane's Defence Weekly*, 'Seeking to replace Sea Kings', 21 Feb 96

125. McDonald S.J., Diary Entry dated 15 Feb 93 and Moon M.J., Diary Entry dated 15 Feb 93

126. Moon M.J., Diary Entry dated 15 Feb 93

127. Mellor W.J.A., Annotations on first draft of Chapter 6

128. Diary entries, letters and interviews with several 1 RAR officers

129. Mellor W.J.A., op cit

130. Worswick R.J., Undated essay, 'Urban Patrolling—A Company Perspective'. 'Visitors on patrols was a big hindrance during the initial stages of the deployment.'

131. Tweedie P. Interview with author on 3 Nov 93 and Martin G.K., Interview with author on 4 Mar 95. Tweedie was a logistics staff officer at HQ Logistics Command and Martin was commanding the Army's Ground Liaison Group who were responsible for the co-ordination of Air Force support to the Army.

132. CAPT G.W. Kosciuszko

133. Martin G.K., Interview with author on 4 Mar 94. Undated AirHQ Brief 'OP *Solace*—RAAF Support' probably written in May 93. This brief pointed out that AirHQ did not know the purpose of the 'monetary supplementation' from the Army for OP *Solace* until late in Apr 93 when AirHQ received LHQ signal SIC E3L 14J 40022/93 dated 20 Apr 93. On the evidence available, Ian Jamieson and his staff did not know that the $2.4 million supplementation, transferred to them by financial managers at Army Office to support OP *Solace*, was to pay for air freight. They appeared to have set the funds aside to pay for redeployment of the AFS back to Australia in May.

134. Martin G.K., Annotations on first draft. Greville R.H., Conversations with author Mar 96

135. LTCOL G.K. Martin

136. Martin G.K., Annotations on first draft

137. Dodds S.J., Diary entry dated 20 Feb 93

138. 'marry up' is an Army term for 'bring together'

139. Harnwell M.D., Diary Entry dated 25 Feb 93

140. Harnwell M.L., Diary entries of 15 Feb 93 and 25/26 Feb 93. On 12 Feb 93 HQ AFS requested that stores should be flown into Baidoa where the C 130 could be refuelled and then loaded with stores that needed to be back loaded to Australia. LHQ K 93 00030 Op *Solace* Service Demands Folio 8. HQ AFS signal to Land, Logistic and Air Command dated 12 Feb 93. Woolnough offered Australian Air Force personnel from Greg Jackson's staff attached to UNOSOM to co-ordinate unload and load of stores in Baidoa.

141. Martin G.K., Annotations on first draft. NGOs unloaded their C 130s carrying humanitarian aid and miscellaneous stores using Somali contract labour. On the evidence available, the 1 RAR Group were never given the option of unloading Air Force C 130s by hand in Baidoa or undertaking a 500 km round trip to Mogadishu to pick up stores.

142. Several diary entries from 1 RAR Group officers, interviews and conversations by 1 RAR Group officers with the author. Woolnough G.T. Interview with author on 10 Nov 93 and 1 RAR Bn Gp Post OP Report— OP *Solace* Annex F Command, Control and Communications.

143. Woolnough G.T., Annotations on second draft dated 24 Jul 96

144. Harnwell M.D., Diary Entry dated 2 Mar 93

145. LHQ K92-00998 Op *Solace* Log Aspects Part 4 Folio 65. Army Office (ASRP-A) did not send approval for use of $ 10 000 to purchase canteen stores, 'As a result of representations from LHQ', until 26 Feb 93'

146. HQ AFS NAM LOG 19 dated 1 Mar 93

147. David Creagh and Mick Fulham at HQ AFS created a data base on a lap top computer that showed the status of each demand. Copies were regularly sent to Vale's staff.

148. LHQ K93 93-00030 Op *Solace*—Service Demands Folios 77-83. This file is full of signals from MLG to Log Comd advising of the numbers of unsatisfied service demands from Somalia. This system of informing Log Comd when service demands could not be satisfied from stocks held at supply depots was known as the Refer Demand Process.

149. A Fleet Manager was a civilian or military officer who had responsibility for managing the acquisition and issue particular items of supply.

150. LHQ K93 93-00030 Op *Solace*—Service Demands Folio 86. Tweedie P., Interview with author on 3 Nov 93. Another interesting feature of the communications about service demands that frustrated Woolnough was the habit by the Moorebank Logistic Group of saying that certain service demands had left 'on consignment.' This gave an impression that they had been satisfied and were on their way to Mogadishu. In reality, it meant that they had been put into Multi Pack containers or on pallets and were now at Warehouse 90 at Moorebank.

151. Harnwell M.L., Diary Entry dated 2 Mar 93

152. LHQ K93 93-00030 Op *Solace*—Service Demands Folio 26. Tyres and tubes were in Service Demand 1/93. The impact of camel thorn on vehicle operations had been identified but possibly not well communicated well before the 1 RAR Group Deployment. LHQ K92-00964 ASC UNOSOM Somalia—OP Iguana Med/Health Aspects. HQ ADF Signal 2204 dated 3 Sep 92. Quote 'A notable feature of the regional vegetation [in Somalia] is the preponderance of thorns, some which reach lengths of 10 centimetres and are capable of puncturing the tyres of a standard two and a half tonne military transport vehicle.' Unquote. The 1 RAR Group had deployed with 12 Ply Landrover tyres that were no match for the camel thorn. McLeod had ordered 18 Ply tyres and additional inner tubes to overcome the problem.

153. Moon M.J., Annotations on second draft in May 96

154. Hurley D.J., Interview with J.D. Simpson dated 20 Jun 95

155. McLeod R.D., Diary Entries dated 25 Feb 93

156. McLeod R.D., Diary Entries dated 1 and 3 Mar 93

157. LHQ K92-00998 Op *Solace* Log Aspects Part 4 Folio 24

158. LHQ K92-00998 Op *Solace* Log Aspects Part 4 Folio 11. The ships correct name was Sabrian. Entry to Kenya from the sea was via the port of Mombassa. It is not clear where King obtained the funds to finance these sea and commercial air freight options for resupply.

159. Harnwell M.L., Diary Entry dated 4 Mar 93 and LHQ K92-00998 Op *Solace* Log Aspects Folios 12-50. During the period 3-20 March numerous signals were sent back and forward from AFS to Australia and within Australia to and from Land, Logistic and Air Commands. They would form an excellent case study of how the resupply system to Somalia operated.

160. McKaskill D.G., Annotations on first draft of Chapter 6 dated Apr 96

161. Greville R.H., Interview with author on 7 Feb 96. Martin G.K., Interview with author on 4 Mar 94. An interesting feature of these negotiations was that Air Force 'operators' at 86 Wing were very keen to gain experience in operational flying while Air Force logisticians were cautious about allocating Air Force training hours to fly resupply sorties to Somalia.

162. Woolnough G.T., Diary entry dated 5 Mar 93

163. LHQ K92-00998 Op *Solace* Log Aspects Folio 27. Though the author has no evidence either way, it is likely that Army Office paid for this air freight bill as a 'one-off' unforeseen cost to OP *Solace*. It is unlikely that AirHQ used part of the $ 2.4 million allocation because the files at LHQ contain no evidence that Army Office, LHQ or Log Comd asked them to pay.

164. Terms such as 'satisfied' and 'satisfaction date' were ways of saying that stores had arrived on or before the date they were required by the unit who had sent the service demand. To satisfy a service demand on its satisfaction date meant that the store had been delivered to the requesting unit on time.

165. LHQ K92-00998 Op *Solace* Log Aspects HQ AFS LOG 19 Facsimile to COLOPS [COL Phil McNamara and COL ADMIN [COL Brian Vale] dated 110301ZMAR93. Mellor attached a list of unsatisfied demands for HQ Logistic Command's urgent follow up by 15 March.

166. LHQ K92-00998 Op *Solace* Log Aspects Folio 23.

167. Harnwell M.L., Diary Entry dated 14 Mar 93

168. Harnwell M.L., Diary Entry dated 15 Mar 93

169. Interview MAJ M.L. Harnwell with author on 5 May 93. The seal arrived through the resupply system several weeks after the forklift operator received his through the mail from a mate.

170. WO2 M.E. Robinson. McKaskill D.G., Annotations on first draft of Chapter 6 dated Apr 96

171. Tweedie P. and Woodward B., Interview with author on 3 Nov 93. The author has concluded that if 50 percent of the demands were overdue on 17 March, as reported by Woolnough, then 'a significant number' were most likely to be in the In trays of Fleet Managers and others were somewhere else in the system.

172. Westworth I.G., Interview with author on 31 Jan 96. Westworth was the QM who took over from Rod McLeod a few weeks before 1 RAR returned to Townsville. He found that service demands sent by McLeod in Somalia had their priorities adjusted by hand by persons in the resupply system.

173. LHQ K92-00998 Op *Solace* Log Aspects Folio 31 and 43

174. LHQ K92-00998 Op *Solace* Log Aspects Folio 29. The last four APC engines and transmissions were transported from Bandiana to Moorebank late in March. If these were required in Somalia, the Army had no more spares. Engines and transmissions would have to be cannibalised from the APC fleet in Australia.

175. Air hours are the 'currency' of air support in the Australian Defence Force and are divided into Category 1, 2 and 3 depending on the nature of their use and priority for use. For example, air hours allocated for operations should have higher priority than air hours allocated for training. Each year the allocation of air hours is the subject of robust negotiation between staff from HQ ADF, AirHQ, Army Office and Navy Office.

176. AirHQ Brief and supporting papers 'Op *Solace* RAAF Support' undated but probably written in May 93

177. Richard Greville, Gary Martin and Jim Wallace had entered Duntroon in 1970.

178. Wallace was able to do this responsibly because high priority was given to counter-terrorist training and special forces contingencies. This training often involved air deployments from Perth to the eastern seaboard States.

Consequently, when training was rescheduled or cancelled, air hours became available. Fortunately, in early 1993 Wallace was able to transfer air hours to support OP *Solace* rather than returning them to HQ ADF for central reallocation.

179. Greville R.H., Interview with author and conversations of Feb 96. Martin G.K. Interview and conversations with author Mar 96.

180. 'OP *Solace*—An Analysis of Command Control and Communications.' Section of Operational Study Team Report P 7-8

181. Ibid

182. LTCOL G. Allan CSC

183. Kevin Taylor later noted that 'Army signallers ashore were extremely reluctant to request communications assistance from *Tobruk*.' HMAS *Tobruk* Minute 5/7/15 dated 4 Jul 93

184. 'OP *Solace*—An Analysis of Command Control and Communications.' op cit

185. LTCOL T.R. Jones

186. MAJ B.A.R. Scott

187. Scott B.A.R., Interview with author on 10 May 93

188. MAJGEN D.J. McLachlan

189. COL T.B. Winter. Log Comd SIC NAM OPS 435 dated 230511ZMAR93 and LOG COMD SIC NAM OPS 449 dated 24055ZMAR93. On 6 April, McLachlan, recalling his experience in the Vietnam War when he had observed service demands being tracked at Army Headquarters through the Directorate of Maintenance, wrote on Ayerbe's minute, 'I am not sure that pushing it [monitoring service demands] down to MLG is the way to go.' McLachlan conferred with Thomas separately and decided that Thomas and his staff were probably in a better position than Winter's staff to improve the resupply link to Somalia. Thomas G.R., Interview with the author 15 Aug 96

190. LTCOL B.C. Cook. Stevenson C.R. Interview with author on 6 Mar 96

191. LT C.R. Stevenson

192. SGT G.G. Weston CPL D.R. Fisher. Interview Stevenson C.R. with author on 6 Mar 96

193. Thomas G.R., Interview with author on 15 Aug 96. Stevenson C.R., Interview with author on 6 Mar 95 and follow up conversations in Aug 96

194. Ibid.

195. Martin G.K., Discussions with author May 96.

196. Greville R.H., Interview with author on 7 Feb 96

197. Mellor W.J.A., Interview with author on 10 Nov 93 and Caligari J.G., Diary Entry dated 19 Mar 93. Numerous interviews and conversations with both Mellor's staff and Hurley's subordinates.

198. Mellor W.J.A., Annotations on first draft of Chapter 6 dated 8 Apr 96

199. Ibid

200. Hurley D.J., Interview with author on 25 Jul 93

201. Mellor W.J.A., Annotations on third draft of Chapter 6 dated 19 Aug 96

202. Caligari J.G., Interview with author on 3 May 93 and subsequent conversations with author in Jul 96

203. Greville R.H., Annotations on second draft of Chapter 6 dated 7 May 96

204. COL J.J.A. Wallace AM. LHQ K92-00998 'Op Solace Log Aspects' Part 6 Folio 35 AirHQ signal dated 26 Mar 93

205. Conversations LTCOL R.H. Greville with author Jan-May 95

206. LHQ K92-00998 Op Solace Log Aspects Part 4 Folios 12-50. Woolnough G.T., Interview with author on 9 Nov 93

207. By this time B Squadron's 36 APCs had travelled over 120 000 km in three months, the equivalent of 12 months of normal operations in Australia. By the end of four months of operations in early May, 8 engines, 4 631 track link, 15 276 track pads, 7 transmissions, 95 road wheels, 3 differentials and 5 final drives had needed to be replaced. B Sqn 3rd/4th Cav Regt, Post OP Report—OP Solace.

208. Hurley had received permission to deploy MAJ Mark [M.C.] Fairleigh, 2IC 1 RAR, CAPT Ian [I.G.] Westworth, QM, and WO1 Dale [J.R.] Sales, 1 RAR RSM, who had been posted into 1 RAR in December to replace MAJ Glenn Crosland, MAJ Rod McLeod and WO1 Greg Chamberlain who had been posted out of 1 RAR, but had been retained for OP Solace.

209. LHQ K93-00030 Op *Solace*-Service Demands Part 3 Folios 9. 17 and 20.
     'hasteners' a term used for correspondence sent to remind a supply agency
     that a demand for stores has not been satisfied on time.

210. Stevenson C.R., Interview with author on 6 Mar 95 and subsequent
     telephone conversations

211. Ibid

212. Ibid

213. Ibid

214. Thomas G.R., Interview with author on 8 Aug 96

215. Ibid

216. LHQ K93-00030 Op *Solace*-Service Demands Part 3 Folios 9. 17 and 20.

217. LHQ K93-00030 Op *Solace*—Service Demands Part 3 Folio 21 Overseas
     Movement Advice dated 2 May 93. Like all of its predecessors it carried
     Multi-pack containers showing the size, weight and priority, but not their
     contents. Harnwell's staff would be none the wiser about the stores being
     carried on this C 130 as they had been on any of the others.

218. Woolnough G.T. 'Brief for Comd 1 Div. Op *Solace*—Supply Matters dated
     26 Apr 93 Woolnough wrote that only five APCs had been prevented from
     operating for 5 days because of a lack of track link and two forklifts had been
     unable to operate for 30 days, awaiting spare parts.

219. McKaskill D.G., Op *Solace* Post Operational Report B Squadron 3rd/4th
     Cavalry Regiment May 93

220. Woolnough G.T., Annotations on second draft dated 24 Jul 96

221. In early March Mr Malcolm Love, the Australian Quarantine Inspection
     Service manager in Townsville had travelled to Mogadishu and Somalia and
     made and assessment of the situation. He reported that a major cleaning
     effort would be required to meet quarantine regulations.

222. Angelatos P.L., ASC UNOSOM Weekly Situation Report 9/93 dated 28
     Feb 93

223. CAPT A.N. Jakab. Westworth I.G., Letter to author dated 27 Feb 96

224. MAJ L. Collins

225. Martin G.K., Annotations on first draft and conversations with author in Apr 96. Collins L., LHQ K93 00082,'An Appreciation on the Return to Australia of HQ AFS and 1 RAR Group dated 19 Mar 93'

226. The author has this assessment based on the subsequent behaviour of senior Air Force officers in trying to secure the support of LandHQ staff for an 'all Air Force' option for the redeployment.

227. Interview LTCOL R.G. Greville with author on 7 Feb 96. By this time Greville knew that all of the 'fat' from the Army's allocation of air hours had been trimmed by Gary Martin for earlier resupply flights. He knew that the Army would be billed for the redeployment and this would have a significant impact on Army operations and training in Australia over the coming months.

228. McNamara P.J., Brief to LCAUST on Redeployment of AFS dated 8 Apr 93

229. Nye T.W. Notes, 'AirHQ JOPG Meeting held on Friday 16 April 1993' dated Apr 93. AirHQ Brief 'OP *Solace*—RAAF Support' undated but likely to have been written in May 93. Later in early May US Vice President Al Gore requested that the 1 RAR Gp's return be rescheduled to the end of May to allow time for a UNOSOM II unit to be found for Baidoa. Senator Evans, the Foreign Minister, refused to adjust the dates as movement arrangements had been finalised and vehicles and stores had begun moving to Mogadishu for cleaning.

230. Treloar R.B., Air HQ Minute 'JOPG Discussion Notes—16 APR 93' dated 22 Apr 93

231. Nye, op cit. Conversations COL P.J. McNamara with author of Aug 93

232. MAJ D.N. Hasson. CAPT A.J. Hall. SGT D.A. Coiro. SGT A.G. Wormington. Hall A.J., Interview with author of 9 Nov 93

233. McKaskill D.G., Annotations on second draft of Chapter 6 of Jun 96

234. Mellor W.J.A., (1993) OP *Restore Hope*—The Australian Experience, Personal Experience Monograph, US Army War College, Carlisle Barracks.

235. Westworth I.G. Letter to author dated 27 Feb 96

236. HQ AFS Sitrep 139/93 of 19 May 93

237. 'Zero Alpha.' the radio callsign and slang term for HQ 1 RAR

238. General Blake had second thoughts about this decision. On 6 April he directed Lieutenant Colonel Paul Retter to investigate how well the command and control arrangements had worked between HQ AFS and the 1 RAR Group. In retrospect, he felt that the tasks of a national headquarters should have been clarified better. He was also concerned that 'cobbling together' a headquarters from a 'makeshift' structure and putting a Colonel in command may not have been the best way to set up command and control arrangements for a battalion group on operations. Retter P.B., Notes—'LCAUST Brief on Tasks of the OST' dated 6 Apr 93

239. Technically, HQ AFS commanded Air Force personnel for about two weeks during the redeployment phase of OP *Solace*. However, the operation to redeploy the bulk of AFS personnel was planned and conducted by the Land and AirHQ, in conjunction with specialist ADF personnel deployed from Australia.

240. Mellor's group was called a National Liaison Team on HQ ADF files and a Reconnaissance Group on LHQ files. This suggests that they were satisfying a strategic as well as an operational role. The Navy took a more conventional approach. They deployed a reconnaissance and a liaison officer. The reconnaissance officer returned to Australia to report back and the liaison officer stayed in Mogadishu with Mellor and his staff.

241. Mellor's background was in Army Aviation, Petersen was an Intelligence Corps officer and Catchlove was a Transport Corps officer, seconded to the Director General of Public Information.

242. In Nov 97 the ADF was authorised to plan contingencies for sending a battalion group to protect the distribution of humanitarian aid to Rwandan refugees in eastern Zaire. This time the national commander and the battalion group commander were scheduled to depart together to affect liaison and to conduct a reconnaissance of the area of operations. However, some of the lessons from OP *Solace* appeared to have been forgotten. Once again the liaison and reconnaissance group was limited to 10 personnel and was to travel in an Air Force Falcon jet. This time the Australians would have been depending on the good will of the Canadians to accommodate them, protect them and move them around.

243. LTCOL C.H. Green. LTCOL I.B. Ferguson

244. Grey Jeffery., (1988) *The Commonwealth Armies and the Korean War*, Manchester: Manchester University Press. Gallaway Jack., (1994) *The Last Call of the Bugle: The Long Road to Kapyong*, Brisbane: University of Queensland Press. Ferguson I.B. Interview with Kit Denton recorded in 1986.

245. LTCOL I.R.W. Brumfield LTCOL A.V. Preece

246. Breen Bob., op cit p. 26-27

247. Group Captain R.L. Perry, Director Joint Plans, HQ ADF commented on 22 July 1993 after reading the LHQ OST Papers that there was an interesting inconsistency. One part of the OST Report stated that 'the DJFHQ concept was validated as a workable option for "national" command of a unit deployed offshore' ... Pointing to another part of the report Perry noted that " Clearly, however, there were problems of unfamiliarity that came from divorcing a unit from its parent formation HQ."

248. Gallaway Jack, (1994) *The Last Call of the Bugle—The Long Road to Kapyong*, Queensland University Press: Brisbane. Gallaway presents compelling evidence that British and American commanders used 3 RAR to attack strongly-held Chinese positions and to act as rear-guard far more frequently than other battalions in the same brigade.

249. Breen Bob., (1992) *The Battle of Kapyong—3rd Battalion, the Royal Australian Regiment, Korea 23-24 April 1951*, HQ Training Command: Sydney p. 10. A quote from a manuscript written by LTCOL I.B. Ferguson MC DSO (RL) about his service in Korea, given to Kit Denton by Ferguson in 1986 and passed on to the author in 1992.

250. Ibid p. 104

251. The American nickname for Jackson was 'Daddy Diaper'. He was perceived to be too attentive to the employment of the 1 RAR Group and media reports, and too sensitive about casualties, which by American standards were deemed to be light. The nickname was meant to evoke the image of an anxious father always changing the nappies of a baby. The inference was that Jackson was too fatherly and the 1 RAR Group was not a 'baby' and could take care of itself. With hindsight, Jackson's attentiveness and sensitivities were well justified.

252. The three major controversies were over poor clothing and equipment issued to the 1 RAR group, the delay and inefficiency of the mail service and the conditions under which a soldier was held in detention

253. Limitations imposed by meeting deadlines for this book and security classifications still applicable to post-operation and post-exercise reports from recent operations and field exercises do not permit a full exposition of why the author is still concerned about the ADF's capability to deploy and logistically support off shore operations efficiently. However, some examples will be given to support the analysis. Some service officers assess that logistics is difficult and that the persistent problems encountered during exercises and operations are the norm. If recent logistical performances are the norm when exercises have been planned months in advance and off shore operations have been relatively benign, then the consequences of replicating the norm for more complex and dangerous operations in the future are potentially quite serious.

254. ACOPS File A 92-34185 Part 1 Australian Contingent to Somalia—OP *Solace* Logistic Aspects Folio 7 "Logistic Assumptions—Options for supply of rations and options for general maintenance", undated but written in Dec 92'. The nomination of *Jervis Bay* and SUCCESS to support OP *Solace* may have been because the planning group were aware that *Tobruk* was undergoing major maintenance in Sydney until 20 Jan 93

255. Army Office DGLWP E3L/I4J DAAPA 1062 of 200123ZDEC92

256. CDF Canberra (Info to DEFARM but not LHQ) SIC E31/I4J of 160703ZDEC92 issues FAD 1 to *Tobruk* and *Jervis Bay*. LHQ signal to DGLWP (for SO2 OPS) dated 22 Dec 93 requests FAD 1. On the same day MAJ R.J. Patrick requests FAD1 from HQ ADF and it was issued immediately.

257. HQ ADF CDF WNG O E3L/I4J of 15 Dec 93 'Phase 3 Conduct of ops— MCAUST be prep to prov strat sea tpt for force resupply. ACAUST 'be prep to prov strat airlift for log resupply and AME from Australia.'

258. 'shadow posted.' Term to describe a person who had a full time appointment as well as another appointment in special circumstances. King only became HQ MC after an Australian force was deployed off-shore and required resupply. In effect, he continued to fulfil his other duties while acting as a one-man joint movements agency for OP *Solace*.

259. The HQ MC concept survived OP *Solace* and was implemented again for OP Lagoon in Sep 94 when Australian ships, a force HQ and specialist personnel were deployed to Bougainville to facilitate a peace conference. HQ ADF reported. 'The responsibilities of HQMC were not well known and required the issue of a HQ ADF message to both clarify their role and overcome difficulties in the movement of sustainment stores and mail. ... due to the ad hoc nature of HQMC, HQ LOGCOMD-A staff are considered to have some difficulty divorcing their day to day Army role from that of a Joint movements organisation.' HQ ADF DJLOP DB 092/94 OPS 94/27354 dated 4 Nov 94

260. There were also consignments of stores sent via commercial and service air to Nairobi in February and by commercial sea freight to Mombassa in April.

261. Logistic staff officers currently serving at HQ Logistic Command who have reviewed earlier drafts of this chapter have taken exception to this generalisation about fleet managers, and by inference those who supervised them during OP *Solace*. They point out that much of the evidence offered to support the conclusion comes from an American exchange officer, the newly-appointed QM of 1 RAR and a junior officer, all of whom dealt with fleet managers directly, but not the fleet managers themselves. Aside from begging the question, 'Why should these officers not tell the truth ?', the author acknowledges that it would have been better scholarship to have interviewed several fleet managers to understand the pressure they were under to satisfy the service demands of units in Australia as well as those deployed overseas. This book may help fleet managers and those who supervised them to decide on whether the priorities for satisfying service demands they accorded the 1 RAR Group during OP *Solace* were justified or not. After reading the book, some logistic staff may be relieved that they were not interviewed and their excuses made public.

262. In September 1994 it was the supply of spare parts to helicopters, not APCs, Landrovers, forklifts and generators that exposed weaknesses in the resupply system to deployed forces. During an overseas deployment of an Australian-led force off Bougainville in PNG, a Sea King helicopter became unserviceable and it took two weeks for the spare part needed to fix it to be flown to Forward Operating Base (FOB) on Buka Island. Later it was found that the part waited at Richmond Airbase for two weeks before being put on the C 130 transport aircraft shuttle supporting the FOB at Buka Island.

Chapter 7

1. LCPL D.J. Miller, Clerk at HQ 1 RAR during OP *Solace*.

2. CAPT J.C. Hill

3. Hill V. Interview with author on 23 May 93

4. WO2 R.J. Vine

5. WO2 K.G. Hudson.

6. Vine S., Interview with author on 28 Jul 93

7. Hudson J., Interview with author on 28 Jul 93

8. WO1 H.J. O'Brien

9. O'Brien H.J., Interview with author 2 May 93

10. PTE D. McLeod

11. MAJ D. Hills

12. McLeod D. Interview with author on 3 Aug 93

13. Mooney A., Interview with author on 29 Jul 93

14. 'Rear Details staff' Units that deploy on operations and field exercises leave some administrative staff behind to be a point of contact for the unit. Typically, Rear Details staff maintain a 24 hour Duty Officer roster.

15. Mooney A. and Hannay E. , Interview with author on 29 Jul 93

16. WO2 N. Walsh

17. CAPT G. Dick

18. MAJ A.E., Smeaton

19. Welfare Co-ordination Cell., Family Information Package, 1993

20. WO2 D.R. Caple

21. SGT M.C., Sullivan

22. CPL D. McCoy

23. Hurley D.J., Interview with author on 27 Jul 96

24. Next of Kin, the closest relative, guardian or friend nominated by a member of

the ADF as the first person to be contacted in case of emergency, injury or death. Hurley L., Interview with author on 27 Jul 93

25. Spouses of Somalia Newsletter No 1 distributed in mid January.

26. Ibid

27. During three visits to Townsville in May, July and August 1993 the author interviewed a cross section of 35 spouses, partners, fiances, family members and close friends of officers, Warrant Officers, NCOs and soldiers who served in Somalia. The interviews were conducted at HQ 1 RAR Conference Room in Townsville in small groups based on the ranks of the interviewees spouses, partners, fiances or boyfriends He also reviewed two collections of correspondence received by a wife and a girlfriend from an officer and a soldier who served with the 1 RAR Group and spoke informally with several spouses and partners. Hereafter these sources will be titled 'Spouses, partners and friends interviews May/Jul/Aug 93' Most interviewees did not want their views to be attributed.

28. Hurley L., Interview with author on 27 Jul 93

29. Spouses, partners and friends interviews May/Jul/Aug 93

30. Ibid

31. 1 RAR Welfare Log., Entries in Jan 93

32. Ibid

33. McKinnon M., 'Shots fired, but Aussie soldiers arrive safely' *Townsville Bulletin* dated 16 Jan 93

34. Hall M., 'Troops could battle Somali warlords' *Townsville Bulletin* dated 9 Jan 93 'Spouses, partners and friends interviews May/Jul/Aug 93

35. *Townsville Bulletin*, 'Somalia Messages' dated 16 Jan 93 *Townsville Independent*, 'Hello Somalia' dated 22 Jan 93

36. D'Hage A.S., Letter to 'family and friends in included in second SOS Support Group Newsletter undated but probably written late Jan 93

37. 1 RAR Welfare Log entries Jan-Feb 93

38. Spouses, partners and friends interviews May/Jul/Aug 93

39. WO1 J.D., Collins

40. PTE D. Bretherton

41. Collins J.D., Interview with author on 2 May 93

42. LHQ K92-00997 Part 1 Folio 24 Welfare Situation Report No 1 dated 15 Feb 93

43. Spouses, partners and friends interviews May/Jul/Aug 93

44. Concluded by the author after over 100 interviews with a cross section of members of the AFS in Baidoa in May 93

45. Spouses, partners and friends interviews May/Jul/Aug 93

46. Ibid

47. Ibid

48. Ibid.

49. MAJ M.J. Fleming

50. Fleming M.J., Letter to Ms Gail MacKay, Director of State Education, Northern Region dated 22 Feb 93, Fleming M.J., Letter to Sr Mary McDonald Director, Catholic Education Office, Diocese of Townsville dated 22 Feb 93

51. Jacovos I., Letter to Chaplain Martin Fleming dated 5 Mar 93

52. McDonald M., Letter to Fr Martin Fleming dated 11 Mar 93

53. Jacovos I., op cit

54. Spouses, partners and friends interviews May/Jul/Aug 93

55. Ibid

56. Ibid

57. WO1 J.D. Collins. PTE D.R.Bretherton. Collins J.D., Interview with author 2 May 93. It became clear from the author's research that Collins and Bretherton worked extremely hard to make the postal system work. Without their efforts the mail would have been even more intermittent.

58. 1 RAR Welfare Cell Log Book Entries Feb 93

59. 1 RAR Group Post OP Report—OP *Solace*. Administration Aspects

60. Ibid

61. Dodds S.J., Diary Entry dated 23 Mar 93

62. Tyler D.J., Letter to wife dated 10 Mar 93

63. Spouses, partners and friends interviews May/Jul/Aug 93

64. 1 RAR Welfare Cell Log entry 31 Mar 93

65. Tyler D.J., Letter to wife dated 9 Mar 93

66. Spouses, partners and friends interviews May/Jul/Aug 93

67. Gittoes G., Journal entry dated 29 Mar 93

68. Stated to the author during interviews and conversations in Baidoa 1-12 May 93

69. Tyler D.J., Letter to wife dated 9 Mar 93

70. Tyler D.J., Letter to wife dated 10 Mar 93

71. The AUTOVON is a US Armed Services telecommunications system that connects American military units in the US and around the world. The AUTOVON system also links to all other telecommunications systems in the countries around the world through switchboards located in the US Wherever US military units go, AUTOVON goes.

72. MAJ M. Fairleigh. MAJ W. Austin

73. Tyler D.J., Letter to wife dated 17 Mar 93

74. Spouses, partners and friends interviews May/Jul/Aug 93

75. Spouses, partners and friends interviews May/Jul/Aug 93

76. A point made in numerous interviews by the author with members of the AFS 29 Apr-11 May 93

77. Spouses, partners and friends interviews May/Jul/Aug 93

78. Blake M.P. Minute LCAUST 97/93 dated 5 Apr 93

79. Australian Associated Press Report dated 31 Mar 93 Typed transcript

80. Tyler D.J., Extract included in letter to wife dated 2 Apr 93

81. Blumer A.D., Diary Entry dated 4 Apr 93

82. Moon M.J., Diary Entries dated 10 and 14 Apr 93

83. Ibid

84. Moon M.J., Op cit. Caligari J.D., Diary Entry dated 14 Apr 93 By 15 April 13 members of the AFS had returned to Australia, 10 from 1 RAR and four from other units. Of the 13, only three were returned for compassionate reasons related to the death of close relatives and two to resolve marital issues. Of the remainder, four had returned for medical reasons, two for disciplinary reasons and two returned because their work as communications specialists was no longer required. Foster M., Thirteen soldiers back from Somalia, *Townsville Bulletin*, 15 Apr 93 Based on a press release form MAJ W. Pickering, Army PR Officer in Townsville.

85. Spouses of Somalia Support Group Newsletter No 7 distributed early Apr 93

86. Hurley L., Interview with author on 27 Jul 93

87. WO2 M.E. Robinson

88. Green J., Diggers launch appeal for Baidoa children, *Townsville Bulletin*, 28 Apr 93

89. Hall M., Companies make sure seconds come first, *Townsville Bulletin*, 10 May 93

90. *Townsville Bulletin.*, 'Top brass laud digger's efforts', 18 May 93

91. *Townsville Advertiser.*, 'Private had family support', 20 May 93

92. MAJ I. Hughes

93. Conversations with author on 22 May 93

94. Hannay E., Interview with author and notes given to author on 28 Jul 93

95. pugarees were distinctive Australian hat bands

96. Visiting VIP Civil: Senator Robert Ray, Senator Eric McGibbon, Mr Tim Fischer, MP, Mr Ted Lindsay, MP, Senator Ian McDonald, Senator Margaret Reynolds, Queensland Premier Wayne Goss, MLA. Alderman Les Tyrell, Mayor of Thuringowa. Visiting VIP Military: LTGEN J.C. Grey AO, CGS, MAJGEN P.M. Arnison AO, Commander 1st Division, GPCAPT R.K. McLennan and CMDR B. Hunt

97. The management of weapons and other personal equipment returning with members of the AFS, both before and after the Welcome Home Parade, was organised by WO2 Don Hunt, who had been the hard-working 1 RAR Barracks Warrant Officer in Lavarack Barracks during OP *Solace*.

98.   Breen Bob, *Australian Diggers* .... op cit, p. 51-52, 256.

99.   Spouses, partners and friends interviews May/Jul/Aug 93

Postscript

1.   Kieseker P., Relationships between Non-Government Organisations and Multinational Forces in the Field', in Smith H. (Ed)., (1993) *Peacekeepting Challenges for the Future*, Australian Defence Studies Centre. Kiesker was the Manager of CARE Australia in Baidoa while the 1 RAR Group and the French were there in 1993

2.   _____, Somalis kill 28 UN soldiers, *The Sydney Morning Herald*, 7 Jun 93 and Kiley S., US diverts Marines to Somalia, *The Australian* 22 Jun 93

3.   Trofimov Y., UN raid leaves problem unsolved, *The Weekend Australian* 19-20 June 93

4.   _____., Italy tells US to stop combat in Somalia, *The Canberra Times* 14 Jul 93

5.   McCarthy P., US moves to limit peacekeeping role, *The Sydney Morning Herald* 7 May 94, p. 19_____ ., UN to reduce troop levels in Somalia, *The Sydney Morning Herald* 27 August 1994

6.   _____ ., UN to reduce troop levels in Somalia, *The Sydney Morning Herald* 27 August 1994.

7.   Four soldiers had been found guilty of participating in the torture and subsequent death in custody of a 16 year old Somali youth, Shidane Arone, who had been caught trespassing in the Regiment's base in Belet Eun on 13 Mar 93

8.   Kiley S and Evans M., Rival clans battle for Somali airport, *The Australian* 28 February 1995, p. 11

9.   Kiley S and Leonard T., Sun finally sets on UN Somali mission, *The Australian* 4-5 march 1995, p. 17

10.   Ibid

11.   Ibid

Index